The Day
KENNEDY
WAS
SHOT

ALSO BY JIM BISHOP

The Day
KENNEDY
WAS
SHOT

★ ★ ★

JIM BISHOP

HARPER ● PERENNIAL

NEW YORK • LONDON • TORONTO • SYDNEY • NEW DELHI • AUCKLAND

HARPER ● PERENNIAL

A hardcover edition of this book was published in 1968 by Funk & Wagnalls.

First Harper Perennial edition published 1992, reissued 2013.

Library of Congress Cataloging-in-Publication Data is available upon request.

ISBN 978-0-06-229059-5 (reissue)

HB 08.23.2023

This book is dedicated to Kelly Bishop,
my wife, my assistant, my life

CONTENTS

CONTENTS

THE EVENING HOURS

THE MIDNIGHT HOURS

For the Record

There is a lingering melancholy, mixed with suspicion, regarding the assassination of John Fitzgerald Kennedy in Dallas, Texas, on November 22, 1963, and it is difficult to tell which is the harder to sustain. The shock waves which radiated from Dealey Plaza on that warm noon day seemed, like some cataclysmic sound, to pass around the world and back again many times, hardly diminishing in intensity as it bruised consciences. He had been a fair prince indeed, bringing youth and sophistication and an air of confidence to the throne. He had had his political triumphs and gaffes, but in either case he was facing forward when they occurred.

His friends, his enemies, his *aparthétiques* had sharp memories for the thick brown hair, the square gleaming teeth, the sailor's eyes, the frame of the dandy. They recalled his wife, too, as First Ladies are seldom remembered. They knew the dark slab of hair, the piquant, brilliantly lighted face, the modes, the moods, the veneer of the well-bred Georgetown girl. She had the clasped hands, white-knuckled ecstasy of a little girl at her first prom, and an appreciation of privacy. The world, it seemed, could conjure visions of little Caroline, fair-haired and sweet, and John-John, with a babyish face which had yet to make up its mind, an innocent in a white suit who could bow gravely from the waist without falling.

The nation and the world had no trouble remembering. The problem was forgetting. Waves of masochistic guilt swept America. Many people felt that, in some vague manner, we had failed Kennedy. Everything about the funeral, including the stiff-legged march to the Pro-Cathedral with the muffled tattoo of drums, was calculated to reduce the nation to tears. President Johnson appointed a committee of distinguished men to investigate the assassination and the subsequent slaughter of Lee Harvey Oswald. The simple became complex; the obvious, obtuse. The death of Mr. Kennedy became the subject of lengthy newspaper articles, magazine series, books, pamphlets, tracts, theses. The more the people read, the more certain they became that they had not heard the facts.

Some of the writers were irresponsible and sensational. These drew the most attention. A single hair could be split in several ways, and they split bullets, assassins, affadavits, and innuendo. There are—according to my count—16,500,000 words of research material on John F. Kennedy and the amount is growing. The tan smiling figure of the President began to take on the aura of a mysterious martyr.

He became a bigger man than he was. A library was proposed. Donations were solicited. The father of the President donated a million dollars. A brother-in-law, Stephen Smith, was appointed chairman of the drive for ten million dollars. A train was sent on a tour of cities with Kennedy memorabilia. The people gave freely of their hearts, their tears, but not their money.

On a warm day in 1959, John F. Kennedy had permitted me to remain at his side so that I might write a newspaper story called "The Day Kennedy Was Nominated." In public he was self-assured, an eager fighter. In private, as we parted, he said: "I have to make it by the second ballot. All of our delegates are pledged for two. If I don't make it by then, the votes will scatter. I'm going to watch it on TV." I shook his hand and said he could make it on the first. "I hope so," he said, and walked up an alley beside a small Beverly Hills apartment house. He hopped a fence to visit his father, who was in the home of actress Marion Davies.

The last time I saw President Kennedy was at the White

House. It was a month before the assassination. I was present to research an article for *Good Housekeeping* magazine to be called "A Day in the Life of President Kennedy." He was cordial and helpful, and he seemed anxious for me to write a book rather than a magazine piece. At one point, feeling that he had not persuaded me, the President addressed Mrs. Bishop: "Kelly," he said, "don't you think he has enough material for a book?" I had, barely.

Our last chat was held in the big Oval Office and he selected the topic. It was assassination. He was in his Kittyhawk rocker, the black shoes gleaming as he prodded himself back and forth, and he said he had enjoyed a book of mine called *The Day Lincoln Was Shot*.

"My feelings about assassination are identical with Mr. Lincoln's," he said. "Anyone who wants to exchange his life for mine can take it." The words were uttered with bland good humor. "They just can't protect that much."

He expatiated on the subject without belaboring it. He told how he avoided church crowds by having the Secret Service drive him first to St. Matthew Pro-Cathedral. If the street was jammed with people, he would ask to be taken to St. Patrick's, in downtown Washington. If that appeared to be too busy, he would go out to Georgetown, to the little church he and Mrs. Kennedy visited when they lived there.

Sometimes, on Sunday, he would leave his pew as the priest left the altar, and genuflect in the aisle. Behind him, two Secret Service men—not necessarily Roman Catholics—would do the same. So would two in the pew in front. The five would turn to face the rear of the church, and, as they walked toward the church doors, the President would start to crouch. As his knees flexed a little more with each step, he would grow smaller and smaller. Then he would whisper to the two in front: "If there is anybody in that choir loft trying to get me, they're going to have to get you first."

While I was at the White House, a blanket order had been issued to all hands to grant interviews and to take me wherever I desired to go. There were many people to see. The President asked, at least three times: "How did you make out with Jackie?" She had been gracious and lovely, as I knew she would be. Still the repetitious question made me wonder why the President asked it. Newspaper reporters in the West Wing had told me that Jacqueline Kennedy would not see me, and would not permit me to see the private quarters of the family.

On the contrary, she did. If she was not friendly, she was a consummate actress. Many of her observations about her husband and children were directed to Mrs. Bishop. Her father-in-law was convalescing from a cerebral hemorrhage and she said she was "mad" about him. She had proposed several times to her husband that they bring Joseph P. Kennedy to the White House to live—"I could put him in the Lincoln Room"—but John F. Kennedy said no. Whatever reason she had for inviting him, her husband's contrary wish was more adamant. At no time did she indicate that the invitation might include her mother-in-law, Rose.

On another occasion, she said: "Next month I'm going to Texas with my husband. It will be my first campaign trip." She did not appreciate or understand politics. She knew that her attractive presence would help her husband secure votes, and Mrs. Kennedy seemed to have feelings of guilt about not accompanying him. She said she was not happy in the White House and, even though she remodeled the interior of the mansion, she was fond of pointing out that, of more than one hundred thirty rooms, only eight were the private apartments of the President.

She was a woman who was pleased and offended by small things. The President who often assured worried political friends, "I can control Bobby" was never heard to say it of his wife. To some observers, they seemed to live as separate entities. Mr. Kennedy applied himself to the hard world of politics;

Mrs. Kennedy enjoyed yachting off the Greek Islands, or booking a ballet for the East Room. The connubial X intersected after dinner, when they sat alone in the living room on the second floor. He smoked a cigar, sipped a beer, and studied his "night work." She puffed a cigarette, put long playing records on the turntable, and examined her correspondence.

Dallas was her first political pilgrimage. It was a crude world of roaring voices, motorcades and inadequate quarters. It consisted of whispered conferences, resounding speeches, and damp, cold dinners. Mrs. Kennedy, I am certain, went to Texas to please her husband. The President was joyful. When she was irritated because of lack of assistance at her toilette, Mr. Kennedy upbraided one and all and yelled: "Get on the ball!"

She represented votes. Many Texans who were not particularly fond of the President turned out to yell: "Hi, Jackie!" When the motorcade left Love Field at 11:50 A.M., the First Lady had reached a point where she seemed to be enjoying the attention. She waved and smiled out the left side of the car, and left the right side to him. The shattering impact of the shots, the slow turn to see the top of his head blown away, the numbing sag of death on her lap was an emotional shock beyond calculation.

So, too, was Mrs. Kennedy's recovery. Within a short time, the First Lady had steeled herself to accept her husband's demise, and in a manner which can be described as the product of a masterful will, Mrs. Kennedy took charge. Nothing was too small to escape her attention.

The funeral was Jacqueline Kennedy's tribute to her husband. She was consulted about every aspect, from public parade to flickering torch. She refused sedation. She examined the guest list, made the determination about who would say the funeral mass, and who would preach. The funeral of Abraham Lincoln became an *idée fixe*.

Before she quit the White House, she had her husband's name and vital statistics chiseled on the marble mantle of the

President's bedroom. She approved a change of name from Cape Canaveral to Cape Kennedy, and she allowed Idlewild Airport to be renamed Kennedy International Airport, but she withheld permission to name hospitals and boulevards and bridges in honor of John F. Kennedy.

My agreement with President Kennedy was that when the slender book about him was complete, I would send him a carbon of the manuscript and he would be free to correct errors of fact. Two months after it was ready, Mrs. Kennedy assumed the license I had given to the President and asked Pierre Salinger to phone me with sixty-odd corrections. So far as I could see, few of them involved errors of fact. She did not want me to say that she never met a politician she admired, which she had said. She did not want me to say that she went to bed in a nightgown. She did not want . . . .

I was surprised that a grief-stricken widow could bear to read the words of a book about a happy day. All of the corrections were made, as she requested. When I sent an inscribed copy, she held it between two fingers over a wastebasket and dropped it.

A note arrived from Nancy Tuckerman, secretary to Mrs. Kennedy, beginning: "Mrs. Kennedy is so grateful to you for sending her the copy of your book, which you have inscribed to Caroline . . ." A similar note arrived from Robert F. Kennedy the next day.

I returned to work on *The Day Kennedy Was Shot*. In February 1964, the word was out that I was working on such a book. Robert Kennedy, at a party, met Robert Bernstein and Joseph Fox of Random House and asked bluntly why they would want to publish a book about the assassination written by me.*

---

* This book was originally under contract with Random House, but for reasons completely apart from the Kennedys' pressure on that publisher, I decided to have Funk & Wagnalls bring it out.

At a luncheon, Bennett Cerf, the President of Random House, was distressed to find that he was sharing a table with Jacqueline Kennedy who begged him, sobbing, not to publish my book. I met Richard Cardinal Cushing and, in chatting, he said that the Kennedys did not want him to talk to me. George Thomas, the President's valet, shook his head sadly and said he was not allowed to discuss the President with me. I was standing in a long corridor listening to doors slam.

When I continued my work, Mr. Cerf became unnerved by a phone call from Mrs. Kennedy. Again she wept. Again she begged him not to publish my book. He told her that there would be many books about the assassination, and that it would not be in the interest of historical accuracy to sponsor one author to the exclusion of all others. He then addressed a letter to Robert Kennedy, stating in part:

"I can understand that you would want to cooperate completely with only one author, but I urge you not to cut other writers off completely. . . ."

A movie magazine appeared with a headline: "The Man Who Made Mrs. Kennedy Cry." I was the man.

A letter arrived from Mrs. John F. Kennedy. It was on mourning border letterhead written in her script: "I cannot bear to think of seeing—and seeing advertised—a book with that name and subject and that my children might see it or someone might mention it to them." She feared never-ending conflicting books about that day in Dallas, she said. "So I hired William Manchester—to protect President Kennedy and the truth." He would interrogate all who could contribute knowledge of the assassination. "If I decide the book should never be published, then Mr. Manchester will be reimbursed for his time . . ." In the next breath, she supposed that she had no right to suppress history.

All the people were asked "not to discuss those days with anyone else—and they have all kept the faith." It did not occur to her that, if all the sources had been shut to her husband, he could not have written *While England Slept* and *Profiles in Courage*.

I wrote a flattering reply to Mrs. Kennedy and ended by saying that, with doors opened or closed, I would continue to research this book. Her reply was a flash of lightning: "None of the people connected with November 22nd will speak to anyone but Mr. Manchester . . . The Manchester book will be published with no censorship from myself or from anyone else . . ." The legal threats, the denouncements, the befouling of that man's work were all ahead of her. The Kennedys, in effect, were trying to copyright the assassination.

The reader will find ninety-two sources listed in the back of this book. I found hundreds of doors still open.

The Kennedys tried to sue their author. Alterations were made in the work. *The New York Times* quoted Robert Kennedy: "Maybe we ought to take a chance on Jim Bishop." It was too late for that.

This book was not stopped, nor even slowed. The prime source for all the years to come is *Hearings Before the President's Commission on the Assassination of President Kennedy*, volumes 1 through 26. It required two years to read and annotate the 10,400,000 words but within the maze of repetition and contradiction, there is a mass of solid evidence which, if used as a foundation, will help any author to build a book of fascinating credibility without rancor, bias, or censorship.

The personal interviews are numerous and they embrace some people who are still friendly with the Kennedys. The most unusual occurred when Judge Joe Brown of Dallas County, with pipe in hand, took me to the fifth floor of the Dallas Police Department jail. I was introduced, as I recall, as a relative from Florida. The judge was so popular with the jailers that we got inside Lee Harvey Oswald's cell and were given a running account of all that happened to him when he was there. The first interviews granted by President and Mrs. Johnson occurred in May 1968 when I was invited to the White House.

An anonymous fan in Dallas sent me a copy of the Zapruder film of the assassination. Who or why, I do not know. I used it for study and, knowing that the original belongs to *Life* magazine, burned my copy. I consulted the books of Mark Lane and R. B. Denson and Josiah Thompson and dozens of others, including the friendly superficialities of Salinger, Evelyn Lincoln, and Maude Shaw, but for incontrovertible fact, I absorbed *The Truth About the Assassination* by Charles Roberts.

This book is as accurate as I can write it. If, anywhere in it, I have given someone eyes of the wrong hue, or if I have in any sentence uttered a wrong phrase or sentence, there is no malice intended—not even toward Lee Harvey Oswald. As it stands, this book comes as close to re-creating every minute of that day of November 22, 1963, as unremitting work on my part will permit.

Jim Bishop
Hallandale, Fla.

NOVEMBER 22, 1963

★ ★ ★

The Morning Hours

Throughout the book, all times given are Central Standard.

★ ★ ★

The
Morning
Hours

Throughout the book, all times given are Central Standard.

7 a.m.

The morning light was weak and somber, seeping in diffused grays across the north Texas plain, walking along Route 80 from Marshall to Big Sandy to Edgewood, not pausing, not hurrying, through Mesquite and between the granite headstones of downtown Dallas to Arlington and Fort Worth, inexorably scouring the night from Ranger and Abilene, walking westward always westward, bringing to focal life the clustered communities, the all-night lunchrooms, the laced highways with ribboned loops, jogging trucks, flat farms with tepees of corn shucks, the quiet, shallow streams swimming to bottomland, the pin oaks huddled in hummocks hanging on to old leaves, the land smelling spongy and good in the warm wind and a mist that matched its gray with the walking dawn.

The clouds were low, kneading themselves into changing figures as they swirled in slate against the red clay below and the sandwich of electric lights between. It was a day that would be much rainier, or much brighter, a capricious time when the glimmering sky flowed on a well-muscled wind, and then, an hour or two later, might be sawed into shafts of sunlight.

At the Continental Trailways terminal a big bus slowed, headlights glowing saffron along the shiny pavement of Commerce Street, and the brakes sighed as the vehicle inched into the terminal, on time. Some passengers slept. A few, sleepless, squinted drowsily at the tall brown-brick Hotel Texas diagonally across the street. As the bus inched into the terminal, the hotel disappeared and the driver said: "Fort Worth, Fort Worth. Fifteen minutes." Time for an egg sandwich and a mug of coffee;

time for a morning paper; time to return to the uneasy sleep of the traveler.

Time.

The elderly lady stared at the ceiling. She had lived in the Hotel Texas a long time. For Helen Ganss, this room on the eighth floor was home. Yesterday there had been much excitement. Liston Slack, the manager, had been conferring for days—maybe weeks—with men who wore sunglasses and everybody on the eighth floor had been moved out. The whole L-shaped corridor had been emptied of guests. All except Mrs. Ganss. She hadn't been shrill about it, but she was an old widow and the men in the sunglasses had been perfect gentlemen. They had thought it over and had told Mr. Slack: "Okay."

The President of the United States was down the hall in the corner suite, 850, but Mrs. Ganss wondered how he could possibly sleep. All night long she had heard the march of feet up and down that green rug with the big flowers, and now, in daylight, the feet had voices. Sleep was impossible. Some feet walked. Some ran. The voices ranged from a loud call the length of the corridor to sibilant whispers outside her door. Sadly, there was nothing exciting about the ceiling. Mrs. Ganss stared at it because a lady of years and frailty has so few options.

The noise in the corridor grew by solitary decibels. One of the three hotel elevators was reserved for presidential traffic and waited on the eighth floor. On the opposite side, Rear Admiral Dr. George Burkley, the President's physician, was up and had phoned for breakfast. He is a short, gray man of considerable reserve, and he looked out the window and then peeked down the hall toward Suite 850. The Secret Service men nodded good morning. The doctor knew that everything was all right.

George Thomas, a chubby valet, came down the corridor with an arm full of clothing. As he was admitted to Suite 850, a Secret Service man picked up a phone near the fire hose and said: "The President is awake." Thomas walked through a small foyer, shifted some Texas newspapers from one hand to the other, and tapped lightly on the door. Inside, there was a moment of silence, and President John F. Kennedy muttered, "Okay."

The word had meaning which only the President and his valet would appreciate. In the White House, when Mrs. Kennedy shared her husband's bedroom, a light tap by Thomas would elicit a small cough as response. The tap and cough were designed not to disturb Mrs. Kennedy's slumber. The word "Okay" would signify that Mrs. Kennedy had slept in another room.

Thomas opened the bedroom door, deposited the clothing on the back of a chair, dropped the newspapers on the bed, and exchanged greetings with the tousle-haired sleeper who was turning the sheets back from the left—and window side—of a big double bed. The President sat up, swung his long slender limbs over to the floor, and picked up the packet of newspapers. Mr. Thomas was already in the bathroom, mixing the water and drawing a bath.

On the mezzanine floor, Master Sergeant Joseph Giordano completed the work of screwing the Seal of the President of the United States to the lectern as Secret Service men, stationed around the big room with its long rows of breakfast tables, watched him. He took another Presidential Seal downstairs to the parking lot across the street. Mr. Kennedy would make two speeches this morning. The formal one would be at the Chamber of Commerce breakfast around 9:30. These people, Mr. Kennedy had learned, were largely Republicans. The Democrats of Fort Worth had protested that the workingmen had not been invited. So the President had agreed to meet them in the parking lot before the breakfast.

The handsome General Ted Clifton, military aide to the President, rapped on the door of 804. The man who answered was The Bagman. "You packed?" Clifton said. The man said he was. Behind The Bagman stood Cecil Stoughton, the White House photographer. He, too, had plenty of gear to pack, and he knew that he had to keep several cameras ready with film of varying speeds. The Bagman, Ira Gearhart, was important. He carried the small suitcase with the safe dial. It was his job never to be more than a few seconds from the side of the President, because inside The Bag was the electronic apparatus with which Mr. Kennedy could call, in code, for a nuclear strike.

It was assumed by all knowledgeable persons in the White House, and the Pentagon, that The Bag would never he used. Still, in the event that the Continental Army Command tracked flights of "birds" coming in across the top of the world and over the DEW line, a decision would have to be made at once. The Bagman was never far from Mr. Kennedy. The function of the man was to remember the combination to the dial; the function of the President was to order one of several types of retaliatory attacks.

In the hotel was a "White House switchboard." This was usually manned by the military. It, too, moved in the wake of the President. It provided instantaneous communication between Mr. Kennedy and Washington. Coded information that the President had awakened was already in Washington. At Carswell Air Force Base, Colonel James Swindal, commander of *Air Force One,* had called in five minutes ago that the craft had been inspected, tested, and was ready.

On the seventh floor, a teletype machine chattered and the daily information report began to come in from the Central Intelligence Agency. The CIA, with its finger on sensitive pulses around the world, was giving the President a morning rundown on the political climate of the world. General Godfrey McHugh, the only American officer of rank with a French accent, signed a receipt for it and walked up to the eighth floor, to be confronted by a Secret Service agent who blocked his path at the head of the stairs until the general was recognized. Then he went on to Suite 850, to be studied momentarily by another man with a key in his hand.

McHugh would wait until summoned, then give the report to the Commander-in-Chief. The general's strength was his weakness. He was a perfectionist in all his work. The general even maintained a record of the precise minute that the report came off the machine, and he would duly note the moment it left his hands for Mr. Kennedy's.

Two Secret Service agents were at Fort Worth Police Headquarters examining two limousines. The cars had been rented for the Kennedys and the Secret Service for the four-mile drive from the Hotel Texas to Carswell Air Force Base. Everything, including ballrooms, parking lots, bedrooms, bathrooms, parade routes, stairwells, lobbies, kitchens, cooks, waiters, telephones, local personnel, from food to forks, had to be "sanitized" by the Secret Service.

Three weeks prior to this visit, Manager Liston Slack was surprised to learn that the Secret Service declined use of the Will Rogers Suite on the thirteenth floor. It would be more difficult to "protect," the agents had said. So Kennedy was now in a smaller suite in a corner of the eighth floor, and the Will Rogers Suite was being used by Vice-President and Mrs. Lyndon Baines Johnson. The President's quarters cost $106 per day, but the management would not send a bill. Even though normal life at the hotel had been cruelly upset, Liston Slack would not send charges to the government.

The measure of Fort Worth's excitement was in the lobby and the parking lot. The first was jammed with men wearing fawn-colored cowboy hats; in the lot, five hundred men and women stood waiting in the misty rain, even though the President was not expected for more than an hour. A half dozen mounted sheriff's deputies patrolled their horses in and out of the growing crowd, herding them toward the lectern.

Presidential assistants Kenny O'Donnell and Larry O'Brien—the first lean and grim with a pulsing mandible, the second a myopic redhead with a gift for solving political puzzles—walked into Dr. Burkley's room and said: "We can't see anything from the other side of the hotel." They raised the venetian blinds and studied the crowd. O'Donnell murmured: "They're waiting for him in the rain. And there will be more of them." They thanked the doctor and left.

O'Donnell went back to his room to shave. He glanced at the presidential itinerary. Two speeches in Fort Worth, one in Dallas, a flight to the capital at Austin, two cocktail parties, a speech at a banquet, a slow motorcade late at night, and a flight to Lyndon Johnson's ranch for a two-day rest. O'Donnell was the watchdog, the harrier. As the man who, except Attorney General Robert Kennedy, was closest to the President, Kenneth O'Donnell managed the show, made many of the peremptory decisions, kept Mr. Kennedy close to his schedule, tried hard to please Mrs. Kennedy, ordered the White House staff to its appointed duties, and, when necessary, compressed his lipless mouth and said "No" to senators and congressmen.

Agent O'Leary, under the marquee of the hotel, kept his eyes roving from the sidewalk to his left, across the Bus Terminal, down the emptiness of the parking lot, across Main Street with its Century Building and Fort Worth National Bank, and down the sidewalk to his right. The eyes began the searchlight progression again, and midway, O'Leary saw a man reclining on a roof diagonally opposite Suite 850. The Secret Service man called a policeman and pointed. "Get him off that roof."

Clinton Hill, assigned to Mrs. Kennedy, had inspected all the entrances and exits to the hotel last midnight. Now he did it again. He reported to Agent-in-Charge Roy Kellerman that everything was all right. On the thirteenth floor, the Johnsons dressed swiftly, and the Vice-President sipped coffee, sans caffeine. Lady Bird glanced out the window at the dismal weather and noticed the people in the lot. She knew that her husband was expected to be at the President's side, but she wasn't informed whether Mrs. Kennedy would be with her husband. If so, Mrs. Johnson should be there, too. She didn't want to phone 850 and ask because, if the First Lady wasn't going, the call would point up her absence.

Mrs. Johnson phoned the Connallys and spoke to Nellie. Yes, the Governor's wife would be at his side in the parking lot. Nellie said that she and John regarded Fort Worth as home because, years ago, he had worked for the rich Sid Richardson in this town. Lady Bird decided to go along with her husband.

In the Arlington Heights section, the stout face of the martyr, Mrs. Marguerite Oswald, peered from behind curtains in her small apartment on Thomas Place. The weather matched her mood. She turned the kitchen light on and puttered with the coffeepot. The gray hair was tight in a bun, but skeins of it hung loose. In common with other citizens of Fort Worth, she was aware that the President of the United States was in town, and she planned to watch the event on television.

Mrs. Oswald was a hardworking saleswoman and practical nurse. She was fifty-six years of age, stout, and full of outraged righteousness. The mouth was thick and pursed. She enjoyed conversation but, except for chronically ill patients, she had no social life. Years ago she had married three times and had three sons. One of the husbands died. The others left her. The sons enlisted in military service early. None of them ever came back.

In Dallas, seventeen men lined up before Deputy Chief W. W. Stevenson. The patrolmen were told that their function would be to "seal" the Trade Mart. None of them could understand why the work had to begin at 7 A.M., but they knew that Chief Jesse Curry and the Secret Service had been in conferences for three weeks and had driven slowly, in squad cars, along several routes to and from Love Field.

Stevenson glanced over the enormity of the interior, where 2,500 persons would greet the President at 12:30 P.M. The Secret Service had studied the overhead catwalks and had shaken their heads disapprovingly. But from this moment on those catwalks would be denied to everyone except the fluttering blue parakeets that darted from the huge fountain at the back of the building to the rafters overhead. The big head table was placed inside the main entrance. The interior "side streets," which featured shops, would be closed off.

The policemen listened to their individual assignments and were told how to recognize Secret Service men by the tiny orange pins in their coat lapels and to deny access to anyone without a luncheon invitation, even if the policemen recognized the intruder. Stevenson placed the last of his men at the receptionist's desk in the big front lobby. This one would assist the ticket takers to screen guests.

The head table had already been "sanitized," flowers and all. The chefs in the kitchen had petitioned the Secret Service to permit them to select a fine marbled steak for the President of the United States. The request had been denied. When the huge platters of steaks began to come from the kitchen, the Secret Service said, one would be selected at random for Mr. Kennedy.

The men posted at the freight entrances and along the sides of the structure were told that no one was to be permitted to enter, unless Mr. Saich, the caterer, came to the door personally and identified the person as an employee. One man stood in the rain on the roof over the entrance. He carried a rifle and had a good field of vision, not only along the feeder lane leading to the Trade Mart, but also behind him, along Stemmons Freeway from downtown Dallas to Parkland Hospital. He didn't have to worry about the freeway. Other men would be patrolling the route. The Dallas Police Department had canceled all leaves, and all personnel except a handful of squad cars and some detectives were working the Kennedy assignment. The dispatcher had been told to keep Channel One open for superior officers with the President and to use police Channel Two for all other business.

At 7:08 A.M. the police chief, a mild, spectacled man who maintained a clean city, appeared on television and announced that the President would be in Dallas today and that Dallas wanted no incidents. He knew that the citizens desired to give the Chief Executive a cordial welcome, but there was always a chance that some "extremist" planned to demonstrate. If so, Chief Jesse Curry was putting such people on notice that the police department would brook no nonsense today. Curry did not mention the whacking of Adlai Stevenson with a placard a short time before, or the shrieking, shouting crowd which once chased Mr. and Mrs. Lyndon Johnson into a hotel lobby. Dallas had an articulate rightist group which was obsessed with the notion that all others in the political spectrum were Communists or "fellow travelers" plotting against the Republic.

The words came off the television screen calmly, but, by the nature of the appeal, they exposed the helplessness of law enforcement in the face of a sneak. Chief Curry concluded by asking all good citizens to please report to the Dallas Police Department anyone who had voiced violent opinions against the President or who had boasted, publicly or privately, of plans to demonstrate today.

The television set in the modern little four-room house at 2515 Fifth Street in Irving was shut off. The owner, Mrs. Ruth Paine, was still in bed. The suburb, off the western edge of Dallas, is a collection of small ranch homes astride Route 183 to Fort Worth. In several thousands of these houses, men were up, preparing to leave for office and plant; children were up, breakfasting on hot cereal for school.

In the kitchen of the Paine home, a young, slender man poured boiling water into a cup with instant coffee and sat at the table. He was alone and he sipped his coffee, as he always did, with the fingers of both hands around the cup. He had pale eyes, thinning brown hair, and a mouth which pursed itself in a permanent pout. Anyone who knew Lee Harvey Oswald was aware that he did not mind being alone and he enjoyed long silences.

He would not turn the television set on to listen to Chief Jesse Curry. Mr. Oswald was having trouble with his wife. She had awakened to feed their infant, Rachel, at 6:30, taken a look at the other little girl, June, and closed her eyes. Mr. Oswald had said: "Don't get up." Marina Oswald thought this was funny, because she never got up to make breakfast for him. It wasn't sarcasm. She was sure of that. He whispered softly, in that throaty, bobbing-Adam's-apple manner, that she should buy shoes for June. She opened her eyes, watching him dress, and grunted before returning to unconsciousness.

The baby had awakened several times in the night. The blonde head on the pillow tried to concentrate on what he was saying, and some of it remained with her, and some didn't get past her ear. Lee told her to buy a pair of shoes for herself. That registered. He donned a tan-gray work shirt, gray slacks, and an old zipper jacket. Without opening her eyes, she could feel him stop beside the dresser, and she knew that he wanted to start a friendly conversation. "Maybe someday June will remember me," he said.

Mrs. Oswald kept her eyes closed. She did not want to be friendly. Mr. Oswald removed his wedding ring from his finger and lowered it carefully into a Russian cup on his wife's dresser. He opened a drawer carefully and placed his wallet inside. It contained $170. He kept $13.87, insufficient for a man who might wish to leave the area. And yet the gesture of the wedding ring and the sum of money for his wife—more than he had ever given her—are symbols of a marital break.

Last night, he had tried to restore the marriage. He had come to Mrs. Paine's house unasked, unwelcome, unexpected. On previous occasions when he visited his wife, he had left his tiny room in the Oak Cliff section of Dallas on Friday evenings and had remained with her until he could get a free lift to the plant where he worked, on Monday mornings. This time he came in on Thursday, played out front with his gladsome idol, little June, and had tried to have a private chat with Marina.

Her respect for him was dead. Sweet words would not resurrect it. He had personality flaws which she could not understand. Marina, a Soviet pharmacist, had met him in Minsk and married him after a short courtship. He was an American defector with ideals unattainable. He was, he proclaimed, a United States marine who wanted to renounce his citizenship and embrace the Soviet Union. In the next breath, he said he was disillusioned with Russia, because the government was deviationist from the principles of Karl Marx. The inference was that his politics was pure communism; Russian socialism was opportunistic and despotic.

When the Soviets denied citizenship and, for a time, even sanctuary, he had cut his wrists in Moscow and, as in most other crises in his life, had failed. He asked Marina if she would like to return to the United States with him—particularly to Texas—and she said yes. He had extolled the virtues of his mother, Marguerite, and then later forbade his wife to see her. He spurned the friendliness of the Russian expatriate group in Texas, and refused to teach his wife to speak English.

He talked big but couldn't hold onto a job. When he had one, he doled out small sums to his wife and told her she would have to get along as best she could. At night he read library books about Marxism and others concerned with history, and there were long silences. He brooded sullenly and appeared to have trouble making love to his wife. The average attempt occurred once a month, and Marina, bristling, told her husband he was not a man.

Sometimes, in his frustration, he beat her with his fists. At others, he became the supplicant and begged her forgiveness. The man who seldom spoke could weep. He bought a mail order rifle and a revolver, and these were anathema to Marina. To a young man whose father had died two months before he was born; to a boy who had slept with his mother until he was eleven years of age; to one who had, of necessity, spent time in orphanages, one who was now accused of lacking manhood, the weapons may have made him as big as the biggest man.

He told her that he had tried to kill General Edwin Walker, an avowed reactionary, but had missed. On another occasion, he announced that he was going out to kill the Vice-President of the United States—Marina had thought of Richard Nixon, although the reference was probably to Lyndon Johnson—and he had permitted her to lock him in a bathroom, supplied with books, until the storm of violence had left him.

Nor had he complained when Mrs. Paine, a student of the Russian language and a dark, pretty Quaker, had offered Marina and June a home until "Lee could get on his feet." It had happened before in other homes. A few weeks ago, a second child, Rachel, had been born. Marina had still felt that the marriage might be "saved" for the sake of the children, but when Mrs. Paine had phoned him at his rooming house the woman who answered said that there was no Lee Harvey Oswald there. They had a young man named O. H. Lee.

Marina, in anger, lost all confidence in her husband. He, in turn, was angry to learn that Mrs. Paine had tried to contact him. His unexpected arrival on Thursday evening did not endear him to his wife. She had busied herself in the kitchen with Mrs. Paine, fed the babies and him, and chilled all his Russian entreaties. In bed she had turned away from him. She was tired. She didn't want to talk.

It is possible that Marina Oswald misjudged Lee. She saw the current situation as another dispute. She might have relented in her own time. The punishing wife was conscious of the needs of her children. But the ring and the money showed that Lee Harvey Oswald was at the end of his tether. Day by day his affection had turned more toward June, and, according to the inexorable law of transference, away from his wife.

He needed someone more helpless than himself. His personal inadequacy was known to him. In school he had shunned the friendship of boys. He played by himself. For years he had submitted to the scourging of his mother's domination and, like John and Robert before him, had left her as soon as the U.S. Marine Corps would take him. The military gave him training, discipline, foreign service and a marksman's medal.

At the age of fifteen, books taught him what the United States symbolized as a democracy, and he chose the role of dissenter. Furthest removed from what his country stood for was the Soviet Union, and he chose that, with reservations. In time, his studies of Karl Marx made Oswald feel equipped to explain it in theoretical terms, but he could draw the attention only of those who did not understand it at all. Friends who had studied political science exposed him in conversation as superficial and for using communist terminology without understanding it.

He had left the Marine Corps as a "hardship discharge" to take care of his mother in Texas. He gave her three days of his time and left for New Orleans and a long trip to Russia. Marina, a shrewd, intelligent girl, was not a helpless person, but he could make her so by returning to the United States. She would be dependent on him just so long as she did not speak English. But she was not compliant. At Texas parties given by Russians, she asserted herself and agreed with those who said that life in the United States was far better than "at home."

Oswald threatened to send her back to Russia and ordered her to write notes to the Soviet Embassy asking for repatriation. His frustrations mounted as he lost job after job. Recently he had taken a bus to Mexico and had appealed to the Cuban Embassy for a visa. He had formed a Fair Play for Cuba Committee in New Orleans, been arrested, told his story on radio, and tried to "enlist" in the Castro forces. They did not want him.

The young man who seldom responded to a friendly "good morning" found himself at the end of his particular blind alley. He was friendless, homeless, "hounded by the FBI," as he said, and now he knew he was a cipher. He aspired above all other things to be big, to be known, to be respected or feared (equal values).

The coffee cup went into the sink. He went out into the garage and turned the ceiling's naked bulb on. He opened a rolled blanket on the floor, slipped a rifle out without disturbing the convolutions of the blanket, and closed the flap. He took some wrapping paper, placed the rifle in it, and wrapped it in such a way that one end appeared to be thick, the other thin. He went back into the kitchen, forgetting to turn the light off. Oswald left quietly.

The President, standing in the green tile bathroom, finished toweling himself and began to shave. He could hear someone rolling up his special hard mattress in the bedroom and, without looking, he knew that the black leather chair with the thick backrest would leave with it. Wherever he went, they went. He saw his plump, slightly jowled, and tan face in the mirror. It was a good strong face—many would call it handsome—but a man seldom dwells on features as he shaves. The pull of the razor is automatic, done without conscious thought, furrowing the white mantle of shaving cream in a pattern which is peculiarly the shaver's own.

It gives a man freedom to dwell on other matters—the day's schedule; the minutiae of business; the problems, if any; the triumphs—if any. Mr. Kennedy had a sturdy, almost youthful, body with patches of hair on the chest, and legs a bit slender for the bulk of the torso. It never got a good grade from the President.

His back was in pain constantly. Long years ago, in a football scrimmage at Harvard, the spine and adjacent musculature had twisted, and it was beyond repair. A delicate and protracted operation did not help. Massages and medication made him feel better, but, as he sometimes said, the pain was never eliminated. It was lessened. It became bearable.

He combed his thick brown hair and moved to his right to the vanity set. There he donned his underwear and a surgical corset. The President had them in different sizes. He put on a large one and yanked the laces tightly. The vanity chair had a concave seat and the President sat to pull a long elastic bandage over his feet. He twisted it so that it formed a figure eight, then slipped it up over both legs. When it was adjusted over the hips, the Ace Bandage supported the bottom of the torso, as the back brace held the lower spine rigid. The figure eight constricted the natural long stride, but today was going to be a "backbreaker"—sitting, standing, walking, making speeches, handshaking, and spine-creaking climbs up airplane ramps.

He accepted help from George Thomas in slipping on a white shirt with a blue pinstripe, a plain blue silk tie, and a gray-blue suit with a half inch of kerchief showing in the breast pocket. Mr. Kennedy nodded toward the window. "How does it look, George?" George Thomas was picking up night gear from the bed, preparing to pack. "It's raining out," he said. The President said, "That's too bad" and left the bedroom for the large sitting room.

The elegant little family dining room on the second floor of the White House was never brighter than when it was filled with the morning chatter of Caroline and John. She would be six next week, and already she was accustomed to the serious business of being a lady. She knew how to keep a pretty frock tidy and unwrinkled in the back, how to wear white gloves and keep them white, how to flick the well-brushed brown hair back off her shoulders, how to apply herself to study.

John would be three years old in a few days, part baby, part boy. He enjoyed running through the White House corridors, hitching a skip in his stride and he could make it at top speed to his father, falling against the parental knees, and wrapping both arms around the tall legs. He understood little about his father's work, but he was willing to extend his complete confidence to the many strangers he saw in his home, men and women who stooped to hug him or to say hello.

The children spent ten minutes with their father shortly after 7 A.M. When he was eating from a tray in his bedroom, he could hear the typewriter speed of the little feet coming down the corridor from their bedroom, and the President of the United States grasped both sides of the tray and held on, bracing against the assault of morning kisses and hugs.

The routine of conversation seldom varied. The President asked his daughter for a report on her current schoolwork. Shyly she would hold out a sheet of paper on which the alphabet had been printed in large block letters. Mr. Kennedy would study it and fall back against his chair in mock surprise. "Caroline," he would say, "did you do this? All by yourself?" The child was girlishly embarrassed by lavish praise and often hung her head and twisted her laced fingers.

Then, noting that John was waiting, Mr. Kennedy would crook his finger at his son and say: "John-John, tell me a secret." This too was a morning ritual and, even before the familiar routine began, the little boy laughed and held his stomach. He went to his father's side, stood on tiptoe and whispered: "Bzzzz-bzzzzzzz—bzz-bzz." The President threw both hands in the air, reared back with surprise, and whispered: "You don't tell me!" The effect on the boy was to cause him to fall to the floor, rolling over with laughter. It was repeated almost every morning.

The Kennedy children were accustomed to having one parent home. When father went on a trip, quite often mother remained with them. When mother flew away for a rest, father was in the White House. This morning, neither was home and they sat in the dining room with Miss Maude Shaw, their British nanny. The lady was slender and middle-aged.

She slept in a small alcove bedroom between theirs, and time and understanding had built a solid edifice of affection among the three. The children were well behaved and tractable. Sometimes, when they awakened before seven in the morning, they would ask respectfully: "Good morning, Miss Shaw. May we get up now?"

She permitted them to chatter for a longer period this morning, and there was still plenty of time. It was 8:15 A.M.* and Miss Shaw said that Caroline had time to wash her hands before going upstairs to the little private school composed mostly of children of old Georgetown friends of the Kennedys. It was a bright room with a ramp leading upward toward the shafts of morning light, and the other students arrived by vehicle at 8:45 A.M. and waited in the front lobby of the White House until schooltime.

Then, said Miss Shaw, she would take John-John for a walk around the White House grounds. His happy hope was to be on the South Grounds when a helicopter landed or took off. The only better one he could think of was to be in one.

The big cellar kitchen of the Hotel Texas was charged with excitement. The chefs and waiters had arrived early, and breakfast orders were being filled quickly and carried up by service elevator to the members of the most important group ever to grace the sedate edifice. An order had come in from 850, and everyone paused to listen. Peter Saccu, the short, dark, jovial man who supervised all the catering and food, took the order.

"The President," he said, "wants a large pot of coffee, some extra cups and saucers, orange juice, two eggs boiled five minutes, some toast and marmalade on the side. Come on now. Let's move." Saccu turned to a tall, dignified Negro waiter. "George Jackson will handle it." Some of the other faces relaxed in res-

---

* Washington time is one hour later than at Dallas.

ignation; Mr. Jackson began to beam. At once, he got a rolling table, a pad, a snowy tablecloth, some napkins, knives, forks, spoons, cups and saucers, and his expert fingers flew as the tools were placed on the table. He kept shaking his head. "Man," he murmured. "I have never even seen a President of the United States. Now I'm going to walk right into the room with him."

In five minutes, the steaming snowy eggs were lifted out of boiling water and placed in a side dish. The table moved off with George Jackson behind it. When he arrived on the service elevator at the eighth floor, a man stood in the doorway of his elevator. He lifted the covers of dishes, stooped to look at the underside of the table, gave Mr. Jackson a cursory study, and nodded for him to proceed.

A silent man outside the door of 850 studied the table and the waiter and gave him a small orange pin to wear in the lapel of his white jacket. George Jackson pushed the breakfast tray inside the small foyer and into the living room to the right. He said, "Good morning, Mr. President" at once, and Mr. Kennedy, chatting with Kenny O'Donnell near the coffee table, said, "Good morning."

The Chief Executive appeared to be bright and forceful to the waiter. A "take-charge" man. Mr. O'Donnell was explaining that the rancorous battle between Senator Ralph Yarborough and his liberal Texas Democrats and Governor John Connally and his conservative Democrats had not been resolved by the President's visit. It was worse, in a way. The Senator had refused to ride in the same car with Vice-President Lyndon Johnson in the San Antonio motorcade and at Houston, in spite of a presidential request to do so.

"Get on that phone," Mr. Kennedy said, pointing a finger at the instrument, "and tell him he's riding with Johnson today or he's walking." O'Donnell asked the President if he had seen the crowds waiting in the rain. Mr. Kennedy strode to a couch near the window and put a knee on it, but he couldn't get a satisfactory look so he went to his wife's room, rapped lightly with his fingernails, and entered.

Mrs. Kennedy, who had promised to sleep late, was awake. Her husband hurried through the room to the window and parted the closed venetian blinds with two fingers. "Look at that crowd," he murmured. "Just look." His wife pulled a robe around her and peeked. It was still raining, and the large parking lot, with its diagonal white lines, was filling with a happy-go-lucky crowd. There were two thousand people down there, jostling and joshing. It was easy to pick the women out; they carried colored umbrellas.

"Take your time," the President said, as he left the room. "The breakfast is at nine or nine-fifteen." Mr. Kennedy was enthused about that crowd. So far, the crowds in Texas had been larger than expected and more cordial. He sat down to his breakfast, cracking the eggs and talking brightly to O'Donnell, when Dave Powers walked in. Mr. Powers was the balding gnome of the Kennedy inner circle.

It was he who had first managed John F. Kennedy's campaign for Congress in 1946; it was he who had taught him the little tricks of choosing topics for speeches; it was Dave Powers who followed his young man on the long run to the White House, telling Irish stories, making the candidate smile, swimming with him in the Executive Mansion pool ("I had to learn to breast stroke because it's the only way to swim and talk"); sleeping in the same room with the President when Mrs. Kennedy was away on a trip ("My family calls me John's Other Wife"); a confidant, a buddy, a lead pony for a race horse, but never a topflight political strategist as Kenny O'Donnell was and as Larry O'Brien was.

"Have you seen the square?" Kennedy said, waving the toast. Dave Powers nodded. "Weren't the crowds great in San Antonio and Houston?" Mr. Powers peeked out at the square again. "They were better than expected," he said sagely. "Listen," the President said, "they were terrific. And you were right—they loved Jackie."

The waiter was in the foyer. He paused a moment to speak to George Thomas. Could Thomas ask the President for some little souvenir? Any little thing that he could keep as a remembrance? George said he would see. He walked back into the living room and whispered to the President. Mr. Kennedy reached into his jacket pocket and arose from the table. In the foyer, he handed George Jackson a PT-109 tie clasp. They shook hands.*

The two confidants who sat in the room with the President were anxious to resolve three pressing problems: How to secure Texas for Kennedy in 1964; how to resolve the fight between Texas Democratic liberals and Texas Democratic conservatives and get both to work for a second term for John F. Kennedy; how to raise money from Texas dinners—half of which would remain in the state, the other half of which would go to the Democratic National Committee.

The President was talking about the contents of the morning newspapers when Mrs. Kennedy came into the room. Even without makeup, she had a dark radiance, a female mystique which attracted men of all ages and forced women to emulate her careless coiffure, her big soft mouth, her street clothes, even her hats. Mr. Kennedy, more than anyone else, knew that Mrs. Kennedy was a co-equal in marriage. There was nothing suppliant about her. Submissiveness was anathema.

She was a woman of will and intellect; a charming conversationalist obsessed perhaps with what she referred to as "good taste"; a wife who tried to draw her husband's attention to fine arts, ennobling music, schools of painting. She professed to distrust the press and her attitude toward politics was that it was a dreary game infested by untrustworthy persons. "I wish," she once said, "that my husband was still a United States Senator. We would be living in Georgetown with our friends."

Still, a trip to Paris had been welcomed, because there Mrs.

---

* Within a year, George Jackson was dead of cancer.

Kennedy drew more attention and more admiration than the President. Jacqueline Kennedy had enjoyed that trip. Before and since, she had expressed feelings of guilt because she managed to remain out of campaign trails. In late October, 1963, she had said, almost happily: "You know I'm going to Texas with Jack. It's the first real political trip for me."*

It was obvious that she was doing this to please her husband. He was so acutely aware of it that he had asked General Godfrey McHugh for a forecast of Texas weather so that Mrs. Kennedy could properly plan a wardrobe. McHugh had contacted the Air Force meterologists and they guessed it would be chilly. The weather was unseasonably warm and McHugh had been dressed down venomously by the President. One of Mr. Kennedy's major considerations on this trip was to help his wife enjoy herself so that she might be cajoled into making further political excursions. She was an asset.

In public, the Kennedys were a happy, gracious family. In private, there was room for disagreement and asperity. This is not to say that it was not a happy marriage, but rather, like others, there were times when the wife disagreed with her husband. Mrs. Kennedy, for example, had once worked as an inquiring photographer for a Washington newspaper, but she felt little empathy for the press and often used her Secret Service guards to prevent newspaper photographers from taking her picture. The President, on the other hand, had once been an International News Service reporter, and cultivated a public aura of patience with his editorial detractors. In private, he was not above writing furious notes to editors and publishers about the "inaccuracies" of certain White House reporters. Now and then, as though to beard the enemy someday, he vowed to buy the *Washington Post* after completing his second term of office.

---

* On October 26, 1963, while showing the private quarters of the Kennedy family to the author and his wife.

The difference between public and private opinions seeped down to the press corps, and they were often at variance. After the Vienna Summit Conference with the Russians, Mr. Kennedy spoke well of the conversations. In private, he said of Nikita Khrushchev: "Why, that son of a bitch won't pay any attention to words. He has to *see* you move." In 1959, when he was in California fighting Richard Nixon for the presidency, he was aroused by motion picture star John Wayne's efforts for the Republican party. On a notepad, he scribbled: "How do we cut John Wayne's balls off?"

The opposition in Congress were often "bastards." He made a mental note to criticize Mary Gallagher this morning. Miss Gallagher, as private secretary to Mrs. Kennedy, might have seemed an insignificant item for presidential attention, but the young lady had volunteered to be Mrs. Kennedy's personal maid on this particular trip, and the First Lady told her husband that she could not find Mary when she was needed. So the President was going to tell O'Donnell, "For Christ's sake, keep Mary Gallagher on the ball." Anything which irritated Mrs. Kennedy aroused the President.

The newspapers of Texas irritated him. Aloud, he read headlines from this morning's *Dallas News*: "President's Visit Seen Widening State Democratic Split"; "Yarborough Snubs LBJ"; "Storm of Political Controversy Swirls Around Kennedy on Visit." The paper was cast aside. He finished eating the eggs, picked the paper up and turned it inside out. "Have you people seen this?" It was a full-page advertisement headlined "Welcome Mr. Kennedy to Dallas." Around the page was a quarter-inch black mourning border. It was signed by "The American Fact-Finding Committee, Bernard Weissman, Chairman." The copy asked twelve questions of the President, each slanted toward the arch-conservative attitude of oil-rich Dallas.

"WHY do you say we have built a 'wall of freedom' around Cuba when there is no freedom in Cuba today? Because of your policy, thousands of Cubans have been imprisoned, are starving and being persecuted—with thousands already murdered and thousands more awaiting execution and, in addition, the entire population of almost 7,000,000 Cubans are living in slavery?

"WHY have you approved the sale of wheat and corn to our enemies when you know the Communist soldiers 'Travel on their stomachs' just as ours do? Communist soldiers are daily wounding and/or killing American soldiers in South Viet Nam.

"WHY have you urged greater aid, comfort, recognition, and understanding for Yugoslavia, Poland, Hungary, and other Communist countries, while turning your back on the pleas of Hungarian, East German, Cuban and other anti-Communist freedom fighters?

"WHY has Gus Hall, head of the U.S. Communist Party, praised almost every one of your policies and announced that the party will endorse and support your re-election in 1964?

"WHY have you ordered or permitted your brother Bobby, the Attorney General, to go soft on Communists, fellow-travelers, and ultra-leftists in America, while permitting him to persecute loyal Americans who criticize you, your administration and your leadership?

"WHY have you scrapped the Monroe Doctrine in favor of the 'Spirit of Moscow'?"

Why, why, why . . . Mr. Kennedy poured a little fresh coffee. "How can people write such things?" he said. To Mrs. Kennedy, he said, with obvious disgust: "We're really in nut country now." To the others, he spoke with contempt about oil millionaires, reactionaries who peddled hate but had no alternatives to the program of the Administration. He had not seen the *Dallas News* of the morning before, in which a sports columnist had written glibly: "If the speech is about boating you will be among the warmest of admirers. If it is about Cuber, civil rights, taxes or Viet Nam, there will sure as shootin' be some who heave to and let go with a broadside of grapeshot in the presidential rigging."

Nor had he seen the handbills which spun across the clean sidewalks of Dallas for the past few days. These had not been signed, nor was the printer's signature on them, but they featured a solemn front and side view of the President with the words, in large type:

WANTED for TREASON

It was a typical sheriff's poster. The copy read:

"This man is wanted for treasonous activities against the United States:

"1. Betraying the Constitution (which he swore to uphold):

"He is turning the sovereignty of the U.S. over to the Communist controlled United Nations.

"He is betraying our friends (Cuba, Katanga, Portugal) and befriending our enemies (Russia, Yugoslavia, Poland.)

"2. He has been WRONG on innumerable issues affecting the security of the U.S. (United Nations—Berlin Wall—Missile Removal—Cuba-Wheat Deals—Test Ban Treaty, etc.)

"3. He has been lax in enforcing Communist registration laws.

"4. He has given support and encouragement to the Communist inspired racial riots.

"5. He has illegally invaded a sovereign state with federal troops.

"6. He has consistently appointed anti-Christians to Federal office;

"Upholds the Supreme Court in its anti-Christian rulings.

"Aliens and known Communists abound in Federal offices.

"7. He has been caught in fantastic LIES to the American people (including personal ones like his previous marriage and divorce.)"

In Section Four of the *Dallas News*, the President read a story by Carl Freund which raised the hackles on his neck: "Former Vice-President Richard M. Nixon predicted here Thursday that

President Kennedy will drop Lyndon Johnson from the No. 2 spot on the Democratic ticket if a close race appears likely next year. Nixon said Johnson is becoming a 'political liability' to the Democratic Party."

Mr. Nixon had been narrowly defeated for the presidency in the autumn election of 1959. Since, he had been defeated in a race for the governorship of California by Mr. Pat Brown. Nixon moved to New York and became an attorney for a firm which bottled soft drinks (Pepsi-Cola). He was in Dallas to promote business and, to keep his personal political ambitions alive, often submitted to interviews which concerned themselves more with "gut and gutter" politics than with bottling beverages.

The President was still fulminating against the press when a Secret Service man said he had a call from the Dallas office, asking if the bubbletop should be put on the car. A negative headshake came from Kenny O'Donnell. Mr. Kennedy said he didn't want it on. Furthermore, he said, he wanted the Secret Service men told to stop running beside the car and hopping on the rear bumpers. His convictions were firm about this and had been restated many times. "The people come to see me, not the Secret Service." Besides, the bubbletop offered no protection except from rain. It wasn't bulletproof, nor would Mr. Kennedy permit himself to use it even if it was.

No one ever had the temerity to introduce the subject of assassination to the President. But there were occasions when he dragged the ugly subject into focus.* Mr. Kennedy's feelings were that a President is conscious of sudden death only when he first assumes office. He learns that he cannot expose himself to crowds without prior warning; he is surrounded, in the White House and out, by silent faceless men who are always looking in another direction; his family cannot go shopping without notifying the agent in charge of the White House detail; the heating

---

* He brought it up to the author at the White House, October, 28, 1963.

units in his office are tested daily for radioactivity. So is his jewelry, his watch, his telephone.

After being in office awhile, the President loses his personal fear and it is replaced by irritation. He feels overprotected. Often, he orders Secret Service agents away. In a slow-moving motorcade, the President sees ocean swells of smiling faces; the Secret Service watch for a sudden movement, a flying object. The function of these men is, when necessary, to place their bodies between the President and potential danger. This becomes difficult in a follow-up automobile.

Mr. Kennedy said that his feelings were the same as President Abraham Lincoln's. "Any man who is willing to exchange his life for mine can do so," he said. Leaving church, with two Secret Service men in front of him and two behind, Mr. Kennedy used to crouch lower and lower. His joke was to whisper to the two men in front: "If there is anybody up in that choir loft trying to get me, they're going to have to get you first."

When he was a United States Senator, in the spring of 1959, Kennedy received a note from Mr. Harry A. Squires of Lakewood, California. The reply is revealing:

"The historical curiosity which you related in your letter of May 4th is, indeed, thought-provoking: 'since 1840 every man who has entered the White House in a year ending with a zero has not lived to leave the White House alive.' . . . On face value, I daresay, should anyone take this phenomenon to heart . . . anyone, that is, who aspires to change his address to 1600 Pennsylvania Avenue . . . that most probably the landlord would be left from 1960–1964 with a 'For Rent' sign hanging on the gatehouse door."

In addition, Kennedy had personal courage. It was something he felt honor-bound to display. A warning from the Secret Service that it would be dangerous to attend a Harvard football game without prior screening guaranteed the President's presence. A reminder not to pause to shake hands with citizens behind police lines was almost always ignored. Nor did he appreciate seeing law enforcement men on rooftops with riot guns. The possibility of losing his life by violence occurred to this bright young man, but it never deterred him nor did he believe that it would happen.

Philosophically, to Mr. Kennedy, death was a state of abrupt termination. It stopped everything: thought, ideals, projects, progress, love, action. There is a Hereafter; there is a Heaven; there is a God sitting in judgment; there is a religious code through which these happy states may be attained, but the President was in no hurry to attain any of them. When Caroline brought a dead bird into his office, Mr. Kennedy averted his head. Against his will, he had shot a deer on the Lyndon Johnson ranch and it offended him to think about it. The news that a friend or acquaintance died brought Kennedy's activities to a halt. The dreadful finality of death stopped his thinking and momentarily numbed him. The previous summer, at Hyannis Port, he had taken an afternoon cruise with his father, victim of a cerebral accident, and when the President returned he had dashed angrily into the bedroom, ripped his tie off, and growled to Mrs. Kennedy: "Don't ever let me get like that."

The room was now quiet for a moment. Mrs. Kennedy returned to her bedroom as the waiter, George Jackson, wheeled in a second table with scrambled eggs and crisp bacon for her. Mr. Kennedy sat sipping coffee. Then, glancing at Kenny O'Donnell, he murmured: "Anyone perched above the crowd with a rifle could do it." The President's assistant slipped off the windowsill and reminded him to phone Mrs. J. Lee Johnson of Fort Worth. She had hung several original paintings in Suite 850. Mr. Kennedy also wanted to phone congratulations to former Vice-President John Nance Garner at Uvalde, Texas. He was ninety-five. It was time to start the business of the day. The phone calls were made.

In the corridor, Secret Service Agent Roy Kellerman spoke to Agent Winston Lawson in Dallas. The bubbletop was to remain off the car unless, of course, there was heavy rain at the time of arrival. Kellerman also advised that the President had again requested that Secret Service men remain away from the lead car. "He wants everybody to remain on the follow-up," said Kellerman.

The Kennedys drew no joy from Suite 850. The management had redecorated and painted the three rooms, but, when the President arrived shortly after midnight, he and Mrs. Kennedy had looked the place over with little appreciation. The first remark was: "Get that damned air conditioning off. I can't sleep with air conditioning." The suite had been selected by the Secret Service because, as it stood in an elbow of the corridor, it was the easiest to protect.

The paintings, the appointments drew no huzzahs from the sophisticated. The rooms seemed small, almost dowdy, to the Kennedys. To Fort Worth, a friendly, old-fashioned cattle town, the suite was lavish. The large bedroom on the left had a double bed, or rather two singles pressed together under a broad walnut headboard. There was a chest of drawers, a pineapple bridge lamp, a green-tiled bathroom with a recessed formica vanity and chair. The walls were painted blue; a portable television set was on casters near the bathroom door. Under a glass top was a message: "Hotel Texas. Check out time is 12:30 P.M. If you plan to stay after this time please contact Assistant Manager."

There was also a green upholstered chair on the opposite side of the room from the bed, a small coffee table, a golden lamp with a shantung shade. A couch reposed against the far wall with checkered upholstery; a phone, an ash tray, and a phone book reposed on the small table. An extra phone had been placed in this room. It was hooked up to the Secret Service men in the corridor. A few colored throw pillows were on the bed. Mr. Kennedy glanced out the window. Fort Worth was quiet. He saw lighted signs proclaiming: "Hotel Texas Official

Parking Lot." "The Fort Worth Press." In the distance, the yards of the Texas & Pacific Railroad and "Bewley Mills."

This bedroom, the larger, had been set up for him. He glanced into the bathroom. The tub, judging from the breadth of the shower curtain, appeared to be small. None of it was spectacular; still the management and the scions of the Amon Carter family, who owned it, had worked hard to display Western hospitality to the First Family of the nation.

The sitting room faced south and east. Mrs. Kennedy, fatigued, walked through it glancing left and right and rubbing her arms against the chill of the place. There were three windows facing south, toward the parking lot. On the opposite side of the room was a recessed bar. Radio music was coming from the ceiling somewhere. A green Chinese cabinet, which had no relation to anything else, held a television set behind gold-ornamented doors. At the corner window was a black-topped table with four chairs upholstered in blue.

In the ceiling were two chandeliers with electric candles. Proceeding toward the second bedroom, there was a long tan couch against the windows. A low serving table stood before the couch. A Gideon Bible and a lamp reposed on a cabinet. Walking slowly, Mrs. Kennedy pushed open the door to her bedroom. It was smaller than her husband's.

The bed was small, with a brass headboard designed like harp strings. End tables on each side were adorned with Oriental base lamps. She saw a blue easy chair, a leather chair, and one window. There were two framed crests over the bed. The closet was small. Two snack tables were folded inside. The bathroom was small. The tub was small. The glass shelf over the basin was not big enough for lotions and unguents.

The caterer, Peter Saccu, arrived in the suite and turned the air conditioning off. Mr. Kennedy asked if a window in his room could be raised "halfway." Saccu lifted the window and noted that the President might be awakened by the slamming of freight trains in the Texas and Pacific yards. Mr. Kennedy smiled for the first time. It wouldn't bother him, he said.

The word flashed up and down the hall that the President and Mrs. Kennedy had retired. The sudden relaxation of tension hit the Secret Service. Stiffly erect bodies sagged. The agents standing out in the drizzle on Eighth and along Main began to chat with each other. In the corridor, the soft whirl of elevator cables started and stopped. Some off-duty agents went to bed. Others looked for an hour or two of relaxation.

They asked the clerk behind the desk if there was any place open after midnight. The clerk pointed up Main. "The Cellar," he said. A group of agents walked up three blocks and saw a flashing sign. They entered, but The Cellar was not in the cellar. It was up a flight of dark stairs with a red light at the top.

A young man asked them for a dollar apiece entry fee. They gave it and he stamped the backs of their hands with invisible ink which shows up under ultraviolet light. This was to identify customers who might depart and return or non-customers who claimed that they had already paid.

The Cellar was a huge square room lighted in perpetual dusk. The walls were painted flat black. In a corner on the right, a combination of young men with long hair hooked their guitars to loudspeakers and lyrics roared. The waitresses, young and slender, with opera-length stockings, and breasts which appeared to float on top of corset stays, took orders for "setups." In Fort Worth no one can buy an alcoholic drink. Customers bring their own in paper sacks and buy ginger ale, Coke, and ice in a glass.

The Secret Service men found seats at a long straight board table in the rear. They shouted their orders as the country rhythm pulsed against the walls. The only lights were small red bulbs in the ceiling. A waitress, noting that they were strangers, tried to tell them that soft drinks would cost sixty cents. No one could understand her.

It was too dark to read a small legend in white on the far wall. It said: "Tomorrow is cancelled."

A sweet, sad face was framed in the window. Mrs. Linnie Mae Randall was washing the dishes in the kitchen sink and looking out at the veiling of rain, at the same time partly listening to the breakfast chatter of her younger brother, Buell Wesley Frazier, her mother, and the children. The house is on West Fifth Street in Irving, but the kitchen faces Westbrook Drive.

She saw Lee Harvey Oswald, bare head down, coming up Fifth Street with a long package in his hand. He held the fat part under his arm; the tapered end was pointing at the sidewalk. The rain didn't seem to bother him. He walked steadily, up Fifth, across the corner lot, toward Mrs. Randall's garage. She kept watching him, a dark, pretty woman with shoulder-length black hair. By rote, she set the dishes upright in the drain.

Mrs. Randall did not know Mr. Oswald, but in a way she had gotten a job for him. In early October her neighbor, Mrs. Ruth Paine, had asked about employment for the husband of the Russian woman, and Mrs. Randall had said that her brother Wes worked at the Texas School Book Depository, and they were looking for people. It wasn't much; if a man earned a dollar or a dollar and a quarter an hour, it was as good as he could expect. The work was digging book orders out of the warehouse on the upper floors and bringing them down for shipment to school districts in Texas.

Lee Oswald got the job. He wasn't at it long, but he didn't like it. There was nothing to it but drudgery. He picked up book orders on the main floor, took one of two elevators to the sixth or fifth floors, dug out the proper number of copies of the right book from the right carton, and brought them back downstairs. Lunch was 12 to 12:45. A ten-minute coffee break could be worked into the afternoon. Mr. Roy Truly, the boss, expected all hands to be on time, 8 A.M., and he was a fair and firm man who expected a day's work and no trouble.

Oswald had been there six weeks. Mrs. Randall watched him walk toward her garage and she wiped her hands on her apron and opened the kitchen door in time to see him open the right rear door of Wes's old car and drop the bundle on the back seat. He stood waiting under the shelter of the overhead garage door. He was alone, but not lonely; friendless and solemn. Wes had helped him get the job, and it was Wes who gave him a free lift to Irving on weekends and a free lift back to Dallas.

Her brother was sipping his coffee and talking about some kiddie program that the little ones had watched earlier this morning. Wes was a tall, dark Alabaman with the twang of the back country heavy on his tongue. He took the easy way with people, work, and problems. He enjoyed getting along with people, and so he was often given to conversation in which he said what he thought the other person might want to hear. It didn't cost anything.

Mrs. Randall was peeking out her kitchen window when she saw that Mr. Oswald had changed his position. He was staring in at her. This irritated Mrs. Randall. "Wes," she said. "Somebody waitin' on you out there." Her brother left the table, donned his jacket, snatched a bag of lunch, and went out to the car. He hopped into the driver's side and turned the windshield wipers on as Lee got into the front seat with him. Frazier kicked the old car and it started. The battery was low, and the teenager knew that it would either start at once or die with a moan.

Wes turned to look behind him. "What's in the package, Lee?"

"Curtain rods," Oswald said, looking at the glistening pavement along Fifth Street. The drive from there to the Texas School Book Depository was about ten miles and required about a half hour in the morning flow of freeway traffic. "Oh yeah," said Wes. "You told me about them yesterday." Lee didn't nod.

Most conversations were closed abruptly. Yesterday, when some of the fellows had lain dozing on the ground floor counters at lunch time, Lee had borrowed a *Dallas Herald* and had read the story about Kennedy's visit. He had seen the chart showing that the motorcade would end as it passed the Texas School Book Depository. After that, according to the newspaper account, the group of cars would go under the overpass, turn up onto the Stemmons Freeway, and get off at the Harry Hines cutoff for the Trade Mart.

Two hours after lunch, Oswald had met Frazier in the rear of the ground floor at the Depository and asked if he could get a ride to Irving. Wes nodded. "Sure, Lee. But I thought you usually go out Friday." Lee said he wanted to visit his wife to pick up some curtain rods for the little room he had in Dallas. "Sure, Lee," said Wes. "Any time." Oswald's room at 1026 North Beckley, across the Trinity River in the nearby Oak Cliff section, had four small windows on one side. The landlady, Mrs. A. C. Johnson, had long ago equipped them with venetian blinds and filmy curtains. She did not permit roomers to make changes.

The car hit speed and swung east on Carpenter Freeway. As the traffic melted onto Stemmons Freeway, the two young men could see the skyscrapers of downtown Dallas. Wesley Frazier stared through the snapping wipers, and he tried to think of something to say. He knew he had to be careful, and he was aware that Oswald felt at ease with children. "Did you have fun with the little ones?" he said. Oswald nodded. "Yeah," he said. Then he smiled. "Yeah we had fun playing around."

Lee always carried a little brown bag with a sandwich and an apple. "Didn't you bring lunch today?" Wes said. Oswald said no. "I'm going to buy some in the lunchroom." Wes did not understand Lee, and he was abashed by the total absence of male conversation. For example, he would not dare ask *where* Lee had a room. Nor did he understand how a man could have a wife and children living with a lady in Irving, while the man had a small place in Oak Cliff. It was none of Frazier's business, and any allusions to Oswald's private life would elicit a dead stare.

Once, about a year before, Mr. Oswald began to write a book about Russia called *The Collective*. It was not finished, although he paid a typist to render one part of the manuscript into the printed word. At the time, he thought that it was "literary" to include a short biography of the author, and Wes would have learned more in the one written paragraph than in all the weekend trips to and from Irving:

"Lee Harvey Oswald was born in Oct 1939 in New Orleans La.," it read, "the son of a Insuraen Salesmen who early death left a far mean streak of indepence brought on by negleck. Entering the US Marine corp at 17 this streak of independence was strengthed by exotic journeys to Japan the Philippines and the scores of odd Islands in the Pacific immianly after serving out his 3 years in the USMC he abonded his american life to seek a new life in the USSR full of optimism and hope he stood in red square in the fall of 1959 vowing to see his chosen course through, after, however, two years and alot of growing up I decided to return to the USA. . . ."

The sun was kind to the pale beauty of the White House this morning. The day was cloudy, but the yellow shafts of light poured through the holes and fingered the great building in Braille. The men in the Situation Room in the cellar of the West Wing worked on the next précis to be sent to the President through the military switchboard. The policemen at the East and West Gates stopped cars, counted and studied faces, examined passes, and waved the vehicles on. Two presidential assistants were in the barbershop. Others, young lawyers, worked at their desks on the multitudinous tasks of legislation, studying it, writing it, rewriting it, condensing it. For Bernard West, the urbane chief usher, it was an easy day in his between-floors office. The President's personal secretaries—with the exception of Mrs. Evelyn Lincoln, who was with him—had a backlog of letters to write, some to be stamped with Mr. Kennedy's signature, others to wait on his desk for his personal scribble.

The Rose Garden, outside the President's office, dozed behind stiff, unyielding leaves, waiting for winter. The swimming pool remained heated at ninety degrees, even though no one, including the Kennedy children, would use it today. The long melancholy corridors, flanked with the faces of history in frames, were empty except for the stations where Secret Service men stood. The huge East Room was in semi-darkness, with an ornate grand piano at one end and a Stuart painting of George Washington at the other.

A crowd queued up outside the East Gate. These were citizens. They would tramp through all the public rooms on the ground floor, and some would pluck a piece of gold fringe as a souvenir, and others would stare gaping at the deep rugs, the damask wall covering, the gold pen and ink sets, and the dishes of many administrations. The White House was, at this time, looking better. Jacqueline Kennedy had assumed the burden of calling in decorators and going over the White House warehouse inventory to see what furniture, what paintings, what bric-a-brac might be restored.

She had brought such color and beauty and life to the old mansion that it was being compared to the elegant palaces of the Old World. The work had been arduous, and sometimes she was forced to beg an owner of an historical chair or bust to please donate it. She had even established a President's personal library in a small room on the ground floor, near the South Grounds. The magic of her work showed in the brightness and dignity of the rooms. There was an historical *lift* to the mansion; visitors who had once shouted carelessly now whispered. Guides passed out ornate pamphlets explaining the significance of the items in each room. President Kennedy so admired the result that, instead of holding all formal dinners in the state dining room, he now preferred to use the Blue Room for small parties and both for big ones.

In the East Wing—"the female section"—a necklace of offices held ladies who responded to social mail and Mrs. Kennedy's personal mail; here the invitations to White House galas were executed in script; seating arrangements were worked out with the care of a good chess game, and the First Lady's ballroom gowns were on display in half-lifesize renderings. On the second floor, there were some men. Not many, but a few.

This was the Secret Service office of the White House detail. It had a female receptionist, but this was the only concession to the frilly wing of the White House. Inside there were three desks. They were plain and uncluttered, unlike the men who used them. Gerald A. Behn, special agent in charge of the White House detail, sat here to plot the safekeeping of the First and Second Families of the nation with the care someone else might plot their undoing. His work began a full three weeks before each presidential trip.

The moment the President made a commitment to go somewhere, Behn's work was under way. In the case of Dallas, he followed procedure by pulling a PRS (Protective Research Section) file on the city, and this card, in Secret Service headquarters, would list any persons in the area thought to be potentially dangerous. All persons who were psychiatrically homicidal were listed; all cranks who wrote threatening letters; all persons who had been involved in political riots or arrested and detained for political violence.

Every street the President planned to traverse in each city had to be "sanitized" long in advance by agents. Every name on the PRS list had to be checked for whereabouts and security. Every building Mr. Kennedy might step into had to be screened and searched. The day before the President arrived, men had to be posted at every entrance and exit to each of those buildings. Through Chief James J. Rowley of the Secret Service, liaison had to be established with other governmental investigative agencies, such as the Federal Bureau of Investigation and the CIA so that, if they had any information which might augment the safety of the President, it would go into Jerry Behn's hopper.

The agencies worked well together. So well, in fact, that Chief Rowley often sent some of his Secret Service men to the FBI to take short courses in investigative procedures and the newer and more bizarre devices of detection. In late October 1963, the word that went out from Behn's office was "San Antonio, Houston, Fort Worth, Dallas." The PRS file didn't have much material. The FBI and the CIA had very little.

The name of Lee Harvey Oswald did not come up. Nor would it. He was a defector who had gone to the Soviet Union and had returned with a wife and child. The State Department had a file on him, but it was a file of insolent correspondence, closing with the department's lending him money to come home. The Navy Department had a short dossier on Lee Harvey Oswald. He was a onetime Marine who, after fleeing to Russia, had been court martialled and his honorable discharge changed to a dishonorable discharge. The young man had protested to the Secretary of the Navy, Mr. John Connally, but the DD was allowed to stand. The FBI was aware of him, but only as a "Marxist" who appeared to be "clean." He had never attended a Communist Party meeting, never consorted with Reds, never tried to get employment in a sensitive defense area, appeared to have considerable trouble with his marital life, and bounded from one cheap laboring job to another.

Most of the people Behn had to worry about were emotionally disturbed. A history of assassins is a glossary of persons sick and obsessed. Lee Harvey Oswald never got drunk, never wrote threatening letters, and once told his wife that if the President was killed, he would be replaced by another man who "thinks the same and will keep up the same program."

What worried Gerald Behn was that the Secret Service has no authority over the actions of the President. They had the responsibility but could not make the decisions. Word had already come over the teletype that it was raining in Fort Worth but that Mr. Kennedy did not want the bubbletop on and asked that the Secret Service men remain on the follow-up car. The first part created no anxiety in the White House. The second part did.

President Kennedy was becoming increasingly irritated with the Secret Service. Behn recalled that the Chief Executive had told him, forcefully, to keep his men away from the lead car. On another occasion, in the midst of a motorcade, he had excitedly waved off the men who trotted beside the car. It had reached a stage where Behn and his assistant, Floyd Boring, were no longer popular with the President. He saw them as the leaders of the intruders. Sometimes he almost bumped into an SS man outside his office door. He felt that he was seeing SS men everywhere.

One afternoon, when the President's eyes blurred, he asked to see an ophthalmologist. The Secret Service asked him to please remain in the White House until they could send men to the doctor's office, clean out the waiting room, study the examining room, the doctor, and his nurse, and "sanitize" the sidewalk and the buildings on the opposite side of the street. After all this was done, Kennedy left the South Grounds with a Secret Service car ahead of him and one behind. For the President, it was beyond bearance. At the doctor's office, he had had to sit in the car until the men in the sunglasses nodded to him that it was safe to emerge.

Gerald Behn had been running beside the President's car in Mexico City in June 1962, amid the din of a full-throated Latin welcome, when a "beatnik" broke the police lines and planted himself squarely before the President's car. When he saw that it would not stop, he skirted the fender with a twist of the hip like a matador avoiding horns. As the car passed, the bearded one approached President Kennedy, and Mr. Behn had knocked him down with a punch. The man had been arrested by the Mexican police and was found to be an American with a police record. The President was angry. He told Mr. Behn that he should not have hit the man.

In a Berlin motorcade, enthused youths broke police lines and the Secret Service agents dropped off the follow-up car to interpose themselves between the President and his admirers. This also incurred presidential wrath. In Seattle, Phoenix, and Bonham, Texas, in November 1961, Mr. Kennedy ordered the Secret Service to stop riding the rear bumpers of his car. Only four days ago, in Tampa, Florida, the President looked over his shoulder and saw Special Agents Donald Lawton and Charles Zboril on the rear steps of his car and he ordered them off. The motorcade was moving too fast, so Floyd Boring radioed the follow-up car and the President's driver to slow down. The Secret Service men got off.

It was not that the President did not appreciate the protection. He didn't want it to be obvious. When he was in a good mood, he said: "Protection is Jim Rowley's job. He has never lost a President yet." Mr. Kennedy knew as well as the Secret Service did that 100 percent protection is impossible. "Any man who wants to trade his life for mine . . ." The percentage of protection decreases with the daring of the "boss." If he waves his personal police force away, he hampers its work. If he departs from schedule, or stops the motorcade to shake hands, or leaves a welcoming group to walk along the edge of a crowd shaking hands, or even if he stands still in a street of tall buildings, his percentage of protection drops to the danger point.

Now Gerald Behn had the news from Fort Worth. He could sit at his desk and worry. He could call his Chief and win understanding and sympathy. Or he could proceed with the small tasks of his office, knowing that Mr. Kennedy had always been proved right before, and the Secret Service wrong. Nothing had ever happened to him that could be called dangerous. In a dozen hours, the President would be at the LBJ ranch for a day or two, and the place was a cinch to secure. It was off the main road, and the entrances and exits were easily sealed. The nearest town, Johnson City, was about fourteen miles away. The two families would rest up, enjoy a Texas barbecue, invite some of the Johnson friends over, then take a plane back to Washington. A simple and safe procedure.

On the wall of Mr. Behn's office hung a framed poem:

Fame is fleeting, fitful flame
Which shines a while on John Jones' name
And then puts John right on the spot;
The flame shines on
But John does not.

The President took a call from the White House. It was Richard Goodwin, an assistant, who said that *The New York Times* was about to write an article about him. Goodwin had told the *Times* "No comment," but he wanted presidential advice in the matter. Mr. Kennedy ordered his man to go ahead and write a release about him. At the same time, Pamela Turnure, the dark, attractive secretary to Mrs. Kennedy, was issuing a press statement in the name of the First Lady that Texas had turned out to be just as warm and hospitable and friendly as she had always heard. Yesterday had been "a wonderful day."

The White House press corps was stacked in rooms all over the big U-shaped hotel. The dean, gray-haired Merriman Smith of United Press International, had filed some overnight copy; Seth Kantor made notes that the crowd in the parking lot had started to collect "before dawn." Charles Roberts of *Newsweek*; Tom Wicker of *The New York Times*; Robert Donovan of the *Los Angeles Times*; Jim Mathis of the *Advance Syndicate*; Jack Bell of the Associated Press; there were correspondents accredited from Washington, from New York, from Fort Worth, Dallas, Chicago, and there were newspaper photographers, television cameramen, radio and TV reporters, Western Union telegraphers and, in some cases, editors-on-the-scene to correlate the efforts of groups of reporters.

The importance of the press was never underestimated by the Kennedys. The President, having served his apprenticeship as a reporter, understood professional jargon such as "overnight," "bulldog," "lead to come," and "folo-up." In a manner of speaking, he was his own press secretary. The post was nominally filled by a stout, jolly man named Pierre Salinger, a onetime investigator for the Senate McClellan Committee, whose counsel was Robert Kennedy. The President dealt with the press through Salinger, and the reporters heard only what Mr. Kennedy wanted them to hear, without exposing himself to charges of "managing" the news.

The attitude of the President was that the press, in a real sense, was akin to a fire: it can warm a man, but it can also burn him. At morning conferences, he and Salinger tried to anticipate the questions—especially "the curves"—which might be asked at Salinger's daily briefings. The increasing importance of the press to presidential aspirations is seen in the fact that, in the Woodrow Wilson administration, his personal secretary, Joseph Tumulty, dealt with the newspapers when he was so disposed, whereas in the Kennedy administration, Mr. Salinger, assisted by Malcolm Kilduff and Andrew Hatcher, occupied a suite of White House offices full of researchers and stenographers.

It is possible that the Kennedys (Mrs. Kennedy feared press coverage and especially unflattering photographs of herself) attributed more importance to the press than it deserved. Mr. Kennedy began his Administration by trying to seal the sources of news and information. He demanded that ministers of Cabinet rank and less, even servants in the White House, agree not to take notes and write tracts, magazine articles, or books about their experiences. He also asked bureau and department heads not to write articles of major importance or make speeches without first submitting the copy to the White House for endorsement.

On the surface, Mr. Kennedy handled the press with urbane wit and a first-name camaraderie. As a minority President, one who had won election by the narrowest of margins, he was aware that he needed the goodwill of these questing men and women who, by the nature of their daily work, had to fear being used by a charming man and his attractive wife. The breakfast speech on this particular morning was of no moment if addressed solely to the 2,000 members of the Fort Worth Chamber of Commerce and their guests. It must be directed more to the press, which could funnel the words and their import to 180 million Americans outside the area.

Under the surface, the President was chagrined to find that the goodwill of the press had to be solicited anew every day. No warm handshake, no inscribed photograph, no off-the-cuff confidence could keep the press loyal to Kennedy. Their words flogged his hide after the Cuba Bay of Pigs disaster. Their attitude after the Vienna Summit Conference was that the Russians had tweaked the young man's nose. The current trip to Texas was assessed as a two-day whirlwind to sweep up the 25 electoral votes of Texas for the Democratic Party. To Mr. Kennedy, the smiling faces he saw everywhere represented 9.25 percent of the 270 electoral votes required for reelection.

The intraparty fight between Senator Ralph Yarborough and Governor John Connally assumed no great importance in John Kennedy's mind until he read the Texas press. He had been misinformed about the depth of the schism and, when he left the White House yesterday, the President had been certain that a presidential knocking of heads together would settle the dispute and align all Democrats behind him. He was pained to find that Governor Connally had arrogated to himself all arrangements for the trip, and invitations, too. The conservative side of the party, which would never support the President with enthusiasm, got all the choice seats, while Yarborough's liberal followers, who would and did endorse Kennedy, were cast in the role of pariahs and outcasts.

Kennedy became increasingly irritated. This morning he had read in the Dallas papers that, far from healing the Connally-Yarborough breach, he was widening it. Governor Connally, who had postponed this visit several times because, even though Kennedy had heeded the intercession of Vice-President Lyndon Johnson and appointed Connally, who was Johnson's onetime assistant, as Secretary of the Navy, the Texas Governor was never a "Kennedy man." As host, he was in the position of a man who could manipulate the luncheons, dinners, and cocktail parties in such a manner that his following sat in the places of honor while Yarborough's liberal wing was either ignored or confined to the back of the hall.

Kennedy found it impossible to bring the Governor and the Senator together for a smile and a handshake, so he settled for asking that Yarborough ride with Vice-President Johnson. In this instance, Johnson was tractable, but Yarborough declined. The President, increasingly incensed at what might become a Connally trap, stopped requesting that the Senator ride with the Vice-President and demanded it. Yarborough rode with the Vice-President and Mrs. Johnson on the final short midnight leg of the trip from Carswell Air Force Base to the Hotel Texas. Few people were in the streets to witness the demonstration of party unity.

This morning, the Washington press corps, which had featured the President's unexpected welcome from Texas, began to read the *Fort Worth Telegram* and the *Dallas Herald*. The sophisticated wire services had been aware of the party fight, but had not pinned their leads on it because, like Kennedy, the press assumed that Kennedy's personal charm would bring the contentious ones together. They were changing their minds.

That is why the President, through Kenny O'Donnell, ordered Yarborough to ride with Johnson "or walk." In effect, he was working the easy side of the street. The proper move would have been to thrash the matter out alone with Connally and Yarborough, but Kennedy was too insecure a party leader to risk a state ultimatum. The Governor, as O'Donnell and O'Brien should have known, was not even sympathetic to the President. Connally, facing reelection in the next year, felt no desire to be seen with the President or to be his host. Connally was a handsome man with a long splash of white backing away from the temples, but he had a stubborn jaw and a thin lip. He and Nellie had come a long way, and he had no relish for sitting in the jump seat of anybody's car.

In Texas, the Governor was accustomed to the comfort of the big seat. He knew his people and he felt that his star was ascending, while that of Kennedy had dissipated its light in a shower of sparks. A national political poll had dropped the President to his lowest point of popularity less than a month ago, and Congress had felt no mandate from the electorate to push the Kennedy reforms. In a test of loyalty, the Governor felt much closer to Vice-President Johnson than to the President. He was also aware that Johnson and Yarborough had fought each other "bare knuckle" for federal patronage in Texas; it was also true that Yarborough had fought the Texas leadership of Lyndon Johnson, even though the Vice-President had a majority of the party faithful in his corner.

At times, it seemed that the President didn't understand Texas at all. "Lyndon thinks we'll carry Texas next year," he had said to the Governor, "but he says it will be hard. Texas is Democratic country; we shouldn't have a hard race in Texas." It seemed pointless to explain the situation again. Instead, the Governor hoped that Mrs. Kennedy would be with her husband in Texas. At the White House, Mr. Kennedy had glanced up sharply from his reverie and said: "I agree with you. I would hope that she would come."

Obviously he was in no position to pledge the presence of his wife. He would consult her about it. But it was Johnson who remained in the most awkward position. He had vowed, when he was nominated, to subscribe with enthusiasm to everything his President proposed, and he had been caught, time and again, doing things and saying things which were opposed to his best political judgment. To keep peace in the family, Johnson had to agree to share an automobile with Ralph Yarborough, even though every student of Texas politics would know that Mr. Johnson, a master assessor of the practical, would not share any part of the state with him. The two men in one car might be worth a photograph, but it would be like tossing two cocks in a pit.

On the thirteenth floor of the Hotel Texas, Rufe Youngblood, the Vice-President's bodyguard, asked: "Anything new from PRS?" There was nothing new from the Protective Research Section of the Secret Service. On the eighth floor, Clint Hill asked the man in front of 850 if Mrs. Kennedy was going across the street with the President. The answer to that was no, so Hill got a cup of coffee and a biscuit and leaned against a corridor wall.

Mrs. Johnson had finished dressing and felt a sudden uneasiness. Her hand was shaking. She had a poetic passion for her native state, but Dallas was the only city which had ever frightened her. Once, just once, an unruly mob of people had chased her and her husband across a street and into the lobby of the Hotel Adolphus, and the lady had never quite forgotten it. No real harm had come to them, but Mrs. Johnson knew that "Big D" took its politics seriously, and the extremely conservative side of it was not timid about demonstrating. She wasn't worried about her husband this time; it was the President. She wished that it was all over and they were at the ranch. She kept thinking: "There might be something ugly today."

On the eighth floor, Mrs. Kennedy had finished breakfast earlier than she had planned. This would give her more time to dress. She might have been happier (Mrs. Connally, too) if the President had decided to put the bubbletop on the car. Men didn't seem to realize what an open car could do to a coiffure, even a slow-moving automobile. The decision had been made, and Mrs. Kennedy had laid out a street outfit the night before. It consisted of a strawberry pink suit with a burl weave and a grape-purple collar. She would comb her shoulder-length dark hair down straight and part it in front toward the left side of her head. She had a matching pillbox hat to be worn well off the forehead. She would wear a small gold bracelet on the left wrist and white gloves. It was a good cool-weather open-car ensemble. In addition, Mrs. Kennedy dropped a pair of sunglasses into her purse.

The nine-year-old Chevrolet came off the freeway. Wes Frazier wished he could think of something to say, but he couldn't. The big trucks had tossed mud and mist at his windshield, and the old wipers smeared the mess so that Frazier had to stare around the arc of the wipe to see. Now he turned them off and said: "I wish it would rain or clear off altogether," but Lee kept his pouty mouth closed.

The car was driven around behind the Texas School Book Depository. The area was flat and open, full of railroad tracks and sidings. Wes pulled the car into a space and put the clutch in neutral. Then he revved the engine to restore a little strength to the battery. Oswald said nothing. He opened the door and got out, reached in back for his curtain rods, and left. The back of the Depository building was two hundred yards ahead.

Frazier was saying something, but Oswald could not hear him over the roar of the old motor. He kept walking toward the loading platform. Wes shut the engine off and followed, calling to Lee, but he didn't pause. On top of the building was a big flashing Hertz Rent-a-Car sign. The clock on it proclaimed the time to be 7:56 A.M. Under the sign, a cote of gray pigeons nested, waiting for the rain to stop, so that they could swoop along the railroad tracks, looking for spilled grain.

Lee Harvey Oswald didn't say "thank you" for the free ride. This was not unusual. But Wesley Frazier kept asking himself why, for the first time, Lee was walking ahead of him? They always walked the last few yards together, watching the diesels yanking and shoving strings of cars back and forth. This time, Lee Harvey Oswald reached the loading platform fifty feet ahead of Wes, and didn't hold the door for him.

8 a.m.

The second platoon lined up. The policemen stood under their metal helmets at attention. Lieutenant William R. Fulghum studied the faces until the three sergeants had the men dressed in two rows. This was the Southwest Substation of Dallas; the men patrolled the residential areas of Trinity Heights, South Oak Cliff, Fruitdale, and Oak Cliff. This was a fragment of a police department, one of the outlying fragments which would not be devoted to the big assignment of the day.

Some of them had already listened to the dispatcher calling the brass at headquarters, at Love Field, the Trade Mart, Stemmons Freeway, on the motorcade route. Others had heard the solemn warning of the chief. These men were glad to be out of it. They did not know that the presence of a President for 180 minutes would require a cancellation of days off, a repetitious dry run of the motorcade route, a host of conferences with Secret Service men—a sapping of the city's law enforcement for the safety of one man.

Sergeant Hugh F. Davis called the roll and the patrolmen hollered "here" to such names as Truman Boyd, Rufus High, J. D. Tippit, and Roy Walker. Dallas had a humorless department. It was composed largely of young "skinheads" with a passion for promotion. The word had been out for years that the men who controlled Dallas demanded a clean city, and these young men, zealous and with little formal education, kept it clean.

They also kept the old jail on top of city hall well populated. Some of the trustees had to peel potatoes all afternoon just to keep the sullen Negroes and loudmouthed whites fed. Every day, batches of new prisoners were being brought in, mugged, fingerprinted, and led upstairs to the dismal rows of cells. Naked bulbs stared down at gray concrete.

Dallas policemen had a deceptive politeness reserved for lawbreakers. They said "yes sir" and "no sir" while writing out a traffic violation, and, if this encouraged a motorist to use bitter and abusive words, he was quickly carried off in a squad car to the prison on an additional charge: resisting arrest. When violence occurred, the Dallas policeman was never afraid to draw his revolver and seldom hesitated to use it. His manner of dealing with known hoodlums who could not be trapped into breaking the law was to make certain that he "fell," thus converting an ambulatory person into an ambulance case.

Lieutenant Fulghum recited the patrol areas with Boyd, Tippit, and Walker, and reminded all hands that they were to tune in to police Channel Two, because One was being reserved for Curry and the motorcade. They were to be especially vigilant, because all suburban squads were short of men today, and, if trouble developed, they would be moved from area to area, and Fulghum wanted an immediate response.

At police headquarters, Captain Perdue W. Lawrence had his huge detail of police motorcyclists ready, and he again read the orders of the day. This group was to report to Love Field and patrol it to keep people away from the President and his party. No strange planes would be permitted near Gate 24, even if given permission by Love Field tower. The entrances and exits to the field were to be sealed; additional men would patrol the passenger terminal.

When Mr. Kennedy was ready to leave the field for Dallas, the time would probably be 11:45 A.M. A line of motorcycle policemen would follow the first car, which would be a quarter of a mile ahead, and it would be followed by Chief Curry and a Secret Service man in the headquarters car. Four men on two-wheel motorcycles would flank the President's car, slightly behind him on both sides.

In the event that the crowd broke the police lines at any intersection, the Secret Service would hop off their follow-up vehicle and fend the people off. If they needed help, they would motion the four motorcycles forward. Otherwise, they were to remain slightly behind the President. Captain Lawrence was still reading the order when a thought occurred to him. His men, on three-wheel motorcycles, would close out the back of the parade, but there were too many of them.

Also what should they do in case someone tried to cut in behind the parade and move up past the three-wheelers? The captain got on Channel One and asked Assistant Chief Charles Batchelor, already at the airport, what to do. No, Batchelor said, no regular Dallas traffic should be permitted to overtake the motorcade from behind. Lawrence asked if he had permission to send some of his extra men up on Stemmons Freeway. He wouldn't need them in the motorcade, and it seemed to him that, when the presidential cars left Dealey Plaza, they would get up on Stemmons for a high-speed ride to the Trade Mart.

Part of the freeway had rises in the roadway and the President's men might go over one and find a citizen dead ahead at low speed. This would constitute a danger. Batchelor agreed, and Captain Lawrence said that he would send some of his men directly from the airport to Stemmons, and that they would be staggered along the freeway so that they could signal each other to get local traffic out of the way as Kennedy negotiated the last part of the Dallas run.

Lawrence made the changes, but to protect himself he put asterisks on the squad chart, showing where the men had been diverted and why. Someone said it was still raining out, and the captain gave the men permission to wear raincoats. "If the weather clears," he said, "take those raincoats off and stuff them in the motorcycle bags."

The department refused permits to all groups which might want to protest anything or anyone on this day. Their spokesmen were asked to return to headquarters on Monday, and permission would be granted. The worst thing that could happen, the superior officers knew, was for some bystander to throw a rotten tomato or an egg at the motorcade. Or someone in a lofty office building might drop an object on the parade. This, Curry knew, would be featured on every front page in America, and on every television news channel. Dallas was jealous of its image as a rich, reactionary city, and it was not going to be baited into cheapening that image by permitting some emotionally unstable adherent to bring the city into national contempt.

The possibility of an assassination attempt had been studied, but this was thought to be a remote possibility. The Secret Service had scanned the police lists of agitators, professional protestors, the insane-at-large, the known communists, and the poison pen writers. Curry's police worked in concert with the Secret Service in running down every lead. The city was as secure as it could be made. Even those violent ones who had left Dallas were checked at their distant residences to make certain that they would not be in the city on Friday.

And yet, in downtown Dallas, quite often one out of every five local citizens carried a gun. The city, part Southern in character, part Western, believed that it was the right of the citizen to carry a gun if he had a permit and did not use it unlawfully. Thousands of permits had been issued, and the Dallas Police Department had no reason to regret the number of arms bearers.

More young than middle-aged men carried firearms. It seemed, symbolically, to be a part of growing up. Gun play came to life in only two types of cases: too much alcohol or a fight between two men over a woman. It was common for police, in heading off a car full of wild boys, to frisk the occupants and find three or four revolvers. They were the badges of manhood; they gave the young men a strutting walk and a consciousness that they were as big as any man in the world, but their use was considered a violation of the police permit.

Shotguns and rifles were even more common. The outlying areas of the city were good for hunting, and men cleaned and pampered their rifles in the manner of little boys shining bicycles. In the off-season, many of the men took their rifles to local ranges to fire at targets and reset the windage markers and the drift of steel-jacketed shells. This was not equated with violence in Dallas, because the constitution guaranteed citizens the right to bear arms, and Dallas was not too many generations removed from its own frontier settlement days, when the Trinity River, a watering place for cattle, frequently was a battleground.

Tippit, a tall, wavy-haired policeman, was a minute late getting out to his car, number seventy-eight, and moving it away from the Southwest Substation for patrol. At 8:01 A.M. he was on his way.

The Hotel Texas elevator took Vice-President and Mrs. Johnson down to the mezzanine with their Secret Service man, Rufus Youngblood. Mrs. Johnson, who was gifted with the animated face of a happy child, composed her features in rare solemnity as the car slid silently downward. Secret Service men watched them get off the car, nodding good morning. In the dimly lighted Longhorn Room, the Democratic leadership of Texas gathered, whooping loud greetings to each other and falling into whispering groups along the wall.

On the eighth floor, Mrs. Evelyn Lincoln finished breakfast and hurriedly recompleted her makeup. She was a small, dark woman, the adoring private secretary of John F. Kennedy for twelve years. At the White House, it was her function to keep the half dozen daily deliveries of mail coming into the President's desk between appointments, just as it was expected that she would keep the covey of stenographers in her office busy with outgoing mail. This consisted of more than the President dictated. If he learned that a Democratic politician was ill, it was Mrs. Lincoln's function to compose a "get well" letter or send flowers or both. On her desk small plates of enticing candies reposed for those who, *en passant*, felt like reaching.

She kept her door to the President's Oval Office slightly ajar unless he ordered it closed. This was so that she could hear him if he called to her. It was she who knew when to admit George Thomas to the inner office with freshly pressed suits of clothes, a tie and a shirt, several times each day. He crossed the office noiselessly when the President worked alone and hung the clothing in the little lavatory.

Mrs. Lincoln was also the liaison between the President and Mrs. Kennedy. She could tell Jacqueline when he was too busy to talk to her; she could judge the time that his daily work would be completed and he would be ready for dinner in the President's eight-room home on the second floor; Mrs. Lincoln kept a secret "log" of every presidential appointment and what was discussed, promised, and refused; she lifted every fiftieth letter from the mass of unsolicited mail and opened it and placed it among the letters the President should read.

She was more than an amanuensis; Mrs. Lincoln was Kennedy's confidante from the time he emerged on the Washington scene as the junior Senator from Massachusetts. She worked long, exhausting hours because she believed in Kennedy's "star." Her office walls were heavy with color photographs of the First Family, and she applied herself with complete loyalty to his every wish.

It wasn't necessary for Mrs. Lincoln to have made this trip. There were younger stenographers in the White House pool, but Mrs. Lincoln, like Mrs. Kennedy, had a desire to make this one quick trip to Texas. She had written happily to her sister-in-law's sister, Mrs. Jo Ingram of Dallas, inviting her to breakfast "very early" at the Hotel Texas in Fort Worth, so that Mrs. Ingram might meet the President of the United States briefly after breakfast. Mrs. Ingram, in a burst of feminine enthusiasm, had replied that she would be there, and she was bringing her cousin and her cousin's daughter.

The four sipped hot coffee and waited for the Secret Service to advise Mrs. Lincoln that the President was ready to emerge from his suite. The anticipation was too much for the ladies. The breakfast was still on the table in the bedroom, and the four rushed up and stood in the corridor near the elevator, watching with excitement as men hurried up and down the hall, screening the waiting elevator and the young Mexican girl who stood at the controls.

In the parking lot, the crowd now numbered more than four thousand. In the rain they looked like a conglomerate of shiny colors. They lit cigarettes; they joshed each other; the mounted police herded them into a large circle around the speaker's stand and the loudspeakers. Some stood off on the next block, leaning against shop windows which had awnings. Many wore the big fawn-colored Texas hats. These were not the elite of Fort Worth, who would be inside at the breakfast. These, until two days ago, were the forgotten supporters of the President: the railroad brakemen, the clerks, the off-duty policemen, housewives, waiters, mechanics, store managers, also the unemployed.

They did not think of themselves as an "afterthought." Fort Worth is a lusty, hospitable Western town, and the local citizens were glad of the opportunity to see their hero person-to-person and to see the beauteous Mrs. Kennedy. These, multiplied many times, were the men and women who had given the electoral votes of Texas to John F. Kennedy in 1960. They asked nothing in return but a hearty hello, a few witty expressions to be remembered later, and a "Godspeed."

At 8:20 A.M. they numbered close to five thousand—double the number sheltered inside against the rain—and some began a good-natured hog-calling shout: "Come on, Jackie. Come on, Jackie."

The coffee was on, and Troy West watched the excited percolation as he wrapped the first early delivery of books. He had a long shiny counter on the first floor and he could pack the books into cartons, buttress them with old newspapers, seal, and twine them while watching that percolator. Mr. West had been a mail wrapper at the Texas School Book Depository for sixteen years and he always arrived early, put the coffee on, and began the wrapping as though his heavy hands could do the work without involving his mind at all.

The air was aged. It had been imprisoned in the archaic building a long time, like Troy. The floors had been scuffed so long they had concave paths between the counters and the two birdcage elevators. The windows, large and dusty, had rounded tops, reminiscent of Ford's Theatre in Washington. The Texas School Book Depository had never lifted anyone's spirit; it was a warehouse of work, with school books coming in by the gross and being mailed out by the dozen. There were seventy-two persons on the payroll, and two were absent this day because of illness.

It was a place where Negroes and whites worked well together. The wits told jokes and perpetrated pranks; the serious ones, fellows and girls, gravitated together and had discussions on their lunch hour; the loners nodded soberly and moved toward the rack with the fresh orders, yanked one out, took the elevator upstairs, filled the order, came downstairs, put it on Troy West's counter, and went back to the rack for another order.

There were thirteen employees working on the sixth floor. More than half of them were laying a tile floor. It wasn't easy, as Bonnie Ray Williams learned. He and the others had finished the fifth floor, which entailed starting at one corner of the huge square barn, moving cartons of books out of the way, cleaning a section and laying tiles, then moving the cartons farther away until they became jammed in another corner. At that time, they had to be lugged back to the freshly finished part—hopefully, dry—while the last corner of the floor was worked on.

On the sixth floor, moving the cartons made the work of the order-fillers difficult. Men like Danny Arce and Wes Frazier, Lee Oswald and James Jarman, Jr., had to sort among the pyramids of cartons to locate histories, mathematics books, and grammars. They worked the sixth floor and knew where to find any book, but the floor crew started work at the west end of the room and, on their knees, were working east. Little by little, the boxes of books were being inched toward the front windows of the building.

Some of the fellows were in the habit of disappearing into a small place on the ground floor called the Domino Room to sip Troy's steaming coffee. Lee Harvey Oswald never chipped in with the others toward the purchase of a pound of coffee and never asked for a cup. He sometimes joined the others merely to rummage through yesterday's *Dallas Times Herald* or the current day's *Dallas News*. No one ever saw him buy a newspaper, but he often sat alone, running through the news items swiftly, seldom pausing to respond to a greeting.

The Depository faced Dealey Plaza with little confidence. There were new glassy buildings around the perimeter, including the Dal-Tex and county sheriff's office. The School Book Depository and the old stone county courthouse were all that was left of old Dallas. The plaza itself was dedicated to Joseph Dealey, founder of the *Dallas News*, and he stood in bronze facing away from what he had wrought.

The grassy square was the funnel for downtown Dallas. Three main streets, Elm, Main, and Commerce, met in the plaza, carrying the bulk of traffic headed south, southwest, west, and northwest out of town. The buildings formed an inverted U at the top of the plaza, and the cars ran downhill for two hundred yards toward a railroad overpass, then disappeared underneath to reach any one of six viaduct speedways.

The final part of the presidential motorcade would come down the middle avenue—Main. When it reached the plaza, it would turn right for one block to Elm, facing the Texas School Book Depository at the top of the hill. The cars would make a left on Elm, pass directly in front of the Depository, and go downhill, the President and Mrs. Kennedy waving to the last few hundred people on the parade route. Once at the underpass, the automobiles would speed up, make a right turn up onto Stemmons, and head for the Trade Mart, less than three miles away.

To those who work hard within small horizons, there is little excitement in seeing a President. There was more interest among the deputy sheriffs on the opposite side of Dealey Plaza than in the School Book Depository. Some of the Negro employees at the Depository said that they wanted to eat their lunches quickly and maybe get a look at Mr. Kennedy, but to the majority of $1.25-an-hour laborers this was a drizzly morning, the day before Saturday's late sleeping.

Some did not know that the motorcade would pass here; some felt no desire to see the Chief Executive; a few said they had too much to do. Roy Truly, the short, bespectacled manager, sat in his half-glassed office inside the main entrance, thinking nothing at all of the motorcade. At lunchtime he had an appointment to meet one of the owners of the business, and this took precedence over any parade. He hunched over his desk, puffing on a cigarette, now and then looking up at his jacket hanging on a clothes tree without seeing it.

Duty was all. Every morning Mr. Truly made a quiet inspection of the building, getting a head count from his foremen, nodding a soft greeting to old hands, glancing at the order rack to make certain that the merchandise was moving, and getting a fresh feel of the plant.

He knew the good workers and the shirkers. He was aware that his presence spun all hands into bursts of activity, and he knew which ones would loaf when he disappeared. The workers knew that Mr. Truly was fair; he gave a man an even break. Old hands broke new ones into the routine. If an employee didn't want to work, he was paid according to the clock and dismissed. Six weeks ago, Mr. Truly had hired Lee Harvey Oswald because he liked the quiet, almost uncommunicative attitude of the boy, and especially the way he prefaced his responses with "Yes sir" and "No sir."

Truly was in his middle years, and he gave his best efforts to everything he undertook. He wasn't much more of a conversationalist than young Oswald, but he was shrewd. When he was thinking, he pulled on his ear, and his philosophy was that "nothing is going to change my life. No matter what happens, I am not going to get an ulcer." He had seen this business start small, and now he was shipping school books to five states.

The foreman, Bill Shelley, told Roy Truly that the boys laying the floor were now working the sixth floor, and, while it would not be finished today, it certainly would be by Monday. The only two employees who had a facial resemblance worked that floor that day. Billy Nolan Lovelady, a crew-cut with a lean pale face and a grim mouth, was setting tiles. At the other end of the room, Lee Oswald was filling orders which involved books published by Scott Foresman & Company.

There was a burst of silent activity on the eighth floor. It was electric without the crackle, just a man leaning over a telephone. His whisper could not be heard: "The President is on his way to the mezzanine." The door to Room 850 opened, the wood gleaming against the lights in the corridor. Mr. Kennedy came out. He looked young and tall and buoyant; his light step made his shoes twinkle.

The elevator was waiting when he saw Evelyn Lincoln and her ladies. At once the President stopped and his secretary introduced them. He shook hands and, in mock surprise, said: "Are these some more of your relatives?" Mrs. Lincoln chuckled and said: "Well, relatives of relatives." He backed toward the elevator, said, "Good morning" to Elaine San Duval, the operator, and turned to see agents Clint Hill and Muggsy O'Leary and complained that Mary Gallagher was not available when Mrs. Kennedy required her services. He wanted someone to "get her on the ball."

In the Longhorn Room, agents hurried to the elevator door

on the left and asked the politicians and their wives to move back. The door swung open and a polite patter of applause rippled across the room as the President walked from group to group, shaking hands, nodding good morning to the Vice-President and Mrs. Johnson; the Governor and Mrs. Connally; a special word to Congressman Jim Wright; a pause before Senator Ralph Yarborough to freeze the smile and growl: "For Christ's sake, Ralph, cut it out!"

He was ready to go, and a few joined the President on the elevator. Others hurried down the flight of stairs. In the lobby, a cheer went up from the group permitted to wait beyond the manager's desk. The President saw the clerks at the drugstore and behind the florist stand staring agape and he smiled and nodded. Lightly he danced down the few steps to Eighth Street, a sizeable entourage behind him, and out into the street. He was one of the few who was bareheaded, and the crowd saw him step into the street and shouts went up: "Here he is!" "Here's the President!" A Secret Service man ran up to Mr. Kennedy with a raincoat, but he was waved away. Mr. Kennedy began to chuckle. He got up on the buckboard truck and saw the immensity of the crowd, more than twice the size of the one expected at the breakfast, and, although he made no comment, it induced laughter in his throat as he studied the density of human beings waiting in the rain, and there was no doubt that Mr. Kennedy was impressed and encouraged regarding the warmth of Texans.

The roar of the crowd broke against the red brick of the Hotel Texas and, up on the eighth floor, the face of Mrs. Kennedy could be seen for a moment, looking down. In another room Evelyn Lincoln and her women crowded at a window to watch. Behind them the door opened and Mary Gallagher came in to announce that she and Mrs. Lincoln should be getting out to the plane. She wanted to be aboard before Mrs. Kennedy.

The Vice-President and the Governor flanked the President on the truck. Their brows were grim in the drizzle. Up in Dr.

Burkley's room, O'Donnell and O'Brien watched Kennedy's arms trying to flag the crowd into silence. They had spoken to Senator Yarborough before the President saw him, and their persuasion had been fruitless. Yarborough had his own reasons for not wishing to ride with Mr. Johnson, he said, and, besides, no one was watching who rode with whom; the people of Texas were out to see the President. Besides, he had ridden into town with the Vice-President last night and that ought to be enough.

Like a receding wave, the parking lot welcome began to soften and fall back. The President, smiling and waving, saw more than the crowd. His eyes caught the policemen on the roof of the Continental Trailways Bus Terminal and on the roof of the Washer Brothers Building with riot guns. The big broad smile did not change; it was an irritant which went with the job, like listening interminably to "Hail to the Chief."

"Mr. Vice-President," the President said loudly, chin a little high before the microphone, and the crowd shouted again. "Mr. Vice-President, Jim Wright, Governor, Senator Yarborough, Mr. Buck, ladies and gentlemen": Another cry went up for Raymond Buck, president of the Fort Worth Chamber of Commerce. "There are no faint hearts in Fort Worth, and I appreciate your being here this morning."

Someone yelled: "Where's Jackie?" and Mr. Kennedy broke into laughter with the crowd. "Mrs. Kennedy is organizing herself. It takes longer, but, of course, she looks better than we do when she does it." The people whooped. "But we appreciate your welcome."

For a moment, his thoughts wandered. "This city has been a great Western city, the defense of the West, cattle, oil, and all the rest. It has believed in strength in this city, and strength in this state, and strength in this country."

He was trying to appeal to their pride and their civic vanity, but the words were not marshaled and clichés boomed through the loudspeakers. In the small bedroom in Suite 850, Mrs. Kennedy could hear the voice of her husband and the shouts of the citizens. She hoped the rain would continue so that the bubble-top would be on the car; even a slow-moving motorcade twisted hair into dissolute tendrils. Besides her mirror told her that her eyes were fatigued.

"What we are trying to do in this country and what we are trying to do around the world, I believe, is quite simple, and this is to build a military structure which will defend the vital interests of the United States." The President's right hand was pointing outward now, then back up, and out again as punctuation for an informal talk. No matter what he said, or what the President left unsaid, the crowd permitted the veils of mist to shine its collective face and it endorsed every pause with open throats and enthusiasm for the man.

"And in that great cause, Fort Worth, as it did in World War II, as it did in developing the best bomber system in the world, the B-58, and as it will now do in developing the best fighter system in the world, the TFX, Fort Worth will play its proper part. And that is why we have placed so much emphasis in the last three years in building a defense system second to none—until now the United States is stronger than it has ever been in its history." He was not telling them anything they did not know; they would have preferred to hear him tell what he was going to do to Senator John Tower and the Texas Republicans next year, but Mr. Kennedy was on a defense topic and found it difficult to separate the bait of federal payrolls and the fish of local workers.

"And secondly," the President said, "we believe that the new environment, space, the new sea, is also an area where the United States should be second to none." This was a thought borrowed from his dedication of the Houston Space Center yesterday, but the crowd whistled approval.

"And this state of Texas and the United States is now engaged in the most concentrated effort in history to provide leadership in this area and it must here on earth. And this is our second great effort. And in December—next month—the United States will fire the largest booster in the history of the world, putting us ahead of the Soviet Union in that area for the first time in our history."

Again he was borrowing from the Houston speech. Mr. Kennedy had fallen into a witty slip of the tongue at the Space Center by calling the booster the "largest payroll in history" instead of the "largest payload." Mist was now shining on his forehead; the Governor and the Vice-President faced the crowd with neutral expressions. Reporters made notes, even though they knew that the speech at the breakfast and the one at the Trade Mart in Dallas were the ones with built-in impact.

"And thirdly, for the United States to fulfill its obligations around the world requires that the United States move forward economically, that the people of this country participate in rising prosperity. And it is a fact in 1962, and the first six months of 1963, the economy of the United States grew not only faster than nearly every Western country, which had not been true in the fifties, but also grew faster than the Soviet Union itself. That is the kind of strength the United States needs, economically; in space, militarily."

"And in the final analysis, that strength depends upon the willingness of the citizens of the United States to assume the burdens of leadership." He was finished. He had to find a thought to get him off the stand. "I know one place where they are, here in this rain, in Fort Worth, in Texas, in the United States. We are going forward. Thank you."

There was a thunder of applause, shouts, and rebel yells. The President and Vice-President hopped down from the truck, and stood before the American flag and the Seal of the President, pumping hands, trading smiles, studying the awestricken faces which would never forget this moment. The Secret Service began to break through the crowd, back toward the hotel.

Mr. Kennedy stopped a moment to reach up and shake

hands with the ponchoed troopers who sat on their horses and kept the lane open for him. Along the corridor on the eighth floor, the word was passed: "He's on his way back to the hotel." The police on the rooftops slid their shotguns under their arms and watched him disappear under the big metal canopy which said: "Welcome to Texas."

The man with the dark wavy hair initialed a paper and placed it in the outgoing box. A cigarette, dying on a tray, was touched to a fresh one and puffed. Gordon Shanklin depressed a button. "Let them come in," he said. Behind his head was a color photograph of J. Edgar Hoover, director of the Federal Bureau of Investigation. Mr. Shanklin, head of the Dallas office of the FBI, was a well-dressed, low-key man. He administered the field office, in the old Santa Fe Building, in the manner of a confident banker who would rather listen to the depositors than talk.

He was ready to start the biweekly meeting with his agents. They came in and said good morning. Some stood and some occupied chairs against the wall opposite Mr. Shanklin's desk. He had their reports before him and he went over them, asking a few questions, making suggestions. This morning he wanted to bring up the matter of the President's visit to Dallas once more. The protection of the President and Vice-President and their families was the province of the United States Secret Service, but there was always a chance that one of his men might have a lead on something, and he had reminded them, at earlier meetings, to mention any they might have in mind, so that such tips could go out at once to Washington and the PRS, and also to Roy Kellerman, who was the Secret Service agent in charge in Dallas today.

They had nothing. "If there is any indication of any possibility of acts of violence," Mr. Shanklin murmured through his own smoke, "against the President or the Vice-President . . ." He glanced along a row of forty faces. "If you have anything," he said, "anything at all, I want it confirmed in writing." Agent James P. Hosty was in the group. He had nothing to offer. Neither did anyone else. Hosty was rated as a solid, non-panic agent, a man who, in the absence of any big cases, kept checking a number of small ones and who often sat in the outer office in the late hours laboriously pecking at a typewriter to keep his reports up to date.

The FBI men listened to Mr. Shanklin's admonitions. He ran down a list of pending files. Each man involved gave an oral report on the status of the case to support the written work on Shanklin's desk. Most of them would not see President Kennedy today. Their work was in other areas and, unless they could arrange to have lunch somewhere in the neighborhood, they wouldn't get to see the reception on television either.

Hosty, for example, had learned by accident last night that there was going to be a motorcade. He was home reading a newspaper. He scanned the story of Big D's welcome to Mr. Kennedy, and noticed that there was a map diagram of the parade route, but he didn't study it. Hosty's path crossed that of the President only indirectly. Yesterday he had seen some street pamphlets with front and side-view pictures of President Kennedy and the words: "Wanted for Treason."

The matter may have been of small moment, but Hosty had carried them over to the Secret Service office and had given them to Agent Warner. A man in James Hosty's squad had some information about someone in Denton, Texas, who had made threatening remarks about the President. This, too, was given to the Secret Service. Last night, there had been a tip about a demonstration against Kennedy at the Trade Mart—picketing, perhaps—and this, for whatever it was worth, was passed from FBI to SS.

The only thing that Hosty recalled about the parade route was that it would come down Main Street about noon, and he thought that, when the Friday morning meeting closed in Mr.

Shanklin's office, he might get a window table and have lunch at a restaurant along Main. He had no reason to think of any of his small "follow-up" cases in relation to the safety of the President.

One of them was Lee Harvey Oswald. Mr. Hosty's most recent report on this matter had been filed with the Washington office four days ago. It said that Oswald had been in communication with the Soviet Embassy in Washington. Oswald was a chronic chore to Mr. Hosty. He had been on it a year. The Federal Bureau of Investigation had interviewed Lee Harvey Oswald when he returned from Russia with Marina and baby June.

Oswald felt he had nothing to hide. He had served his country as a United States marine in foreign service, and his country had rewarded him with a dishonorable discharge. The Soviet Union turned out to be a disappointment because there was no freedom for the workers. He had been employed in a parts factory in Minsk; the trade union meetings had turned out to be dull and doctrinaire. At one time he couldn't even get permission to leave Moscow. At another, he couldn't leave Minsk to join his wife at a vacation resort.

The exploitation of the workers, Oswald said, was even worse in the United States, but here he could go where he pleased, and he did not have to answer Mr. Hosty's questions. If Hosty intended to inform Oswald's boss of his defection to Russia, then he would be harassed out of work. All Lee expected was to be left to work in peace and support his family. Yes, he was a communist, but he didn't expect an FBI agent to understand the word. He was a Marxist in the purest sense; not a socialist-despot like Stalin and Khrushchev; a true communist.

Mr. Hosty checked the Oswald case regularly. Oswald was never home. Hosty spoke to Marina Oswald, who resented him, through Mrs. Ruth Paine, who interpreted. To the agent, Mr. Oswald was a chronic complainer who lost jobs regularly. He had no friends. He had no admiration for the Russian expatriates who tried to befriend him and his wife. They detested communism.

There was a small communist cell in the Dallas area, and the FBI had an undercover agent in it, but none of them knew Lee Harvey Oswald and the young malcontent had no desire to join. He felt superior to them and, in the time Hosty kept him under surveillance, Oswald vacillated between wanting to stay in the United States; sending Marina back to Russia with the baby; getting a Soviet visa for himself so that he could get to Cuba and the communist bloc countries. From day to day, he seemed to change his course abruptly, so that even his wife could not understand him.

Hosty was aware that the newest job Oswald had was in the Texas School Book Depository. He worked filling book orders, but there was nothing sinister in this. Another thing: Oswald was not a violent person; he was never seen with firearms; never walked a picket line; never wrote hate letters to newspapers; he never even went to a motion picture.

If Hosty had followed the newspaper diagram of the parade route and noticed that it would pass the Texas School Book Depository, it would have been witless to draw the attention of his superiors to the presence of the defector, because he represented no physical danger to anyone. Hosty made the trips out to the little house in Irving as a matter of duty, but he never met his man. The investigation disclosed Oswald as a sullen braggart—nothing more.

At Love Field, the captain of American Airlines Flight 82 asked for taxi clearance. In a few moments, he would be headed for Idlewild Airport in New York. One of the passengers, a stewardess had told him, was Richard Nixon. Apparently the former Vice-President was not going to remain in Dallas to watch the presidential parade.

9 a.m.

The chefs stood motionless beneath the gleaming ranks of hanging pots and kettles as the Secret Service men burst into the kitchen. One man ran ahead and placed a chair in the doorway leading from the cooking ranges to the Grand Ballroom. The President strode through the kitchen, walking fairly fast between the counters where, in the past hour, over two thousand breakfasts had been cooked and carried on huge trays. He wasn't smiling. He patted his forehead and thick hair with a handkerchief and glanced back at the entourage of political chieftains.

It nettled him to know that he could not settle intraparty disputes by fiat. Over the years, Kennedy's personal loyalty to party, and especially to party hierarchy, had been constant, and it seemed to him that when the President of the United States said, "Do this," that all hands should do it, not merely as a matter of unity, but in obedience to the wishes of the Chief.

The President sat in the chair and commanded a view of the double-tiered head table, with the big "Fort Worth Chamber of Commerce" banner hanging behind it. He got to his feet, looked through the assortment of faces waiting in the kitchen and summoned Agent Duncan. "Where is Mrs. Kennedy?" he said. "Call Clint Hill and tell him I want her to come down to breakfast." He would have preferred to escort her to the head table, but, like many husbands, he had little understanding of the natural feminine contempt for time.

Loudly he said: "Everybody set?" and again sat. The men and their wives began to move toward the doorway, skirting the chair and listening to the rising sound of applause from the Grand

Ballroom. The President chatted again with Ralph Yarborough; no one heard the words, but the face of Mr. Kennedy was stern and frowning, and the index finger pointed and probed whatever point was being made. The Senator didn't appear to resist the President; his features appeared to be in shock, as though he could not credit the words or ideas he was hearing. Ralph Yarborough went on into the ballroom, a man unconvinced that his political quarrel with the Connally-Johnson forces had any bearing on the President's future.

A moment later, as the Governor passed, the President said: "John, did you know that Yarborough refused to ride with Lyndon yesterday?" The Governor knew. The President didn't seem to understand, or perhaps appreciate, that Governor John Connally was a prime mover in this battle and that he had an interest in war rather than peace. In every case he had thwarted the Yarborough group, even to the point of denying them seats of honor in the presidential party.

The rich, the affluent, the oil money of Texas backed Connally and it was important for the Governor to display his antagonism to Yarborough and, in a more subtle manner, his lack of enthusiasm for Kennedy himself. It was the Governor who had postponed this visit to Texas several times; it was Connally who felt that a Kennedy-Johnson ticket might be defeated in his state in 1964, and there was considerable risk in being seen with Kennedy.

Yes, the Governor knew that the Senator had refused to ride with the Vice-President. "What's the matter with that fellow?" Kennedy said, as though the political schism was obviously the fault of Yarborough. The Governor said he didn't know. "Well," the President said, "I'll tell you one thing: he'll ride with Johnson today or he'll walk."

The Governor moved into the big ballroom with Nellie; a burst of applause greeted the handsome couple. For a minute, the President would sit alone, except for the agents who stood near him, before making his entrance. The chefs still remained immobile behind the chopping blocks, the counters, the gleaming steel sinks. In that minute, the President's mind, like a waterbug on a big summery lake, could dart in many directions. His speech was already under the yellow light at the lectern. He could hear Ray Buck's booming voice and the crest of laughter, but he couldn't decipher what was being said.

In the seat of power there is sometimes a lofty loneliness, and this was one of those minutes. Like primordial man, he desired most of all to leave a good deep scar on the wall of the cave, to be remembered as a leader with high purpose and firm resolution. He was at his best when he was politically, economically, and inspirationally far ahead of his people, beckoning over his youthful shoulder for everyone to follow him. His heels were fleet, but the veterans of Congress studied this man through other prisms and often viewed him as an opportunistic son of a rich and merciless man. Beckoning to the people, Kennedy learned, is good publicity, but not good politics. His party had a clear majority in the legislative halls, but the President could not command it.

Sometimes he revealed himself in flashes, as when a friend asked him why he wanted to be President. "Because it's the seat of power," he had said. "I don't know anybody around can wield it better than I. Do you?" In the White House swimming pool, Kennedy had floated on his back and said he wouldn't want the job more than eight years. "Look at it. Laos may go to hell again next week. There's this nuclear testing thing. Berlin, Vietnam— all that. Yeah, I know that's what makes it exciting, that's what makes it challenging. But eight years seems enough."

After the first year in office, he had said: "This job is interesting but the possibilities for trouble are unlimited. It's been a tough first year. But then they're all going to be tough." He was the scion of Irish forebears who placed a premium on adversity. None of them desired an easy victory; none would admit to one. Men fought, won or were beaten, but they never wept.

Mercy, forgiveness, compassion—these were the pious pity of the effete. "I run for the presidency of the United States," he had shouted in the Boston Garden, "because it is the center of action. . . ."

On his wedding night, he dismayed his bride by locking the door to their suite at the Waldorf-Astoria Hotel and sitting at a desk to note his speaking engagements for the following two months. Emotions, like ablutions, are best concealed. The family had a horrifying history of sudden death and disaster, but the Kennedys learned to steel themselves against whimpering. Always they were grim pallbearers turning in unison to face the Roman Catholic Church.

The President stood, compressed the knot in his tie with his fingers, and strode into the ballroom to a standing ovation. Grinning, the arms waved for the people to sit, to return to their breakfast. In the back of the huge room, a small red light began to glow on a television camera. Overhead, wagon wheels festooned with small lights served as pioneer chandeliers.

The gray head of Monsignor Vincent Wolf of Holy Family Church was too full of excitement to do justice to the food. He turned toward the back of the head table and motioned to Peter Saccu, the catering manager. Mr. Saccu took an envelope from the priest and carried it to the President. The note read: "We, the school children, the nuns and priests of Holy Family Church in Fort Worth are happy to offer one thousand Masses for the spiritual and temporal welfare of you and your family, and to show our love and devotion to the President of the United States of America. . . ."

Mr. Kennedy looked down the length of the table, caught the eye of the monsignor, and nodded his thanks. The message went into his jacket. To the Roman Catholic endowed with full unquestioning faith—and the Kennedys are such—there is both solace and protection against the unknown in the holy sacrifice of the Mass. It is the most direct appeal to God, and

Masses which are said by innocent children are, to some Catholics, the most inspiring of all.

And yet death was as common as the veils of rain which drifted up Eighth Street outside the hotel. In California at this moment a great writer, a mind of compassionate sophistication, was in final coma. Aldous Huxley was dying. At Hobe Sound, in Florida, a man who exerted considerable influence on John F. Kennedy had died of cancer, and no one had bothered to tell the President. The Rev. George St. John, headmaster of Choate School, who often exhorted students like Kennedy: "Ask not what your school can do for you, ask what you can do for your school," was dead at the age of eighty-eight.

Nor had the presidential eye paused at Page 14-A when reading the *Dallas Times Herald* this morning. There had been twenty-five death notices, a normal complement, ranging from James Meek, who had struck it rich in oil, to a relatively unknown man named Carl Lucky. Ten miles south of Norwalk, Ohio, brisk winds fanned a blaze in the Golden Age Nursing Home. Sixty-three persons would die there this day, none of them quickly or easily.

The sky over Irving brightened and darkened and brightened again. Marina Oswald, tidying the breakfast dishes, was interested in the weather. She and her hostess, Ruth Paine, always washed their own laundry; nothing was sent to a cleaner's; no diaper service truck ever stopped in front of 2515 West Fifth Street. Today was to be a laundry day, but it was too threatening to start washing clothes now.

Besides, the television set had been left on. Marina never turned it on without asking permission. This morning Ruth had left early for a dental appointment with one of her children. Mrs. Paine remembered that the President of the United States would be in this part of Texas and had left the set on before Marina arose from bed.

The laundry could wait. Marina sat on the couch in the little ranch house, so that she could watch June, playing on the floor, hear the baby if she cried, and watch America's First Family. Mrs. Oswald did not dress. Household laziness was a prerogative she exercised at will. The babies could not be taken out in the rain anyway. She heard the commentator's words, but she understood only the camera.

She had not looked in her grandmother's Russian teacup; therefore, she had not seen Lee's wedding ring. Nor had she examined the money in the dresser drawer. If she had dwelt on her husband at all, Marina Oswald recognized the situation for what it was: a problem which could be postponed at least until next week. The young woman with the straight brown hair, the piquant Nordic face, the Soviet peasant philosophy that it was a husband's right to beat a wife, wanted to remain in America. She would return to Russia, if it was Lee's will, but it would not be hers.

All of her roots, her relatives, her culture, were in Leningrad and Minsk, and yet the seductive influence of the United States, with its air-conditioned shops, its TV, its millions of small homes with gardens, the benevolent ease with which travel was permitted, the lack of suspicion among neighbors who owned automobiles and would give one a lift into town or home, the gatherings of friends over beer and cheese, the lack of restrictive forces in private or public life all superseded the natural affection of this woman for her homeland.

Her husband seldom saw anything as she saw it. To the contrary. At this moment he was on the ground floor of the School Book Depository, picking up a book order. He was looking across the counters to the front entrance of the building. Lee asked James Jarman, Jr., a checker, what the people were crowding the front step for. They would be waiting to see the President, Jarman said. He didn't know what time, but he guessed it would be late this morning.

Oswald looked surprised. "Which way do you think he's coming?" he said. Jarman was surprised, too, because it seemed as though everybody in town knew the President was coming, all except Lee. Then, too, it was rare to elicit this much conversation from the silent man. The checker said he had heard that Kennedy was coming on down through Dallas on Main Street, then right on Houston, and left on to Elm, right out in front of that door.

"Yes," Oswald said. "I see." The early arrivals were few. There were vantage points for seeing the motorcade all over Dealey Plaza, and, as this was the place where the President would pick up speed for the run to the Trade Mart, it wasn't considered good enough to attract crowds. Most of the curbside watchers would be employees from buildings around the square, in addition to some who chose to remain free of crowds. The few on hand were trying to keep out of the rain.

Weather was a subject of interest to Special Agent Sam Kinney of the Secret Service. Twice he had left the underground parking lot at Love Field to study the sky. And twice he had returned to tell his co-worker, Agent George Hickey, that he didn't know whether to put the bubbletop on or leave it off. Right now, it wasn't raining hard enough to put it on the big special Lincoln. It would be better to wait awhile.

The decision, which was of concern to so many people this morning, could be decided when Forrest V. Sorrels picked up Kinney and Hickey at 11 A.M. Mr. Sorrels was special agent in charge of the Dallas office. This, for him, was a big day. He was the man in charge, even though he shared his decisions with and could be countermanded by Jerry Behn at the White House, Chief Rowley at Secret Service Headquarters, Kenneth O'Donnell, and the President. For three interminable weeks Sorrels had worked this assignment, doing the field assignments and reporting to Washington.

It had turned out to be one of those slowly accelerating responsibilities, moving faster and ever faster, entailing more and more work until now—on the day of days—a thing like the bubbletop could not be decided, finally and unalterably, until 11 A.M. Sorrels had his own men well briefed, but he had also to meet the "advance man," Mr. Winston Lawson of the White House Secret Service detail, apprise him of the meetings with the Dallas Police Department chiefs, the route selected for the motorcade, the security of Love Field, the selection of the Trade Mart as the site of the luncheon speech, even though it had overhead catwalks (difficult to secure), the interviews with the Federal Aviation Administration officials at Love Field for landing patterns and gates for presidential aircraft, the screening of luncheon guests, the motorcade from the Trade Mart back to Love Field, the probable time of takeoffs to Austin, even the arrival of the two automobiles on an Air Force C130 at 6:05 last evening.

Kinney and Hickey had accompanied 100-X, the President's car, and 679-X, the Secret Service follow-up automobile. The cars could not be taken for granted. They too had to be secured. Both had been taken to the basement garage at Love Field last night. Dallas policemen stood guard all night. At 8:30 P.M. Kinney and Hickey were helping Sorrels and Lawson check, once more, the speaker's stand at the Trade Mart, the seating arrangements, the kitchen, and the exits. It had been done several times before, and, in the morning, it would be done once more by Deputy Chief Batchelor of the Dallas department.

Hickey and Kinney had checked out of the Sheraton Hotel at 7:30 this morning. They had breakfast, met Warrant Officer Art Bales of White House Communications, who had already established instantaneous communication with Washington from a private switchboard in the Sheraton and arrived at Love Field at nine. Both men relieved the Dallas police and began the customary job of getting the automobiles ready for the motorcade. This consists of examining the engine section, trunk, and chassis, removing the seats to look for detonating devices, checking oil, water, batteries, and then polishing the cars.

• • •

The special phone on the eighth floor had a light which flicked on and off. Agent Clint Hill, whose assignment was to guard the First Lady, picked it up at its station beside the hotel fire hose. Agent Duncan said that the President was waiting at the breakfast for Mrs. Kennedy, and wanted her downstairs at once. In the ballroom, a boys' choir sang "The Eyes of Texas" and President Kennedy led the applause. He kept smiling and tapping the tabletop with the fingers of his right hand and glancing surreptitiously toward the wings.

It was 9:22 when Clint Hill advised Mrs. Kennedy that she was expected downstairs. She hesitated over two pairs of white gloves, selected one, and left with him. Of all the First Ladies, this one was the most naturally beautiful, the most romantic, and the most dedicated patron of the arts. She had poise, presence, and a smile that reduced statesmen and commoners to the absurd and speechless. When the elevator stopped at the mezzanine, and a door opened on the back side of the elevator, Mrs. Kennedy said: "Aren't we leaving?" Mr. Hill shook his head. "No," he said. "You're going to a breakfast."

When the boys' choir completed a second number, Raymond Buck glanced toward the wings and shouted: "And now, ladies and gentlemen, an event I know you have all been waiting for." Clint Hill nodded to Mrs. Kennedy. She was a vision of pink confusion as she stepped out into the talcum glare of the klieg lights. The audience got to its feet, and so did the President. He saw his wife blinking in the lights and beckoned to her.

At the head table, Nellie Connally was standing and applauding with the others when she saw that pink wool ensemble and looked down with dismay at her own pink wool suit. An embarrassment of this sort happened only in comedies. The First Lady stood beside her husband, and Texas realized that all it had read and heard about this remarkable young lady was true. The men stood on their chairs to whistle with their fingers between their teeth. The women stared archly at their men.

In the back of the room, there was some commotion when General Godfrey McHugh tried to pass in, and a Texas ticket taker tried to keep him out because he didn't have a ticket. This occasioned a little pushing and shoving with both sides shouting: "Do you know who I am?" A Secret Service man explained the matter to the ticket taker, and the President's military aide was permitted to proceed, feathers askew but dignity intact.

At the microphone, Buck was presenting gifts from Fort Worth to the First Family. It would be doubtful that Mrs. Kennedy would ever wear those ornate hand-tooled boots, but Buck wasn't going to permit the President to get off as easily. He gave Mr. Kennedy a big fawn-colored cowboy hat and motioned for him to put it on. The television cameras were on the scene; still photographers down front trained their Rolliflexes and their Nikons and got ready for the big moment. Those who knew the President had heard him speak of such "baloney pictures" with contempt. He thought that former Presidents sacrificed something when they adorned themselves with broad-brimmed hats or Indian feathers.

He stood to accept the gift, smiling sheepishly and glancing appealingly around him. A few guests yelled, "Put it on, Jack!" and the President held the hat out in front of him and, at last, in a burst of laughter, he said: "I will put this on Monday in my office at the White House. I hope you can be there to see it." As the President sat, the master of ceremonies began his introduction, and Kennedy felt in his pocket and showed his wife the Mass card sent by Monsignor Wolf. Mrs. Kennedy looked down the table and smiled her gratitude to the priest.

In a moment, the President was grasping the lectern with both hands, looking out at the enthusiastic crowd, and, as one who had become accustomed to eyeglasses in private, his eyes darted to the large type on the pages to make sure that he could read it. The press out front studied its copies of the same speech to check whether the President would depart from it. He did at once. Instead of opening with "I am glad to be here in Jim Wright's city," the tribute to the local congressman came second to one he reserved for his wife:

"Two years ago," he said, "I introduced myself in Paris by saying that I was the man who had accompanied Mrs. Kennedy to Paris. I am getting somewhat that same sensation as I travel around Texas." The crowd loved it. So did Mrs. Kennedy, who covered her laughter with her white glove. "Nobody wonders what Lyndon and I wear," he said lugubriously.

His desire in this speech was twofold: to pander to the aircraft payrolls the government maintained in this region and to make the speech a major pronunciamento on defense. The President said that Fort Worth was always in the forefront of national defense. A hundred years ago, he said, the town had been a fort, a shield against marauding Indians. In World War I, Canadian pilots had been trained to fly in Fort Worth; in World War II, the President's brother Joe had picked up his B-24 Liberator in Fort Worth to fly it overseas. The B-58, "finest weapons system in the world," was built in Forth Worth; also "the Iroquois helicopter from Fort Worth is a mainstay in our fight against the guerrillas in South Vietnam."

Mr. Kennedy had the attention of his audience, and he began to speak forcefully, almost stridently, as though challenging anyone to deny the veracity of his statements. He espoused the cause of the controversial TFX—Tactical Fighter Experimental—which was designed to save enormous sums by being available in slightly modified form for both the Air Force and the Navy.

In appealing to local pride, no statement was too outrageous for the President, no allusion to government spending overdrawn. "In all these ways," he said, "the success of our national defense depends upon this city in the western United States, ten thousand miles from Vietnam, five or six thousand miles from Berlin, thousands of miles from the trouble spots in Latin America and Africa or the Middle East. And yet Fort Worth and what it does and what it produces participates in all these great historic events. Texas as a whole and Fort Worth bear particular responsibility for this national defense effort, for military procurement in this state totals nearly one and one-quarter billion, fifth highest among all the states of the union."

He drew a laugh when he pointed out: "There are more military personnel on active duty in this state than any in the nation save one—and it is not Massachusetts—any in the nation save one, with a combined military-civilian defense payroll of well over a billion dollars."

He was at a tender moment in history when a President could boast about his spending, rather than apologize for it.

"In the past three years we have increased the defense budget of the United States by 20 percent; increased the program of acquisition for Polaris submarines from 24 to 41; increased our counterinsurgency forces which are engaged now in South Vietnam by 600 percent." The President decided that he could afford to take a laughing shot at the rich oil millionaires of Texas who plaster the highways with billboards demanding a stronger America and lower taxes. "I hope those who want a stronger America and place it on some signs will also place those figures next to it."

In the field of foreign affairs, the President had felt that he failed one test—the invasion of Cuba by Cubans—and had triumphed in another—the confrontation with Khrushchev over Intermediate Ballistic Missile sites in Cuba. Now he proposed to tell all Americans that, for good or ill, they had firm commitments all over the world.

". . . As I said earlier, on three occasions in the last three years, the United States has had a direct confrontation. No one can say when it will come again. No one expects that our life will be easy, certainly not in this decade, and perhaps not in this century.

"But we should realize what a burden and responsibility the people of the United States have borne for so many years. Here, a country which lived in isolation, divided and protected by the Atlantic and the Pacific, uninterested in the struggles of the world around it, here in the short space of eighteen years after the Second World War, we put ourselves, by our own will and by necessity, into defensive alliances with countries all around the globe.

"Without the United States, South Vietnam would collapse overnight. Without the United States, the Southeast Asia Treaty Organization alliance would collapse overnight. Without the United States, the CENTO alliance would collapse overnight. Without the United States, there would be no NATO. And gradually Europe would drift into neutralism and indifference. Without the efforts of the United States in the Alliance for Progress, the Communist advance into the mainland of South America would long ago have taken place."

It was not a new military posture for the United States. The rich and powerful democracy had been on the world scene since 1917, but its stand as an Atlantic power and a Pacific power had seldom been stated so succinctly, and never had a President made it so painfully clear that, without the United States, most of the treaty organizations would have collapsed. The cheers of the Chamber of Commerce were ample proof that, though the members may not have supported John F. Kennedy politically, they certainly admired any statement which made their country look stronger while others appeared to be weaker.

"I am confident," he said on a softer note, a time to conclude, "as I look at the future, that our chances for security, our chances for peace, are better than they have been in the past. And the reason is because we are stronger. And with that strength is a determination to not only maintain the peace, but also the vital interests of the United States. To that cause"—he flung up both hands in farewell—"Texas and the United States are committed. Thank you."

The President remained standing through the tumult of cherished sound. It was obvious that he would not remain. The Secret Service men fell in ahead of him and along the flanks. Mrs. Lyndon Johnson, a cameo in a pale gown and triple strands of pearls, stood with her husband and the Connallys, applauding as long as the crowd did. One agent remained behind to pick up the black leather prosthetic chair Mr. Kennedy always used. He would also pack the Seal of the President.

In a room on the eighth floor, O'Donnell and O'Brien watched the President catch his wife by the arm and lead her across the breadth of the speaker's table to the wings, where they disappeared into the kitchen. O'Brien shut the set off. "She seemed to enjoy it," Kenny O'Donnell said. Mrs. Kennedy had enjoyed it. O'Donnell must have sensed a satisfaction in this, because it had been whispered with political venom that Jack Kennedy's wife felt that politics was beneath her. O'Donnell and O'Brien both knew that she had been pregnant through a presidential campaign and again in a bielection year. Earlier in 1963 she had lost a baby, Patrick Bouvier Kennedy, to a natal edema of both lungs. She had left the White House for a rest in the Greek islands.

When they got off the elevator at Suite 850, Mrs. Kennedy preceded her husband into the sitting room and appeared to be refreshed. The President picked up the phone and called O'Donnell. He asked Kenny to come to 850 at once. "We leave here at 10:40," he said. Mrs. Kennedy sat on the gold couch and began to remove her gloves. "We have a whole hour?" she said. To a vivacious woman who must acknowledge that the presidency maintains priority over her personal wishes, this is a golden time of privacy. It wasn't quite an hour. She asked the question at 9:55.

He reflected her good spirits. Subordinate to the political considerations of the trip, but never far from his consciousness, was the certainty that this was a trial political run for Jackie. She would enjoy it or she wouldn't. If she didn't, he would feel awkward asking her to join him on these political forays, where everything was counted by the minute. The crowds, the bands, the motorcades, the handshaking, the bedding down in strange places, the whispered conferences between men, the eternal rushing to be ready on time—Mrs. Kennedy couldn't remember the last time she had laid out a suit, shoes, stockings, a hat, and some gloves before retiring, but she had done it last night.

It was tiring, but she knew now that she could help him. Mrs. Kennedy—without saying a word—was a vote getter. Women admired her style, and men admired the woman under the style. The President took a chance. Casually, he asked her if she was enjoying the trip. "Oh, Jack," she said, "campaigning is so easy when you're President." Kenny O'Donnell came into the sitting room as the President said: "How about California in two weeks?" His wife nodded happily. "Fine," she said. "I'll be there."

There was nothing on the Texas trip that could lift the spirits of Mr. Kennedy more than those four words: "Fine. I'll be there. . . ." He was proud of his wife, of her obvious "class" as the President called it, and he was not averse, in private, to comparing her with other "dames." His heart's desire* was to show her off to the nation, but to protect her from the buffeting of the sweaty crowds and the hearty backslapping of the politicians.

The President asked to restudy the heavy schedule for today with Mr. O'Donnell. But the Boston Irishman, who seldom smiled unless he was studying an adversary's teeth embedded in his own fist, was staring across the room at Mrs. Kennedy, his face wreathed in a grin. He didn't hear his boss.

---

* Mr. Kennedy referred to this matter obliquely several times during the author's stay at the White House.

10 a.m.

The breakup of a presidential encampment is dramatic. There are thirty minutes of elasticity between the time the vanguard begins to check out and the departure of the President. The red-brick front of the Hotel Texas was glistening with rain when it began. Newspaper reporters wearing crumpled rain slickers, carrying portable typewriters and suitcases were in the lobby queued up at the cashier's window waiting to sign their overnight bills or pay them. Some had to report long-distance phone calls, and the telephone operators were requested to get the charges at once. The bellmen waited for service calls near the palm fronds. Texans lounged on the settees waiting for a final look at President John F. Kennedy.

Two Secret Service men had already checked out. They had packed the Presidential Seal, flags, and the President's prosthetic chair, and sound equipment in a car and had started out from the Main Street side, asking the doorman, "Which way to the Dallas-Fort Worth Turnpike?" They were headed for the Trade Mart in Dallas to set these things up before Mr. Kennedy arrived. They had a run of thirty-six miles in rain. Twenty minutes were needed at the Trade Mart to set everything in order. Roy Kellerman, the special agent in charge, forgot to notify the Secret Service men at Dallas, who were waiting with an automobile at Love Field.

The corridors of the hotel were jammed with people in controlled panic. The three elevators—deducting one which must always wait at the President's floor—were busy passing floors on the downward journey as operators listened to the knocking and said: "Next car, please." There were sixty-eight persons assigned to fly the short trip on *Air Force One* and *Air Force Two*, and another fifty would fly the leased Pan American plane which would carry the press. Some of the Texas reporters and photographers pooled their cars and left in groups, swinging up the ramp to the turnpike in an effort to get to Love Field first.

The police had a problem in front of the hotel's main entrance. A good part of the crowd which had laughed and cheered with the President in the parking lot was now on both sides of Eighth Street. The local police, assisted by state troopers, had to get them back on the far side, and no one wanted to move. The mounted men urged their horses up onto the sidewalk and moved them sidewards against the mass of citizens. Some had to squeeze between the bumpers of limousines already waiting. Men and women checking out of the hotel came out the door and went back in to try leaving by the Main Street entrance.

The Secret Service walkie-talkies were busy on several floors. One, from the Will Rogers Suite on the thirteenth, called the men standing in front of 850 to announce that the Vice-President was on his way down to see the President for a moment. Other people, worried by the clock, used the stairways to hurry down to the lobby. Doors to many rooms were ajar. The occupants ransacked drawers and closets to make certain that nothing was forgotten in the packing. Others yelled: "What time do we leave?" and the answer was: "Now. Now."

Negro women in neat white aprons stood openmouthed in all the corridors. They had cleaned rooms and made beds for years, but none had seen anything like this. One or two patrons paused to drop a dollar tip; others packed, dressed, and ready, tried the switchboard once more to get a line to home or the office.

An elevator opened at the eighth floor and two Secret Service men, noting the red light over it, blocked it. Inside was

Vice-President Johnson with his sister and brother-in-law, Mr. and Mrs. Birge Alexander. They had never met a President of the United States, and Lyndon had promised. In the summer of 1959, they had thought that the Democratic Party might nominate a Texan. Kennedy had beaten their Lyndon, and many voters in the Lone Star State felt that the Democratic ticket was "wrong-side-up." The nominee should have been the man with the broadest political experience: Lyndon Baines Johnson, the youngest majority leader in the history of the United States Senate. The polished, almost foppish Kennedy, they felt, was no more than the junior Senator from Massachusetts and should have been happy with the vice-presidential nomination.

After the election, Johnson spent considerable time discouraging the residue of rancor. He told his intimate friends that he tried to make a good run for the nomination, but in his heart he knew that he would never be President of the United States. A week before this day in Texas, he had dinner with friends at Chandler's Restaurant in Manhattan. Between the lobby and the bar stood a screen made of squares. Each one held a portrait, cased in glass, of the Presidents of the United States.

Johnson left his table and his friends to pluck his glasses from a breast pocket and examine the screen. The owner, Mr. Louis Rubin, saw the Vice-President crouching, looking from one glass face to another, and he smiled and pointed to the empty glass square next to the youthful head of John F. Kennedy. "When will I put your picture in there?" he said. The Vice-President straightened up, and anger darkened his face. "Never," he said. "You'll never see it." His loyalty to his President was so absolute that it was the subject of jokes among Kennedy's assistants.

The Secret Service opened the door to 850, and Johnson led his relatives inside. Kennedy, who was chatting with O'Donnell, turned and hurried to shake hands. Johnson presented the couple, uttered a cheerful word or two, and started to back out, motioning the Alexanders to follow him. Kennedy appeared to be elated. "Lyndon," he said, following the guests to the door, "I know there are two states we're going to carry in 1964—Texas and Massachusetts." The Vice-President grinned and said: "Oh, we'll do better than that." There was a gentle irony in this, because, until this moment, Kennedy had never stated that he wanted Johnson on the ticket with him again. "We're going to carry" was an encouraging phrase for the Vice-President, and he turned and left, reminding his sister that he and Lady Bird had to hurry so that they would be in the motorcade before the President.

Mr. Kennedy ordered all visitors to be kept out of the suite. He started to yank off his tie. Johnson's presence reminded the Chief Executive of Yarborough. It was an irritant which he could no longer countenance. As he started toward his bedroom, he told O'Donnell to phone Larry O'Brien and to tell him that Senator Yarborough would ride beside Johnson today, even if the Senator had to be shoved into the automobile. *I want him in that car!* the President said.

He went into the bedroom. George Thomas was ready. The suitcases were on the double bed. Mr. Kennedy changed to a blue-striped shirt, a solid blue silk tie, and a fresh lightweight gray and blue suit. In the other bedroom, Mrs. Kennedy readjusted her coiffure and her hat, and took one more glance out the window. The skies seemed to be clearing and reclouding. She hoped that the bubbletop would be on that car today.

In the corridor, that question was propounded for the final time. The special phone rang. It was Special Agent Winston G. Lawson in Dallas. He asked to speak to the boss, Roy Kellerman. The weather appeared to be clearing at Love Field, Lawson said, although it was still "drizzling." Kellerman, in charge of the Secret Service, did not want to make the decision. "One moment," he said, "and I will check with you one way or the other."

He went into 850 and asked Mr. O'Donnell. The President's assistant had concluded his conversation with O'Brien, in which he advised Larry to go the limit with Yarborough. "Mr. O'Donnell," said Kellerman, "the weather—it is slightly raining in Dallas. We have predictions of clearing up. Do you want the bubbletop on the President's car, or should we remove it for this parade to the Trade Mart?"

O'Donnell glanced around the room to make certain that he had left no memos. "If the weather is clear," he said, "and it is not raining, have that bubbletop off." Kellerman thanked him. It was not the unequivocal response he desired, but it was obvious that the Kennedy group, given a choice, would rather have the top off than on. He returned to the phone and repeated O'Donnell's words verbatim.

It was still "iffy," but Lawson read the message correctly. He would leave the top off unless, at the last moment, a downpour made it advisable to put it on. It was never a simple matter to put it on or take it off. The car was a special 1961 Lincoln Continental, accommodating seven passengers: a chauffeur and agent, two persons on jump seats, and three on the rear leather seat. The bubbletop was in four pie-shaped sections and covered the area from the back of the rear seat to the front windshield.

The bottom sections had to be bolted and screwed to the car. There was a metal bar, attached to the rear of the driver's seat, which stood fifteen inches above the headrest. This was used when the President and a visiting Chief of State wished to stand, holding on to the bar, to respond to crowds. On the back of the trunk, at the bumper, were two fixed stirrups with metal bars on the trunk so that agents could hold on when the car moved at speed.

The two jump seats were three inches lower than the rear seat at the lowest position. The President, by flexing a small metal arm, could lift the back seat ten and a half inches. However, after several experiments, he decided that the bottom position, which placed him and his wife a little more than three inches above their guests in the jump seats, was about right.

Lawson told his men to leave the top off.

The capital was always serene, impassive, and majestic from the streets. The broad boulevards, spokes in an unbalanced wheel; the huge edifices of granite; the bronze figures of statesmen, standing to declaim, sitting to ponder, riding to battle on horses green with the droppings of pigeons; the fountains arcing iridescent in the morning sun; the feel of history under the pedestrian's foot; the saucer of green hills curling away from this American Mecca; this Parthenon; this Vatican City; and, withal, the strained dignity of the courtesan who is not aware that she has lost her soul.

The motivation of the city of Washington has always remained the same: self-righteousness. The American posture, which pours from this heart through the provincial arteries, is one of nobility, giving away money, merchandise, and counsel to a retarded world. Behind the granite façades of many public buildings, the imperturbable regality is lost; the outside and the inside have no more relation to each other than the dignity of an opera house lobby has to the turmoil onstage.

This was a routine day. The United States Supreme Court was not prepared to turn the country from its natural course with a shattering decision; Congress, in an hour, would convene, listening to two well-paid chaplains invoke the favor of God on this nation above all others; the President pro tem of the Senate would step down to permit the new junior Senator from the State of Massachusetts, the Honorable Edward Kennedy, to sit behind the gleaming gavel and make parliamentary rulings; the Oval Office of the White House was empty, but the East and West Wings would continue to grind out the wordy projects of new laws, speeches major and minor, the hand-scripted invitations to a gala, the stereotyped responses to missives from the citizens, the publicity handouts to the few journalists who were not with the President; the logging of phone calls to and from Texas; the situation reports from the Central Intelligence Agency, the Federal Bureau of Investigation, the Department of Defense, and the Department of State.

The gentlemen sat around the spacious and darkly subdued office waiting. They had been summoned by the Attorney General of the United States and, when he burst through the door with an assortment of papers in his hand, they got to their feet. Robert F. Kennedy waved them back down, and the toothy smile was turned on. This was the most energetic of the Kennedys, the most belligerent and, in the same set of scales, the least tactful. He was the President's brother, campaign manager, buffer, and hatchetman.

In childhood, it was John who protected Robert. The older brother, Joseph, who "minded" the family when the parents left the house, was a martinet not above administering corporal punishment to the younger members of the family. In these situations, Robert—younger, smaller, weaker—hid in the dark at the head of the stairs as John diverted Joe's attention and sometimes took the blows intended for Robert. The situation was now reversed. Robert, as Attorney General, often diverted attention from his brother and faced the wrath of the public himself.

The President felt that Robert would make a good Attorney General, but he found little support for the appointment, even among his followers. It amounted to arrogant nepotism and, one night in Georgetown, the young President-elect said: "I think the best way to announce Bobby's appointment is to wait until late some night right here and then go out front and look up and down the street. If nobody is around, I will take my brother by the hand and lead him out on the porch and shout: 'I appoint my brother as Attorney General of the United States!'"

As the legal counsel to the nation, Robert Kennedy had little experience in courts and even less in the field of political compromise, but the President was pleased with his work and more than pleased. He appointed Robert to the National Security Council, where the secret decisions were made; he sought "Bobby's" advice in all matters and often listened to propositions from outsiders only to ask the rhetorical question: "What does Bobby think about it?"

As counsel to the McClellan Committee, young Robert fought organized crime, an appellation pinned to those (usually of Italian ancestry) who had permanent committees and boards of directors in many cities for the enrichment of all in the fields of vice, narcotics, and gambling. He had also applied himself to the exposure of union racketeers, notably James Riddle Hoffa of the Teamsters Union. Exposure turned out to be easy, with the assistance of renegade witnesses and the cameras of television, but conviction in court was seldom achieved and the devising of new statutes by the committee was lax and ineffectual.

As Attorney General, Robert F. Kennedy found that the Federal Bureau of Investigation belonged to his Department of Justice. This opened a new avenue of investigative procedure, a broad one encompassing the use of thousands of trained agents in many cities across the land. The free use of this weapon, Kennedy found, was blocked by the massive presence of John Edgar Hoover, who had been prosecuting interstate felons since 1924, the year before Bobby was born.

Hoover had enjoyed the confidence and respect of presidents from the administration of Calvin Coolidge onward. Now, in advanced years, the old tiger and the young wildcat were in the same hunting preserve. One of the least appreciated of Kennedy's virtues was his habit of stepping on the polished shoes of other public servants. In some cases, fear of the President kept the victims from protesting. In others, notably Hoover and the FBI, the schism became a gaping wound, unhealing and suppurating.

The Attorney General wanted to take charge of the FBI. Hoover did not relish being "summoned" by an inexperienced "boy." In this contretemps, John F. Kennedy could not help his brother. Hoover was a national hero; his FBI had never been tainted by scandal and permitted no encroachment by other departments. John F. Kennedy, elected president by a margin of only 118,000 votes out of more than 70 million cast, could not risk the wrath of the people by retiring Hoover. The wildcat was stuck with the tiger.

This morning, Bobby was making one of his periodic moves designed to keep a needle in the hide of Hoover. He had a group called the Organized Crime Committee. Some were federal prosecutors, beholden to the Attorney General. Others were officials in other offices. A few were investigators and public relations men. Their work was to expose and prosecute the American Mafia, or Cosa Nostra. This, of course, was high on the agenda of the FBI, but Robert Kennedy hoped to take the play away from his own FBI.

On the surface, the Department of Justice and the FBI worked well together. The attitude of subordination was maintained by Hoover, and the departmental amenities flowed in memoranda between the wings of the big doughnut-shaped building on Pennsylvania Avenue. But, in law, Robert Kennedy could issue unpalatable orders and force their execution.

This was the second of two meetings of the Organized Crime Committee. United States attorneys had been called in from many parts of the country to attend. Robert Kennedy's desire was to exploit a break in the Cosa Nostra—an FBI prisoner named Joseph Valachi, a minor member of the Mafia group, had secretly revealed more about his organization than the authorities had known. He had, under FBI persuasion, named names and places and events. In return, the Bureau had promised to protect his life, even if they had to arrange security in a federal prison.

Kennedy sat behind his desk, dropped the papers on the blotter, and removed his jacket. He loosened his tie from its collar moorings and unsnapped his cuff links and rolled his sleeves up. Before him was a long list of members of the Cosa Nostra who lived in many cities. Now he wanted to know what these federal prosecutors and assistants had been able to do with the information he had given them.

The city of Washington was bland with a cool majesty of purpose. The sun was strong now. The morning work, from Pentagon to Post Office to Patents, was orderly and routine. Perhaps it seemed dull because this was a time for football games and chilly air, a hoar frost on the grass and invitations to the galas of the social season. In hundreds of offices, workers sipped coffee and wondered whether it would be worthwhile—this being Friday—to take the family to the shore once more.

The dark sedan made the turn off Harry Hines Freeway and onto the small service road. Detective R. M. Sims did the driving, and he pulled by the main entrance to the Trade Mart and stopped it in the east parking lot. Captain Will Fritz, head of the Homicide Division of the Dallas Police Department, was not assigned to murder today. He and the two men who drove with him, Sims and E. L. Boyd, had nothing to do from 10:15 A.M. until 2:30 P.M. except watch a table.

Fritz was a big bifocaled man with hyperthyroid eyes who wore a cowboy hat. He was shrewd as captain of Homicide, but he had the potential pensioner's attitude of obeying the boss without question. No one ever found fault with Will Fritz, because the captain lived by the book. He was not a man for browbeating prisoners or bludgeoning confessions from the sullen. It is characteristic that his office on the third floor at police headquarters had walls of glass.

He walked back to the front of the Trade Mart with his men and, once inside, stood behind the President's table to study the layout. He saw the flowers, the speakers table spreading across two thirds of the Trade Mart, the solid ranks of long tables with their snowy tablecloths, the waiters, armed with knives and forks, dogtrotting from the kitchen to the tables, the overseers, who shouted orders to the waiters, the green and yellow parakeets, excited by the noise, flitting from rail to rail on the overhead crosswalks, the fountain, swelling to show the color red, receding into a pale blue.

Agents Grant and Stewart saw Fritz and came over to brief the Dallas cops. Fritz was to safeguard the President's table. He knew almost every person of note in the Dallas area; he also knew most dangerous persons at sight. If Sims could flank one side of the table, and Boyd the other, then Captain Fritz would be free to cross behind the table to assist either of them. Of course, Secret Service men would also be assigned to the head table.

The captain noted that directly behind him was the main entrance. He asked the Secret Service who would meet the motorcade out front, who would escort the presidential party inside, and who would clear the big lobby before the arrival. "Now," said Fritz, "when he gets inside, how does he approach this head table. From which side?"

The Secret Service answered all the questions and gave the police little orange buttons for their lapels. Robert Stewart said that the back and sides of the building were already secured. Local policemen were all over the place, under Deputy Chief M. V. Stevenson, and, except for a possible emergency delivery of more bread and rolls, no one would enter the block-and-a-half-long building except by the canopy at the front. The men stationed there, uniformed motorcycle police and Secret Service agents, would funnel the guests through the main door into the lobby where, as a matter of course, they would be properly screened by ticket-takers, local politicians, and more agents.

As the guests passed the back of the head table to find their places in the gigantic room, it would be up to Fritz and his men to keep them outside the roped area of the head table. No one, even if properly identified, was to be permitted to pluck a flower from one of the many vases at the table nor to leave an object or package for anyone.

The front of the head table would be guarded exclusively by Secret Service agents from the President's party, but Stewart wanted to make certain that the flanks and the rear were secure at all times. He asked Fritz if he and his men had experience with explosives. The captain nodded. Stewart asked if Fritz would make a complete inspection of the entire head table and under it in about forty-five minutes and again when the word reached the Trade Mart that the President was five minutes away.

Fritz said yes. The assignment was going to be easy. There was a shout from the lobby and a few of the men ran out. Lieutenant Jack Revill of the Dallas Subversive Unit, had been checking Trade Mart merchants in and out with the assistance of Detective Roy Westphal. It was Westphal who decided to frisk a merchant at random and found a small Cuban flag in his pocket. This had led to a complete fanning of the merchant, who protested. He insisted that he was anti-Castro. Chief Stevenson came running and settled the matter by reminding the merchant that the Dallas Council had just passed a special ordinance about signs, picketing, protests, thrown objects, and threatening language during the visit of the President. This would include anything, he said, which might tend to embarrass or intimidate the President. The merchant would have to surrender the flag temporarily or leave the building. He gave up the little banner.

Up at Elm and Houston, Officer W. B. Barnett was a lonely man. He had nothing to do but stand in uniform against a wall and try to keep the drizzle from soaking him. Barnett was a traffic cop. Today he was to stand at the corner where the motorcade would make its final turn. His superior, Captain Lawrence, had told the traffic men that they were to divert all traffic five or ten minutes before the motorcade passed, and they were to scout the crowds and the windows overhead for thrown objects.

Barnett stood on the corner opposite the School Book Depository, looking down at the railroad overpass, where donkey engines, small and energetic, shoved strings of freight cars back and forth. A man came over from the School Book Depository, trotting in the rain, and asked the police officer what time the motorcade would be passing. Barnett asked him why he wanted to know. "Because," he said, "our building is full of people who would like to see President Kennedy go by."

"Tell him to come out around 11:45," the cop said. The stranger dogtrotted back to the front entrance, and Barnett walked toward the statue of Joseph Dealey to get a better look at the old red brick building. All the windows were closed.

At police headquarters, Lieutenant D. H. Gassett strode down the cross-hall on the third floor and entered the dispatch room. The big antennae on top of the municipal building would carry heavy traffic today. At 10 A.M. three operators had reported for duty. Two were on Channel One of KKB 364, and one was on Channel Two. Gassett went over the situation with them once more, explaining that the two operators on Channel One, facing each other with the console between them, would handle the heavy traffic. The operator on Two had a separate console and faced the others. There was to be no foul-up, men reporting in were to do so by number, and Gassett would tolerate no friendly conversation. The motto was: "Get 'em on, take the message, get 'em off."

As he left the room, Lieutenant Gassett whacked one of the time stamps. There were three, one for each operator, but they never agreed on the time of day. One operator, looking at the little clock from a low stool, read the time one way; another, sitting high or leaning over the clock, saw it a minute later or a minute earlier. It was odd, Gassett thought, that in a business where time was so important, that the clocks of the three radio operators, if stamped simultaneously, would probably be a minute apart.

At the Sheraton Hotel, crouching in the shadow of the huge Southland Life Building in downtown Dallas, the President's communications headquarters was now complete. Anyone who called in from the parade route, or the plane, could be hooked in with the White House in Washington or, for that matter, to anyone in the world. This was the most sophisticated telephone equipment to be found anywhere. No conversation would go through the Sheraton switchboard; no operator could listen because there were voice scrambler attachments on both ends. A master sergeant, who was also a master electrician, sat in a small room and said: "Ready in Dallas" and asked for a few tests.

The code names for people and places were before him. President Kennedy was Lancer; Mrs. Kennedy was Lace; Vice-President Johnson was Volunteer; Mrs. Johnson was Victoria; the White House was Castle; *Air Force One* was Angel; the President's car was SS-100-X; Chief of the Secret Service James Rowley was Domino; the LBJ Ranch was Volcano; The Bagman was Satchel; the Pentagon was Calico, and the FBI was Cork.

It seemed involved, but Colonel George McNally and his communications men made it function almost instantaneously. The President, in his car, could lift a phone and say "Lancer to Lyric" and, in a breath, his daughter Caroline, at class in the White House, would be on the phone. The Vice-President, smiling to crowds, might hold a phone to his ear and say: "Volunteer to Daylight" and be talking to Secret Service Agent Jerry Kivett at the ranch hundreds of miles away.

At the moment, the sergeant at the switchboard was listening to Colonel Swindal, command pilot of *Air Force One*, call in from Carswell that he was ready to go and had cleared flight plans with Love Field, Dallas. The long checks had been run through by the crew; the fan jets had been tested; fuel was aboard, and the colonel had alerted the brass at Carswell that the President of the United States was expected at the base in half an hour. An honor guard was sent to the entrance gate.

Not all communications were as wrinkle-free. Mr. Jack Ruby, a worn face in a disorderly apartment, had dialed his sister's number. He could hear it ring and he knew she must be home because she had been ill. At last the receiver in Apartment I at 3929 Rawlins Street, Dallas, was lifted, and a small voice said hello. Mr. Ruby did not say: "This is Jack." He assumed she recognized her brother's voice. He was upset, he said, over a full-page ad in the *Dallas News*. Mrs. Eva Grant said she was sorry, but she had not seen it.

Her brother began to shout. She should look on page fourteen; it was awful. It asked the President of the United States a lot of insolent questions. Worse, it was signed by Bernard Weissman. This sounded Jewish. Mr. Ruby was a Jew. Eva Grant, his sister, was a Jew. All the difficult tenement growing up in Chicago was not enough; Adolf Hitler was not enough; now a Jew had to get fresh with the President.

"He's a son of a bitch," said Mr. Ruby. "The *News* was wrong to accept the ad." Jack knew a lot of nice guys in the ad department at the *News* and he was going to ask them about this so-called Weissman. Who was he? Where had he come from?

Mrs. Grant listened. She was too ill to argue or to soothe ruffled feelings. Of her brothers and sisters, this one was strange. He had never married, and now he was in the middle years, talking about exercising in the YMCA, taking sauna baths, making a full-time hobby of being friendly to policemen, trying to run a brace of cheap nightclubs with strippers and dirty filthy masters of ceremonies who seldom heeded Mr. Ruby's warning to "clean up them jokes—especially the ones with the Jewish dialect."

The listening continued. Lately, the conversations between the two had been one-sided. Once, when Eva and Jack were full partners in the nightclub business, she had been shrill and inexhaustible and she kept track of every dollar that came in or went out. But they had bounced from one place to another, always failing or tempting failure, always a step ahead of the bill collectors, trying to fight competition which was using "amateur stripteasers," which seemed to draw more male customers and heavier drinkers than the professionals. Now the woman who, perhaps was more mother than sister, more *hamisher* than Jack, was tired.

Tuesday morning—three days ago—he stopped in, not to ask her how she felt after hospitalization and an operation, but to show her a newspaper photograph of President John F. Kennedy and his son John. The more Jack Ruby studied the photo, the more emotional he became. "That man," he had said, choking, "doesn't act like a President. He acts like a normal everyday man with a family."

Mr. Ruby seldom permitted anyone to get off a telephone easily. He had phoned the *Dallas Times Herald*, he said, and they told him that they had refused the ad. That, thought Mr. Ruby, was class. Further, he thought that the ad should not have been addressed to "Mr. Kennedy." They might at least have called him "Honorable Mr. President." At least, Mrs. Grant was small and middle-aged and patient.

He had phoned the *News* and, after making certain that he had a minor advertising executive and not an editor, had asked: "Where the hell do you get off taking an ad like that? Are you money hungry or something?" Mr. Ruby groused a little more and then told his sister: "If that guy is a Jew they ought to whack the hell out of him."

In this, Mr. Ruby, average citizen, was crying aloud against the slurs and slanders which had smashed against his sensibilities in Chicago and in Dallas. This was his real gripe. His admiration for President Kennedy was genuine, but his fear of being a defensive Jew was paramount. Sometimes, in the company of friendly *goyim*, he had to force a wry smile when he was patted on the back and referred to as a "white Jew." Always Mr. Ruby had tried to be twice as nice to them as they were to him, but on occasion, especially in his strip joints, his temper deserted him and he lashed out with his fists, knocking a customer to the floor and kicking him down a flight of stairs. Or listening to one too many Jewish jokes and yanking the master of ceremonies off the tiny stage and hurling him across the floor.

Why did such a thing have to be signed "Weissman"? Why, Eva?

The President strode back into the sitting room and gave his wife a big smile. Yes, she was ready. Mary Gallagher had gone on ahead with Mrs. Lincoln and some of the others. Mr. Kennedy told the Secret Service he was ready to leave. The word was whispered through the partly open door of Suite 850, and it passed down the hall and men became even more alert. On the telephone near the fire hose went the final message. "Lancer is leaving the Hotel Texas." And, as it always did, the message spread downstairs to men with walkie-talkies, to others on rooftops, to Carswell Air Force Base, to Washington, D.C., to Love Field in Dallas, to the communications center at the Sheraton, to interested parties in many places. Lancer was leaving.

Eighth Street was choked with vehicles and mounted policemen. The crowd had been herded to the far curb and, for a moment, a shaft of sunlight brightened the scene. The automobiles were in three rows—congressmen's cars along the outer edge; press buses and staff cars in the middle; the big limousines rented for the President and his personal party at the curb.

Between the Secret Service follow-up car and Lyndon Johnson's limousine, Larry O'Brien stood bareheaded, entreating Senator Ralph Yarborough to please get in the vice-presidential car. The reporters in the buses could not hear the conversation, but they could see O'Brien's hands making the plea, and they could see the small silky wisps of red hair lift and fall on

O'Brien's head, and they could see Yarborough, studying the curb and shaking his head negatively.

Mr. O'Brien could not exert the blunt, brutal pressure which was available to the President. He tried persuasion and the Senator found that tack easy to resist by stating that nobody cared where he rode or with whom. To the contrary, O'Brien said softly, nodding toward the press bus with its array of eager faces pressed to the windows. Yarborough barely looked up. He knew that his presence in that car, or his absence, could be the big story in the nation's press tomorrow morning.

It could darken the Kennedy triumph and hide it in shadow. The trip, thus far, had been bigger, warmer, friendlier than Kennedy had expected and was a surprise to the knowledgeable and conservative Governor. Now it was threatened by personal pique. The man to send to balm the raw sensitivities was not Lawrence O'Brien. The redhead was a peacemaker for those intelligent elements which did not set themselves against peace; he was a compromiser, a friend, a buddy, a favor-doer before becoming a favor-maker, a collie dog trotting along the perimeter of congressional sheep, urging the stragglers onward, bringing the wanderers back into line, looking to the shepherd in the White House for a compass heading aimed toward greener pastures for all.

But this was a personal vendetta. O'Brien was not a man to threaten. Mr. Kennedy might have sent Mr. Kenneth O'Donnell, who could have turned on his Humphrey Bogart peel of lips and who might have whispered: "You will get in that car, Senator, or you will wish you were dead. The President says that if we have to get a few guys to lift you up and toss you in the back seat, we're to do it. Which way would you like to have it?" If Mr. Robert Kennedy had been in Texas, he would probably have summoned Yarborough to his presence yesterday, in San Antonio, and he might have said: "Senator, we are going to have unity in this goddamn party and we're not going to have the boat rocked by you. We demand that you sit with the Vice-President on every occasion for the next two days. You don't have to make love to him; just sit beside him so that you are not in the position of handing ammunition to the press. After we go back to Washington, if you want to continue your childish quarrel, go to it, but while my brother is in this state you and Lyndon and John Connally are going to smile like brothers. If you don't, we think your support is too expensive for us and we may have to dump you."

Yarborough desired to compromise with O'Brien. All right, he said, I won't ride in the car but I'll issue a statement. Larry O'Brien wagged his head no. The sun was back behind the billowing, slate sky, and the faces at the bus windows were less distinct. The Vice-President and Mrs. Johnson emerged from the hotel and, smiling at the applause, got into their car. Yarborough felt that the situation must be abrasive to the President to warrant all this attention, and with head down he said: "Well, if it means that much——." O'Brien permitted himself a grin of relief.

The Johnsons were comfortable in their car, but it had no jump seats; two Secret Service men were in front; the Johnsons used most of the rear seat. Two cars ahead, they noticed a commotion. The Kennedys had discovered that their car also held but five passengers. The front seat had William Greer, Secret Service agent assigned to drive the President's car, and Roy Kellerman, head of the Dallas office. Governor and Mrs. Connally were to ride with the Kennedys, but there was room for only one of them on the back seat.

The President was apologetic. Nellie Connally said it was understandable, and that her husband should ride with the President. For the short trip to Carswell, she would go back with the Johnsons, who were old friends. Up ahead, at the level of Main Street, the police motorcycle escort had started the assortment of explosions which always signified the imminence of departure. Mrs. Connally hurried back to squeeze in beside the Vice-President.

Sunlight flicked on bright and hot. Seth Kantor, a journalist, found the Eighth Street side too crowded for leaving the hotel. He hurried out the Main Street entrance and down the side street, looking for the press bus. A raincoat was on his arm, and his notes were clutched in his hand. People in the motorcade watched him step off the curb between cars and one foot came to rest in fresh manure. Mr. Kantor went down, left arm trying to break the fall, and the hand braced itself in the manure. Whatever earthy humor there is in such a situation is beyond analysis, but laughter swept the motorcade.

Two congressmen left their cars (Henry Gonzalez and Olin Teague) to help the reporter to his feet. They brushed his clothes and both observed that, journalistically, no matter where Kantor went, he managed to step in this substance. Malcolm Kilduff hurried to Kantor's side to inform him that, unless he could find a way of diminishing the residual odor, he would have to ride on top of the press bus. Kilduff, assistant press secretary to Pierre Salinger, was the man in charge of the press on the Texas trip.

O'Brien escorted Yarborough to the Johnson car and helped him in beside the Vice-President. Under his breath, Lyndon Johnson muttered, "Fine" to O'Brien, and Mrs. Johnson, in the middle, turned a friendly smile to the Senator. On the far side of the seat, the door opened and Nellie Connally was trying to squeeze in. Gallantly, the Senator began to back out. Larry O'Brien saw the broad hips he had just assisted into the car backing out toward him. He glanced at Mrs. Connally in mute appeal. She, gracious and confused, backed out as O'Brien's shoulder jammed the Yarborough hips back into the car. Mrs. Connally got out of the car, walked around the back, and got in the front seat between a policeman and Secret Service man Rufus Youngblood.

No one, except O'Brien and Johnson, seemed to comprehend this complex game of musical automobiles, but the Secret Service men nodded to each other from the back of the motorcade to the front, and the Fort Worth police stopped all traffic on Main and slowly began the run to Carswell. The crowd cheered, the Kennedys waved, and the excitement around the Hotel Texas died in a pall of blue smoke.

The people of Fort Worth were confused by the route chosen by Kellerman and O'Donnell. The direct way to Carswell would have been Route 20, the east-west freeway, between Trinity Park and Forest Park and then the residential section of Arlington Heights to White Settlement and the base. It was 10:40 A.M. and the decision-makers wanted to use a half hour, so the motorcade swept up Henderson Avenue and onto Jacksboro Highway in a north-northwesterly direction, although Carswell was west of the hotel, then a big swing left onto Ephriham Avenue, which melts into General Arnold Boulevard, passing the airport on the opposite side, to the south.

Someone in authority must have thought that this would be good exposure for the President, but, as the morning breeze cupped the ears of the riders, they saw but a thin rime of citizens standing on the curb. In the press bus, reporters saw a few shapely women out on the sidewalk in housecoats. One said: "Hustlers?" A second said: "This early?" Seth Kantor, looking ahead, saw a big automatic farm forklift at the side of the road with two men sitting high in the shovel waving to the Kennedys. They were old friends: Harry Rubin and George Levitan. Mr. Kantor had been married in Levitan's house eleven years ago.

The sun, still tentative, seemed a bit more positive and flashed light and heat across the lush fairways of Rockwood Park and Shady Oaks. President Kennedy squinted at the sky and smiled. He was a man who liked to make his best showing, and a good portion of truculent, independent Dallas might come out on the streets to swell the attendance if he could have sunshine as an ally.

His wife, flicking the lower strand of dark hair back from her eyes, held onto the little pink hat and said the weather was warm for November. It was, but in this matter her husband would not attribute it to the vagaries of high and lows and un-usual isobars. It was General Godfrey McHugh's fault. Kennedy had asked him specifically to get a good advance reading on the weather, and McHugh had the pooling of several meteorologists at his command: the Dallas Weather Bureau; the Air Force; Carswell; Fort Worth; the national meteorological projection for all areas, issued at the observatory in Washington, and McHugh had come up with "cool." It wasn't cool; it was hot, and Mrs. Kennedy was now in cold-weather wool. Before noon, the tem-perature might climb to the seventies. If it did, the President's temperature was going to climb higher, especially in the pres-ence of Godfrey McHugh.

11 a.m.

The pace of the clock was deliberate and unhurried, the large second hand flicking a leg like a British soldier executing a slow march. It is the most exasperating of man's inventions because it must prevail and can never be dominated. The timepiece displays an impassive face and man moves at a pace faster than he wishes. He fights this index to events and movement and must always lose the final engagement.

It was so with the Fort Worth motorcade. Surprisingly, there were a few bands tooting music as a welcome to the President, and he waved, turning backward in the seat to grin as his thick hair lifted high over his head, but the speed began to accelerate as the O'Donnells and the Kilduffs noticed that the schedule was six minutes late.

The air was damp and uncomfortable, and, as the dwellings and curbside watchers thinned, the cavalcade turned off onto Meandering Road to the gate at Carswell Air Force Base. The guard of honor, sparkling in the sporadic sunlight, stood at present arms as the cars inched through the security checks. The commanding officers stood in array, the white laurels on their peaked caps proclaiming rank.

Mr. Kennedy turned on his youthful smile, but he was in a hurry. The post band began the solemnly punctuated "Hail to the Chief," and everyone remained at rapt attention until the last note. There were greetings to be exchanged with generals and some wives who wanted to be presented, and there was no way to hurry the takeoff to Dallas. Mrs. Kennedy, gracious and smiling, met the ranking officers and their ladies with the poise that was always hers. The President, fingering the top button on his suit jacket, murmured to O'Brien: "This certainly has been an interesting and pleasant morning."

The President was sure that he could board *Air Force One* at once, because Carswell is so security-conscious that young officers who left the base for a movie often wondered how they would get back in. When Major General Montgomery was the commanding officer, he assigned new lieutenants to live in Fort Worth and ordered them to try daily to "sneak" into his base without proper credentials.

The President and First Lady, with the accouterment of generals, began to move toward *Air Force One*. They almost reached it. Suddenly, a group of mechanics in coveralls burst through the lines and engulfed the party. There were yells and hurrahs and grimy hands were thrust forward. High in the flight deck of 26000, Colonel James Swindal looked down and saw the Kennedys disappear in the enthusiastic maelstrom.

Lyndon Johnson and his group hurried over to *Air Force Two* and boarded at once. The press, with a Pan American Clipper 707, was ready to go. No one could move until the President was ready, and Secret Service men tried to insulate the First Family against all those hands waving near their faces. But the President, pleased at the outburst, began to lend himself to it and he held both arms above those nearest to him to reach and shake a wheel of arms.

Secret Service men waved everyone else aboard, urging the passengers to hurry. Mrs. Lincoln, who had stopped the President at the elevator with her relatives' relatives, waited to get a picture of the Kennedys with a new camera. The captains of the three aircraft received clearance to taxi. The breeze was out of the northwest and slight, and Swindal—even though his plane was traveling light—chose to taxi south and use the long runway north over Lake Worth.

The doors of the press plane closed, and the ramp was pulled away. Aboard *AF-2* were thirty-two passengers, including ten Secret Service men. Of the remaining twenty-two, Lyndon and Lady Bird Johnson had as guests seven congressmen and Texas Attorney General Waggoner Carr and Mrs. Carr. General Ted Clifton, The Bagman's boss, was aboard, but The Bagman was manifested on *Air Force One*. Elizabeth Carpenter, the stout, jolly, and observant secretary to Mrs. Johnson, put on her seat belt, blew a sigh, and hoped that Dallas would behave itself.

The Kennedys inched their way toward the rear ramp of *Air Force One*. Military police were now trying to pry the well-wishers from the couple, but many of the workers had waited a long time for this opportunity, and they had no intention of going home tonight saying: "No, I did not shake hands with the President."

There were thirty-six aboard *Air Force One*, not counting the crew. This was the powerhouse group. No matter how friendly the relationship between the people on *AF-1* and *AF-2*, there was always an aura of condescension from the first to the second. The planes were the same model, same size, almost the same accommodations, but rank and precedence are important. The personnel on both planes shared a spirit of camaraderie, but, when opinions diverged, it was *AF-1* which prevailed.

Besides the President and Mrs. Kennedy, *AF-1* had Kenneth O'Donnell, Lawrence O'Brien, and David Powers, the Athos, Aramis, and Porthos of the President's private circle. General Godfrey McHugh was aboard, proud and officious. Mary Gallagher, who was not *en rapport* with Mrs. Kennedy this morning, sat quietly with Mrs. Lincoln and hoped that the First Lady's squall of displeasure would blow away with the rain.

Malcolm Kilduff, bright and tough, was in charge of the press but he rated *AF-1* in case the President might wish to commune about stories in the Dallas press. Dr. George Burkley, rear admiral and physician to the President, walked up and down the aisle, a mild gray man who felt that he should be close to the President at all times, but who was often ordered to the back of the line by Kenny O'Donnell. It seemed pointless to Burkley to have a doctor at a distance from the President of the United States—of what use would he be in an emergency?— but he was also thankful that the President, other than a chronically bad back and a chronic cortisone deficiency, was a healthy animal.

Governor and Nellie Connally had to be on *AF-1*. So, too, did Ralph Yarborough. Roy Kellerman, the tall, wavy-haired man who was in charge of Secret Service on this trip, was also aboard, sitting at a breakfast nook in the back of the plane. So was Clint Hill, whose function was to protect the First Lady.

Four members of the press, representing all the others on the Pan American plane, were permitted aboard *AF-1*. The dean was Mr. Merriman Smith of United Press International, who would, if the opportunity arose, flash a story to UPI and, on leaving the only telephone in the area, yank it from its moorings.

Practically, this flight was ludicrous. Three giant jets were about to fly thirty-three miles. There had been considerable surprise among the crews when the order had come down that all the infinite care and preparation for a flight overseas would be put into operation for a small commuter run between twin cities. Automobile drivers on the expressway made the trip in thirty minutes. The commercial air companies referred to Love Field as the Dallas–Fort Worth Airport, as though both cities were one.

The situation had already created a traffic problem. No bombers were taking off from or landing at Carswell. The B52s on the hardstands stood like tired birds in pools of rain. Any aircraft en route in was ordered to remain clear or proceed to Bergstrom Air Force Base. The small private planes at Luck, Oak Grove, Plyon, Arlington, Blue Mound, Grand Prairie, Red Bird, Greater Southwest International, and White Rock had been ordered to clear the air. For weeks the big commercial companies which had flights coming into Love Field, or out-bound, had been warned to cancel, postpone, or move ahead. The FAA monitors at Love Field had been told to "pick up" the three presidential planes on radar the moment they lifted off Carswell.

The soft drawling Alabaman, Colonel James Swindal, filed a flight plan asking for 5,000 feet altitude. *AF-1* could make the climb in two minutes, but a jet is inefficient under 24,000 feet, and Swindal would be swinging into the Greater Dallas area before he could get higher. He could climb north to the outer perimeter, swing slowly to starboard, remaining mostly in the turbulent cloud cover, pass over Grapevine and on toward Richardson and Garland, then southeast to Mesquite, turning northeast over the Trinity River, dropping slowly as he passed the skyscrapers of downtown Dallas, then up the alley between Hines Boulevard and Lemmon Avenue to runway 31, which is close to 310 degrees on the compass.

The colonel saw his prime passengers climb the ramp, and, when the exit door light flicked off on his panel, he lit number three pod, and then the others. The first officer passed the word that *Air Force One* was ready. The President and Mrs. Kennedy proceeded to their private cabin, slightly aft of the big silver wing. The seat belt sign went on. Colonel Swindal watched the press plane move out, a broad, ugly, rocking thing on concrete but a flash of white swan in the sky.

Behind it, the Vice-President's Boeing 707 whined at a slow pace as it moved to the bottom of the two-mile strip of concrete. The air began to crackle with repeated flight plans, courses, and handoffs. Dallas came in to warn two commercial jets, now twenty miles northeast of Love Field, to please hold at their respective altitudes and positions for the next fifteen minutes.

The crew of *Air Force One* flew the world's most pampered airplane. The exterior, blue and white trim, had been selected by President Kennedy. The interior had been designed by him, with assistance from Mrs. Kennedy and a decorator. The paired seats in the front of the plane are turned backward because it is more conducive to safety; at the wing section there is a bulkhead with the Seal of the President of the United States. Behind it is a combination office and lounge, with a desk, a presidential telephone, a long settee, an overhead television set, deep pile rugs, an electric typewriter and, behind this, a narrow walkway on the port side of the plane. Facing the walkway is the private cabin of the President and First Lady, with twin beds, a makeup table, a desk, and flowers. Here, the current newspapers and magazines were laid out in rows for the Kennedys.

Behind the private cabin is an aft area with a breakfast nook and facing benches, a few stuffed seats against the opposite bulkhead, and a storage compartment in the tail of the aircraft. Up front, behind the flight deck, is the most elaborate communications equipment ever devised for air travel. The President may, if he chooses, speak to anyone in the world who has a telephone. Should any part of the system fail, there are "backup" facilities. At the press of a button, soft music floods the plane from ceiling loudspeakers. A motion picture can be shown. A Secret Service man may depress a button on a separate telephone system and speak to his confreres anywhere.

The plane cost the American taxpayers $6 million with fan jet engines; communications ran to $2 million; there was even an original painting of a French farmhouse over Mr. Kennedy's bed. News came in on a teletype machine and the sheets were ripped off and placed before the Chief Executive as they were enunciated. The plane had a small device which would encode a message and decode it. The rug underfoot was blue with a taloned golden eagle and stars representing the original thirteen states.

Swindal turned his four big engines up, and concrete began to slide under the nose. He watched the press plane pass him, going the other way, and he saw it lift off into the low billowing gray over Lake Worth. Its gear was tucked in at 11:17 A.M.; *AF-2* roared north one minute behind it, still in the turbine turbulence of its predecessor. Colonel Swindal waited until 11:20, completed all his last-minute checks, and poured the JP-4 to the engines. The few in the passenger area who would not sit when the stewards requested it, felt themselves hanging onto seats to keep from sliding back down the aisle.

In a moment, 26000 was airborne. Swindal asked for seven thousand feet instead of five and got it. He could have asked for anything because nothing else was permitted to fly in the area, and only three blips showed on the radar screens. He throttled down to 225 knots, and started his big lazy swing to the right. The Colonel didn't know it, but the FAA monitors at Dallas had him on their scopes before he cleared Fort Worth.

For John F. Kennedy, landing at Dallas would complete 75,682 miles of travel in *Air Force One*—approximately three times around the world in one year.

Dispatcher Henslee at Dallas Police Headquarters noticed a swelling volume of traffic. He was handling Channel Two and the voices were coming in from all parts of the city to ask questions or to contact other officers. A deep voice came in calling "30." This was Sergeant R. C. Childers' number. "When you start receiving information from the tower on that plane," he said, "advise 531 (radio dispatcher's telephone extension)." Sergeant Gerald Henslee announced his agreement. "Ten four," he said. "Will be on Channel Two." He called Deputy Chief N. T. Fisher at the airport. "Four. Will you advise as to crowd estimate and weather condition." Fisher was not willing to predict weather. "Ten four," he said. "It's not raining now and we have an estimate of a crowd of eleven hundred people."

Officer Murphy called in from the opposite end of Dallas. "Could you send a city wrecker to the triple underpass, just west of the underpass on Elm, to clear a stalled truck from the route of the escort?" Dispatcher McDaniel knew that this was between the Texas School Book Depository building and the railroad trestle. There were three traffic lanes, but that truck would have to be hauled out. The dispatcher complied by saying, "Ten four."

Patrolman R. B. Counts cut in to say: "I've got an Air Force car here that has the President's seal and flags and he's got to get to the Trade Mart before the President does. . . . I'll escort him out there about code two." Code two permits speed but caution at intersections with siren and red flashers. Murphy came back on Channel Two: "Disregard the wrecker at the triple underpass. We got a truck to push him out of here."

Captain Glen King, an articulate officer who supervised public relations for the police department, sat in his small office. He was adding up the roster for the 8 A.M. to 4 P.M. tour of duty. The chief would be with the President. A deputy chief and fifty-four men were at Love Field. On the parade route, one hundred seventy-eight men were assigned to the critical intersections. Fifteen police motorcycles would be with the motorcade. An additional two to four men were stationed at each point where the cars would make a turn. Extra men were standing on every bridge and railroad trestle. Others waited inside tunnels and underpasses.

Detectives and off-duty patrolmen in civilian clothes were assigned to mingle with crowds and man rooftops where crowds were expected to be heavy. Sixty-three policemen surrounded the Trade Mart, and 150 more were already inside the big building. The Dallas Police Department totaled 1,100 men. Of these, 850 were now working this one tour of duty, and most of them were within the presidential area of four miles. Captain King was satisfied. The superior officers knew from experience that the visit of a VIP always brings the hotheads and a few mentally irresponsible people to the front. Dallas had seldom entertained a President of the United States, and Glen King wanted no trouble, no irate citizens, no threats to the motorcade. The department was literally breaking its back to make certain that Dallas would "come up smelling like a rose."

There was additional assistance which Captain King did not add to his list. Off-duty firemen were impressed into service. Sheriff's deputies were on duty. The State Police Department of Texas sent men in big, curling hats. The Texas Rangers were stationed from airport to Trade Mart. Governor Connally's own Department of Public Safety sent its best men from Austin. Chief Jesse Curry issued a pronunciamento empowering anyone—in case of riot or disturbance on this day—to make a citizen's arrest.

Henslee was monitoring Channel Two and Assistant Chief Batchelor came on in a worried tone. "Nine," he said. "Have you received information that his arrival is twenty minutes late?" Henslee said: "I have not received that information." The President was due at the Trade Mart at exactly 12:30. The Secret Service and the police had made the dry runs four or five times and it required forty-five minutes, allowing for a speed of fifteen to twenty miles per hour in the rural sections and ten to twelve in the crowded streets. Twenty minutes amounted to a great deal of time. It would stretch the travel between the Hotel Texas and the Trade Mart, a distance of thirty-three miles, to two hours and a half, including jet plane.

Batchelor was wrong. The schedule was seven minutes slow and could be improved by allowing a shorter time for official greetings at Dallas and a faster run from the triple underpass to the Trade Mart. Secret Service agents at the airport were lining up the cars for the motorcade, and chauffeurs were moving ahead or back in line as ordered. City police held a large crowd of sightseers behind the steel fence. A group of fire engines, foam trucks, an ambulance, and a jeep with a tailgate marked "Follow Me" moved out on runway 31.

The sun came out, not peeking, but staring. The temperature was 76 degrees Fahrenheit. Dallas, the narcissistic city, found itself impressed by an alien. The people not only were lining the curbs from the airport to the underpass, but a second wave, seeing the hot sun, now jammed buses and cars on the roads. Almost everybody was headed for a part of downtown Dallas. Each one, so it seemed, felt that there would be plenty of room for watching the motorcade. Cedar Springs and Main and Lemmon and Mockingbird and the roads between were lined and triple-lined with people.

The porches of homes were heavy with people who stood looking. Others leaned in shop doorways. Some sat on gravel roofs. In the skyscrapers venetian blinds were pulled up. Faces appeared in windows. Secretaries tore old scrap paper and watched it float, like confetti, toward the crowds below. The Lamar Street viaduct was choked with cars coming in from Oak Cliff, trying to find a place near Main to park. Some left their cars along the curb between the *Dallas News* and the railroad station, even though police citations would be tied to the windshield wipers. The city had decided, with the advent of the sun, to stage a rip-roaring hell-bent-for-election Western-style welcome. For this day, this moment, Dallas did not want to be reminded of the politics of John F. Kennedy. He was the President of the United States, and, if San Antonio gave him an "Olé!" and a big, sprawling Houston tossed its Stetson in the air, and Fort Worth waited patiently with rain on its face to roar, "Where's Jackie?" then Dallas was about to deafen this Easterner with a roar of sound and a bear hug that would send him to the LBJ Ranch dancing with joy.

Those who could not hurry downtown turned the television sets on and left them on. In the *Dallas News*, E. M. Dealey, publisher, permitted the young secretaries to use his television set to watch the proceedings. In Irving, Ruth Paine hurried from the dental appointment to find Marina Oswald excited and still in her bathrobe, explaining in Russian that the President and his pretty wife had left Fort Worth but there was nothing on television at the moment because the plane had not arrived in Dallas.

In City Hall, a portable set was on and too many faces tried to watch. In shopping centers, crowds anchored before radio stores to watch the proceedings on TV sets in the windows. At Parkland Memorial Hospital, sets were turned on. Starched nurses, en route from room to room with tiny paper "cocktails" of pills and water, watched the excitement. In the rectory of Holy Trinity Church, Father Oscar Huber sat upstairs in the recreation room with Father James Thompson. Huber, the pastor, was a small, alert man with a wry Missouri humor.

The blue veins in his snowy hands were pulsing with excitement. He watched the jets lift off from Fort Worth and he said to his younger confrere: "I have never seen a president. Want to walk up to Lemmon and Reagan with me and watch him pass by?" Father Thompson did not. He would be content to watch the proceedings on TV. The little pastor sat for a moment, watching, then got up and put on his black jacket and left. He had heard that the nuns at Holy Trinity Church were escorting students up to Lemmon and Reagan, and, if it was good enough for them, it was good enough for him.

The clouds, like swirls of mud in a blue lagoon, jarred the three jet planes as they approached the glide path to Dallas. General Ted Clifton, on *AF-2*, could not understand the reasoning behind O'Donnell's decision to fly rather than ride. Between clouds, the flat prairie of Texas, modified by thousands of white ranch houses, lush green golf courses, and damp river beds, came into focus and disappeared in the next cloud. The seat belt sign went on and most passengers were of the opinion that they had hardly left the ground.

Colonel Swindal's first officer, wearing a headset, could hear Love Field repeating the warning for all aircraft to stay away: "D.V. Arriving. D.V. Arriving. D.V. Arriving." The Distinguished

Visitors, even though locked inside these flying aluminum tubes, were busy, each with his own problem. Senator Yarborough felt that he had made the peaceful move by riding with Vice-President Johnson; now it was up to the President to see that Yarborough was treated like a United States Senator at Austin tonight. That, unfortunately, could not be solved by a seat in a car. Governor Connally, a fellow Democrat, had no intention of honoring Yarborough at the party affairs this evening. Nellie Connally was more spiteful. She would not invite the Senator to the cocktail party at the Governor's mansion, although it belonged to the people of Texas. Yarborough might surrender to persuasion, but the Connallys were going to maintain the split in the party, and the President could not persuade them to be civil. The Governor felt little enthusiasm for Kennedy.

Mrs. Lincoln, with a handful of memoranda which had been typed the previous night, was trying to steady herself as she went aft to ask the President to sign them. Chris Camp and Pamela Turnure split parts of the Austin speech and tried to type some of it before the plane reached Dallas. For a moment, Mr. Kennedy was in the main cabin, sitting at his desk, facing Texas politicians. They were boasting about the fine receptions, and the President conceded with gratitude, but kept coming back to the scurrilous attacks of the newspapers, and the congressmen reminded him that they had no more control over the press of Texas than he had in Massachusetts.

Without reading, he signed the memos Mrs. Lincoln placed on his desk, and he returned to the papers. He had a new one. It was a column published in the *Dallas Times-Herald* two days ago. "Why Do So Many Hate the Kennedys?" was the title. It was written by A. C. Greene. As Mr. Greene analyzed it, antipathy was felt toward all the Kennedys, the President, his wife, father, brothers, "his daughter, Caroline, and to some extent, even the little tyke, John Jr." The President was referred to as a man whose "money still stinks."

Mary Gallagher was in the private cabin, helping Mrs. Kennedy to get ready for the automobile ride. In a few moments she was out, sitting beside Mrs. Lincoln, who urged the young lady to join the motorcade. But Miss Gallagher, who had worked eleven years for the Kennedys, was afraid of incurring the President's wrath twice in one day. "Remember Adlai Stevenson!" said Mrs. Lincoln. "We may run into some demonstrations." She also thought that the President seemed to be nervous. Mary decided to take a chance and join the parade. There might be some perverse excitement.

A brilliant burst of sunshine came into the plane and President Kennedy arose from his desk and walked aft. "It looks as though our luck is holding," he said to the congressmen. In the narrow alley between the main lounge and the aft breakfast nook, Mr. Kennedy noticed that O'Donnell was chatting with Governor Connally. The President must have known at once that Larry O'Brien was forward chatting with Yarborough. Keeping the Governor and the Senator apart on one plane was not easy.

Kennedy did not want to listen to the persuasions and evasions. He called Mr. Dealey, the publisher, a bad name and went into his cabin. He motioned Congressman Albert Thomas, the only stowaway on *AF-1* (he was on *AF-2*'s manifest), to step inside with him. Mr. Kennedy sat with the legislator, and Kennedy, who always ate before keeping dining dates, asked a steward to bring some fruit and sugar.

Powers came in briefly to display the large-type copy of the Trade Mart speech. The Congressman, shaking his head dolefully, said: "I'd be careful what I say. Dallas is a tough town." Kennedy did not answer. He was scanning the speech, rocking with the turbulence of the aircraft, and preparing to take care of his personal ablutions before the landing. In the speech, he was going to rock the right-wing element of Dallas and he was as adamant about it as the Connallys were about Yarborough. This equated the President with the people he criticized because, if they were stubborn and refused to see, so was he.

He detested the fanaticism he detected in himself. The rich Hunts and Richardsons and Murchisons of Texas represented the radical right—the Dealeys, too—and Mr. Kennedy saw them as implacable and hateful, preaching a dead doctrine to an ignorant people. The reactionaries saw Kennedy as a spiteful liberal who had seized the dammed-up political power of America and had opened the valves to flood the country with the venom of civil rights and who drowned all opposition with fancy phrases which painted them with decay and himself with youthful nobility.

The rich of Texas remembered Kennedy's words with bitterness:

"I believe in an America where every family can live in a decent home in a decent neighborhood—where children can play in parks and playgrounds, not the streets of slums—where no home is unsafe or unsanitary—where a good doctor and a good hospital are neither too far away nor too expensive—and where the water is clean and the air is pure and the streets are safe at night. . . .

"Terror is not a new weapon. Throughout history it has been used by those who could not prevail either by persuasion or example. But inevitably they fail, either because men are not afraid to die for a life worth living, or because the terrorists themselves come to realize that free men cannot be frightened by threats and that aggression would meet its own response. . . .

"I don't think, really, in any sense, the United Nations has failed as a concept. I think occasionally we fail it."

It was a standoff of the obdurates and the obstinate; he in Washington, they in Dallas. The affluent Texas begged for less government; Kennedy gave it more. Texas asked for lower taxes; Kennedy tried to seal the tax loopholes. Texas was narrowly nationalistic; Kennedy was a sophisticated European. They put millions of dollars into their propaganda; Kennedy nullified their work with a White House press release.

Now he had come to their land, their home. The millions of Texans who do not matter were awestruck by the presence of youth and power wrapped in one aluminum smile. The Texans who did matter sat behind their money and wondered if he had come as a friend, or if he was about to start more trouble. They backed Governor Connally, and the Governor had assured them that he had tried to postpone the visit into oblivion, but the President had insisted and, if he must come to Texas, it was better to have some control over the stops, the parades, the luncheons and dinners than to divorce oneself from the headstrong young man.

Jack Ruby was one of the millions who do not matter. The lumpy body carried the tired face across the gleaming floor of the *Dallas News* lobby on his way upstairs to see Toni Zoppi, the amusement editor. He noticed a classified ad clerk, Gladys Craddock, and the features hoisted themselves toward congeniality and he yelled, "Hi!" Then, en route to the elevator, they slid again toward his chin as he contemplated the phrasing of the weekend advertising he wanted for his nightclub, and the cutting words he was saving for the person who accepted the "Welcome, Mr. President" ad.

There is a warm effusion of righteousness when a little man can indict a big one. Jack Ruby, under threat of dropping his one-column two-inch ad, proposed to "dish it out" to the *News* advertising department for ridiculing the President of the United States. Ruby also wanted to know who Bernard Weissman was. As he waited for an elevator, it is doubtful that he heard the whine of jet aircraft.

Now he could see the first one. It was under the floe of broken clouds on final. Jack Jove stood in the glass octagonal tower and watched the Clipper coming, head on, with about 20-degree flaps, like a dragonfly trailing a dirty veil. It was over the skyscrapers of downtown Dallas, dropping steadily. The overhead speaker crackled and the first officer gave his call letters and said: ". . . on final, 31 right." Jove, chief of FAA at Love Field, watched five of his men work the consoles. A radar man below had the plane on scopes and said: "Okay, Pan American Clipper 729."

It was obvious to Jove that, in flight, the three presidential planes had spread farther apart. They had taken off from Carswell at two-minute intervals but radar showed that the Clipper had come in a little faster; the Vice-President's *AF-2* had swung wide to the right which gave it additional space; and Colonel Swindal had climbed to 7,000 feet to ease the moderate turbulence and had held the speed down to a minimum.

Mr. Jove liked that. A little spacing never hurt any aircraft. He was a dark, wavy-haired man with an office below the tower, but he wanted to be in the tower this time. Two weeks ago, the Air Force had sent Major Charles Nedbal to Jove to work out the entire procedure. It was a nothing flight, from nowhere to nowhere in no time, but Nedbal wanted to know where the fire trucks and ambulance would be stationed, whether they could follow the plane on an apron runway without being hit by jet blast, who would man the "Follow Me" jeep, what turns would be made, and where each of the three planes would come to rest.

He also had to know and report on departure procedures when the President had finished his Trade Mart speech; what gate; what minute of what hour; what commercial jets might be taking off or landing at or near Love Field at 2:45 P.M.; when he could have a weather projection for November 22, 1963, and what would be the three favored flight plans direct to Bergstrom Air Force Base at Austin, Texas.

Jove answered the questions. Even service trucks with fuel for airliners were to be ruled off the runways, aprons, and roads from a half hour before the flight from Fort Worth until Colonel Swindal brought his plane to the parking area at Gate 24 and called an all clear after he stopped his engines. Jove was proud of the airport. It was modern and as flat as a good billiard table. It ran over two miles between Mockingbird Lane and Bachman Lake and a mile and a half in the other direction from New Lemmon Avenue to Denton Drive.

The terminal was in the middle, between runways. There was a huge lobby, full of shops and restaurants, and a bronze statue of a Texas Ranger. There was a parking area for 1,666 automobiles, but it was not enough and there was contention between the police and the residents. Love was used by several airlines; the main ones were American, Braniff, Delta, and Eastern. Mr. Jove used his binoculars and could see mechanics on the opposite side of the field dropping work to come out on the apron.

The Clipper threaded the line down Lemmon Avenue, came in over the fence, and used a little runway before dropping with a screech of rubber to the concrete. The time was 11:29 A.M. It ran a little way down and then the skipper closed his "clams" and threw the engines into reverse. No problem. The shriek overpowered all other sound, then died to a whisper as the skipper made a left onto taxiway 14 and back up on 16. He was moving past the tower as Jack Jove picked up AF-2 coming in on final and heard his men, now divided between telling the Clipper where to park and working the Vice-President's plane down over Lemmon Avenue.

It made an ideal landing at 11:33 and was parked before AF-1 came in at 11:39. The Johnsons and their guests had disembarked and been greeted by the Dallas mayor and council before the striking blue and white of aircraft 26000, gleaming in sunlight, came to a stop on runway 31 and watched the little "Follow Me" truck lead it to the nest. The crowd behind the turnbuckle fence was bigger than anticipated, and Channel One of the police department was busy reassigning men to the congested areas.

All of it seemed to be leisurely and jolly, but there was a

surge of excitement which permeated even the jaded. This was the event, and it was a predictable one, but it carried a thrill which could not be suppressed. The crowd behind the fence was screaming and the people were jumping up and down. In the tower, Jove told himself that it was an uneventful landing, but his stomach convulsed when he thought: here is the President of the United States. The press hurried to places behind a roped area on the field. Secret Service men coming off the first plane were met by Winston Lawson, who sent some directly to the Trade Mart.

Along the airport fence was a long line of silent limousines, peopled with chauffeurs. The three monster birds stood almost together as they disgorged people, some of whom ran while others walked. Inside the rear of *AF-1*, the President fingered the knot of his tie and Kenny O'Donnell threaded his way toward him with good news. He whispered that Governor Connally had been impressed by the crowds and had said of Kennedy: "If he wants Yarborough at the head table, that's where Yarborough will sit."

Mr. Kennedy was pleased. "Terrific," he murmured. "That makes the whole trip worthwhile." Mrs. Kennedy, hearing the news, smiled. If it was good news for Jack it was good news for her. Down on the hot concrete, Vice-President Johnson led the state and local delegation over to *AF-1*. A runner rug was rolled to the ramp; Mrs. Johnson had a bouquet, and all faces turned up to the curved door of the plane. A flick of the hand from the White House advance man, Winston Lawson, and the drivers started the engines of the automobiles.

The door opened and Mrs. Kennedy, radiant in pink, stood in view. Air Force personnel at the foot of the ramp came to attention and saluted. The crowd shouted from behind the fence as she inched carefully down the thirteen steps, a gloved hand on the railing, the other holding the pink handbag, the dark face alight with appreciation. Behind her, the President stepped slowly, glancing down at Lyndon Johnson with a "What? You again?" grin. There was no smile on the face of Connally as he held his big cowboy hat in his left hand and assisted Nellie with his right.

Behind them, Agent-in-Charge Roy Kellerman preceded the congressional delegation. He had to remain close to the President and, at the same time, establish immediate contact with Lawson. Mr. Kennedy was shaking hands with the Johnsons. The press was protesting that the ropes represented too small an enclosure to see anything; television cameras, like boxy turtles, slowly followed the action; Mrs. Kennedy was presented with a bouquet of velvet-red roses; Governor Connally walked swiftly ahead of Mr. Kennedy to shake hands with the Dallas politicians; Secret Service agents moved from group to group, saying: "Please get in your car. Please get in your car." Mrs. Connally smiled her sweetest as she accepted a bouquet.

Kennedy was given two charcoal portraits, one of himself and one of his wife, and he glanced at them with a studied esteem and gave them to Paul Landis, an agent. Congressman Henry Gonzalez, from the San Antonio area, walked around patting his chest and studying the crowd: "I haven't got my steel vest yet." Dignitaries began to hurry to the automobiles and sit in the wrong ones. Kellerman got to Lawson, who said: "Your program is all set. There should be no problem here."

Police Chief Jesse Curry stood beside the lead car, calling in that there was a slight delay in starting, but it would be less than estimated. The President drew his wife's attention to an elderly woman who sat in a wheelchair, and they paused for a moment, stooping to chat. Mr. Kennedy looked up to see the people jammed against the other side of the steel fence and he led his wife toward it, bowing and smiling. It was a friendly crowd, shouting to be seen and acknowledged, but some of the members of the press sensed hostility and a few reporters and still cameramen tried to follow the Kennedys.

Congressmen and local officials joined the President and,

in a moment, he was lost to view. Roy Kellerman elbowed his way through the throng; Clint Hill was pushed away from Mrs. Kennedy. Hundreds of hands were sticking through and over the fence and it was obvious that Mr. Kennedy, far from feeling a sense of danger, was surprised and elated at the warmth of the greeting. He "walked" his hands along the fence. His wife felt that the people were friendly but that they were "pulling" her hands. Some of the writers assumed that the President was trying to show the press that he was not afraid. Vice-President and Mrs. Johnson were part of the eddy which swirled along the fence, shaking hands. Johnson, tall and solid, looked over heads to make certain that the President had not left the fence for the car.

He was now fifty feet ahead of the motorcade, and Secret Service driver William Greer, a stocky veteran who derived pleasure from driving the President, began to inch the vehicle forward so that the Kennedys would not have far to walk. The President was tiring of the handshaking; the reaching through and over the fence would sharpen the ache in his back. He could not see that, beyond the foremost fringe of people, some high school students were holding aloft unfriendly placards: "Help Kennedy Stamp Out Democracy"; "In 1964, Goldwater and Freedom"; "Yankee Go Home"; "You're a Traitor."

The elevator came down with the noise of an off-center disc rotating slowly. Some of the workers on the ground floor were stalling. It was almost lunchtime and they could see the people gathering all over the sunny plaza. Often, a game of dominoes proved to be exciting at lunch, but today was special. White and Negro, they took a few orders, filled them hurriedly, and prepared to finish a sandwich and coffee before the parade arrived. The elevator arrived at the ground floor and Lee Harvey Oswald got off.

He did not join the group nor seem to notice the people gathering outside the Texas School Book Depository. He scouted the order box and picked out three. The first was from Mrs. Hazel Carroll of the Reading Clinic at Southern Methodist University. It asked for a copy of *Parliamentary Procedure* at $1.40. The second was from M. J. Morton of the Dallas Independent School District asking for ten copies of *Basic Reading Skills for High Schools, Revised,* at $1.12 per copy. The third came from M. K. Baker, Junior High School, Reynosa, New Mexico, requesting one copy of *Basic Reading Skills for Junior High Schools.* All were published by Scott, Foresman & Company.

They were snapped onto the clipboard Oswald carried and, with the monotonous attitude of a mine mule, he went back to the elevator and started up to the sixth floor. Some of the employees went to the second floor to the little commissary to get bottles of Coke and cookies from machines. The small office force on that floor had practically quit, with the exception of a woman clerk, because the excitement of the Kennedy visit now permeated areas which had been immune to it. As Bonnie Ray Williams, Negro employee, said: "We always quit five or ten minutes before lunchtime, but today, well, all of us is so anxious to see the President—we're quitting five or ten minutes ahead of that so that we can wash up quick and not miss anything."

Some of the fellows played the daily game of manning the two elevators—which were back to back in the middle of the Depository—and racing each other to the main floor. Today Charlie Givens had the east elevator and Bonnie Ray had the west one. At the sixth floor, each called to his friends. Givens saw Lee Harvey Oswald on the fifth floor and he yelled: "Come on, boy," and Lee shook his head negatively. "It's near lunchtime," Givens said.

The sullen clerk said: "No, sir. When you get downstairs, close the gate to the elevator." This was the only elevator which could be called back up if both doors were closed at the ground-floor level. Charlie looked to make sure that his group was inside the car, and he could hear Bonnie Ray's car moving down ahead of him, picking up other clerks. "Okay," he said to Lee Harvey Oswald. "Okay."

The two elevators completed their race to the main floor, and Givens patted his pockets and found that he had left his cigarettes in a jacket upstairs. Alone, he rode back up. When he reached the sixth floor, he saw Lee Harvey Oswald walking along the panel of windows facing Elm Street and the crowds below. There was nothing uncommon about it, except that Charlie Givens thought that, a moment ago, Lee had been on the fifth floor. It made no difference and they did not exchange greetings.

When Charlie got his cigarettes and ran back to the elevator, the fifth, sixth, and seventh floors of the Texas School Book Depository were, for a time, empty except for the presence of Oswald. The foreman, William H. Shelley, who had been busy supervising the laying of the sixth floor, was down in his small office next to Roy Truly's eating half his lunch. It was his habit to eat part and to finish it in mid-afternoon.

Shelley had ordered the workers to move the small book cartons along the Elm Street side of the sixth floor but had left some *First Grade Think and Do* cartons along the back wall. These cartons were four times as large as the others. Lee Harvey Oswald now moved the big cartons diagonally across to the Elm Street side, making a wall about four and a half feet high. Smaller cartons were placed inside the "wall" so that, if a man chose, he could sit on them, recessed from the window but be able to look out. As a spyglass on the motorcade, this sixth-floor window in the east corner of the building was one of the best in Dallas.

Looking straight out, across the top of Dealey Plaza, a Kennedy enthusiast could see the motorcade approaching head on along Houston Street. Then, immediately below the window, the cars would make a left turn and run westerly, down the slope to the triple underpass, and, from the window, the happy couple in the car would veer neither to left nor right—merely grow smaller. This window was one of the rare places where a citizen could look down with a grand view of the parade on two streets, one northbound, one west.

Lunch had never been suffered with such silent speed. The men ate and hurried outside to find elevated positions on the Depository steps or on the lawn at the opposite curb for an un-obstructed view. Bonnie Ray Williams thought he heard some of the fellows, like Danny Arce and Billy Lovelady and Charlie Givens and some of the other sociable "guys" say that they were going back upstairs to watch the parade.

Williams got his sandwich, and ran to the second floor and got a bottle of Dr. Pepper, then back to the elevator and up to the sixth floor. He looked around the vast barn-like floor and saw no one. It was silent. The sun sifted its beams through the rows of double windows and picked up motes of dust in the air. Bonnie Ray looked around, but none of the fellows was on the floor. He saw the everlasting cardboard cartons stacked in mounds here and there, and the cleared space where the floor-workers had knocked off for lunch, and it just seemed funny that he had beat everybody back up.

He saw the new high wall of cartons near the Elm Street windows and he walked beyond them to another set of win-dows. He laid the bagged lunch on the dusty windowsill and took the top from the bottle of soda. The sandwich wasn't much. It was a piece of chicken with the bones still in it, imprisoned by two pieces of spongy bread. "Just plain old chicken on the bone," Bonnie Ray called it. He looked out and down and saw the warm, friendly sun and the green of the plaza, and the po-lice herding people this way and that, and the whistles blowing for cars to get moving before all traffic was closed, and couples on the grass adjusting cameras and looking up at the Depository building, and a final diesel locomotive yanking a small string of freight cars across the overpass as a policeman trudged to the top to keep unauthorized personnel off the property.

It was a good view, but Williams didn't want to see the parade alone. Directly below him, a few order clerks watched from the fifth floor. Bonnie Ray finished the soda, and left the remains of his sandwich on the windowsill. If anyone else was in that big room with him for those ten minutes, Williams did not hear him.

All along the tawny concrete, men were running to and from vehicles in controlled panic. The President had told Kellerman that each person should have the same seat in the same numbered car as in Fort Worth, but now there were twenty-four vehicles and some occupants did not bother to count; some, finding an empty seat, tried to squeeze in with friends; others, consigned to the buses in the rear, tried to use empty cars closer to the President's position. Lawson and his men were urging one and all to please be seated. Mr. Kennedy had broken away from the crowd at the fence, saying, "Thank you. Thank you. I'm happy to be in your city today. Thank you. Thank you. This is a real Texas welcome. . . ."

Few heard him because the crowd shouted its individual exhortations and imprecations. The congressmen were elbowing each other for preferred positions, and the Dallas mayor and council felt its lack of sympathy for Kennedy congeal when the Secret Service pointed to the seventh car in the line. The President helped Mrs. Kennedy into the big Lincoln. She sat hard on the left side of the rear seat and dropped the bouquet of roses between them. This was a hot day. The hypocrisy of politics was that the young lady had to smile while sweltering in a merciless sun.

The President, grinning at one and all from the right side of the car, assisted Mrs. Connally to the jump seat in front of his wife, and watched John Connally, still dour, unfold the seat in front of the President. Senator Yarborough was spotted, over the Kennedy shoulder, walking past the Vice-President's car and the President caught Larry O'Brien's eye and pointed. At once, the President's special assistant ran along the line of cars and came abreast of the senator at the time that Special Agent-in-Charge Kellerman flagged the motorcycles in front to start. Yarborough was brought back, protesting a little, and was shoved by the back of the trousers into the Johnson car. The door slammed behind him and Yarborough dropped into the left side, next to Mrs. Johnson. O'Brien, watching the cars start out, studied the faces and seats and hopped into the first one which had space for him.

Two motorcycle policemen swung their vehicles through a hole in the fence and motioned to patrolmen to keep the people back. The two cops moved slowly at first, glancing back over their shoulders toward the pilot car. The motorcycles and that car should open a lead of at least a quarter of a mile in front of the procession. Deputy Chief G. L. Lumpkin, in the pilot car, had policemen Jack Puterbaugh, F. M. Turner, and Billy Sinkle with him. If there was going to be trouble of any kind, this was the car to raise the alarm.

None was expected, but four pairs of trained eyes would be watching the curbside crowds and the overhead windows all the way. If trouble did come, the policemen expected that it might be in the form of a crowd at a particular intersection which would break through police lines and engulf the President's car. At worst, some fanatic might carry an insulting sign or shout a curse at the Chief Executive of the nation. Lumpkin was in touch with Chief Curry all the way, and he kept reporting monotonously everything he and his men saw.

The two motorcyclists were ordered by radio to increase their speed coming out of the airport, and Lumpkin and his men lengthened the distance between them and the rest of the motorcade. In police headquarters, it was announced that, at 11:55 A.M. the President was leaving Love Field. The word was passed by walkie-talkie radio among Secret Service men. It reached the White House switchboard at the Sheraton Hotel and was passed to Washington, D.C. Jack Jove in the tower heard it on short wave and returned to his office. Colonel Swindal heard it. The word was hammered into newspaper offices on the police radio band; it went to the radio stations, and the local television directors heard it and placed camera crews at strategic areas to pick up the glistening cars as they turned off the airport road at Mockingbird Lane. The President, for the first time, could see the huge cluster of modern buildings which comprised rich, conceited Dallas.

The main section of the motorcade was led by Chief Jesse Curry in a white car. He drove it. The radio speaker was open and the volume was turned up. The chief was never a shirker. He was a hardworking, eye-blinking martyr to his job. He wanted to stay on as chief, and he knew that he could stay just as long as he executed the will of the Dallas Citizens Council. The city fathers wanted no incidents today, and there would be none because Curry and his men had researched, planned, and rehearsed this assignment so deeply that today it seemed anticlimactic.

The chief sat with Sheriff Bill Decker, an old political fighter, Special Agent Forrest Sorrels, and Agent Winston Lawson. This was called the "lead car," and Curry had four motorcycles in front of him to trim the curbside crowds. Three car lengths behind was the big Lincoln. Agent William Greer had the presidential standard and the American flag snapping from the foreward fenders. Beside him sat Roy Kellerman, and they listened to Lumpkin and Curry on the police channel. The Kennedys and the Connallys nodded and smiled passing the filling stations and the big Coca-Cola plant and the restaurants where lunchers poured out, wiping their mouths and waving.

Behind the Lincoln were four motorcycles, one on each side of the rear bumper. They were ordered not to pull up on the President unless he was endangered. Next was the big Secret Service car, full of men in sunglasses. Sam Kinney drove; Emory Roberts manned the communications set, which called this particular automobile *Halfback*. Mrs. Kennedy's guardian, Clint Hill, stood in the forward position of the left running board, directly behind her. John Ready had the opposite position on the right. Behind them stood Bill McIntyre and Paul Landis. Glen Bennett and George Hickey occupied two-thirds of the back seat. The seat also cradled a powerful automatic rifle. It was in a better position than Kenneth O'Donnell and Dave Powers, who occupied two jump seats. In front of their knees was a compartment holding a shotgun.

Next came a rented Lincoln convertible, occupied by the Johnsons and Yarborough. In the front seat Rufus Youngblood, the Vice-President's agent, sat beside Hurchel Jacks of the Texas Highway Patrol, who drove. The Lincoln was followed by another Secret Service car called *Varsity*. Next, a Mercury with Mayor and Mrs. Earle Cabell. Behind it was the press pool car. The majority of reporters were in a bus farther back in the procession, but this one consisted of four men who, if a news story broke, would get it on the wires as a "flash."

The President's press representative on the trip to Texas, Malcolm Kilduff, rode in front on the right. A driver furnished by the telephone company manned the radio transmission set. Merriman Smith of United Press International sat between the driver and Kilduff, his knees hunched under his chin. In the rear seat were Robert Clark of the American Broadcasting Company, a Dallas reporter, and Jack Bell of the Associated Press. The ninth car was a Chevrolet convertible for White House motion picture photographers. It was impossible to take pictures in a position so remote from the President. Behind it were two more automobiles with photographers.

Three cars were assigned to congressmen. Number fifteen, a Mercury station wagon, was for "unplanned guests." Behind this was a huge Continental bus for the White House staff; then a second one containing the White House press. The nineteenth car was most important: this was White House communications, the traveling radio car through which the President, in his automobile, could address himself to anyone in the world. This one was the link between Kennedy and the Sheraton military board. There was a Western Union car, with operators who could take stories from reporters and get them on the wire quickly. In addition, toward the back of the procession, there were two Chevrolets for "unexpected developments," a local press car for newspapermen and television, and, closing the ranks, Captain Lawrence's specially assigned police car and motorcycles.

The motorcade was spread over a half mile. Leading it was Deputy Chief Lumpkin with his "pilot car." All drivers had been instructed to remain tuned to police Channel Two, which would be manned by the chief himself. All other police matters would be handled by Channel One. The press was displeased with its place in the parade. Some felt they could have reported a better story watching the motorcade from any of the buildings downtown. Even their wire representatives—AP, UPI, and American Broadcasting—sitting forward in a special car, were six hundred feet behind the Kennedys and could see little except the mayor of Dallas directly ahead.

The Secret Service men were not pleased because they were in a "hot" city and would have preferred to have two men ride the bumper of the President's car with two motorcycle policemen between him and the crowds on the sidewalks. Kennedy was "wide open," but the SS recognized him as "the boss" and he dictated his protection. Had they been able to exercise power over his decisions, the Secret Service would have forbidden the parking lot speech in Fort Worth and the fraternizing at the

Love Field fence. These were explosive situations, but Kennedy had survived many of them with a happy smile and, as always, the Secret Service was made to appear overly protective, overbearing in insulating the President of the United States from his people, and unduly alarmed.

Dr. Burkley was unhappy. O'Donnell had relegated the President's physician to the sixteenth car. It had happened before, and this time the admiral had protested. He could be of no assistance to the President if a doctor was needed quickly. He was reminded that this was a sunny day encompassing a friendly crowd. Mr. Kennedy was in reasonably good health, and no doctor would be required. In that case, Burkley felt, there was no need for a doctor to be present. Mr. O'Donnell sent Burkley back, and ordered Mrs. Lincoln back with him.

No one was pleased, Kennedy was unhappy with the Dallas newspapers; Mrs. Kennedy, dressed for cool weather, was sweltering in a pink suit; the man who sat closest to Kennedy, Governor Connally, might have exchanged places easily with Dr. Burkley and felt better for it; Chief Curry was tired of the security burden placed on him and his city; Kellerman counted the minutes until they would all start the trip to the privacy of the LBJ Ranch; even the cop on the motorcycle behind the President's left ear, B. W. Hargis, had been on duty since 7 A.M. and didn't look forward to an additional hour of strain. This could also be said for B. J. Martin, the cop on the other side of the rear bumper, who watched Mrs. Kennedy wave. He squinted into the sun ahead of his bike.

On that day, one man assumed an assignment without an order. This was Captain Perdue W. Lawrence, who got in his car and left Love Field before Chief Lumpkin. To Lawrence, the event suddenly appeared to be bigger, greater, more important than he had thought, and he wanted to precede the pilot car by a half mile to make certain that his sergeants and their patrolmen were holding the crowds back and blocking the cross-street traffic. The captain was a cautious and thorough police-man. He listened to Chief Curry on Channel Two. En route alone, Lawrence displayed himself to his men at the crossings so that they would be alert to their work.

Lawrence became more and more surprised the farther he proceeded toward the Trade Mart. This was no average Dallas crowd. It was a metropolitan mob. The people were numer-ous even in the outlying area off the airport. At Lemmon and Atwell, the captain could not believe the number of people who waited. He recalled that, at some of these remote streets, he had not assigned a policeman to shut traffic off. Now the captain saw that it was not necessary: the pedestrians were stretched solid athwart all side streets.

At the first turn off Mockingbird, Governor Connally grabbed the metal bar in front of him and raised himself up and then sat back again. He looked incredulous. Unless all the early signs were wrong, about a quarter of a million people had turned out to see Kennedy in Dallas.

It was late in the season for a cookout. The sun was radiant over the lush hills of Virginia, and the men sat in the back of the big house, around the swimming pool. The beeches had shucked all but a few leaves from their branches, the red maples hung onto an assortment of shiny copper leaves, stiff russets, and some apple yellows. A mist hung over the hills, and yet it was not a mist. It was as though blue wood smoke had gathered in the low places to hang quietly, like old nets drying in a cove.

The Attorney General was in a buoyant mood. He had broken up his Organized Crime meeting and had taken two of the visitors to his home in McLean for lunch. The thin, at-tentive face of Robert Morgenthau, United States Attorney for the Southern District of New York, was at his side. At Morgen-thau's side was his deputy, Silvio Mollo. The lunch, according to Mrs. Kennedy, would be creamy New England clam chowder, some crackers, and coffee. This was a Catholic Friday—even for Morgenthau.

The Attorney General was a vital wire of a man, a sort of combat Kennedy. He relished the prosecution of crooked union leaders and members of the Sicilian Mafia. The latter had grown rich in an era when America had enforced the Volstead Act, which prohibited the sale of spiritous liquor of any kind. Since the repeal of that act and the return of licensed liquor in the United States, the criminal element, still organized into "families" with assigned terrtiories, had worked their way into legitimate businesses. The Attorney General was fascinated with the notion that he could drive them out of business and into prison, or out of the country.

He was a man who dared. His experience in matters of law was not extensive, but he was enthusiastic about his new role as the righteous prosecutor battling a world of evil. Robert Kennedy was designed to play the part of David.

Someone remarked that the wives always looked bright and refreshed in the early hours. Perhaps it was the plumage, or the makeup, but they smiled and brightened a scene which, to the men, might be grim. The time was 6:55 A.M. at Honolulu, and the ladies had arrived from Washington the night before. It wasn't often that they had an opportunity to make a trip with their busy husbands, and this one was going to Japan.

The men had been in Honolulu for two days of conferences, at the direction of the President. He had been possessed of a suspicion that American military involvement in Vietnam was beginning to stick to his fingers. Mr. Kennedy, as was the case with Mr. Eisenhower, his predecessor, would like to be out of Vietnam. Each day the military boots of the United States sank a little deeper in this Oriental rice paddy.

Mr. Kennedy had become disenchanted with the Vietnamese President, Ngo Dinh Diem, and Diem distrusted Kennedy.

The American military advisers complained that supplies sent to Vietnam were being diverted, that American suggestions regarding the conduct of the war were ignored, that Diem and his "dragon lady" wife, his brother-in-law, a Catholic archbishop, his brother, a general—all of them spent more time fighting the Buddhists than in ridding their land of Vietcong terrorists and North Vietnamese soldiers.

Less than two weeks ago, President Diem and his brother had been assassinated in a military coup. The Catholic archbishop was out of the country; the "dragon lady" had fled, presumably to Italy. The situation, as far as the Americans were concerned, should have been good, but it wasn't. There were whispered charges that President Kennedy had agreed to the assassination of Diem, that he had been aware of the palace plot and had kept the American ambassador, Henry Cabot Lodge, from warning Diem. True or untrue, there was an odor of American chicanery in his assassination, and world newsmen spent the better part of a week trying to piece the story together.

Kennedy had ordered five Cabinet members and his press secretary, en route to Japan for a state visit, to pause in Honolulu for conferences with Ambassador Lodge and General Paul D. Harkins, who flew east from Saigon. Pierre Salinger, the press secretary, sat in as an observer, and he felt that the Vietnamese generals who were not in control "were doing a good job." To some of the other conferees, this was a secondary aspect of a larger problem. The United States was spending blood and treasure beyond its means in Southeast Asia. Further, it had lost the sympathy of the world.

At the conferences were Secretary of State Dean Rusk; Secretary C. Douglas Dillon of the Treasury Department; Robert McNamara of Defense; McGeorge Bundy, presidential adviser on Foreign Affairs; Luther Hodges, Secretary of Commerce; Orville Freeman, Secretary of Agriculture, and Stewart Udall, Secretary of the Interior. It is hardly likely that this much of the Cabinet, in effect, the United States government, had met in Hawaii to hear Lodge and Harkness tell how well the new regime was doing. They could appreciate that from the daily précis in the Situation Room of the White House. It is more likely that they were worried over public interest in Kennedy's involvement—if any—in the assassinations and how best to divert attention from it. They may also have discussed how best to extricate the U.S. from Saigon; in fact, it was a probable topic and the President may have asked the military for a timetable of withdrawal.

The conferences were made important by the status and number of the conferees. The meetings had broken up last night, and Secretary McNamara and McGeorge Bundy had taken a plane back to Washington. At 6:55 A.M., the remaining Cabinet members and their wives walked out onto the strip at Hickam Airport and boarded a presidential jet for Japan.

In Tokyo, Rusk, Dillon, and the others would discuss trade agreements with their Japanese counterparts. It was a beautiful day. Pierre Salinger got aboard with his wife Nancy and remembered that a good breakfast had been promised on the plane. The press secretary had one additional assignment: he was to sound out the Japanese secretly on the possibility of John F. Kennedy's making a state visit to Japan in 1964. The President didn't want any of the student riots which had turned Dwight Eisenhower back from a similar visit at the halfway point.

The Japanese would have to maintain public order if they wanted Kennedy. It was Salinger's job to find out. The door closed and the wives were thrilled to be making the trip. It would take all day, even flying with the sun, and there would be one stop at Wake Island for fuel. From the air, Oahu looked like a fresh salad in mint aspic.

* * *

The
Afternoon
Hours

12 noon

The sun was high and steady and the few remaining edges of gray in the sky changed to snow-blinding white. This was a day to match the springtime of a man. All the threats of the heavens had been dissipated by a band of light which warmed the concrete of Dallas. The welcome had been hesitant; a little stiff, like offering the courtesies of the house to a policeman. From Inwood Road on, the faces old and young, stern, senile, congenial, analytical, and apathetic became infused with warmth and the smiling eyes all blinked the same message: "Hello, Mr. President. Glad to see you." It was cordial and a little more than that. They looked upon him with favor. Their sound, provincial judgment told them that he was a handsome young man with a friendly grin and blue eyes that drifted from face to face, laughing and glad to be alive.

His little woman was perky, too. Bright and sweet, a girl who didn't have her hair all frizzed and was not so made up you couldn't tell what she was like. Liked horses, too, this one. Rode some good hunters when she wasn't busy having babies. And lookit that pink two-piece suit with the little collar and cuff trim. Couldn't cost more than forty or fifty dollars down at Neiman Marcus. The President's wife—nothing put on about her. Nice couple with no scandal running their marriage into the ground. Almost too young—wouldn't you say?—to be the First Family of the entire United States. Like a couple of zippy kids ready to kick up their heels at the Grange.

Dallas, slow to admire, to enjoy, to give affection; quick to suspect, to indict, to distrust; this giant of Texas which was the end of the South and the beginning of the West, which was neither and was both, this multiphrenic city sitting alone on a hot prairie like an oasis spouting a fountain of silvery coins gave its elixir to John F. Kennedy. The decision was made somewhere along Lemmon around Mahanna and the throats began to open with a continuous roar which spread from street to street and ran ahead of the motorcade. It swept over the sound of the motorcycles and made them run, as it were, silently. None of it affected the men with offices high up in the Southland Life Building or the oil men in the mahogany chambers with the deep pile rugs, the men with the cowboy boots and the pendulous bellies; none of it altered the opinion of the monarchs of Big D. They watched Kennedy on their color television sets and snapped him off. They had millions of dollars and they wanted additional tax breaks and write-offs and they wanted offshore oil drilling, too. They recognized the face of the man who would stop them.

The people did not matter. Dallas could buy and sell people. The metropolitan area had a population of 1,125,000 and the cheapest, meanest millionaire had more money than that. The city was so new it squeaked. One hundred and twenty-two years earlier, John Neely Bryan had built a log cabin on the confluence of three forks of the Trinity River. It wasn't a good choice, but the settlement was named after George M. Dallas, Vice-President of the United States. Dallas became the runt of the northern plain.

It was a stop for pioneers; a railhead; a haven for wholesalers; a cotton broker. The state government, in 1908, passed a restrictive law making it mandatory for all Texas insurance companies to make their headquarters in Dallas. The city grew at once. Oil companies followed the insurance organizations. Banking followed both. The biggest commodity in Dallas to buy, to sell, to exchange or trade was cash. With riches and growth Dallas developed a constitutional inferiority complex.

Everything it did had to be civically bigger and better than anywhere else. For the advancement of poetry appreciation, five Browning Societies were organized. It had more air conditioning per cubic foot than any city in the world. The rich women dressed richer; the Neiman Marcus store featured "His" and "Her" planes. The Symphony Orchestra, the Civic Opera Guild, the Museum of Fine Arts were oversubscribed annually.

The city wasn't crude. It was isolated. It was a wealthy hypochondriac looking away from himself; a kept woman living in a florist shop; a sultan with a hat full of diamonds begging for a glass of water; a tower in a tunnel. Dallas consists of five main sections, but North Dallas, lying between downtown and Love Field and Highland Park, was worthy of special attention.

To a Dallasite, the ideal was always to narrow everything to the ultimate. The best country is America; the best state is Texas; the best city is Dallas; the best section is North Dallas. Across the Trinity River to the south were the 280,000 people of Oak Cliff. One of them, in a rooming house on North Beckley, was Lee Harvey Oswald. Oak Cliff was used car lots and movie houses, supermarkets, filling stations, and unpainted porches; it was clerks, cops, and warehousemen; crooked flagstones, bargain stores, and buses. The damp dark bed of the Trinity might as well have been the Great Wall of China.

West Dallas and South Dallas are slum areas. Shacks and unpainted tenements abound; automobile cemeteries line the expressways. Religion is fundamental Protestant; the churches of North Dallas are more sophisticated, but for old-fashioned Bible-whacking and brimstone, the outlying districts are preferred. Oak Cliff supported 215 churches. Sin was the secret pleasure of the rich. The ladies of North Dallas were expected to flirt with their friends' husbands, but this was cocktail polo. The men all knew a motel clerk on the Dallas-Fort Worth Turnpike who specialized in keeping his mouth shut; each one knew an eager stenographer or an airline stewardess; the rich men

143

exchanged gifts five feet five inches tall; a welcome stranger was always asked: "Want me to take care of you?" The question was asked of John F. Kennedy on his first visit to the area in the presidential campaign of 1960. Many men recalled the question; no gentlemen remembered the answer.

Downtown Dallas supports all the others. The money is here. It consists of three parallel avenues: Elm, Main, and Commerce, which run from east to west, ending at Dealey Plaza. They are crossed by a dozen side streets. From the sky it looks like a broken banjo. Within this small embrasure are the tall office buildings, the courthouse, city hall, the library, the airlines offices, banks, insurance companies, county jail, hotels, smart shops, and fine restaurants. The freeways and viaducts lead into and out of Dallas within these three avenues and the power downtown is almost absolute. Any notions which are not approved by the powers are denounced as "creeping socialism" or "communism." Here, the Chief Justice of the Supreme Court of the United States, Mr. Earl Warren, was a traitor. The living thing, the treasured spirit was the frontiersman with his Conestoga wagon, his mules, his woman, his horse, and his rifle. The enemy, once the hostile Indian and the rattlesnake, was now the government in Washington.

Physically, downtown Dallas is as clean and unpolluted as any business area in the country. The leaseholders are rich in natural gas, and everything but the sewers are air-conditioned.

Old, sunbaked buildings come down and edifices with tinted glass go up. Dallas will decline a government loan for downtown improvement but accept $50 million or $100 million from one or two of its own families, such as the Murchisons and the Hunts.

Nightlife in Dallas is superficial. There are three choices: a cultural binge at the symphony in a black tie and strands of pearls; a private party at a palatial home where husbands and wives mingle only at dinner; strip joints, where the biggest thrill is amateur night and the moist-eyed customers sit at darkened tables pouring drinks from a bottle which remains in a brown paper sack. Of the latter, Jack Ruby's Carousel Club was one of the poorest. By midnight, downtown Dallas is dead, except for the lights used by the women who scrub the office floors and the drunks who stagger along the dark walls looking for a taxi.

Dallas is out of bed early and to work early. The executive and the elevator operator are often in that skyscraper before 8:00 A.M. and the men work hard to find new ways of forcing money to make more money. Nor are the giants of industry above quarreling over the price of a steak or a tip to a waiter. The desire not to be swindled forced one man to buy eight automobiles for his family so that he could postpone trading the vehicles in at a loss. The same man would donate a million dollars for a new laboratory at the University of Dallas or buy a $100,000 ranch house as a wedding present for his daughter.

The group he represented, The Establishment, built two small cities within Dallas—Highland Park and University Park. They abut influential North Dallas and have their own taxes, police departments, and fire departments. No Negro lives among the 35,000 residents, but Negroes commute to Highland Park and University Park as servants. Separating the two communities is a boulevard called Lovers Lane.

The Establishment could be vague, but the Dallas Citizens Council never is. It has been denounced as the political instrument of the rich, but it works hard and unselfishly for a bigger, better Dallas. A long time ago, men who could make decisions without recourse to corporate stockholders or to voters took over. These men could fire a lax police chief or allocate funds to an opera company. Each man had to be powerful enough in his own right to win the endorsement of the Citizens Charter Association, which is the political body which selects candidates to the council.

It is easy to refer to the Citizens Council as the rulers of Big D, but they are individually and collectively responsible for maintaining the pride and the self-respect of a community which is defensively egocentric. The membership comes to about two hundred men, and there are no doctors, clerics, writers, or lawyers among them. All of them are corporate heads and their hobby is Dallas. Once, when Chance Vought Aircraft was contemplating a move from Connecticut to Dallas, it was found that Love Field's runways were too short. Mr. D. A. Hulcy called a special meeting of the Citizens Council the same afternoon, voted $256,000 to lengthen the runways, and the aircraft company moved to Texas and employed twenty thousand Dallasites.

The city government, oriented around the dollar, must, to be consistent, be conservative Republican. It was anti-union, anti-centralized government, anti-liberal, anti-welfare state, anti-foreign involvement. The political pendulum swung so far to the right in Dallas that the city symbolized conservative extremism in the eyes of the nation. Actually the description was inaccurate but the Birch Society members and the adherents of Major General Walker—small in numbers—were loud in public relations. The Dallas Council, which thinks of business first in relation to politics, was annoyed when Adlai Stevenson, Democratic liberal and nominee for the presidency, was hit on the head by a woman holding a rightist placard.

The story hurt Dallas. Too late, the Citizens Council tried to dam the extreme right. The community was alive with thirtyish to fiftyish couples who were hospitable but who, once the topic of politics was introduced, became hysterical fanatics. They shouted intemperate extremist doctrine and would not listen to a dissenting voice. The women, as Warren Leslie pointed out in his book, *Dallas City Limit,* were louder and more violent than the men. To them, the enemy was Washington, D.C. When they paused for breath, they decried and denounced any conversation which might support another view. These were the scores of thousands of voters who infused fascist hysteria throughout Dallas in 1963. This is what John F. Kennedy meant when he said: "We're in nut country." The council endorsed the extremists with the left-handed animadversion: "They go a little too far."

The cloistered attitude of Dallas, which makes its citizens feel alien and separate even from Fort Worth, extends indeed to its legal values. Here drunken driving can be most reprehensible, but murder is often condoned. Some European countries do not average a hundred homicides per year, but Dallas sustains more. A stripteaser was sentenced to fifteen years in prison for possessing marijuana; several murderers won suspended sentences.

It is legal for a Dallas husband to kill a rival if he believes the other man "is about to commit adultery on his wife." And yet, in comparison with other communities, Dallas is a law-abiding city. Crooked policemen are few and are dealt with mercilessly. Henry Miller's *Tropic of Cancer* did not violate any law, but it was never sold in Dallas because the authorities *requested* the bookshops not to sell it.

One must return to the *Dallas News* to capture the metropolitan conscience. The news columns cover the same stories as the evening *Times Herald* and both are as fair in the treatment of national and local news as can be expected from writers who know that the publishers are conservative. It is on the editorial page that the *News* becomes cutting, sarcastic, inaccurate, and unfair. Here the drums beat in unison. Warren Leslie, who once worked for the paper, said: "Its editorial page has been not just dissenting, but insulting."

At a publishers conference at the White House, most of the owners of newspapers accorded respect, if not admiration, to the President. Only E. M. Dealey of the *News* criticized Mr. Kennedy face-to-face. He said that the country was looking for a man on a horse to lead it but that Kennedy was trying to do the job "on Caroline's tricycle." Kennedy seethed. Later, he asked government agencies to "check up on Ted Dealey."

A perusal of *News* editorials discloses another facet of its Dallas character: the writers seldom approve of anything or anyone. Their motif is negation and anger. The United States Supreme Court is a kremlin; the State Department harbors "perverts"; the White House plays the Soviet game; this morning, it published the full-page hate advertisement: "Welcome Mr. Kennedy to Dallas . . ." The *Times Herald* spurned the same ad.

An exultation began to take hold of John F. Kennedy. The crowds grew heavier, street by street. All along, the Texas receptions had been bigger and friendlier than the predictions. San Antonio had been great; Houston was greater. But Dallas— Dallas was incredible. In the backup car, O'Donnell and Powers turned on a pair of Gaelic grins to match the sun. They knew that, if this continued, one million Texans would have seen the President by the time he reached the LBJ Ranch. One million persons who left home or business to see JFK. And this, of course, not counting the other millions who would see it on television or hear it on radio.

The motorcade was at Craddock Park when the President saw a long, limp sign held by little boys and girls. It read: "Mr. President, Please Stop and Shake Our Hands." Mr. Kennedy leaned between Governor and Mrs. Connally. "Let's stop here," he said to Bill Greer. The car stopped. The motorcycle cops braked down and swung away. The President leaned out and shook a lot of little hands. The bigger boys, yelling: "Our sign worked! It worked!" followed the little ones, and then the police tried to step in between. The Secret Service men on the backup car got off the running boards to herd the children away.

Curry, who was at the junction of Lemmon and Turtle Creek Boulevard, received word of the unexpected stop and brought his vehicle to a slow walk. He asked if there was any problem and a voice said: "No." The motorcade started again. Once, near Reagan, the President tried to turn around, probably to see whether O'Donnell and Powers were reacting to the welcome. Powers was standing, half crouching, making home movies. O'Donnell was racking up votes mentally.

The bank clocks flicked to 12:08, Temp: 68. 12:09, Temp: 68. 12:10, Temp: 68. In Washington, D.C., the clocks read: 1:08; 1:09, and John F. Kennedy, Jr., was not interested in them because time means little to a child who will enjoy his third birthday in three more days, and whose sister Caroline will be celebrating her sixth the day after tomorrow. Their British nanny permitted a great deal of birthday talk at lunch because their cousins, Teddy and Kara, the children of Uncle Ted Kennedy, were luncheon guests.

Miss Shaw made certain that the children ate well, then she slowly turned down the lamp of excitement so that little heads might nap on downy pillows. There was no nap for J. D. Tippit this day. He was alone in a squad car, touring Post 78 in Dallas. He had Fruitdale and Cedar Crest to himself. It was a quiet sector, with two country clubs, some sparsely settled streets leading to Trinity River, and a veterans hospital.

So quiet that Officer Tippit reported on Channel One that he was going to drive home for lunch. If he had any interest in Kennedy, Tippit was sure he could be seen on television in the house. But this assumption was mistaken, because local television coverage did not include the major portion of the motorcade; the sound portion described the welcome to the President, but the camera remained rooted to the interior of the Trade Mart.

In thirty-five minutes, Tippit was back. He moved across Kiest Boulevard toward Illinois on his lonely vigil. Tippit was the good cop—big, duty-conscious, a robot who put in his time without complaint, listened to the dispatcher, and was seldom reprimanded. His boss, Sergeant Owens, thought of Tippit as a reliable policeman who would not advance in position. The rumor had gone around that Tippit may have taken the tests for promotion to sergeant more than once, but lack of sufficient formal education tripped him in the written examination.

Tonight, after supper, he would go to Stevens Park Theatre and work a part-time job keeping order among the youthful moviegoers. At other times, he put in hours at Austin's Barbecue to make a few extra dollars. Mrs. Tippit and the children needed everything he could bring in. The rest of today's schedule was monotonous: keep patrolling 78, keep a lookout for stolen cars. If more patrol cars were needed for the downtown festivities, a few outlying patrols would be called in closer to the city. Other than that, keep the car moving slowly along the curbs until 3:50 P.M. Then turn in to Oak Bluffs substation and report off duty.

Others were disappointed in the television coverage of the parade. Marina Oswald and Ruth Paine, in Irving, sat watching, but there wasn't anything to see. They listened to the vocal report, and Mrs. Paine translated the substance of the story into Russian. Mrs. Oswald felt excited about this because, as she said, her husband was a chronic critic of all things American, but he had read several favorable articles about Kennedy, aloud and translated them in his pidgin Russian. When the rare mood to read was upon him, young Mr. Oswald was fond of reading a section and then giving his wife what he called the "real truth" about the story.

Now, as she waited for the Trade Mart speech, Mrs. Oswald recalled that, when speaking about the President of the United States, her husband had once observed that eliminating him would do no good, because the American system was so devised that the man who took his place would continue the same political policy. Marina's interest was not in Mr. Kennedy's politics. It was in his family.

Jack Ruby had a similar interest. Politics, at its simplest, consisted of top dogs and underdogs. Mr. Ruby had an inordinate interest in any "underdog." He had been one, a ghetto Jew from Chicago, as he described himself, and Mr. Ruby became emotional when an underdog was hurt. President Kennedy was an underdog. He was Roman Catholic and a regular guy with a nice wife and two youngsters, but the *Dallas News* had insulted this wealthy and powerful young man. Ruby would like to have fought someone with his fists to protect the President's good name, but he was not interested to the point of watching a parade.

He was in the advertising department of the *Dallas News* and he had told his sister Eva that he was going to repeat what he had said on the phone to someone at the *News*: Were they hungry for the money that dirty ad brought in, and who was this Bernard Weissman? Don J. Campbell, salesman, returned from early lunch and conferred with Ruby. Somehow Kennedy was forgotten. Business was a more imperative topic, and Ruby told Campbell that, for his information, business was lousy.

The space salesman listened. The account wasn't big, but it was regular. Ruby showed Mr. Campbell the copy for two ads, one for the Carousel Club, the other for the Vegas Club. Ruby wrote a check for $31.87 on the Merchants State Bank (leaving a balance of $199.78) and began to boast about the fistfights he had engaged in to keep obstreperous customers from his nightclubs. He told Campbell that, when he saw trouble coming, he always carried a gun.

The President's car was at Reagan, and the crowds bulged from the building line to a point off the curb. The President held his right hand up and out, kept grinning at the crowd, and flexed his wrist in a gesture half flirty and part benediction. Softly, he kept saying: "Thank you. Thank you. Thank you." Governor Connally could barely hear it from the back seat, and he wondered if the people knew what the President was saying. Back in the crowd, the small figure of Father Oscar Huber hopped up and down as he strained to look over many shoulders to see the President of the United States. So many people were waving back to Mr. Kennedy that the task became impossible. After the President had passed, Father Huber jumped once more and saw the back of the familiar head and the right side of the face. The priest waved, but he felt ridiculous because he knew that the President could not see him, so he turned away from the crowd for the walk back to the rectory. Father Huber reminded himself that he had not seen Mrs. Kennedy at all.

At Hood, the cars made a slow right turn onto Turtle Creek. The Kennedys were now in the section of North Dallas where greenery and parks and a statue of Robert E. Lee created a different and aloof world. Some people lined the righthand side of the street, but they were neither deep in numbers nor loud in enthusiasm. They watched; they squinted in the hot sun. Some waved. Some did not.

Mrs. Kennedy foraged in her purse and withdrew sunglasses. She put them on and smiled. The President, without losing his smile, said: "Jackie, take your glasses off." She seemed surprised, but they came off, and rested on her wool skirt. Nothing, the President felt, masks a face and its individual personality more than sunglasses. He never admonished the Secret Service men, who were addicted to them, but they were not present to be seen in any case. A few moments later, Mrs. Kennedy absentmindedly slipped the glasses over her eyes. Her husband did not notice it at once, but when he did, he turned to her and said: "Take off the glasses, Jackie."

A nun in black habit stood on a corner with some small children, and the President ordered Greer to stop the car again. He didn't get out. The long arm reached, the smile was polite and deferential, and he said a few words to the nun and waved the motorcade into motion. For Mrs. Kennedy, it was a long, hot ride, made a bit more aggravating by the fact that the breeze was out of the northwest at ten miles per hour, which matched the speed and direction of the car and left the occupants in a breezeless vacuum. Unless the honored guest feels a kinship for the people, a motorcade can be fatiguing, and the incessant need to smile becomes an aching monstrosity.

Someone held up a homemade sign that spelled: "Kennedy Go Home!" The President nudged the Governor. "See that sign, John?" he said. "I see them everywhere I go." The voice became tinged with bitterness. "I'll bet that's a nice guy," he said. Riding through the Turtle Creek area, Greer moved the car up to twenty-five miles per hour and a cool breeze curled around the corners of the windshield. The First Lady held one hand on her hat. The spectators were sparse.

Mr. Kennedy said: "John, how do things look in Texas?" He did not expect an optimistic response because he knew that the state had been split into three political entities: Republican; Democratic conservative; Democratic liberal. The President was also aware that John Connally, in spite of a Kennedy appointment as Secretary of the Navy, was not an ally. The Governor turned to his right, and squinted back toward the President. "There will be a *Houston Chronicle* poll out tomorrow," he said. "That should give us some ideas."

It was a noncommittal reply, neutral at best. Mr. Kennedy was in a mood to press Mr. Connally. "What's it going to show?" he said, aware that if the Governor placed any credence in the poll, he would have advance knowledge of its findings. "I think it will show that you can carry the state," the Governor said guardedly, "but that it will be a close election."

The President seemed to be pleasantly surprised. Knowing that Connally was also running for reelection next year, he said: "Oh? How will it show you running?" The Governor grinned. "Mr. President," he said, "I think it will show me running a little ahead of you." Kennedy said, "That doesn't surprise me" and closed the conversation. The President did not address the Governor again.

• • •

From the sixth-floor window, Bonnie Ray Williams had seen the people collecting on the green below and had known that the excitement would begin soon. He had finished his soda pop and his bony chicken sandwich. The remains were on the windowsill. It was strange, he felt, that the fellows had talked about coming upstairs and watching the parade from the window, but where were they? Billy Lovelady had said he was coming up. And that Spanish boy—what's his name?—Danny Arce said he was coming up. Bonnie Ray saw a wall of cartons stacked around the easternmost window, so he leaned against them and finished a bag of Fritos while watching and waiting.

Once, he thought he heard some conversation below. The floors were so worn that cracks of daylight could be seen between the ribs. Mr. Williams crumpled the empty bag and took the elevator down to the fifth floor. If no one was there, he would try the fourth. As he emerged from the elevator on the fifth, he spotted two Negro co-workers, Harold Norman and James Jarman. They had a whole array of windows to themselves, and there was a lot to see.

If you leaned out far enough, you could see the people directly below, on the Elm Street sidewalk. Billy Lovelady was down there. So were Wesley Frazier and William Shelley, the foreman. Frazier didn't know where Oswald was, and Wes was too keyed up to ask. Lee wasn't too friendly driving to and from Irving, so there was no use chasing him to ask if he wanted to see the President. Frazier did, and he kept moving up and down the front steps of the Depository as the time of the motorcade approached.

Roy Truly left for lunch with his boss, Ochus V. Campbell. They stopped on the sidewalk and decided to watch, hoping that the motorcade wasn't too far away. They saw the people gathering, as family groups and couples might at a picnic on the green. Some sat on the grass. Others stood with arms folded watching the late traffic move up and down the three streets which funnelled into the underpass: Elm, Main, and Commerce. Still others were involved in the complexity of threading fresh film into balky cameras. On the opposite side of the square, prisoners in the county jail gathered at a window, like caged birds on a perch. In a post office building window, an inspector used his seven-power binoculars and leaned on a fifth-floor windowsill to bring the individuals up close.

As the Hertz sign on the depository roof clicked to 12:15, the pigeons jumped in flight and circled Dealey Square, a peerless view unencumbered by people.

Channel Two was busy. Captain Souter called Deputy Chief Stevenson at the Trade Mart: "Advise three that the ambulances have arrived and are standing by." Chief Curry: "One. Just turning off Turtle Creek." The conversation was almost incessant and, on the loudspeakers of the cars, it sounded strident and flat as it cut through the surging sound of the people on the sidewalks. Channel One was busy with a report from Detectives J. R. Leavelle and C. W. Brown. They had been looking for a man named Calvin E. Nelson, who was wanted for armed robbery. At 12:15 P.M. Leavelle said that they had arrested "this subject" at 2421 Ellis Street and were bringing him into headquarters for questioning.

Patrolman Chance came on. He was at Fairmount and Cedar Springs. The parade had passed, but he had a problem: "There is a V-shaped piece of land out here," he announced, "with no improvements on it. Someone during the parade backed over a water faucet and it is shooting water into the air. Wonder if you can contact the water department and have them come out here and turn it off." Dispatcher McDaniel came on: "Ten-four," he announced as a token of compliance.

"Come on. Come on. Get out of there." The policemen in front of the Trade Mart had orders to keep the canopied entrance clear. The 2,500 guests had been told to be in the Trade Mart at noon, and some were still coming up the feeder road from Harry Hines Boulevard. Police cars were standing around, parked awkwardly, their speakers still on broadcasting the voice of Chief Curry from the motorcade. Time was running out here and the guest cars had to be hurried to a position behind the Trade Mart at once.

"Come on. Come on. No, I got no time for questions. Get that car out of here." Superior officers of the Dallas Police Department watched from the curb and ordered all traffic cleared away when the President reached Dealey Plaza. Deputy Chief Stevenson wanted no problems, and if a few latecomers missed lunch, let them steam. It was no fault of his. Inside, the huge arena buzzed with the indefinable sound of mass conversation. Byron Melcher, the organist, was warming up a special symphony organ, which had been brought in from California, by playing "Hail to the Chief."

Special dispensation had been given to the Catholics present to eat steak. The fruit cup and the tossed green salad, along with the sprays of flowers, already adorned all the long tables which stretched across the width of the Trade Mart like stripes. In the kitchen, white-hatted chefs perspired over the green beans amandine and the twenty-five hundred broiled steaks, which flamed and subsided sporadically. The Crotty Brothers firm of Boston, who had catered the affair, supervised the waiters who streamed in one door and out another laden with small plates of rolls and butter. Others were cutting up huge apple pies and spinning them away to make room for more. An array of stainless steel percolators chuffed steam, permeating the kitchen with the odor of hot coffee.

The city of Dallas had announced that the carrying of placards "peaceably" would be permitted, but the Texas State Police had not heard of it. Without rancor, they picked up three youths by the neck and trousers and carried them and their signs off in squad cars. In the lobby, Dallas police officers checked the last of the guests heading inside, and Secret Service men stood by, watching the checking. Others patrolled the catwalks with walkie-talkies. Captains of waiters pleaded with the guests to please take their places. Overall, there was an elite atmosphere throughout the room. The best of Dallas had been invited and, with only twenty-five hundred tickets to satisfy all, there had been some belated scrambling to buy them. The best of Dallas society, Dallas civic welfare, the leaders of the Dallas County Democratic conservative wing, the rich, the leaders of the arts were all present. Many of them may not have been in sympathy with Kennedy, but none had come to jeer. He was about to be accorded the hospitality of the house. Deputy Chief Stevenson asked a lieutenant to check with the dispatcher on Two and find out where the motorcade was.

The parade was off Cedar Springs, heading slowly across Harwood toward Main and the center of the city. Deputy Chief Lumpkin, now a third of a mile ahead, radioed back to Curry: "One. Crowd on Main in real good shape. They have them back off the curb." Curry, driving his Ford slowly, studying the crowds which now spread up walls to pop out windows and off rooftops, picked up the microphone and murmured: "Good shape. We are just about to cross Live Oak. Curry, speaking to the motorcycle escort: drop back. We will have to go at a real slow speed from here on now. One to escort—hold up escort. O.K. Move along. . . . Check and see if we have everything in sight. Check with the rear car."

The crowds, with whole sections screaming "Jackeeeee!" began to close in on the limousine. Greer knew what to do. He turned the car toward the left side so that motorcycles could come up on the President's side and protect him. The Secret Service men dropped off the running board of the follow-up car and trotted beside the President and First Lady. Two helmeted policemen on motorbikes moved up with roaring engines to drive the people back from the President's side.

In a moment, they were at the head of Main Street. On the left stood Police Headquarters. Duty officers stood in the windows, waving. Prisoners on the top floor strained to look down. Chief Curry made a right turn and could see Lumpkin's car far ahead. Greer turned the President's car behind the Chief. For the first time, the Kennedys could see the real welcome that Dallas was tendering to its Chief of State. Except for a center lane of pavement, the entire gourd of skyscrapers was covered with people. The heads were solid for twelve blocks straight down and up the sides of the buildings. It was as though all the rest had been prelude; now the curtain had been raised and, in a flash, the big uninvited part of Dallas was ready to tender its respects.

Governor Connally saw it and thought: "They're stacked from the curb to the walls." Paul Landis, agent on the follow-up car, saw a boy leave the crowd, evade the policemen, and head for the President. He nudged Agent Ready, who dropped off. The boy had a big smile and held his hand before him to be shaken. Ready ran and headed him off. He pushed the boy back into the crowd. A reporter in the press bus made a note that there were "a whale of a lot of people." Lawrence O'Brien, squeezed between Congressmen in the wrong car, heard the representatives say that the crowd was large, but too reserved. Congressman Walter Rogers leaned out of his car and shouted to the people to "smile and look perky."

Malcolm Kilduff, the press secretary, looked up in time to see three letters pasted in three windows. They spelled "BAH." Kenneth O'Donnell, adding and subtracting an election a year ahead, thought: "They are not unfriendly nor terribly enthusiastic." Still, he was pleased. The President's domestic political expert was glad that the congressmen who had been reading in Texas newspapers how unpopular Kennedy was could be along to see the size of the cordial crowd. Greer, driving, thought: "They are very close to us. And very large crowds." As he favored the left side of the street, the motorcade slowed. Greer always allowed four or five car lengths between himself and the lead automobile so that, if an emergency occurred, he could depress the accelerator and swing to left or right. The maneuver is called "getting the hell out of here."

On the press bus, Charles Roberts of *Newsweek* listened to the conversation of other reporters. He felt that there was a consensus that "John F. Kennedy would be unstoppable in 1964. . . . He has everything working for him, including his wife." In Curry's car, Secret Service Agent Winston Lawson listened to the Channel Two chatter and swung his head back and forth over the crowds, looking for an unusual sign, an unusual movement. Forrest V. Sorrels, a fellow agent, looked out the rear window of the sedan and said: "My God, look at the people. They are even hanging out the windows." He and O'Brien, in separate parts of the motorcade, felt that the welcome had changed character suddenly. It was now bigger than ever, more boisterous, more enthusiastic. A shock wave of sound roared down Main Street like a bowling ball down an alley.

Lawson, besides looking dead ahead, had to look backward at Greer because the speed of the motorcade was controlled by the President. This requires swift and certain eyes. As the advance man on this trip, Lawson worked well with Chief Jesse Curry and he kept murmuring: "Move the escort up a little, chief. There. Hold them there. Now, back a little. A little more. Right." When the crowd bulged out toward the center of the street, Lawson requested the chief to use the motorcycles as a wedge.

The President was at Field Street. He was halfway down Main, in the center of a canyon roaring greetings. In the crowd stood an Army intelligence agent and James Hosty, of the Federal Bureau of Investigation. Both were pleased with the big turnout. Hosty, who had worked on the Lee Harvey Oswald case, never gave the sullen young man a thought as the glittering cars swept by. The President had eight more blocks to go, then a right turn on Houston, a left on Elm, and down through the underpass to the luncheon.

He had won the endorsement of the people in spite of their masters.

The sixth floor of the Texas School Book Depository was dead quiet. Some of the windows were up about a third from the bottom. Black pipes and naked electric bulbs marred the gray ceiling. The beige boxes were stacked like small forts in a child's game. A remnant of the breeze puffed lightly through the windows. The sills were only four and a half bricks from the floor so that, if a man stood behind those glass frames, he could be seen clearly from his head to his shinbones.

A man sat there. He was thin-lipped, brown-haired, a flat-bellied man with a permanent pout to his mouth. At the easternmost window, he sat on a small carton, screened completely by a brick wall and two heating pipes near his back. He could look diagonally westward, down the slope of Elm Street from where he sat. He could see all the way down to the underpass and could see the people collecting on the tracks above it. This was a patient man who was compelled to do the thing he planned to do. It is doubtful that he asked himself why. Twice before he had hoped for immortality, aspired to it by plotting against the life of Major General Edwin Walker, an extreme rightist, and again by telling his wife that he would kill the Vice-President of the United States, a Texas liberal.

Now he had the whole sixth floor to himself. He had erected a small enclosure of book cartons, although there was no one to look over them, no one to challenge him. He was alone with his curtain rods, and he and they were about to make history. Never again would he be regarded as a human cipher; a dollar-an-hour book clerk; a U.S. marine with a dishonorable discharge; a baby whose father had died two months before he saw light; the lonely kid who slept with his mother; the renegade who slashed his wrists in Moscow; the hero who had rescued a Soviet maiden from despotism to earn her contempt in Texas; the non-hero in Russia who hadn't even been asked to broadcast his hatred of the United States of America; the boy who at sixteen told a friend he would like to kill President Dwight D. Eisenhower; the silent, sullen psychotic.

The rifle was across his legs. The man and the rifle made a combination. Neither was quite accurate. The Mannlicher-Carcano Italian military rifle fired a 6.5-millimeter jacketed shell. This one, serial number C2766, had been manufactured and tested at an Italian army plant in Terni. It was twenty-three years old; he was twenty-four. On top of it, the man had bolted a four-power scope. The crosshairs were a bit high—not too much—but a little bit. The gun, when fired, had a tendency to bear slightly to the right. The young man knew this; it was like windage. All he had to do at, say, three hundred feet, was to aim a bit to the left of the target. Not much. A little bit. A foot. No more than two.

Lee Harvey Oswald got up, rifle slung under his right arm, and stood near the window. He did this when he wanted to see behind his position, to look at Houston Street, which crossed between Main and Elm. Below, there was a knot of people around a sergeant's three-wheel motorcycle. A spectator had fainted. The sergeant had called for an ambulance some time ago. The man was lying near the curb. Bubbles were coming from his lips and a spectator had tried to stick his crooked finger down the man's throat. An Oneal ambulance came down Main and stopped. The epileptic was rushed to Parkland Memorial Hospital.

As the ambulance raced down Elm and beneath the triple underpass, some of the spectators at Main and Houston looked up and saw the blinking red light of Chief Lumpkin's advance car. The motorcade was coming. Policemen began to blow warning whistles. Automobiles were diverted. Traffic was stopped. Citizens were ordered to get back on the curb. The word was passed. A thrill passed through the thin rime of people. Up in the county jail, Willie Mitchell, known to the Dallas deputies as "a colored boy," pressed his face against the bars in a silent shouldering battle with other prisoners.

They had read in the papers that the motorcade was going to pass here, and some had asked permission to congregate in the big tank on the north side to watch the parade. Mitchell was serving a sentence for driving while intoxicated. He had big eyes and excellent vision. The dark eyes, with a shading of brown in the whites, roamed Dealey Plaza. These were the free people; free to watch; free to ignore; free to roam or stop or say no to somebody. Willie Mitchell, elbowing and being elbowed, kept his hands on the bars and saw the array of citizens on the grass; the pencil line of pedestrians lining Elm Street; the cops pushing cars back and making them disappear against the will of the drivers. Mitchell missed very little.

At the head of Dealey Plaza stood an ornate white memorial pavilion. It was low along the curving edges, around a shallow pool and a fountain. Here, on the Elm Street side, Howard L. Brennan sat. He was forty-five, a good family man, a steamfitter by trade, and a cautious human being who was easily frightened. He had finished his lunch in a nearby cafeteria and had some extra time. Mr. Brennan staked out the low white wall and knew that, when the motorcade came by, he could stand on the wall and look over the people in front.

He was only a few feet from Elm and Houston. Facing him was the front door of the School Book Depository, one hundred and seven feet north. There, had he known them, stood Wesley Buell Frazier, Danny Arce, Billy Lovelady, and, fifteen feet to the left, near the lone V-shaped oak tree, Mr. Roy Truly and

Mr. Ochus Campbell. Farther down, on a slight rise of grassy knoll, stood an elderly manufacturer, Mr. Abraham Zapruder, who was nervously focusing his 8-millimeter zoom lens camera, warning the secretary behind him: "If I back up to you, don't think I'm being fresh."

Brennan knew none of them. His gaze flitted across the faces and back to the School Book Depository several times. He crossed one leg over the other and studied the fire escape on the Depository. It wasn't much. Then he saw faces at the windows and his eyes conned the floors and the windows, roaming without pattern. On the fifth floor he saw three Negroes leaning out, chatting in the bright sunlight, and laughing. Over them he saw a youngish man at a partly opened window. The man held a rifle.

Breenan saw nothing unusual in this. "He is just sitting there," Brennan thought, "waiting to see the same thing I'm going to see, the President." Brennan studied what he could see of the man. He appeared to be sitting. He might be, thought Brennan, in his early thirties, a slender man of perhaps a hundred sixty-five or one hundred seventy pounds. His clothing was light colored but not a suit. The steamfitter noticed that the man with the rifle was on the floor under the top floor, whatever number that might be. Mr. Brennan was a man for looking and minding his own business.

Around the plaza were twenty-two persons with cameras. Ten had motion picture cameras. Six found vantage points halfway down Elm Street, near the grassy knoll. Mary Moorman, with her friend, Mrs. Jean Hill, paced up and down the center triangle of grass, swinging a Polaroid camera. They reminded each other that, with an automatic one-minute developing process, they would be able to shoot one photograph, no more. Miss Moorman would aim the camera and shoot it as the President's automobile went by. Mrs. Hill would yell "Hey!" or something if Mr. and Mrs. Kennedy weren't looking toward the camera. The two friends had been making photographs ever since the sun came out. But this one would be important.

Mr. and Mrs. Arnold Rowland stood close together on the grass. They were a very young couple; Arnold was eighteen, a high school student. To them, it was an enormous thrill to see the President, to be able to say that they had seen him. The young husband was busy glancing everywhere. He studied the Hertz clock on the roof of the Depository, watched the last freight train inch across the trestle, and saw a man in the Depository with a gun.

"Want to see a Secret Service agent?" he said. Mrs. Rowland, turning, said, "Where?" He pointed. She looked. She was nearsighted. She saw no one. Arnold said he must have stepped back, because he saw a man with a rifle. He said the man appeared to be thirtyish. Arnold knew that there was protection all over the city for the President. It didn't surprise him that there was a Secret Service man up in a window. He even noticed that the man had the rifle sloping across the front of his body, pointing downward toward the left foot. It was bigger than a .22 rifle, Mr. Rowland said. He knew.

Some saw the man in the window. Some did not. Some looked up. Some held transistor radios to their ears. A group stood in front of the statue of Mr. Dealey, looking up Main Street. The motorcade was plainly in view now, because it was coming down a slight grade. The stomachs of some tickled with the approaching vision of the President and the First Lady of the land.

The President was close to the new County Courts Building, a steel skeleton reaching for the sky. Lamar Street, then Austin. Two clerks from the county auditor's office, Ronald Fischer and Robert Edwards, had been given permission to remain out—"at lunch"—until the motorcade passed. They were excited and could hear the approaching phalanx of motorcycles. Edwards elbowed his friend. "Look at that guy there in that window," he said pointing. "He's looks like he's uncomfortable." Fischer looked. The only thing he found interesting about the man in the window was that he appeared to be a statue. He never moved his head or his body.

Fischer thought the man was lying down, facing the window. The man had light, close-cropped hair. He wore an open-neck shirt. As the motorcade came closer, Mr. Brennan glanced up again. He had done this several times, and sometimes he saw the man with the rifle, at other times he wasn't there. This time he looked up and wondered why the man seemed to be crouching in the bottom part of the window. Suddenly, the crowd's attention was diverted to Deputy Chief Lumpkin. His pilot car had paused in front of the Depository, and he warned the policemen working traffic that the motorcade was only two minutes behind him.

Someone, anyone, might have approached the car and asked: "Say, who's the man up there with the gun?" Someone, anyone, could have asked the question—no matter how ridiculous it seemed—of any policeman. The deputy chief left the scene and headed swiftly for the underpass, then up onto Stemmons Freeway for the Trade Mart. He warned Captain Lawrence's men that the President was right behind him and to clear traffic from the freeway.

A few minutes earlier, Mrs. Carolyn Walther, who worked as a cutter in a dress factory, walked to a point opposite the School Book Depository with Mrs. Pearl Springer to watch the parade. Mrs. Walther saw a man at the end window of the fourth or fifth floor. Both his hands were on the ledge and in his right hand he held a rifle, pointed downward. The stranger was staring across Houston, toward the edge of Main, where the parade was expected momentarily. Mrs. Walther was sure that she saw another man "standing" in the same window. Because the window was dirty, she could not see the head of the second man.

165

Traffic was stopped in the area, and the streets were clear. Commerce Street, with inbound traffic, was still open. A few cars and trucks began to slow down, the drivers hoping to catch a glimpse of the President. The police did nothing about it, so some additional cars stopped in the left lane. James T. Tague's car, half under, half out of the underpass, stopped. He put on his parking brake and cut the switch. He got out. He would spend a few minutes watching, even though he was standing next to the Commerce Street abutment, farthest across Dealey Plaza from the turn at the Depository. A couple of deputy sheriffs and a policeman stood near Mr. Tague. They, too, were watching.

The lead car popped out of the foot of Main Street onto Dealey Plaza as a cork might leave a bottle. Suddenly, there were no dense crowds, no tall buildings, no diffused roar of throats. Chief Curry edged the car well out into the open and turned slowly to his right. He saw a panorama of fountains and green lawn, clusters of friendly faces, and some buildings, old and new, forming a square horseshoe.

The chief squinted through his glasses and caught sight of the President's car coming down Main directly behind him. The motorcade was over. One more block along the foot of the horseshoe to the right on Houston, then a slow left onto Elm, and a downhill run to the underpass. Curry could afford to breathe easily. There had been no incidents of disrespect to the President. To the contrary, staid old Dallas with its ironclad politics had taken the Kennedys to its heart. The chief could not recall any political event in Big D which matched the hospitality of this one.

Forrest V. Sorrels sat in the back seat on the right-hand side. His job was to study crowds and buildings. The Secret Service agent needed a convertible to do his work correctly. From Love Field onward, he had been straining, swinging, gawking, sticking his head out of the open window in an effort to look up at windows. He sat on the edge of the back seat, and he saw the knot of people to the left, around the statue of Mr. Dealey, then the head swung to the right, along the sidewalk of Houston. No unusual activity anyplace.

He glanced up and down the sides of buildings, moving farther and farther ahead until he spotted the last edifice on this trip. His eyes fanned the old red brick façade of the Texas School Book Depository. Sorrels saw some Negroes leaning out of an upper window. No unusual activity there. The eyes moved up to the roof. A Hertz Rent-a-Car sign indicated 12:29 P.M. in electric lights which could barely fight the sun. Then quickly down to the people ahead who lined Elm Street, and Sorrels saw nothing untoward.

Not many, even in the plaza, noticed the group of girls squealing with anticipation on the fourth floor of the School Book Depository. They clasped and unclasped their hands with delight as the lead car approached. The office belonged to Vickie Adams. She had invited her friends, Sandra Styles, Elsie Dorman, and Dorothy May Garner to watch with her. The girls were thrilled because of the exceptional view, looking downward into the car, and the possibility of seeing the youthful, attractive First Lady and what she was wearing. The girls were prepared to discuss Mrs. Kennedy's shoes, gloves, hat, coiffure, even the roses.

The big Lincoln swept out into the top of the plaza, sunbeams spangling off the fenders and sides. From Elm to Commerce, from the old courthouse to the overpass, attention was on that automobile. Greer was down to eleven miles an hour and he made the turn wide, into the center of Houston. Kellerman, at his side, glanced from side to side, noticing the sudden thinness of numbers, and looked ahead to the Depository. The car could have gone straight down Main, which would have kept it in the middle of the plaza, but it could not have made the turn up onto Stemmons Freeway without inching over an eight-inch concrete curb that separated Main from Elm. It was easier to turn right to Elm, and go under the trestle and up onto Stemmons Freeway.

In the jump seats, the Connallys maintained the crystal smiles to left and right. Mrs. Kennedy, looking left and finding few people, was waving to the right side. So was her husband. He did not know where he was in Dallas, but he knew that the motorcade was at an end, that this was a small thread on the edge of an impressive piece of fabric. He maintained the shiny smile and ran his left hand back across the thick brown hair. The eyes were sailor's eyes, cracked in Vs along the edges. He must have seen the old gray turreted courthouse on his left as Greer made the turn, and he could not have avoided seeing the faded old Depository above the windshield.

Somewhere on the retina of disinterested eyes the images of the four girls jumping with excitement, the "colored boys" leaning and watching, and the crouching figure of the stranger and the rifle may have appeared between blinks. They were present and visible. The blue eyes, which had seen so much that was flattering in the past forty minutes, would hardly have paused on one building, one assortment of faces. Like an infant in a perambulator, John F. Kennedy had been the center of attraction of all the faces which leaned over the sides to look, to approve, to adore. The moment of approval is important to infant and President. For Mr. Kennedy, it was a smashing triumph because, if he could earn the plaudits and the endorsement of the people in the camp of the political enemy, then the fight for other states next year was bound to be easier than he thought.

The car glided noiselessly across Houston. In the sixth floor window, the mediocre marksman could have had Mr. Kennedy in his sights and probably did. From Oswald's perch, the President of the United States was coming directly toward him. He could fix Kennedy in the crosshairs so that, at four hundred feet, the victim appeared to be one hundred feet away. There was one shell in the chamber; there were three more in the clip below, ready to jump to duty.

Oswald could have fired all four into the face of the President at this moment. The target moved neither right nor left as Greer came down the middle. It just grew larger in the sights, the tan, smiling face growing bigger in the telescopic lens with each fifth of a second. Why not fire now? No one knows. No one will ever know. Is it possible that he feared that a missed shot would cause Greer to slam the car into high speed, running out of sight straight on Houston behind the Texas School Book Depository? It is possible. Then, too, if he fired now, from in front and above, all heads in Dealey Plaza could easily turn back up the trajectory of the shell to the window and see the assassin. Those Secret Service men in sunglasses faced him; he was facing marksmen. If he missed, would he have time for a second shot before Greer could make the heavy car leap out of sight? This, too, is possible.

But suppose a patient man could afford to decline the head-on shot in favor of the cul-de-sac? Suppose, just suppose, the President could be placed in a position from which he could not back up in time to save his life and, if he moved ahead, would become a more exposed target? This would be an improvement for a man who permitted himself four shots. He had at least thirty minutes to crouch in this window and study the aspects of murder. As a result of his military training, Lee Harvey Oswald understood the components of ambush. Surprise is a necessity, and the victim must be caught in a position from which it is impossible to escape. In this instance, the car could not back up into the Secret Service car behind it. There was insufficient room on Elm to make a U-turn; the forced move was to continue ahead toward the underpass, exposed to the rifle with the telescopic sight.

Greer watched the pilot car make the left turn onto Elm, but he misjudged it. The Secret Service driver thought of it as a left turn, but it was more than left, curving more than ninety degrees. Instead of getting in the middle of the three lanes, the big car was now edging into the righthand lane, close to the people on the curb. The driver, swinging the heavy car toward the middle, saw the overpass ahead and people on it. Kellerman saw it, too, and so did Lawson and Sorrels in the pilot car. There was a policeman on the trestle in the middle of the group of people. Some of the Secret Service men waved to him to get the people away, out of a position where they would have the President directly below them as the car went into the underpass. The policeman didn't see the arm-waving.

In the back, there had been a long silence. Mrs. Connally flashed a smile over her shoulder and said: "Mr. Kennedy, you can't say that Dallas doesn't love you." The President glanced at her, wearing his crowd smile, but said nothing. In the lead car, Sorrels said to Curry: "Five more minutes and we'll have him there." Someone said: "We have almost got it made." Lawson called the Trade Mart and gave the Secret Service the five-minute warning. Mrs. Kennedy, sweltering in the sun, kept waving and smiling to the thin knots of people, but, looking ahead to the dark of the underpass, she thought: "It will be so cool in that tunnel."

The clock on the roof clicked to 12:30. Cameras were clicking, too. A schoolboy, Amos Lee Euins, sixteen, was impressed by the friendliness of the President, so he waved. The President waved back and Amos Euins, standing opposite the Depository, rolled his eyes upward with pleasure and was surprised to see a piece of pipe hanging out of an upper window. From where the schoolboy stood, it seemed to be hanging in midair, pointed toward the car. He looked back at the automobile. A ninth grader rarely gets an opportunity to look at a President. The pipe was unimportant.

Diagonally down from the sixth floor window, the gleaming car moved toward an open V in the branches of a piney oak on the sidewalk. The curved windshield and the metal trim picked up flashes of sunlight and cast it to the sky. The man in the window followed the man in the car and perhaps led him a little. Then the crosshairs of the telescope sight and the smiling face met, for an instant, in the space between the big branches.

The pigeons on the roof lifted in fright to swing in an aerial covey. Tiny chips of concrete sprayed upward from the right rear of the car. On the sidewalk, Mrs. Donald S. Baker saw the spray and pulled back. A jacketed bullet, striking the pavement at 1,904 feet per second—almost three times the speed of sound—was deflected slightly upward, headed diagonally across Dealey Plaza, hit a curb and broke the shell into fragments, and the spent grains peppered James Tague on the cheek. Then the sound spread across the plaza. It was like dropping a board; like snapping a bullwhip; a sharp intrusive sound; a sound to make every being within range pause in its pursuits, every mind to ask the same question: "What was that?"

Time froze, as though an eternity of things could occur between this and the next second. Faces, still smiling, turned apprehensive eyes on the President. Some hands fell on breasts while the minds murmured: "A motorcycle backfire." "A salute." "A firecracker." "A railroad torpedo." Some cameras kept whirring. Some stopped. The sharp sound slammed around the awkward billiard railings of the buildings on Houston, Elm, Commerce and the railroad trestle and, to some who listened, it came from here, from there, from behind, in front. There was one sound. There were two. Royce Skelton, on the trestle, saw grains of concrete arc upward from the right rear of the big automobile. Tague felt a burst of sand hit his cheek. The President of the United States, feeling the tiny grains hit his face, began to lift both hands upward in fright. He, perhaps better than anyone in the Plaza, understood the import: he had felt the sandy grains on his skin and he had heard the sound he feared. In slow motion, a stunned expression replaced the boyish grin. The hands kept coming up, up, and the face began to turn slowly, an eternity of time, toward his wife.

Governor Connally, a Texas hunter, felt no grains of road concrete, but he knew the sound. His head, his body, began a slow-motion swing to the right. The stern expression under the pale cowboy hat began to change to openmouthed disbelief. His mouth was forming words not yet on his tongue: "Oh, no, no, no." Howard Brennan's mind tripped; it said: "Backfire. No, firecrackers." On the fifth floor of the Depository, Hank Norman had an instantaneous reaction, almost as swift as the pigeons: "Someone is firing from upstairs right over my head."

S. M. Holland, a signal supervisor, standing on the overpass, heard what he thought was a firecracker and saw a puff of smoke come from the grassy knoll parking lot at his left. James R. Worrell, 20, heard the crack of sound and looked straight up at the Depository. He saw part of a barrel of a rifle sticking out of a sixth-floor window. In the Vice-President's car, just turning at the top of Elm, Agent Rufus Youngblood heard the loud pop, sat up in the front seat, saw the uncertainty of the crowd, and yelled, "Get down!"

Roy Kellerman, in the front seat of the President's car, thought he heard Kennedy speak and turned to see both hands coming up toward the face. He said to Greer: "Let's get out of here. . . ." Slowly, agonizingly slowly, the Lincoln continued on past the piney oak and out the other side. Perhaps three seconds had elapsed. Mrs. Kennedy, not sure whether to be disturbed by the sudden sound, turned slowly toward her husband as he held his hands up, pulling the jacket up a little with them and turned his eyes toward her with a dazed, uncomprehending expression. The Governor in the jump seat began to turn the other way, left, to see the President, whose right hand began to come up in front of his body.

The young man in the window had the car in plain sight now. The tree was behind his quarry. The bolt action was turned, the spent shell was ejected onto the floor, and a new one was in the chamber. The crosshairs held the back of the President of the United States in fair focus. In the telescopic sight, Kennedy was about eighty-five feet away. This time the trigger was squeezed with more care. The car was moving away at eleven miles per hour and the bullet overtook it at 1,300 miles per hour. It was aimed diagonally downward and it went through the clothing between the bottom of the neck and the right shoulder, separating the strap muscles, cutting through the trachea, nicking the bottom of the knot in the tie, moving out into sunlight, drilling through Governor Connally's back, coming out the front of the rib cage just in time to shatter itself against his raised right wrist and deflect downward to furrow the left thigh and die against his leg.

"We are hit!" Kellerman said. Greer, bewildered, slammed on the footbrake and the car slewed slightly to the right and almost stopped. The Governor, fearful of that explosive sound, had a sensation of being punched in the back. President Kennedy, with hands no farther than his chin, reacted by trying to clutch his throat. He was conscious and he heard both shots. Slowly, almost sedately, he began to collapse toward the roses and his wife. With the hole in his throat breathing as he breathed, it is doubtful that he could have uttered an articulate sound.

Rufus Youngblood hopped from the front seat of the Vice-President's car, yelling, "Get down!" and shoved Mr. Johnson's right shoulder, pushing him toward Mrs. Johnson and Senator Yarborough and jumping up high enough to sit on his man. The men on the follow-up car turned backward toward the door of the Depository, as though, without consultation, they knew the sounds came from there. The faint scream of a woman came from somewhere. A man yelled, "Duck, for Christ's sake!" One of the drivers hit a siren, and the wail started low and moved up to a pitch to terrify bystanders.

Tague, farthest away on Commerce Street, put his hand against his cheek and saw blood. Governor Connally looked down at his shirt and saw a massive amount of gore and was sure he had been killed. He began to sag toward the car door. Mrs. Connally, alarmed, reached quickly and pulled him toward her. Agent Hickey, in the follow-up car, got to his feet with the AR-15 rifle. He had no target. He was staring wildly at the green of sweet grass and the blue of a warm sky.

The final horror is always reserved for those least prepared. Mrs. Kennedy, the sheltered person, turned to see the agony on her husband's face and she screamed: "What are they doing to you?" He could not tell her and, with his last conscious effort began to slump toward her—who knows?—maybe to protect her from what was still to come. He was conscious and there was time to fall forward, between the jump seats—seven seconds to be exact.

The man in the rough work clothes and the gray steel helmet, Mr. Howard Brennan, looked up in time to see the rifle being made ready again. On the fifth floor, Jarman heard a second empty shell drop. He and his friends began to look uneasy. On the street, a policeman roared: "Oh, goddamn!" Spectators who had struggled to approach the President began to flee. Like the pigeons now circling the plaza, they had been frightened and each moved awkwardly and precipitously away from danger. In the center of the plaza, some ran. Others fell and cupped their hands behind their heads. On the sidewalk near the grassy knoll, those in front knocked down those behind in the scramble to reach safety at the top of the knoll.

The aged Abe Zapruder kept the film in his camera moving, though he had heard the sharp reports. The policemen on the motorbikes flanking the presidential limousine turned their gleaming hard helmets this way and that, looking like inquisitive monsters behind the giant sunglasses. Off to the left, Postal Inspector Harry D. Holmes kept his binoculars on the car. He was sure someone was throwing firecrackers at it because he distinctly saw dust fly up from the street with the first crack. Then he noticed the car pull to a halt, and Holmes thought: "They are dodging something being thrown."

Sorrels yelled: "Anybody hurt?" and a motorcycle policeman, pulling up on the lead car, nodded. "Get us to the hospital." Curry picked up his microphone and said: "Surround the building!" He didn't tell Channel Two what building. The Secret Service saw the President's car pull up to a stop behind them and Lawson yelled: "Let's get out of here!" They would trap the President if Curry's car blocked the Lincoln. S. M. Holland, frozen with fright on the overpass, heard three more shots after the solitary puff of smoke and saw the President slump.

The terror spread. A man on a bench got up, picked his wife up, and slammed her onto the turf. Then he protected her body by falling on her. Officer T. L. Baker slowed his motorcycle and began to run it toward the curb with his feet dragging. Agent Clinton Hill left the running board of the follow-up car and began to run toward the back of the President's car. William Greer hit the accelerator as Kellerman roared into a microphone: "Take us to a hospital, quick!"

The hard-boiled O'Donnell began to bless himself. Powers murmured: "Jesus, Mary, and Joseph . . ." On the sidewalk, Charles Brend, holding his five-year-old son up to see the President, fell on the little boy. Senator Yarborough thought he smelled gunpowder. "My God," he said, "they've shot the President!" Mrs. Johnson, the sensitive one, the apprehensive wife and mother, shook her head negatively and said: "Oh, no. That can't be." In the press pool car, assistant press secretary Malcolm Kilduff straightened up and said: "What was that?" The young Negro student, Amos Lee Euins, was still watching the man in the window. He told himself it was a white man and he didn't have a hat on. Euins could see some boxes behind the man. Mrs. Earle Cabell, wife of the mayor, was in a car still on

Houston Street. She heard shots and jerked her head up toward the School Book Depository. She could see something sticking out of one of the windows.

Mary Moorman was on the grass. She had taken the one Polaroid shot. She pulled at Mrs. Jean Hill's leg, screaming: "Get down! They're shooting!" Brennan, on the low wall in his steel helmet, watched the rifle pull back into the window before the next shot. In the press buses, men were asking each other if that could be rifle fire. A driver said: "They're giving him a twenty-one-gun salute."

The great head was slumping slowly to the left. It came up in the rifle sights big and steady. The trigger was squeezed. As before, the rifle jumped, the bullet split the air, and the slower sound swelled through the plaza, tumbling in its echoes. Mrs. Kennedy was staring at her husband. The shell entered the right rear of the skull. A large portion of the head left the body in two chunks. One flew backward into the street. The other fell beside the President. Dura mater, like wet rice, sprayed out of the brain in a pink fan.

This one the President did not feel. The light had gone out with no memories, no regrets. After forty-six and a half years, he was again engulfed by the dark eternity from which he had come. For good or evil, his work, his joys, his responsibilities were complete. The heart, automatically fibrillating, pumped great gouts of blood through the severed arteries of the brain, drenching the striped shirt, blending on the petals of the flowers, puddling the rug on the car floor. It would stop in a few moments, when blood pressure dropped to zero.

Shock froze the mind of Mrs. Kennedy. She saw a flesh-colored piece of her husband's head turning in air to drop behind the car. Jack's expression reminded her of when he had a headache. Her voice sounded stiff and unnatural and she said: "They have shot his head off." Mrs. Kennedy began to climb out on the trunk of the automobile. "I have his brains on my hand," she said.

The heavy car leaped. "Bill," Kellerman shouted to Greer, "get out of line." Connally felt the car jerk. The Governor was sure he was dying; he was conscious and he was certain that the party had been ambushed by two or three marksmen. As he breathed, the wound in his chest sucked air. "My God," he screamed, "they are going to kill us all!" He heard the sound of the third shot, heard the shell hit something soft, and he knew, reposing on Mrs. Connally's lap and unable to see, that it had hit the President of the United States. His wife, as protective and determined as the Texas frontierswomen of old, cradled his head in her hands and murmured: "Don't worry. Be quiet. You are going to be all right."

The prisoners in the jail strained to see it all. Willie Mitchell, the "colored boy," shouted that the President had been hit from behind by a bullet. "His head burst," he said, excitedly. "It was like throwing a bucket of water at him." Channel Two came on loud and clear:

Henslee: "12:30 P.M. KKB three six four." Chief Curry in the lead car: "Go to the hospital—Parkland Hospital. Have them stand by. Get a man on top of that triple underpass and see what happened up there. Have Parkland stand by." The voices were becoming strained, unnatural. Sheriff Decker picked up a microphone: "I am sure it's going to take some time to get your men in there. Pull every one of my men in there." Dispatcher Henslee: "Dallas One. Repeat. I didn't get all of it. I didn't quite understand all of it." Sheriff Decker: "Have my office move all available men out of my office into the railroad yard to try to determine what happened in there and hold everything secure until Homicide and other investigators should get there." Henslee: "Ten four. Dallas One will be notified." The dispatcher hurriedly addressed Chief Curry. Henslee: "One. Any information whatsoever?" Curry: "Looks like the President has been hit. Have Parkland stand by." Henslee: "They have been notified." Deputy Chief N. T. Fisher came in: "We have those canine units in that vicinity, don't we?" Curry: "Headed to Parkland. Something's wrong with Channel One."

Lumpkin (now up on Stemmons Freeway in the pilot car, using motorcycle policemen to divert traffic): "What do you want with these men out here with me?" Curry: "Just go on to Parkland Hospital with me." Patrolman R. L. Gross: "Dispatcher on Channel One seems to have his mike stuck." Curry: "Get those trucks out of the way. Hold everything. Get out of the way."

The Lincoln was bucking with too much acceleration and Agent Clint Hill barely got a foothold and reached for the small handrail. The car swung hard left and then split the clear air as it picked up speed. Hill felt the forces try to pull him off the car. He hung on, reached forward with one hand, and shoved Mrs. Kennedy backward, into the seat. Agent Ready, jumping off the other side of the follow-up car, was called back. The word, the ugly, shocking news, was in each official car on Channel Two. *It was rifle fire. The President has been wounded.*

Hill pulled himself forward as the car hit the cool darkness of the underpass. He kept pulling up and up. The agony in Mrs. Kennedy's face turned full upon him. The wind whipped her straight dark hair back and forth across her forehead. The agent at last stretched over the seat and she shouted: "They have shot his head off." Hill looked down. The President was on his left side. His head was in the roses. A large part of the right side was missing. The eyes, wide open, stared at the back of Mrs. Connally. One foot was lifted and hung over the door on the right side. Brain tissue was scattered. Governor Connally, down between the jump seats, began to scream louder and louder, and Mrs. Connally began to wail with him.

Agent Hill held onto the back seat and beat the trunk with his hand. The President's wife was looking up at him as the car moved up onto Stemmons for the race to Parkland, four miles away. Pathetically, she held up an arm. "I have his brains in my hand." Hill saw a piece of the President's head lying beside him. The agony on his face was screened by the big sunglasses. He looked back and shook his head no, and turned a thumb downward.

Agent-in-Charge Emory Roberts, in the follow-up car, picked up the phone: "Escort us to the nearest hospital," he said, "fast but at a safe speed." Roberts repeated the words. The President had been lost. Mr. Roberts did not want to risk the life of the new President. He waved the Johnson car closer and yelled, pointing ahead: "They got him. They got him." He pointed to Agent McIntyre: "You and Bennett take over Johnson as soon as we stop." It was the cold intelligent move to make.

Dealey Plaza was quiet. The running had stopped. People sat up on the grass. Policemen were at the School Book Depository door. Some were up the incline, as ordered on Channel Two. In a press car, Bob Jackson pointed: "Look up in the window," he said. "There's the rifle." Tom Dillard, chief photographer of the *Dallas News,* raised his camera at once and took two shots of the red brick façade. The man in the window remained long enough to make certain that his victim had convulsed, then slumped to the left. He had seen the people run; he had heard the screams.

Slowly, the rifle retreated through the window. Euins watched it; so did Brennan. Jackson saw it clearly; then he saw nothing. The automobiles in the motorcade began to pick up speed. From the sixth floor, they looked like derailed cars in a toy train set. The new President was under the crushing weight of his bodyguard. The assassin's heart must have leaped with excitement because he had dared to do a thing which no boy, dominated by mother, sneered at by wife, would have the nerve to do: Lee Harvey Oswald had killed the most powerful man in the whole world; he had done it as casually as a boy might shoot a tin can in an empty lot.

He looked over the little fortress of cartons. The sixth floor was still empty. He stepped over the empty shells, ran diagonally across the big, dusty room and set the gun down between rows of cartons. The fourth shell was still in it. If one of the "colored boys" had remained on the sixth floor, the fourth shot might have become a necessity. Quickly, Oswald walked down the steps toward the second floor. There were eighteen steps to each floor and the young man was sure-footed. All he had to do was to get in that commissary before the wave of excitement came up from the street. Seventy-two steps—one at a time or two?

At the front of the building, Roy Truly was elbowing his way through a group of shocked people. At the first step, he met officer T. L. Baker, who had jumped off his motorcycle with revolver drawn and had run to the entrance of the School Book Depository. Truly, trotting, announced breathlessly that he was the manager. Baker waved the gun: "Then come with me." Reporters were asking to be let off the press bus, but the driver picked up speed and headed for the Trade Mart. Some photographers dropped off another bus, spinning with their equipment, and ran back toward the School Book Depository to cover the story—if there was a story.

It required all of seven seconds for Howard Brennan to get off the wall—having watched the rifleman get off three shots—and become sufficiently frightened to run to the Houston Street side of it and crouch for protection. Jean Hill kept saying: "His hair stood up. It just rippled like this." Wesley Buell Frazier, the boy who was worried about his weak automobile battery, had watched the panic and he remained stock-still. In his mind, he was thinking: "When something happens, it is always best to stand still because if you run that makes you look guilty sure enough." Ronald Fischer, who had seen the gun in the window, ran down the grassy center of Dealey Plaza and back up again. It seemed to him that many policemen were running up toward the tracks.

On the fifth floor, Bonnie Ray Williams' face worked itself toward fear. He and Harold Norman and James Jarman had heard all the shots, had watched the windows rattle, had listened to the empty shells bouncing on the floor above; now dust and plaster was sifting down. Harold pointed to the ceiling and said the shots came from there. Bonnie Ray blurted: "No bullshit!" Norman said: "I even heard the shell being ejected. . . ." Jarman began to edge toward the staircase. "You got something on your head," he said, pointing. "Somebody was shooting at the President." Bonnie Ray began to edge away, and he said again: "No bullshit!" The three dashed for the exit.

In the middle of Elm, Officer B. J. Martin was sickened. He had been riding left front of the limousine. He was wiping blood and brain tissue from the right side of his windshield and his helmet. Another policeman, on the grassy knoll, bent downward in nausea. James R. Worrell, the young man who watched each shot fired from the sixth floor and had glanced at the car to see the effect, realized that he would never forget the screams of so many persons saying, "Duck!"

In the press pool car, Merriman Smith of United Press International took a calculated risk. He lifted the pool phone from between his knees, got the Dallas bureau and said: "Three shots were fired at President Kennedy's motorcade in downtown Dallas." Jack Bell of the Associated Press, sitting in the back seat, knew that this terse message would be heard around the world in a few minutes, and that the world would be waiting to know the destination and fate of those three bullets. He was sitting with Baskin, chief of the Dallas News Bureau, and Clark of ABC.

Jack Bell demanded the phone. He stood. Smith began to dictate additional material in "takes." Over the scream of sirens, he jammed a finger in his ear, squinted his eyes, and asked that his words be read back to him. Each ticking second gave UPI an additional exclusive lead over AP. Bell reached across his adversary for the phone. Smith tried to yank it loose. Bell, swaying in a speeding car, took a swing at Smith and hit the driver.

Kilduff, the President's press representative, tried to pacify the journalists. He didn't know where the car was going or what had happened. But the word was out.

On the first floor, Roy Truly led Officer Baker to the shaft. "Turn loose the elevator," he shouted up the well. Baker could not wait. He asked about stairs. Truly led him across the ground floor and up the steps. The policeman, a big dark man with crew haircut and cleft chin, followed holding his gun downward. At the second floor, Truly made the turn to the third. The cop thought he saw a movement out of the perimeter of his eye. He stopped. Through the glass door, a man stood empty-handed near a soft drink stand. The officer went into the commissary. "Come here," he said.

The young man walked toward Baker. Truly, halfway up the flight of stairs, returned. "Do you know this man?" the cop said, holding the gun close to the belly of Lee Harvey Oswald. "Yes." "Does he work here?" Truly nodded impatiently. "Yes," he said. "He works for me." They left. Oswald dropped a coin in the soda machine. He got a Coca-Cola. This was nervousness because he invariably drank Dr. Pepper. Truly and Baker worked their way up to the fifth floor, found an elevator, ran it to the seventh, examined the roof, and came back down overlooking only one floor: the sixth.

The police dispatcher moved more patrol cars to the School Book Depository, and inched others forward from the outlying districts to protect the city. From the sheriff's office across the street, unassigned deputies poured in to help. The word from Chief Curry and Sheriff Decker had mentioned the overpass; as a consequence, much of the running was up the grade. On top stood Patrolman Earle V. Brown, a fourteen-year veteran. He had screened the men sitting on the overpass, making certain that they were railroad personnel. He knew that no shots had come from his area, but, still carrying his yellow raincoat, he met the oncoming rush of policemen and deputies and helped them to scour the area. No one on the overpass tried to run. Deputy Constable Seymour Weitzman, running up the grassy knoll, heard someone say the shots came from the parking lot behind. At once he hopped the wall, but no one and no automobile was seen leaving the lot. Weitzman canvassed the parked cars, sitting in cool dead rows.

Behind the lot, the railroad towerman, Lee E. Bowers, Jr., could see the parking lot, the railroad tracks, the overpass, and the back of the Depository, without moving from his big window. He had heard no shots, seen no smoke, seen no one leave the area. As the police flooded the trestle and backlots, Bowers threw red-on-red block signals from the switchtower, effectively stopping all trains.

On Mr. Johnson's right shoulder, Agent Rufus Youngblood sat awkwardly, his dark walkie-talkie chattering the code of the secret service. "Dagger to Daylight. Shift to Charlie." "Dusty to Daylight. Have Dagger cover Volunteer." "Lancer may be critically wounded." "Dandy still back on the street." Senator Yarborough, wedged in the left corner of the seat, became fretful. "What is it?" he shouted. Rufus Youngblood tried to stretch his legs to the floor. He leaned toward the big figure of Lyndon Johnson, crushed beneath him and bent toward Mrs. Johnson. "When we get where we're going," he said against the whip of the wind, "you and me are going to move off and not tie in with other people." Johnson, who didn't understand but who recognized the voice of authority, said: "O.K. O.K., partner."

Mrs. Lyndon Johnson, tilted toward the frightened face of Senator Yarborough, was afraid that the President might have been hurt. She, who had always felt that Texas was the finest state, felt a crushing weight. Her stomach began to cramp with visceral anger. The wind of speed whipped at her head. "This can happen somewhere else," she thought, "but not in my country."

Mrs. Robert Reid was emotionally wrung out. She was a clerk on the second floor of the Depository and had hurried down to Elm Street to witness a pleasant and rare scene: the President and his lady moving past in an automobile. She heard the shots, heard the bedlam, and, without thought, had hurried back in the front entrance and up the stairs to her office. It was an orderly refuge from madness. As she opened the little gate enclosure to her desk, she saw Lee Harvey Oswald coming out of the glass commissary, a bottle of soda in his hand. For a moment, her mind told her that it was strange to see a warehouse boy in that room at this time. Then she shook her head sorrowfully, and said: "Oh, the President has been shot, but maybe they didn't hit him." Oswald kept walking. He mumbled, but she did not understand what he said, nor did it interest Mrs. Reid.

He passed her station, walking diagonally across the floor to the front. Then he went down the steps, still holding the soda, and, at the entrance, a man accosted Oswald, flashing an identification card, and said: "Secret Service." Oswald paused in the doorway. This could have been the end of the road. The man said: "Where is the phone?" Lee Harvey Oswald pointed inside the little half-gate. The "Secret Service" man was Robert MacNeil of the National Broadcasting Company.

The time on the roof clock stood at 12:33. The pigeons were still swinging broad circles over the plaza as Oswald brushed through the excited groups, heading east up Elm Street. There was time to think. Of what? Escape? In all the things he did, Oswald was a rational plotter. At each crisis of his life, the surly young man seemed to know what he wanted to do, and how best to do it.

Escape? He had left most of his money with his wife. In this gesture of generosity, he must have known that he had severely limited his chances of ever getting out of Dallas. He had thirteen dollars. He had left the empty shells on the sixth floor. The gun was up there, hidden between packing cases. They could be traced to him. He had purchased them and had used an alias: "Alex Hidell." The signature was in his handwriting. The rifle and a revolver which now reposed in his room on North Beckley Street had been mailed, prepaid, to his post office box in Dallas.

Then, too, when the police began a head count at the School Book Depository, Oswald would be the missing man. He would be the missing man from the sixth floor. His logical mind must also have told him that some of the people in Dealey Plaza had seen his rifle, partly out of the window. When the cops got out to Irving, would Marina stand by him and say that he had never owned a rifle? Would she? Oswald knew that in a crime of this magnitude, he could expect loyalty only from his mother. The rest, he must have known, would ally themselves with the law.

No one in the warehouse could swear that Oswald had been seen, here or there, while the shooting was going on. Everyone saw him some time before the crime, or two or three minutes afterward. He had no credible alibi, and he knew this, too. No money, no alibi, no sanctuary—is it possible that this young man wanted to be caught and tried? It is not only possible, but probable, that the most important circumstance, to Oswald, was that the world must know the name of the doer of the deed. For years, he had fought against the anonymity of the human cipher. What good would it do the ego to escape into additional anonymity? Oh, no. The world must know. The world must appreciate. The world must debate—while he remained silent in the prisoner's dock—whether he did it or didn't do it. And, could the world prove he did, to the exclusion of all doubt? His best course lay in getting caught in a manner of his own choosing and starring in a propaganda trial.

Supreme cruelty is reserved for the defenseless. Mrs. Kennedy was imprisoned in a speeding car with her personal horror. There was no way out, no one to help. Bending low, she cradled her husband's head on her right thigh. The handsome face, once tanned and buoyant and alive with ideals, was blue-gray. The eyes, which had once belonged to a young Senator who had fastened them on a Georgetown society girl and never released her again, were wide open, seeing nothing, never to see anything again. The mouth, which had tenderly sealed a wedding vow in Newport, hung open. Now and then, a snore of sound escaped from it and startled her. She could look down into his brain through a hole big enough for her fist. His right leg, hanging over the door of the speeding car, twitched. The wound in the throat made an irregular, sucking sound, bubbling.

The agony was not John F. Kennedy's. It was Jacqueline Bouvier Kennedy's. He was serene in the cool darkness of death. She, the least prepared for violence, sat with it, prayed over it, crooned at it on her lap. Agent Hill, off the edge of the back seat, could not help her. He hung on, the gun in his right hand, and he saw the knots of people on the street corners; the joyriders pulled to one side and parked with people leaning out. The faces were beaming; the people waved; they saw the pink of Mrs. Kennedy, and they yelled greetings.

They didn't appreciate the high speed of the motorcade, and some said, "Where was the President?" and others said, "Didn't you see him, sitting up there like a king?" There was a blur of buildings, street corners, fuzzy faces, and mouths forming cheery greetings. The sustained moan of the siren was ahead of the car, telling the world to get out of the way; it was weeping for a sturdy young giant who had been shot in the back while his hand was outstretched in greeting.

The kaleidoscope of thought moves swiftly and aimlessly in shock. Often it leaves no footprints. What are its capabilities in six minutes—360 seconds? Would the dark, tearless, shocked eyes see only the hole in his head? Or would it scramble through the files searching for all the tender moments? Would it balance the sheer terror of sight with visions of the children, growing up in his image and hers? Would it encompass those domestic arguments, when her will was opposed to his, when she defied him? Would the mind block all of it out and go back five minutes in time to watch a triumphant President waving to his Texas constituents, the mouth saying: "Thank you, Thank you." The mouth saying: "Take off the glasses, Jackie." The mouth saying . . . It would say no more. "Ask not what your country can do for you . . ." The hair still had that reddish brown thickness, the jawline had the fighter's bulge. But the mouth hung open, and small strings of saliva hung from it. "We're in nut country now. . . ."

The roar of the onrushing wind, the speed of thoughts wanted and unwanted, the omnipotent sight of the sturdy oak felled—by whom? for what reason?—had to be lived and faced squarely by a sensitive young lady unprepared. The last of his blood was running down her stocking.

Gordon Shanklin read the accumulation of reports from his FBI agents. The office was quiet, and the pages on the desk were turned, one at a time, by thumb and finger at the lower left-hand edge. There was a knock on the door and Mr. Shanklin said, "Come in." His secretary said that the young clerk who had been monitoring the Dallas police frequencies would like to see him. The boss nodded.

"Some shots were fired at the President's car," he said. "They're headed for Parkland." Shanklin is a man who thinks fast and speaks slow. He picked up a phone. "Tell Vince Drain to come in here," he said. Then he dialed the private number of the Director, J. Edgar Hoover. He waited a moment, his eyes on the clerk. "Gordon Shanklin," he said. "Dallas office. Let me speak to the Director." In a moment, he heard the familiar voice. "Gordon Shanklin," Mr. Shanklin said. "The President has been reported as shot in Dallas." There was a momentary sucking of breath on the line. "He's on his way to Parkland Hospital. The first word we have here is by police radio. . . ."

Hoover, always under emotional control, asked a few crisp questions. The personal protection of the President was not his business, nor the FBI's. He had a good working arrangement with James Rowley and the Secret Service and no desire to tip the tactical boat. "Offer the full services of our laboratory," he said. "Find out how badly he is hurt and call me back."

Hoover hung up. His secretary dialed the home of the Attorney General in McLean, Virginia. No matter how painful the news, Robert Kennedy should hear it first and, even though no one knew whether the President had been hurt, it would be the Attorney General's function to brace the rest of the Kennedy family against the explosive impact of press and radio.

In the Sante Fe Building office, Shanklin stared at the forgotten reports. He looked up at the bulk of Vincent Drain. "Get to Parkland at once," he said, "and offer our laboratory facilities if we can help." It did not occur to Shanklin, of course, that he had another agent, Jim Hosty, who, in a few hours, would turn out to be the man with a knowledge of the suspect. The only man.

The microphone on Dallas Channel One became unstuck. It was 12:33 and Mrs. Kinney, a dispatcher, ran to the office of Captain W. R. Westbrook to tell him that shots had been fired at Kennedy and the limousine had just passed the Trade Mart on its way to Parkland. Westbrook, in charge of personnel, staring at Mrs. Kinney. If this was another joke . . . Her face began to lose complexion. The eyes were disbelieving. Westbrook asked no more questions.

She had notified Parkland. Mrs. Kinney had also called all patrol craft on the route to the hospital, asking that they seal off traffic and permit the big Lincoln and its wildly swinging satellites to get through. Channels One and Two now were handling Kennedy traffic. They were asking everyone else with routine Dallas business to please keep off the air. Westbrook was dazed.

He stood and then walked away, still listening but no longer hearing, no longer digesting the words. The President had been shot at in Dallas. Shot in Dallas. Right here.

He left the office, meeting patrolmen and detectives and uttering the same words. "If you're not busy, report down to the Texas Depository building. Chief Curry has been on Channel Two, asking for help down there. If you have no big assignment, drop it and go down right away. Sergeant Stringer—yes, you. Joe Fields. Carver. McGee." They saw the deep shock, and they said, "Yes sir," to the captain and hurried to the basement to get into cars. When Westbrook got to the basement, there was no car left. Mechanically, he walked up the ramp to the street, and out into the bright sunshine. His shoulders were squared and he began the long walk to Dealey Plaza, his stride bucking the tide of pedestrians who had witnessed the motorcade and were still talking about what a handsome couple the Kennedys were.

The motorcade, careening wildly and approaching eighty miles per hour, passed the big Levitz Furniture Store on the left, the P. C. Cobb stadium, the Trade Mart, brilliant with snapping flags, and was passed by the big jet airliners letting down slowly to Love Field, two miles ahead. In the convertibles, the wind snapped the eyelids shut. A billboard proclaimed the "Smart Smooth Way to Go." Some of the trees along the edge of Stemmons held onto their leaves even though the season was over the long sleep had arrived. Red berries, like holly, confettied the bushes bordering the lawns.

The grass had turned off its chlorophyll and adopted the beige of the lion's mane. Some boys in empty lots played with kites, getting them aloft and paying out white cord from a stick of wood. A sign said: "Roller Skating Time." There was a shimmer of heat haze on Stemmons Freeway, so that, looking back, the skyscrapers, standing alone on the big plain, shivered a little.

Mr. Stemmons and Mr. Crow, who were co-owners of the Trade Mart, stood with David B. Grant, a Secret Service agent, and asked how to greet the President of the United States. Mr. Grant took them out front, under the canopy, and told them to wait until both the President and the First Lady alighted, and then to present themselves as co-owners and to bid the Kennedys welcome to the Trade Mart. Then both men would please step aside to permit the Secret Service to escort the President and Mrs. Kennedy to the head table.

Agent Grant had received his five-minute warning from the White House switchboard in the Dallas Sheraton. Three of those minutes had gone by. Then he had heard the faraway sirens, getting closer, and the Secret Service man shaded his eyes to look at Stemmons Freeway, immediately behind the Trade Mart. He saw the motorcade disappear at high speed. It was not turning off for the luncheon. And who was that lying across the trunk of the car?

Kilduff, in the press pool car, said to the driver, "What's that large building?" and the man said: "Parkland Hospital." This was the first real clue. Everyone knew that rifle shots had been fired; all hands were aware that the motorcade had pulled away from the plaza at high speed. Now they knew that someone had been hurt. It didn't have to be the President. It could be almost anyone in that head car.

The nurses' station in major surgery was ringing. It was picked up by a plump dimpled woman in starched white. "Nelson," she said. One of the telephone operators at Parkland Hospital, Mrs. Bartlett, said that President Kennedy had been shot and was on his way to the emergency entrance. Doris Nelson, R.N., said: "Stop kidding me." Mrs. Bartlett, almost weeping, said: "I have the police dispatcher on the line."

The nurse, in charge of the extensive emergency section of the hospital, asked Dr. Dulaney, resident surgeon, to report to Trauma One at once. She called Miss Standridge, who said that Trauma One was already set up. A "stat" call was placed for Dr. Tom Shires. Mrs. Nelson inspected the green-tiled room referred to as Trauma Two, directly opposite Trauma One. She opened a bottle of Ringer's lactate. In the hospital restaurant, Dr. Malcolm Perry listened to the emergency call for Dr. Shires.

He also studied the salmon croquettes on his plate. Strange, nobody ever called Tom Shires on "stat." He was the hospital's chief resident in surgery. Another thing: Shires was not in Dallas today. Dr. Perry walked to a phone and picked it up. "President Kennedy has been shot," the operator said. The croquettes began to chill as Perry ran through the long warrens of the hospital to the emergency area. Young Dr. Charles Carrico, a specialist in gunshot wounds, was examining a patient for admission to the hospital. He got the news, left the patient, and hurried to Trauma Two.

Some, out on the emergency dock, could already hear the sirens. Within two minutes, the hospital was going to be a busy place. The Oneal ambulance which had brought the epileptic from Dealey Plaza had dropped him for admission. A policeman came up on a motorcycle and requested that no vehicle move. "Stay right where you are," he said. All of them heard the approaching sounds now, but few knew what they meant.

In newspaper offices across the United States, a small bell began to tinkle. In the wire rooms, the UPI machine was chattering about a murder trial in Minneapolis, Minnesota:

```
DETECTIVES WERE THERE AND THEY "ASKED
HIM TO LOOK IN THERE (THE BRIEFCASE)
FOR SOMETHING."
     THE CASE WAS OPENED AND AN ENVELOPE
WAS FOUND CONTAINING 44 $100 BILLS,
THE WITNESS SAID. THE STATE HAD SAID
IT WOULD PRODUCE THAT PIECE OF EVI-
DENCE BUT IT HAD NOT LISTED IT AS ONE
"OF THE SEVEN LINKS." THE DEFENSE HAS
IMPLIED IT WILL TAKE THE LINE THAT
```

```
CAROL'S DEATH AFTER A SAVAGE BLUDGEON-
ING AND STABBING IN HER HOME WAS THE
RESULT OF AN ATTEMPTED
    MOREDA 1234 PCS
    UPI A 7N DA
    PRECEDE KENNEDY
    DALLAS, NOV. 22 (UPI)—THREE SHOTS
WERE FIRED AT PRESIDENT KENNEDY'S MO-
TORCADE TODAY IN DOWNTOWN DALLAS
    JT1234PCS
```

The walk across Commerce and Main wasn't much for James T. Tague. His car was still dead, half in the underpass at Commerce Street and half out. No one seemed sure of what had happened, and he saw a man near Elm, on the grass, speaking excitedly with a policeman. Tague walked over, still feeling the sandy spray on his cheek. He heard the man say that he had been watching the President and it "just looked like his head exploded." The policeman, Clyde A. Haygood, tried to calm the man down. The man said he had seen a piece of the President's head fly off behind the car. Tague joined the conversation and pointed to his cheek. He had been hit by something, probably bits of a bullet or grains of concrete from a curb. The officer observed flecks of blood on Tague's face. The other man insisted that he had seen the shots and he was certain that they had come from the end window of the School Book Depository.

Howard Brennan, who had watched all of it, was dismayed to see the police "running in the wrong direction." He convinced a policeman, speaking almost desperately, that the whole thing had come from that window up there. The pipe fitter pointed. Quickly, the policeman counted from the ground floor upward, and decided that the shots had come from the fifth floor. Mr. Brennan gave him a description of the man behind the gun.

Officer W. E. Barmett wrote the words: "White male, approximately 5 feet 10 inches tall, weighing 165, in his early thirties."

It was the first "make" on Lee Harvey Oswald. Brennan, slow to become aroused, was now nervously communicative. Sergeant D. V. Harkness, who had been studying the area behind the Depository, stopped to listen. Harold Norman and James Jarman, Jr., walked up to listen, and Brennan pointed at them excitedly and said he had seen both of them leaning out a window on the fifth floor. The people who had not been frightened away from Dealey Plaza were now trying to draw the attention of policemen and sheriff's deputies to individual versions of what had happened. Everyone, it seemed, had a story to which he would swear.

Haygood got on Channel Two and said: "I just talked to a guy up here who was standing close to it and the best he could tell it came from the Texas School Book Depository building here with the Hertz renting sign on top." Dispatcher Henslee said: "Ten four. Get his name, address, telephone number there—all the information you can get from him. 12:35 P.M." A few moments later, Sergeant D. V. Harkness was on Channel Two. He had been approached by a "little colored boy, Amos Euins." The student was appalled when he saw all those policemen—fourteen was his estimate—running toward the overpass to find the rifleman when he, Amos Euins, had seen him plainly in the Depository building. "He was a kind of old policeman," Euins said of Harkness, "so I ran down and got him and he ran up here."

The schoolboy convinced the sergeant. Harkness was reporting on Channel Two, not realizing that he was corroborating what Officer Clyde Haygood had just concluded. "I have a witness," Harkness said, "that says it came from the fifth floor of the Texas School Depository store." It was at this point that Harkness began to seal the Depository building. He went to the back of the building, near the freight loading platform, and saw two men lounging. They identified themselves as Secret Service agents. Harkness did not ask for identification. He acknowledged their authority and went deeper into the railroad yards, where he found some hoboes on freight trains. The sergeant arrested all of them.

Police cars pulled up all along Inwood and on Harry Hines Boulevard to clear the way. The dispatchers had done a good job, pulling them in from nearby areas to open the two vital roads to Parkland Memorial Hospital. They stood awkwardly, at crossroads, their red blinkers flashing, their men in the middle of the road flagging the lead car and the presidential car onward at top speed. Rubber squealed as the cars made a right turn and then another, now heading back toward downtown Dallas. Off Butler Street, Curry swung in onto the service road, and Greer followed, tipping the big car. The follow-up was directly behind them, then Johnson and the press pool car. Merriman Smith handed the phone over to Jack Bell of the Associated Press, and, at this moment, the line died.

At the little emergency overhang, the cars skidded to a stop in attitudes of disarray and men began to tumble out, all running toward the Kennedy automobile. As the car slammed to a stop, Governor Connally hung between the jump seats, his head on his wife's lap, his feet on the other seat. He felt a twinge of pain and, for the first time, hoped he might live. He looked across at the other jump seat and saw on his leg a piece of the President's brain about the size of a man's thumbnail.

Men were running and yelling everywhere. Emory Roberts, agent in charge of this shift of Secret Service, ran from the follow-up car to the Kennedys to learn whether he still had a President to protect. He opened the door on Mrs. Kennedy's side, saw the President face down on her leg, and said: "Let us get the President." Mrs. Kennedy, bending over her husband's head, said, "No." It was firm and final. He turned to Kellerman, nominally his superior, and said: "You stay with the President. I'm taking some of my men to Johnson."

This was the second time in one day that many things would happen swiftly, and yet, in retrospect, they tumbled over each other in slow motion. Sometimes, as the men of government and law sped this way and that, they seemed to stop, frozen in flight. Two men hopped on the Oneal ambulance and ordered the driver to remain where he was. Three agents—McIntyre, Bennett, and Youngblood—hustled Vice-President Lyndon Johnson through the emergency door. He was flapping his arms and trying to get back to the Kennedy car. Youngblood said, "No," and kept pushing. "We are going to another room and I would like you to remain there. . . ." Other agents surrounded Mrs. Johnson, who was looking at the Kennedy car and saw a blur of pink and the edges of some red roses.

The moment was hectic, hysterical, and historical. The nation had a new President, but he did not know it, although the men around him did. Two Secret Service men ran up the hall, with its arrow in the center of the floor to point the way to Trauma One and Trauma Two. The nurse at the triage desk didn't know the situation—no one had told her—and she winced as she saw the men with the guns. They demanded to know where the hell the carts were. Chief Jesse Curry had asked that the hospital be alerted; now where were the carts?

Outpatients sat on benches in the chocolate-tiled corridor, or limped on their way in or out. They were rudely pushed aside and told to stand against the wall. There was no time to explain; just get the carts and clear this damn hall. Greer, who had been within six feet of the President all the way, slid out of the driver's seat and got his first look at the carnage in the back. The tears came and he looked at Mrs. Kennedy and kept mumbling: "I'm sorry. I'm so sorry."

To die so suddenly; to die at the peak; to die in an alien place. William Greer looked up and saw the vast array of medical and surgical buildings, new and bright like the rest of Dallas. He knew that things would never be quite the same here again. There was a man in the back seat with a chest that seemed, every few moments, to convulse. He was dead, just as Mrs. Kennedy and Clint Hill had said, but some of the parts of the body fought the inevitable without any brain to direct them.

The loudspeaker called for ambulance carts. The call was repeated. In the emergency section, doctors by ones and twos got off elevators, bounded down and up stairways, all heading for Trauma One and Trauma Two. The only patient waiting for treatment was Julia Cox, 14, who was admitted for an X-ray. Others were resting after treatment, or on their way out of the numberless little sheet-covered cubbyholes against both sides of the wall. Jack Price, hospital administrator, had heard the ugly news and rushed down from his office to expedite matters and to lend a hand if necessary. In the long corridor, as he passed personnel, he issued orders calling for additional skilled assistance. The call for carts was heard by Diana Bowron, a young British nurse, and she asked orderly Joe Richards to help her run one out to the ambulance port. The press pool car was parked, and Merriman Smith jumped out to take a look. Baskin and the others followed. They saw the carnage, the huddled pink suit, the Governor sagging between jump seats, the moans from the mouth of Mrs. Connally, and the whispered sibilations from Mrs. Kennedy to her husband.

David Powers hurried to the automobile, gasped, and cried: "Oh. Mr. President!" and burst into tears. O'Donnell, the general of the palace guard, did not come. He went inside, looking for carts, came out and ordered the police to cordon the area off, to keep everybody out unless they could present White House credentials, to put special guards over the Lincoln and permit no one to touch it. Senator Yarborough was weeping. Mayor Earle Cabell beat his fists against a wall, roaring: "Not in Dallas! Not in Dallas!"

The cart was beside the car but no one could get over Governor Connally to reach the President. Mrs. Kennedy did not want anyone to take her husband. Clint Hill whispered to her, "Please let us remove the President." She said, "No," Hill removed his jacket and dropped it gently over President Kennedy's head. A security policeman, with his radio on, patrolled the front entrance of Parkland and heard ABC's Don Gardiner cut into another program to say: "Dallas, Texas. According to United Press International, three shots were fired at President John F. Kennedy's motorcade today."

Inside, Mr. Johnson was being hustled to a remote part of the emergency area. He followed the phalanx of Secret Service agents without question. He kept rubbing his sore right shoulder, which had sustained Youngblood's weight, and passing nurses saw it and spread the rumor that the Vice-President had sustained a heart attack and was in the emergency area for treatment. In downtown Dallas, people were saying that two Secret Service agents had been killed in Dealey Plaza.

Outside, police officers were roaring: "Clear this area!" On Channel Two, other patrols were assigned to Harry Hines Boulevard to keep automobiles off the service road. Roy Kellerman snatched the first phone and dialed White House-Dallas, and asked for Jerry Behn, agent in charge of the White House detail, in Washington. He started his conversation by saying: "Jerry, look at your clock. . . ." CBS was rushing a flash to Walter Cronkite in New York, while officials were calling for a cut-in on all CBS affiliates throughout the country. Cronkite hadn't seen it yet, but the announcement read: "In Dallas, Texas, three shots were fired at President Kennedy's motorcade. First reports say that the President is seriously wounded." In Washington, David Brinkley saw the teletype flash, but was powerless to use it. His boss couldn't be found.

At the hospital, men ran at top speed for precious telephones, to have and to hold. Merriman Smith, after a long heartrending look into the back of that automobile, had skidded inside, found a man hanging up a phone, and said: "How do you get outside?" He was told to dial nine. Smith said: "The President has been hurt and this is an emergency." This time, in a couple of sentences, he dictated a bulletin which said that the President had been "seriously, perhaps fatally, injured by an assassin" in Dallas.

The stretchers were going by, almost at a run. First there was Governor Connally; behind him was President Kennedy, on his back with a coat over his face. On his chest were a few bloody roses and a pink hat. Kellerman told Behn in Washington: "The man has been hit. He's still alive in the emergency room. He and Connally were hit by gunfire. Don't hang up. This line should be kept open, and I'll keep you advised." Mrs. Kennedy, as forlorn as the bloody roses, trotted beside the cart, her fingers trying to maintain contact with her husband, while visitors leaving the emergency area bumped into her. Her head was back, her dark hair swinging behind from side to side, the mouth was open in anguish, and the eyes begged for the assistance no one could give.

A nurse found an emergency room to satisfy Roberts and Youngblood. It had one patient—a Negro. He was taken out at once. The room, closer to the emergency entrance than Traumas One and Two, had a small window. The shades were drawn. It could hardly be called a room. There were a dozen or more cubicles in one blue-tiled room. This was the one remotest from the door, and it was screened by sheets on poles. Roberts told Rufus Youngblood to remain with the Vice-President, and guards were posted at the door.

Revolvers were drawn there and outside. Roberts convinced Youngblood and the Vice-President that, at the moment, no one knew whether this was a widespread plot to assassinate the leading men in the United States government. It could be. If it was, they would be after Johnson as well as Kennedy and the Governor. No one knew the ramifications of the plot—assuming there was a plot—and no one knew whether anything had occurred in Washington or in Hawaii, where a big part of the cabinet was. Under the law, Lyndon Johnson was next in line for the presidency; Speaker of the House John McCormack was next. The Vice-President was entreated to please do as he was told, promptly, until the matter could be cleared up. He said: "Okay, Partner." He began to understand that this could be a broad plot. For awhile, he understood fear. Youngblood and Roberts agreed that perhaps it would be best to get Johnson out of the hospital at once and hurry him off to *Air Force One* at Love Field.

The smooth continuation of government depended on Johnson. They had to keep him alive. The stark reality was that, apparently, they had lost their man in spite of the most extensive precautions. Even if he lived, could he reassume the burden of the presidency? When, if ever?

Those who saw the head wound, who looked inside at the scooped-out brain, were doubtful that Kennedy would ever be President again. No matter how they looked at it, the burden reverted to Lyndon Baines Johnson, a huge, rough-tough master politician from this state. The republic was in his hands, and, no matter how, they had to protect this man from all harm and get him back to Washington. Kenny O'Donnell came in, head down, took a look at the Vice-President and the guards around him, and nodded. "Not good," he said.

Lyndon Baines Johnson's guards told him little. He kept asking for the President, and asking if it was all right to go see him, and he received suggestions in reply. Emory Roberts said, "I do not think the President can make it. I suggest we get out of Dallas." Youngblood asked Mr. Johnson to "think it over. We may have to swear you in." The Vice-President held his wife's hand, trying to infuse her with a courage he no longer had. Only she and Cliff Carter, his executive assistant, knew that

Lyndon Johnson, in spite of his 1959 speeches to the contrary, never really expected to be President. He heard heels clicking in the corridor, and saw the SS men run. Mrs. Johnson saw the runners, the frantic faces, and they seemed to her to be frozen motionless.

At the triage desk, the nurse asked both carts to stop. She wanted a history of the injuries and at least the names of the patients. The Secret Service paused for a moment, then went on. The arrow in the floor seemed endless, pointing past scores of people on their way out, swinging left and then, a little way farther, to the right. The stripe went through a door, changed color, and stopped in front of two square-tiled rooms which faced each other across four feet of hall. The one on the left said "Trauma Two"; on the right, "Trauma One."

The coat had slipped off the President's head. His eyes were askew, and the jacket hung over the bottom of his nose. Mrs. Kennedy looked smaller than she had. She kept her hand touching her husband's side, and her eyes appealed mutely. She did not beg or scream. The last words she had said had been addressed to Clint Hill, out in the car: "You know he's dead. Let me alone."

The hat was on her husband's chest. The flowers looked soggy. Damp blood penetrated the white gloves and dried in the tiny swirls of her fingers. The pink wool suit was soaked blackish down the right side. The stockings of the impeccable First Lady were wrinkled, and blood matted them to her skin. The utmost in cruelty had assailed her, and more awaited her.

The Governor was wheeled into one room, the President into the other. Dr. Charles J. Carrico, two years a physician, was ready. Nurses Diana Bowron and Margaret Hinchcliffe looked at the doctor and saw his nod. At once they took surgical shears and began to cut the clothes from the President. Carrico reached down for a pulse. There was none. The doctor tried a blood pressure cuff. There was no pressure. A huge inverted soup bowl of a lamp stared down at President Kennedy, and the young man stared back at it.

The tie was snipped off adjacent to the knot which had been notched by a bullet. The jacket came off in sections. Each item was thrown on a chair in the corner. The striped shirt was plastered with blood from the edge of the collar all the way down the right side to the shirttail. Dr. Malcolm Perry, surgeon, hurried in. He saw a blood-spattered young woman kneeling inside the door. His impulse was to tell her to leave, but as the socks were peeled from the patient, the snowy skin, the cyanotic face, the dura mater leaking out of a massive hole in the head told him to do something and do it quickly.

Dr. Marion Jenkins, anesthesiologist, came in. Two student nurses came in. It was a small room, pale tile, a sink in the floor, a cabinet for sterile instruments, a clock which showed the time to be 12:37. Other doctors hurried in to help. Some were directed across the hall to help the Governor. He was screaming: "It hurts! It hurts!" and he could be heard down the hall. It was the only healthful sound in the hospital wing.

Dr. Fouad Bashour arrived. Dr. Richard Dulaney was working Trauma Two. Dr. Gene Akin rushed into the room. Dr. Kemp Clark was there. There had been no carts waiting outside in the hot Dallas sun; now all the medical help possible was jamming the two small rooms to a point where some must volunteer to leave. Dr. Don Curtis; Dr. A. H. Giesecke; Dr. Jackie Hunt; Dr. Kenneth Salyer; Dr. Donald Seldin; Dr. Jones; Dr. Nelson; Dr. Shaw; Dr. White; Dr. Robert McClelland; Dr. Paul Peters.

The President was down to his shorts and his back brace. The Ace Bandage he wore was permitted to remain between the thighs. One doctor was making a cut down on the right ankle; a nurse was doing it to the left arm. The skin was cool to the touch. The work was professional. Doctors, when necessary, mumbled requests or orders. The electrocardiogram had shown a faint palpable heartbeat, hesitant, irregular, and weak. Then it stopped and the automatic pen behind the glass on the wall began to trace a steady straight line. A doctor tried to assist breathing by doing a tracheotomy and found that a bullet hole was in precisely the right spot. He enlarged it and thrust a cuffed endotracheal tube through and down into the bronchial area.

Hunt had the President on pure oxygen. Nobody stopped working. Everybody knew he was dead, but the work went on in silence as though something magnificent was about to happen. Dr. Burkley, on the wrong bus and taken to the Trade Mart against his will, came into the room. Someone said that this was the President's physician. Burkley had skin which matched his graying hair. Now it seemed paler. He had a black bag with him and he took out the hydrocortisone used to correct the President's adrenal deficiency.

The voices were soft and unhurried; priests at a White Mass, responding acolytes; the sacrifice prone on the altar. Science was trying to impose its will on God.

Agent Lem Johns, the thin man with the basso voice, arrived at the hospital. He had been dropped off in Dealey Plaza, and the motorcade had left without him. Now he was bumping into people in a long corridor, and one he bumped was Art Bales. "Are you The Bagman?" Johns said. Bales said no, Gearhart was. Lem Johns found him and ordered him to hurry to the side of the Vice-President. It was ironic that, in the past eight minutes, no one knew where The Bagman was or who he was; and The Bagman didn't know where the President was, or who he was. If there was a time when the United States could not retaliate instantaneously to a nuclear attack, these were the minutes.

The Bagman hurried to Mr. Johnson in Booth 13, but the Secret Service men didn't know him and couldn't identify him. They saw him with the satchel and shoved him into Booth 8, where he remained under the watchful eye of an agent until Emory Roberts came in and okayed him as The Bagman. All day long, he—Warrant Officer Ira Gearhart—would be lost and found and lost again. Four Congressmen had more luck getting to the Vice-President. Thornberry, Brooks, Thomas, and Gonzalez, all of Texas, walked in glumly, studied the Vice-President awhile, and walked out glumly.

Miss Doris Nelson asked Mrs. Kennedy to leave Trauma One. She led the First Lady out, obviously against the woman's will as the surrender of her husband's body had been against her will, but she left the room, the door closed noiselessly, and someone got a chair for her and for Mrs. Connally. The two women sat with their hands in their laps, studying their fingers. There was no conversation between them; no mutual commiseration. Mrs. Connally was embittered, feeling that the Secret Service was eager to climb over her husband to get to a dead President.

The Governor had been nearest the car door. They should have taken him first. At last they did, but only because they could find no way to reach Kennedy. Mrs. Connally felt that she, too, was a First Lady and was possessed of the feeling that she was the only one in that car who wanted to help John Connally. Now her man was inside. Sometimes, when the door of Trauma Two opened, she glanced up beggingly at a doctor or a nurse for a good word, a kind word. William Stinson, administrative assistant to the Governor, came out and said that Connally had said: "Take care of Nellie." This was worth more than all the medical opinions. If John said that, then for sure he was going to recover. Someone handed her one of her husband's gold cuff links. She turned it over in her lap, and the tears came. They rolled down freely. They were good ones.

She studied the cuff link and put it into her pocketbook. Then she looked up, and saw Mrs. Kennedy staring at her dry-eyed. The women averted their glances and looked down again. An hour ago, they had been pleasant companions, both eager for the gala ball in Austin tonight. Now they had only two things in common: both wore pink suits, both were bloodstained.

It was fish again. Father Oscar Huber did not relish it. He sat in the downstairs dining room in the rectory alone. His curates weren't lunching with him. They were upstairs in the "rec" room watching the Kennedy welcome. The pastor had been the only one with the initiative and energy to walk up to Lemmon and Reagan and hop on tiptoe to wave to the first live President he had ever seen.

He was pushing the shreds of fish from one side of the plate to the other, thinking about all the affluent Roman Catholics at the Trade Mart who had been given a dispensation from fish today. They—la de da—were dining on filet mignon, no less. Father Thompson stood in the doorway. "There will be no dinner for President Kennedy today," he said. The old priest didn't care for riddles. "No?" he said with asperity. "Why not?" Father Thompson sucked in a long breath. "He's been shot, Father." "I don't believe it." The pastor was hurt, personally hurt. He had never seen a live President. Now his triumph was marred by news which no sane person could credit. He was outraged. Father Thompson's voice became soft, coaxing: "Come on upstairs and see." Father Huber, yanking the black trousers up a little, started up the steps. He could hear the voice on television; he could catch the excitement in it: "Several shots . . . No one seems to know . . . Parkland Hospital . . . The Governor fell . . . Bulletins as quickly as they come in . . ."

Father Huber tugged at his young confrere. "Get the car," he said. "Get the car, Jim. You drive. Parkland Hospital is in our parish. Come on, now. Hurry."

The organ music was soft, the tunes were sweet. Sometimes the deep timbre of the tones caused the parakeets to fly off the overhead railings, squeaking as they swooped over the waiters, a half dozen steaks steaming darkly on each platter. Over two thousand diners chatted across the tables, and the steady decibel of conversational noise could be heard over the songs. A few diners brought out bottles of bourbon or Scotch and mixed the liquor in half-empty water goblets.

Captain Fritz and his men of Homicide made a final inspection of the head table, lifting the drapes to look underneath. The lobby was cleared. The ladies were pleased to see a huge spray of Yellow Roses of Texas at the head table. Some of the politicians said that, as the steaks were being served now, this meant that President Kennedy would not eat; he would come in at dessert, wait for the tables to be cleared, and get to his speech. Others assured each other that he wouldn't dare, in Dallas, to make a pronouncement on civil rights or the welfare state all of them feared. "If Lyndon and John have any influence, that boy is going to be moderate today," they said.

Deputy Chief Stevenson had heard something vague on Channel Two about shots fired in Dealey Plaza. One of his men reported that the motorcade was going to Parkland Memorial Hospital; there was a story that President Kennedy had been shot. Stevenson got on the radio to ask the dispatcher what the story was. He wanted to know if the President had been wounded. If he had, would he be coming back to make the speech at the Trade Mart or was he going to send someone to make it for him? All those people were sitting around, waiting.

The big press bus waddled onto the grounds, and thirty-five reporters hopped off and ran pell-mell into the Trade Mart. They knew that something had happened—something. They ran, not knowing where the temporary press room might be, but realizing that they would have to get there quickly to find the story about Dealey Plaza. The diners saw them, and laughter spread through the vast edifice. This was what the public saw in motion pictures: reporters running. An official grabbed one writer by the wrist and said: "Hey, you can't run in here."

The man broke loose and ran. Other diners shook with laughter. One yelled: "Somebody get shot?"

They headed for an escalator. A police officer told them that the press room was on the fourth floor. The escalator was sedately slow. They hurried and, at the fourth floor, they scrambled looking for the proper desks and the right phones. Marianne Means of Hearst Headline Service found one, talked into it a moment, and hung up staring like an alabaster statue. She stood looking vacantly at a wall. "The President has been shot," she said matter-of-factly. "He's at Parkland Hospital."

The scramble reversed itself. None of them could believe it, but all of them ran. They ran downstairs, knowing that they had no car, no taxi, to get to Parkland, and most of them didn't know where the hospital was. If Kennedy was shot, even superficially, it was going to be the biggest story Dallas ever saw. They ran and ran, and when they got to the ground floor they broke for various exits, and the hearty laughter of the diners began again, louder this time, as might be expected when a ridiculous situation is seen for the second time. A waiter, carrying a tray of steaks and vegetables on his fingertips, was caught and spun by a reporter and the dishes clattered and the steaks skidded across the floor. Outside, the journalists begged rides from anyone. Some were lucky. Some were not. The hospital was one mile west.

Inside, the diners moved on to dessert. The handsome head of Eric Johnson, president of Texas Industries and chairman of the luncheon, was alone at the head table. There was a rapping of a spoon for silence. The music faded in mid-tune and the chatter eased until silence prevailed. Mr. Johnson stared with controlled shock at the waves of faces and then began: "Ladies and gentlemen, I regret to inform you . . ."

Dealey Plaza was no-man's-land. The police stopped running. The sightseers began to gather. Little groups told over and over, with succeeding amendments, what had happened. Listeners retold the stories to policemen who grabbed witnesses who, in turn, denied their boasts. Deputy sheriffs, whose duties were not related to the Dallas Police Department, interrogated witnesses who had already been interrogated. Sergeant Harkness was the ranking police officer at the scene for seven minutes. Uniformed men with drawn guns searched the building, rooted through the parking lot behind the grassy knoll, ranged over the underpass, and questioned citizens waiting on the grass for something to happen.

Inspector Herbert Sawyer, who had been a mile north of Dealey Plaza, heard Chief Curry and Sheriff Decker order men to search the railroad trestle, and, without orders, Sawyer inched down through the crowds until he parked his car in front of the Texas School Book Depository. When he found that there was no one present who outranked him, the inspector, a veteran cop of the old school, took charge. He saw that witnesses were being questioned and turned loose. A deputy sheriff volunteered to make an office available across the street in the county building for interrogation, typing of affidavits, and holding prisoners.

Sawyer took advantage of it and started a chain of command, so that cops who found persons with something tangible to contribute could escort them over to Commerce Street. He asked Harkness if the Depository had been searched and sealed. It had been searched and was being searched by new groups of officers as they arrived. No one had ordered it sealed. At 12:37 P.M. Sawyer ordered two guards posted at the front door and guards at the loading platform behind the building. The orders were simple: "No one is to enter; nobody is allowed to leave."

Someone put in a call for fire engines, and they fought their way downtown—hook and ladder and pumpers—to further block traffic in the plaza. The firemen raced around, uncoupling hose lines, but Sawyer told them they were not needed. Deputy Constable Weitzman saw a man lift something out of the gutter of Elm Street. It was the missing part of President Kennedy's head. The deputy constable was mystified, because he did not know that anyone had been hurt.

Officer Clyde Haygood, at the rear of the Depository, saw a Negro standing near the loading dock. "How long have you been here?" the cop said. "Five minutes or so." "See anyone come out this door?" "No, sir. Nobody has come out that door." Out front, Sawyer called to newly arrived detectives and ordered them to go to the sheriff's office to type the statements of witnesses. In all, there would be thirty-seven in the first group. Sergeant Gerald Hill arrived and he asked Sawyer: "Are you ready for us to go in and shake it down?" The inspector said: "Yes, let's go in and check it out."

This time the search of the building would be done floor by floor, counter by counter, box by box.

The daily run was not difficult until Cecil McWatters left St. Paul. Coming down from the Lakewood section, he had kept his bus on time all the way, but now he hit a cloud of cars and pedestrians who appeared to be choking Elm Street. All he saw ahead of his Dallas Transit Company bus was a field of red lights. Pedestrians were filtering through them like lava on a bed of coals. Mr. McWatters, an eighteen-year veteran, was headed for Oak Cliff. He had four or five passengers, and he moved his bus slowly down Elm, the brakes sighing, McWatters leaning over the wheel.

He could see the policemen in the middle of the street, whistles blowing and arms waving like symphony conductors. They weren't getting much music. In time, the Marsalis bus made it to Field Street. He was seven blocks from Dealey Plaza, but it might as well have been seven miles. The density of cars and people was like the end of a game at the Cotton Bowl. McWatters got to Griffin, and the bus stopped halfway across the intersection.

Someone rapped on the front door. There is a city ordinance against picking up a passenger anywhere except a bus stop, but traffic was stalled, so the driver opened the door. He didn't pay much attention to the passenger. He watched him drop the coin in the box, and McWatters saw that he was young and wore "work clothes" and sat next to the window in the second seat on the right.

Lee Harvey Oswald looked out at the people passing. The man with the enormous tolerance for silence was heading back into the danger zone. The bus would eventually negotiate the few streets to Dealey Plaza, and then make the fast run across the Trinity River to Oak Cliff. No one would stop the bus, of course. He was safe, even if the bus paused at Houston, and Oswald stared out at the excitement around the Depository.

The time was 12:40. The assassin was interested in time. Wherever he was going, whatever he planned for himself, time was a factor. It would appear that he was heading to his little room on North Beckley. It would not be a refuge, because, within an hour or two, the police would either discover the rifle and bullets or discover that one employee had not returned from lunch. Either way, the fragments of evidence would be sifted and resifted until someone said: "Now that Oswald fella. Lemme see. That Oswald boy he lives at 1026 North Beckley. Yep."

After the furnished room, what? A change of clothes, a chase leading to where? To Irving? To the arms of Marina? To get there he would have to get on Davis and hitch a ride all the way to Loop Twelve and then make a right turn and get out to Walton Walker Boulevard. No, this would be a long chase for nothing. By the time Mr. Oswald arrived, the police would be waiting. He could remain on the bus and go south out of Dallas along Zangs Boulevard, but wouldn't the word be out, wouldn't the police perhaps have roadblocks out of Dallas, wouldn't they be at the airport soon and the bus terminal? Where to?

Could a man continue to hide in Dallas? Yes he could for an hour or two. Once the identity of the assassin was known, and the police put a photograph on television, the task of remaining free would square itself with difficulty. A day, perhaps? Two? Then what? Then the assassin might have to make a choice: go out in a blaze of glorious gunfire, or contact a left-wing attorney and walk into police headquarters saying: "Are you looking for me?"

McWatters stopped the bus between Poydras and Lamar, four blocks from the School Book Depository. A woman with a suitcase got up from her seat. She had to make a one o'clock train at Union Terminal. She was fretful about missing it. Would the driver give her a transfer, please? She could start walking from here and, if the bus caught up to her, she could reboard. McWatters said he was sorry he couldn't make it any faster, but she could have a transfer. A man got out of the car ahead and walked back and knocked on the bus door. The driver opened it. The stranger was excited. "I just heard over the car radio that the President has been shot," he said.

Oswald moved up behind the harassed woman. "Transfer," he said. Cecil McWatters looked up and gave it automatically. The assassin got off the bus, crossed in front of it, but there was a flaw in his anonymity. Mrs. Mary Bledsoe was one of the five passengers on that bus. Before he took the room on North Beckley, Lee Harvey Oswald had been a roomer in Mrs. Bledsoe's house.

She was a woman with strong feelings for and against people. Within two days, she knew that she did not like Lee Harvey Oswald, although she could not put her dislike into words. She permitted him to finish out his time and then she put it to him bluntly: "I am not going to rent to you anymore." The slender young man with the icy eyes and the pout had stared at her without rancor, and then he left. Now, from a front seat, she had watched him board the bus and her venomous assessment began anew: "He looks like a maniac. His sleeve is out to here . . . His shirt is undone. Is a hole in it, hole, and he is dirty, and I won't look at him. I don't want to know I even seen him. . . ."

The little bells in the wire rooms tinkled again. This time editors—not copy boys—came running. They had seen Merriman Smith's first flash, and it could be a mistake. Editors scanned the Associated Press machine, dreading to place credence in one wire service against the other, but while the AP was rippling through run-of-the-mill stories. . . .

```
UPI A8N DA
  URGENT
  1ST ADD SHOTS, DALLAS (A7N) XXX
DOWNTOWN DALLAS.
  NO CASUALTIES WERE REPORTED.
  THE INCIDENT OCCURRED NEAR THE
COUNTY SHERIFF'S OFFICE ON MAIN
STREET, JUST EAST OF AN UNDERPASS
LEADING TOWARD THE TRADE MART WHERE
THE PRESIDENT WAS TO MA
  FLASH FLASH
  KENNEDY SERIOUSLY WOUNDED
  PERHAPS SERIOUSLY
  PERHAPS FATALLY BY ASSASSINS BULLET
  JT 1239 PCS
```

In the same minute, the Cabinet plane was fleeing the morning sun over the Pacific Ocean. Far below, the men of Kennedy's government could see the white fleece of clouds and the occasional inky glimpse of the sea. Breakfast was over. Some of the wives changed seats and talked of shopping in Tokyo. Some of the Cabinet ministers sat together and chatted; others, like Pierre Salinger, studied the briefing manuals so that they would better understand the functions and purposes of the trip. Robert Manning, Assistant Secretary of State for Public Affairs, came out of the forward cabin and told Salinger that Mr. Rusk would like to see him at once.

He looked up from his reading to find that only the wives were still in their seats. Udall of Interior was with Rusk; so was Agriculture's Freeman, Commerce's Hodges, Labor's Wirtz. Salinger required time to leave a comfortable seat. He went forward and was told that they were waiting for Myer Feldman of Kennedy's staff and Walter Heller, chairman of the Council of Economic Advisers. Pierre Salinger did not understand why anyone was summoned, but he bent over Secretary of State Dean Rusk's polished head to read a sheet of paper which appeared to be badly scrambled in transmission. It was a teletype bulletin but the operator who sent it must have been confused or upset:

```
UPI—207
     HANNOVER, GERMANY. NOV. WW (UPI)—THE
STATE PROSECUTOR
     BUST
     BUST
     QMVVV
     UPI—207
     BULLET NSSS
     PRECEDE KENNEDY
     X DALLAS. NTEXAS, NOV. 22(.708 LAS
THREE SHOTS WERE FIRED AT PRESIXENT
KENNEDY'S MOTORCADE TODAY IN DOWNTOWN
DALLAS
     HSQETPEST
     VVU PLF208
     HANNOVER, GERMANY NOV WWKVUPI)—THE
STATE PROSECUTOR TODAY DEMANDEJ AM QI-
```

```
AMONTH PRISON TERM FOR WEST GERMANYS-
JS ZSTZRILIZATION DOCTOR."
X.X.X.X X,XNXLKDN, VOGEL TOLD THE
THREEJU THAT HANDSOME DR. ALEL DOHRN.
%% WAS N IDEALIST BUT BROKE THE LAW IN
AT LEAST IP OF THE QNEPP STERILIZATION
OPERATIONS HE HAS PERFORME ON LOCAL
WOMEN
     MORE
     HS137PEST
     RV
     SSSSSSSS
     FLASH
     KENNEDY SERIOSTY WOUNDED
PESTSSSSSSSSSS
     HS 138/
     SSSSSSSSSSS
     MAKE THAT PERHAPS PERHAPS SERIOUSLY
WOUNDED
     HSQEOPEST
     SSSSSSSSSSSSSS
     GJ OWHL W WOUNDED BY
HQ139PESTXXXXXXXXXXXXXXXXXXXXXXXXX
     KENNEDY WOUNDED PERHAPS FATALLY BY
VASSASSINS BULLET
HS139PESTSSSSSSSSSSSSSSSSSS
```

Mr. Rusk glanced up and saw Feldman and Heller come into the cabin. In his crisp, flat baritone, he read the last two bulletins. Salinger, who had worked as an investigator when John Kennedy was a member of the McClellan Rackets Committee, felt the shock race across his mind, numbing it. Orville Freeman gasped and murmured: "My God!" Luther Hodges grasped the edge of the desk and began to sag to the floor. Hands helped him to an easy chair. Empty faces stared at empty faces. The tragedy was so enormous, so unexpected, so fraught with danger to the country, that the role of these gentlemen became obvious to all. We . . . have . . . got . . . to . . . turn . . . back . . . right . . . now.

Rusk entertained a small hope. "We have to verify this, somehow," he said. "Get us in communication with the White House and see if you can get Admiral Felt at CINCPAC." The communications men on the presidential special were Sergeant Darrel Skinner and Sergeant Walter Baughman. Within forty seconds, they had the Situation Room in the White House. Salinger, his voice breaking, took the phone and said: "Situation Room, this is Wayside. Can you give me the lastest situation on Lancer?" The voice was clear: "He and Governor Connally have been hit in a car in which they were riding."

That was it. The last doubt died. It was official. The President had been wounded. The White House didn't say he was dead. The fat, jolly man of the Kennedy group said: "Please keep us advised. Secretary Rusk is on this plane headed for Japan. We are returning to Honolulu. Will be there in about two hours. We will need to be advised to determine whether some members should go direct to Dallas." The heart began to drag a little. Salinger had been a clever and adroit jester for John F. Kennedy; he had been a buffer between the President and the press; he had been a good cigar smoker, a poor poker player, a moon of a man who beamed on the Chief Executive he had helped to create. Now, in a matter of two minutes, high over the Pacific Ocean, all of it was swirling down the drain in Dallas.

Salinger started out of the communications shack. One of the operators said: "AP bulletin is just coming in. President hit in the head. That just came in." The Secretary of State made the decisions. He instructed Admiral Harry D. Felt at Pearl Harbor to have a fueled 707 jet ready when this plane got back to Honolulu. The other plane would take Salinger, Rusk, and

Manning nonstop to Dallas. That is, if the President was alive. The plane they now were using would refuel at once and take the members of the Cabinet back to Washington.

Now, someone would have to go back into the other cabin and tell the ladies.

The body on the table was stripped of dignity. It was supple, nonresistant. There were eight doctors left in the small room, two registered nurses, and two aides. All of the proper medical procedures had been dutifully instituted and exhausted. Dr. Kemp Clark, slipping in the watery blood on the floor, tried manual chest manipulation, pressing down hard, holding, releasing, pressing down hard. He asked for a stool. The table was too high, the patient was almost out of reach. A stool was placed under his feet and the doctor worked desperately to preserve life.

Oh, my God, I am heartily sorry for having offended Thee

There was a tube sticking in the throat. There was one in the right ankle. Another one was in the left arm. The doctor's strong hand depressed the chest cavity and, when he lifted up, the dead man breathed a loud sigh. The other doctors, busy with their individual functions, quit silently one by one. No one said he was dead. It was as though everything had been tried, and nothing had worked, just as each man knew all along that nothing would work.

and I detest all my sins because I dread the loss of Heaven

A nurse slipped a watch off his wrist. It went into her uniform pocket. His blood had congealed in the bracelet. Outside, Mrs. Johnson stooped over the withdrawn figure of Mrs. Kennedy and found herself beyond tears as she clasped those hands in hers. Mrs. Kennedy looked up, the drawn, dead expression still on the features, the dark eyes searching Mrs. Johnson's face for something. The new First Lady began to tremble violently. Two men took her arms. She turned to an old friend on the other side of the tiny hall. "Oh, Nellie! Oh, Nellie!" The women embraced, as women do when grief is too dark and deep for expression, or when happiness bubbles in the throat.

and the pains of Hell, but most of all because they offend Thee

The cruelty continued for Mrs. Kennedy. Somehow, she contained herself when it was beyond forbearance. He was dead. She knew it. She said it. Twenty interminable minutes ago she had said it. They could do nothing, none of them. All she asked was to sit in the car with his head in her lap—where surely he would want to be—to say her own farewell in her own way. To say prayers for the repose of his soul. To feel the final communion of man and wife before sturdy hands lifted him away forever.

My God, who art so good and deserving of all my love;

She had pushed that young nurse away rudely. The girl had stepped into the car, speaking in a clipped British accent, and had tried to lift his head. Mrs. Kennedy had looked up, glaring, and shoved those alien hands away. She had said "No" to Emory Roberts. When Clint Hill had dropped his jacket, Mrs. Kennedy had folded it tenderly around her husband's head—tenderly and slowly because she, above all, knew that there was no reason to hurry. Now she sat waiting for a priest. A priest had been requested. It would be unthinkable to permit his soul to leave for an unseen place and an unknown judgment without absolution.

I firmly resolve, with the help of thy grace, to confess my sins,

They came to her, sitting demurely on the little chair, and they murmured the sympathetic words, and sometimes, in kindness, she roused herself to respond. Mostly she sat in silence and nodded. The gloved hands manipulated the fingers and she watched them lace together and part. She saw doctors leave the room where her love lay dead, and she looked up to get a word, a report. Failing all else, Mrs. Kennedy would have settled for permission to step inside and hold his hand. They had nothing to say. They swept out, glanced briefly at her, and fled with the dignity of busy men. She arose and moved to go into the room. Mrs. Doris Nelson, the supervisor, said: "Please, Mrs. Kennedy." It did no good. Mrs. Nelson's hands were pushed away, and Mrs. Kennedy leaned against the door and went inside.

to do penance, and to amend my life.

She looked so small. A nurse was bunching the bloody sheets on the floor. She made a mopping motion, saw Mrs. Kennedy, and hurried outside with the bundle. Dr. Burkley moved over and took Mrs. Kennedy's arm. The doctors who remained in Trauma One retreated toward the back wall. There was nothing more that science could do or say; the shocking scene could not be softened; they stood under the clock. One moved forward, as though he sensed an indecency. He took a fresh sheet from the pile and drew it over the body of the President. He stopped a moment, and, with his thumbs, he tenderly closed the eyelids and patted the jaw closed under the chin. Like putting a baby to sleep. The sheet was pulled up over the head. It wasn't long enough. The shinbones and feet of John F. Kennedy gleamed white under the overhead light.

The unhurried hands of the wall clock were at 12:46. The dark head of Roy Kellerman came into the room. He studied the backs of Mrs. Kennedy and Dr. Burkley, looked at the doctors against the back wall, and the white-hooded body on the table. He went out into the hall and ordered Clint Hill to get back on the phone with Jerry Behn at the White House. "Clint," he said, "tell Jerry that this is not for release and not official, but the man is dead."

He had served as the thirty-fifth President of the United States for one thousand thirty-six days. John F. Kennedy had sought the post because "that's where the action is."

The fat man was getting just enough warm sunlight into the left side of his taxicab to induce a half doze. The huge belly was almost imprisoned by the wheel. Mr. William Whaley edged his taxi driver's cap off his forehead and, for a moment, took in a haze of sight and sound. He was parked at the Greyhound Bus Terminal at Lamar and Jackson, and the big buses were pulling into Dallas and pulling out for the long hauls.

He could hear the scream of sirens around Dealey Plaza, a few blocks away, and Whaley vaguely wondered why they made so much noise. Then he saw the young man walking toward his cab and, with the instinct of the veteran operator, reached over and began to open the back door. To Whaley he was "this boy." The boy shut the back door and asked if he would take him to the 500 block of North Beckley. The driver told him to hop in. The boy didn't want to get in the back. He went around the taxi and hopped in beside the fat man.

Lee Oswald might have taken a bus. Any bus to anywhere. They were pulling in and pulling out and roaring across the north Texas plains to other states. He could have bought ten or twelve dollars worth of geography and been out of Dallas with ease. This was the place to do it. Instead he chose the fat man's taxi and chose to go to a furnished room.

Whaley was ready to pull away from the curb, had pulled the meter down, when a woman looked in and asked if she could take this taxi. The fat man said there would be another one right behind him. The vehicle made its turn and headed for the Houston Street viaduct. The driver was a veteran and had adopted the conversational ploy of taxi drivers and barbers: if the customer wants to talk, talk; if he doesn't, keep quiet.

He asked Oswald what all the sirens were about. There was no response. Whaley turned to look at the profile beside him. "The boy" kept looking straight ahead, as though he had heard nothing. The fat man swung left and passed the *Dallas News* and began to creak and squeak his way across the viaduct. On the other side, he turned up Zangs, swung onto Beckley and, in the 500 block, the fare said: "This will do." Whaley pulled to the curb. Oswald dug into his trouser pocket and brought out a dollar bill. The meter read ninety-five cents. "Keep the change," he said and slammed the door. He started walking south, away from his room.

Oswald was four blocks beyond the little white cottage. Whaley, in a burst of blue exhaust smoke, started out again for the Greyhound Bus Terminal. Oak Cliff was quiet. The array of small homes, lopsided flagstones, young mothers with baby carriages, the high-crowned roads with intermittent traffic were familiar to Oswald. Looking around, it was as though nothing had happened. No sirens could be heard; no running policemen with drawn guns; no screaming citizens falling on lawns; not even a puddle from the morning rain.

The word was out. The shattering news snapped and crackled around the world. No one knew the man was dead except the privileged few at the hospital. The word said that shots had been fired at the President; there were additional flashes of information: he had been wounded; it was thought that he was wounded; he had a head wound; it was alleged that he had a head wound; he might not live; he probably would not live. The shocks met the preceding shocks on the far side of the world and men, great and venal, paused. The sun was down in Berlin. Tokyo waited for dawn.

The sturdy men in the Kremlin sat in bowls of office light not believing. It was almost 2 P.M. in Ossining, New York, and the young woman driving south on Route Nine would not walk again. She hummed with the music on the car radio, and then the flash came and the road spun in gray dizziness and her car tried to climb the front of a building. She was paralyzed from the hips down, a permanent memento of the day.

The news swept through Boston like a breathless hurricane; in Baltimore, shoppers began to weep. In the nation's capital, the antidote to shock became the telephone. All the circuits were tied up. Much of the majesty of the United States government was at the mercy of a busy signal. At the Trade Mart, the Texas politicians decided that, on account of the shooting, their women had better be sent home, but the men would get on that bus and drive down to Austin for the presidential ball. A commitment is a commitment.

New York stopped dead. It came to a stunned pause, as though the subways would not run roaring through the bowels of the metropolis; as though Wall Street would not take the economic pulse again; as though buses and elevators and private cars and jet planes and traffic lights had frozen. The city, lying under its charcoal blanket of smoke, stopped breathing. Everybody told everybody else that it could not be; not in this enlightened century, this cultured era of cocktail party sophistry.

The powerful transmitters of the world picked up the news and city editors everywhere called for reporters to drop everything and get on the next plane to Dallas. In the newspaper morgues, the filed obituaries on "Kennedy, John Fitzgerald" were yanked and updated. The picture editors demanded everything on the motorcade from the wire services in Dallas. Two women on television were discussing fashion in front of a curtain and a man with a drawn, frightened expression came up behind them and they kept glancing at him, trying to keep the conversation in motion, and he was trying to say: "I'm sorry, ladies, but the President has been shot" and the words kept coming out across their discussion of necklines and hemlines, so that neither could be understood.

The phone near the swimming pool at McLean, Virginia, had rung and Robert Kennedy had heard the first word from J. Edgar Hoover. In the United States Senate, the strong jaw and good face of Edward Kennedy sat at the rostrum as president pro tem until someone edged up to him and whispered. Senator Kennedy's world of opulence and politics cracked like old ice, and the voices below, pompous and pedantic, rose and fell like separate sounds not to be hooked together to make a measure of sense. The Senator excused himself. The polished mahogany gavel of authority was placed on its side, tenderly, and the clerks below turned faces up to the young Senator to ascertain what had happened.

The bedlam was worse at the hospital emergency entrance. In the warmth of the early afternoon, windows were open in the several buildings, and patients in pajamas and robes pressed pale, inquisitive faces to the glass. Below, a broad stream of automobiles, the sun glinting off hundreds of windshields like spangles on a stream, were heading for the hospital. The police, on Harry Hines Boulevard, at Inwood, at Butler, became testy and shrieked their whistles of authority and flailed their arms to turn the tide away from the hospital. A good part of Dallas, which had bid the President a warm welcome, now had its automobile radios on loud and had come to the deathwatch.

Deputy Chief George Lumpkin managed to squeeze his car into the emergency area, and Curry ordered him to report at once at the School Book Depository and "take charge." Captain Fritz, not required to examine the President's table at the Trade Mart, reported in, and Chief Curry sent him on to Dealey Plaza with his Homicide detectives. It was a homicide. There was no federal law against killing a President, but there was a local law against killing a person, and this was the law which would apply.

Secret Service Agent Forrest V. Sorrels was the first federal man to return to the Depository building. The police and the sheriff's department had worked closely with him in all the advance work; now he wanted to work with them to clean up the tragedy. He drove back, listening to Channel Two all the way:

Patrolman L. L. Hill: "Get some men up here to cover this School Depository building. It's believed the shot came from, as you see it on Elm Street, looking toward the building, it would be the upper right-hand corner—second window from the end." Dispatcher Henslee: "How many do you have there?"

Hill: "I have one guy that was possibly hit by a ricochet from the bullet off the concrete and another one saw the President slump." Henslee: "Ten four." Patrolman E. D. Brewer: "We have a man here who says he saw him pull the weapon back through the window from the southeast corner of the Depository building. . . ."

The Oneal ambulance was still impounded at the emergency entrance. A nurse's aide promised to mop the blood out of the presidential limousine and forgot it. Two Secret Service men put the bubbletop on the big car. Inside the door, an FBI agent was phoning Gordon Shanklin and rubbing his jaw at the same time. Doyle Williams had hurried into the emergency area, and two overwrought Secret Service men, one with a machine gun, punched him against a wall before Mr. Williams had time to reach for his government identity card.

Father Huber and Father Thompson arrived. Reporters, held back by police, saw the two men, their stoles folded between their fingers, escorted through the emergency entrance. Did this mean that the President was dying? Or do Catholics call a priest in any case? Some said it might mean he was dying. Others rushed back to radio cars to report: "Two priests arrived at Parkland and hurried inside. 12:49 P.M." Father Thompson had parked the car and run to catch up with his pastor.

The little black bag was in Huber's right hand. It contained everything but the Blessed Sacrament. A wounded man, the priest reasoned, is in no condition to swallow a desiccated wafer. The press relations man, Steve Landrigan, broke a path for the two priests through a trail of weeping children with dressings on hands and on eyes, running nurses, and Secret Service men who fringed the long corridor with weapons, examining everyone who tried to pass. They turned right, walked another long corridor, swung right again into the trauma section, and, as Landrigan held the door open, Father Huber stepped inside. Father Thompson was behind him.

The priest lifted his eyes and saw a long table under a diffused glare of light. On it was a figure covered to his knees. Father Huber looked at the snowy feet and thought: "There is no blood in this man." He crouched to open the bag and remove the holy oils, the cotton batting, a prayer book, and to put the thin stole around his neck. He glanced around and saw Mrs. Kennedy standing with a gray-haired man. "Mrs. Kennedy," the priest whispered, "my sincerest sympathy goes to you."

His eyes lingered on her face a moment. It was beautiful and empty of expression. It was the face he had missed when the motorcade went by. Father Huber stepped toward the body. The floor was slippery with blood. He peeled the sheet back from the head to the bottom of the nose. The eyelids were closed. For the first time he thought of the sheet covering the head, the closed eyelids, the doctors against the rear of the room. Father Huber had seen his first live President an hour ago. Now he was staring at his first dead one.

The face appeared to be tan and peaceful. In Latin, Father Huber said: "I absolve you from your sins in the name of the Father and of the Son and of the Holy Ghost. Amen." The priest lifted his eyes and saw that a large part of the back of the head was missing. The violence of it brought its own compensation: this man died instantly; he felt no pain. Roman Catholics have always concerned themselves with the rhetorical question of the soul leaving the body. Does it leave at the moment of death? Or does it remain a few minutes? What is death? Is it the moment the heart stops, even though other organs—liver, bladder, intestines, brain—may go on in reduced function until they stop?

The Church maintains that the sacrament of Extreme Unction is not valid if the soul has departed. The thumb of the priest dipped into holy oil and traced the sign of the cross on John F. Kennedy's forehead. "Through this holy anointing," he said softly, "may God forgive you whatever sins you may have committed. Amen." With the power he had, Father Huber gave the departed

Chief of State a special blessing: "I," he said, louder and in English, "by the faculty granted to me by the Apostolic See, grant to you a plenary indulgence and remission of all sins and I bless you. In the name of the Father and of the Son and of the Holy Spirit. Amen." The priest remained standing because, had he knelt in the blood on the floor, the body on the cart would have been too high for him. Mrs. Kennedy and Admiral Burkley and Father Thompson stood, their voices repeating part of the prayers.

Someone said, "Please pray, Father," so he began to recite the prayers for the dying, although this was pointless. However, it gave the widow and some doctors an opportunity to respond in English, to be a party to the pious adieu, and so he went through the Lord's Prayer, the Hail Mary, and the "Eternal rest grant unto him, O Lord . . . " Then he tapped the oily forehead with cotton, placed the sheet back over the head, and turned to leave, as depressed as he could remember. Mrs. Kennedy bent over the corpse, as though kissing her husband. She hurried after Father Huber and took his arm.

"Father," she said, obviously frightened, "do you think the sacraments had effect?" "Oh yes," he said. "Yes indeed." Out in the hall, where the brown metal chair still waited for her, the once dauntless spirit of the woman was crushed by the finality of death. "Father," she implored, "please pray for Jack." Father Huber agreed. He had already decided to have a Solemn Requiem Mass in his church that evening, before the President's body could return to Washington.

Two Secret Service men took the priest by the arms. "Father," one of them said, "you don't know anything." He understood. No one pretended to know why the death had been kept secret this long, but he promised not to tell. As he and Father Thompson emerged into sunshine, walking toward their parked car, the reporters engulfed them. "Is he dead?" "What time did he die?" "Tell us what he looked like." "Who was the doctor who took care of him?" "Did Mrs. Kennedy say anything?" Father

Huber rubbed his mouth and begged God's forgiveness. "He was unconscious," he said, and hurried into the car.

The voice on Channel One, that of Sergeant G. D. Henslee, carried an unusual pitch of excitement: "Attention all squads," it said. "Attention all squads. At Elm and Houston reported to be an unknown white male, approximately thirty, slender build, height five feet ten inches, weight 165 pounds, reported to be armed with what is believed to be a thirty-caliber rifle. Attention all squads, the suspect is believed to be a white male . . ." Henslee repeated the description slowly. All over the city, men in prowl cars repeated it to themselves or took notes.

"No further description or information at this time," he said. "12:45 KKB-364, Dallas." An unknown voice came on: "What is he wanted for?" Dispatcher Hulse replied: "Signal nineteen (shooting) involving the President." A great deal had occurred within the span of fifteen minutes and Henslee's announcement that there was a suspect lifted the morale of the men patrolling the far reaches of the city. A policeman can, in an emergency, move up closer to the scene. In Dallas, it is the custom not to report these moves because it would clutter the radio channel with men moving out of assigned areas.

John Tippit, cruising the quiet streets of Fruitdale in area 78, swung his car northward to Area 109 and parked at Eighth Street and the Corinth Street viaduct. This was solid thinking on Tippit's part because he had effectively sealed off one of the seven ways of getting out of downtown Dallas to the south. He remained at the last street in town before one crosses the Trinity River. Then, hearing no additional alarms, he turned west into the big Oak Cliff section. He was now in Patrol Area 109, which embraced Zangs, Beckley, and the Houston Street viaduct.

He put his car into low gear and cruised the curbs.

• • •

The hiatus arrived. The energies of the protagonists flagged. The fight was lost; the battle was over; there was time to think of next things next. In the little corridor between trauma rooms, William Greer stood guard over Mrs. Kennedy as Clint Hill phoned the White House. He held the wire open. The operator cut in. "The Attorney General's office wants to speak to you." A small, tense voice came on. "What happened, Clint?" "There has been an accident." "How is the President?" Hill knew the President was dead. "The situation is bad," he said, "we'll get back to you." Mrs. Kennedy sat again on the brown chair. Her grief was consumed in flames of bitterness. Doris Nelson asked her to wash, and Mrs. Kennedy said "No." Someone else asked her, and she looked at the bloody gloves, the soaked skirt, the mixture of brain and blood on the stockings and said: "No. I want them to see what they have done."

Who is they? The world? The nation? The city of Dallas? A young man now walking back to a rooming house? Who is they? Within the grief, there was rancor. The Secret Service could have summoned a change of clothes from *Air Force One*. The hospital was less than two miles from the airport. She would not change, no matter who suggested it. The world was about to get a shocking view of the defenseless widow, bedraggled, bereft, bloodied all day and all night so that they could see what they had done.

For a moment, she roused herself and looked at Mrs. Connally. The eyes of the two women met. Mrs. Kennedy asked softly how the Governor was doing. The hard stare was in the eyes of Mrs. Connally. "He'll be all right." Four words. No more. She did not ask how the President was—she knew. Nor was there any sympathy to offer, not even a hand clasp. "He'll be all right."

In spite of the roars of pain, Governor John Connally did well. Dr. Jackie Hunt attended him; so did Dr. Duke; others hurried into Trauma Two and, on orders, some left quickly for additional equipment. The clothes were cut away and the work of restoring the patient from the ashen skin of shock to a pink of resistance began. The room was less disorderly than the other; the procedures were buttressed by optimism. This one was going to live.

There were cutdowns and X-rays and sutures and the injection of intravenous fluids. The cries of the Governor were encouraging. He was neurologically alert; he could feel pain and could protest. The doctors agreed that, after the first emergency procedures, Governor Connally should be taken upstairs to an operating theatre. Dr. Giesecke made the arrangements and hurried back to help with the cart.

The duty of all doctors in both trauma rooms was to determine the extent of injury and to repair and preserve life. None were detectives; none were acting as pathologists. Each, in his work, had his individual opinions of the wounds he saw, but they would have no weight in law. Most doctors who saw Kennedy's head wound thought that it came from the rear. The same doctors, studying the exit wound in the neck, thought the bullet came from the front. The Governor might have been hit at least twice: once through the back along the frame of the fifth rib, which was partially shattered; once in the wrist.

The doctors at Parkland stuck to their primary province, the preservation of life and the restoration of health. Carrico and Clark remained in Trauma One for a few moments. Professionally, they had no right to feel depressed, but they had lost a President and they had seen the stricken face of his widow. They could have examined the body at their leisure, but as Carrico said: "No one had the heart." The State of Texas, under law, must perform an autopsy in all homicides. They ordered Miss Hinchcliffe to clean up the body.

The police herded the School Book Depository employees on the ground floor, back between the elevators and the order box.

Roy Truly was at a trot, trying to assist by rounding them up and counting them off. Some police were on the roof and these men established that the wall was too high for anyone to fire over it and down. The pigeons, fatigued from taking wing as shots were fired and as police popped out onto the roof, were still circling the plaza. Other policemen were on other floors, looking up among the sprinkler pipes, delving between book cartons.

Truly told the police to check off the name of Charles Givens. He was a Negro employee who was absent. The manager walked around, looking at faces, and he said: "Where is Lee?" Police were taking names and addresses, and no one turned when he asked the question. The foreman, William Shelley, was asked the question: "Have you seen Lee around lately?" and Shelley said no. The manager did not want to get an innocent employee in trouble, so he asked *his* boss, Mr. Campbell, about it. "I have a boy over here missing," he said. "I don't know whether to report it." Campbell threw the question back at Truly. "What do you think?" he said. Truly picked up a phone and got the warehouse. He asked for the Oswald address and telephone number. The warehouse gave him Fifth Street in Irving and Mrs. Paine's telephone number.

The manager was aware that "Lee" might not be involved in any trouble, but he relayed the "missing" data to Deputy Chief Lumpkin, who took Truly upstairs to Captain Fritz. The Homicide division was fine-combing the sixth floor. Fritz, still wearing his cowboy hat, said: "What is it, Mr. Truly?" The name Lee Harvey Oswald was given to him as missing. "Thank you, Mr. Truly," the captain said. "We will take care of it."

At the little house in Irving, Mrs. Paine was making lunch for the children while Mrs. Oswald sat in the living room facing the television set. Ruth could hear the commentary in the kitchen, and kept shouting brief Russian translations. When the shooting was related, Mrs. Paine wiped her hands and came into the living room. The news was as incredible as it was in homes all over the world. As soon as it was verified, Marina went into her bedroom and wept. The unexpected violence, the possible loss of a Chief of State so near home developed into an emotional wrench.

A few minutes later, Marina came into the living room, wiping her eyes, holding the infant. "By the way," Mrs. Paine said, translating as she listened, "they fired from the building where Lee works." For a moment, Mrs. Oswald's heart seemed to stop. Without a word, she put the baby on a couch and went through the kitchen into the garage. She switched on the overhead garage light and started to breathe again. The blanket roll where her husband kept his rifle was intact. She could see the contours of a long object inside. For Marina, it was a great relief.

When she returned to the house, Mrs. Paine was placing candles on a table and lighting them. "Is that a way of praying?" Marina asked. Her friend nodded. "Yes," she said. "My own way." Did Mrs. Oswald think of FBI Agent James Hosty? She didn't like this man; she felt that his sporadic visits to Irving were badgering. Her husband had done nothing wrong. Sometimes, when the FBI man arrived, Lee was not at home and Marina resented the sly questions coming from the calm man with the pencil and paper, then listening to Ruth's translations, then replying curtly in Russian, and hearing Ruth retranslate to English and watching the man write something.

Hosty was eating lunch downtown. He had watched the motorcade go by and had felt the pleasure of seeing an obviously delighted President. A waitress came to his table, scribbling the bill on a pad, and said: "Just came over the radio. The President and the Vice-President has been shot." James Hosty didn't wait for verification. He stopped eating, paid his check, and dogtrotted back to the Sante Fe Building, one block away.

He was ordered out again, told to get in his car and listen for radio instructions. The seconds on the clock were both precious and hectic. Hosty was ordered to Parkland. As he arrived, the radio ordered him to return to the office at once. He was out of breath when his supervisor saw him come into the outer office. He was told to go over the Dallas files carefully and see if he could develop any possible leads in the assassination.

Jim Hosty started on the file immediately. The name Oswald never came to mind.

At Idlewild Airport, in New York, a group of reporters and photographers had been waiting for the American Airlines plane to come back off the landing ramp. It waddled back up the apron strip, whistling like a banshee, and, after some delay, pulled up to its blocks. The ramps were adjusted and passengers disembarked from Dallas. Among them was the briefcase-swinging former Vice-President of the United States, Richard Nixon.

He had a lucrative law practice; he was an officer of a soft-drink corporation. He had been close to the seat of power in Washington once, and he was young enough to think that he would live down the narrow defeat by Kennedy in 1960 and try again. His personal plans were to assist the Republican Party in 1964 by assisting in the nomination of someone else—preferably Barry Goldwater of Arizona—to oppose Kennedy. At the same time, Nixon would run against Pat Brown of California for the post of Governor. If Nixon won his race and Kennedy won his against Goldwater, then Nixon would reach for the presidency again in 1968.

His mood was to keep his political future alive, but not to the point of being nominated in 1964.* As he got off the plane, he thought that he would give "the boys" basically the same interview he had granted to the reporters in Dallas. He wore his smile of camaraderie, related a few facetious opinions about John F. Kennedy and the Kennedy administration, and closed

---

* Nixon confided these plans to me at the Key Biscayne Hotel, Miami, Fla., in January 1963.

on a note of division. "The President may have to drop Johnson as his running mate," he said. "In the fight for civil rights, Lyndon Johnson has become a liability to the ticket. He may be more of a hindrance than an asset."

Nixon posed for a few pictures, then kept walking, the microphones under his nose. He walked out front, waved good-bye and got into a taxicab. He was barely out of the airport when one of the reporters got a message: "The President has been shot in Dallas."

President Kennedy's death was a secret. It was known to a select few, such as Jerry Behn, in the White House fifteen hundred miles away. It was not known to Lyndon Johnson, thirty-five feet away. A brace of doctors and a few nurses knew it. The Secret Service agents whispered the information to each other. In the corridor, Chief Curry saw Stephen Landrigan and said bluntly: "Is he dead?" The press relations man said: "Yes, chief. He's dead."

A few minutes before, Kenneth O'Donnell had peered inside the drapes of the small cubicle in which Lyndon Johnson and Mrs. Johnson huddled on orders of the Secret Service and said: "It looks bad. Perhaps fatal. I'll keep you informed." O'Donnell was issuing the orders. The chieftain had fallen; the palace guard took charge. O'Donnell saw Clint Hill. "Order a casket," he whispered. "Find some place nearby. We want to take him back to Washington."

Clint Hill found Landrigan. He said he needed counsel on the matter of a casket for the President. The press man said that Oneal was reliable and nearby on Oak Lawn. They couldn't get an outside line. In time, they went upstairs to the office of C. Jack Price, administrator of the hospital, and used his private line. Steve dialed LA 6–5221. He got Mr. Vernon B. Oneal Sr. and turned the phone over to Hill. "This is the Secret Service calling from Parkland Hospital," Mrs. Kennedy's bodyguard said. "Please select the best casket you have and put it in a coach and arrange for police escort and get it here as quickly as you can." He listened. "Yes," Hill said, "it is for the President of the United States."

He handed the phone back to Landrigan, who talked to Mr. Oneal. "Wait a minute," said the press agent. Hill was leaving the office. "He wants to know what kind of a casket you want." Hill was still walking. "Tell him to send the best he has and to send it right away." Landrigan relayed the information, and Mr. Oneal started to say he had a bronze casket for $3,900, but he was talking to a dead phone.

Downstairs, Trauma One was a quiet room. Nurse Margaret Hinchcliffe was given a depressing assignment. She was told to wash the President's body and prepare it for travel. Miss Hinchcliffe got the assistance of Nurse Bowron and Orderly David Sanders. All of the clothing, sheared off, was placed in a paper bag and given to the Secret Service. In the jacket pocket was a Mass card given by Monsignor Wolf in Fort Worth four hours ago. It was for the health of the President and his family. Nurse Bowron forgot to include the watch she had in her pocket.

The body was sponged carefully, the legs and arms still pliant. The cart drapes on the right side were heavy with brain matter. This was cleaned up and the edges of the massive wound in the head were wiped. The brown hair was slicked back. The body was lifted off the carriage and white sheets were placed underneath. Enough loose material was allowed to hang off the left side so that, when the President was placed in the box, his head and neck wounds would not soil the white satin interior.

In the hall outside, O'Donnell and the Secret Service and Mrs. Kennedy conferred. Malcolm Kilduff was told he would have to announce the death. He wanted to know the time, the exact time. Mrs. Kennedy and O'Donnell wanted to know what time it was now. It was a minute or two before 1 P.M. The widow wanted the time of death to come after the time the priest had given her husband conditional absolution. The heads began to nod. Dr. Malcolm Perry was called. He was asked if 1 P.M. would be all right. Yes, that would be all right. The death certificate would so state.

At 12:59, Mrs. Kennedy went back into the room. She kissed her husband's ankle and reached under the sheet for his hand. Miss Hinchcliffe and her assistants stood back. They watched. They were professionals, and professionals are not supposed to weep.

It is doubtful that Earlene Roberts ever knew a great joy. She was fat and unpretty and middle-aged, a housekeeper who wheezed when she walked. Even the small pleasures—a gumdrop—were denied to her because she had diabetes. She wore oversized house dresses and spent a great deal of time alone in the little house at 1026 North Beckley, in the Oak Cliff section of Dallas. If she lifted a curtain from the front window, she saw a few struggling shrubs and a sign: "Bedroom for Rent." If this was not enough, Mrs. Roberts could look diagonally across the street at the filling station.

She maintained the little house for Mr. and Mrs. A. C. Johnson. They had a small restaurant which kept them busy all day, so Earlene took care of the dusting and cleaning and counted the towels and face cloths the roomers turned in. Mr. Johnson seldom had much to say. He worked hard and kept his mouth shut. "Mizz" Johnson was alert and in her middle years and could look through a person if she had a mind to. The roomers were mostly men who worked for a while in Dallas; then, one at a time, they dropped off and new ones saw the sign on the lawn.

It was exactly 1 P.M. when Earlene Roberts heard the phone, and she got herself up from a chair by degrees and went to it. There was a girlfriend on the other end. "Roberts," the voice said with the pretentious tone of one who has a secret, "President Kennedy has been shot." The housekeeper was never short of words. She had lots of them if there was only someone around to use them on. And yet all she said was: "Oh, no." The woman said: "Turn on your television set." To Earlene Roberts nothing bad could happen to the mighty. "Are you trying to pull my leg?" she said. Her friend had no patience. "Go turn it on," she said and hung up.

The legs were slow, and Earlene had to walk around the curving couch in the living room, because the furniture was grouped around the square opaque eye of the television set. She turned it on and backed up to sit and then, when the sound came, it was all a babble of excitement as though too many people were talking at the same time. She stepped forward to adjust the volume and the front door swung open and one of the boarders came in.

She seldom saw on in the middle of the day, and she never saw this one in such a hurry. His name was Mr. O. H. Lee and sometimes she said hello and he said nothing. She backed up to the couch and glanced at him and said: "Oh, you *are* in a hurry." Mr. Lee didn't look at her. He strode swiftly across the living room area to the left, where he had a small room. The picture came on the set and the camera kept switching from a hospital to people who were babbling about what they saw, and a young woman and a baby got on and Earlene could see that the woman was excited as she told about shots and where she had been standing and how awful it was.

The roomer had double doors leading into what once must have been an alcove. He opened one and disappeared inside.

The space was five feet by twelve, and an iron bedstead occupied most of it. The walls were pale green. Four windows adjoined each other. They were screened by venetian blinds and lace curtains. The bed had a chenille spread. One window held an air conditioner; the floor had space for a small heater.

There was a pole for hanging clothes, but the roomer didn't have much apparel. He yanked a white zipper jacket from the pole and put it on over his work shirt. On the wall was a solitary naked electric bulb. A fresh towel was lying on a chifforobe. He took his revolver and jammed it down inside the belt of his trousers. It was a .38 caliber snub-nosed weapon, seven and a quarter inches from barrel to butt. He thrust a few extra shells into his pocket.

He came out of the tiny room and closed the door. Mrs. Roberts looked up from the television and might have spoken, might have communicated a fragment of the mass shock radiating out of Dallas, but, as Earlene thought, Mr. O. H. Lee "zipped" out the front door. She had never seen this particular boarder move so fast, so, a moment or two later, she got up, walked to the front window and drew the curtain back. There he was, down on the corner where Beckley, Ballard, and Elsbeth meet, standing at the bus stop. Mrs. Roberts was inquisitive, but the shooting of the President was much more exciting than watching Lee, so she returned to the set. She kept thinking that she never saw him come in and go out so fast.

The lean and pale face of Maude Shaw pulled itself up into a smile when she heard the voice of Nancy Tuckerman on the phone. Like Earlene Roberts, Miss Shaw felt lonely at times, especially when the Kennedy family was away. Her job involved the care and feeding of Caroline and John, and what made it bearable to this British nanny was the innate good manners of the youngsters.

She left both of them in the family sitting room on the second floor of the White House to answer the phone. "Yes," she said when Mrs. Kennedy's secretary said: "Miss Shaw?" There was a silence on the phone, as though Nancy Tuckerman was trying to think of a way of saying something. "I have some bad news for you," she said. "I'm afraid the President has been shot." Sometimes, when words induce horror, the mind refuses to accept and assimilate and, like an overloaded fuse, it shuts down all service. "Will you repeat that, please?" Miss Shaw said quietly.

It was repeated. "Oh, dear," Miss Shaw said. "I do hope it isn't serious." Miss Tuckerman, who was in the East Wing, said:

"That's all we know right now. I'll call you back as soon as I hear how he is." Maude Shaw put the phone on its cradle and stood. She walked back down the long corridor with the uneven floor, passing the dimly lighted portraits of past Presidents and their ladies, and walked through the double door to the living room. Her eyes swept the array of odd chairs, the couch, the end tables with silver-framed portraits of the great men of the world, the small cathedral window, and, under it, Caroline, the prim, willful child who was just learning how to ride a pony. At the moment, she was reading a small book with big block capital letters and small words. On the floor, John was on his stomach. Before him he had a crayon book and an assortment of penciled pigments.

He had the patience to begin placing the right colors on the right flowers and the docile animals, but, if the crayon slipped outside its appointed place, he tended to scribble carelessly. The woman watched the children for a moment. She knew something they didn't. The eyes blinked, and she thought that a nap would be a good thing. It would be a good thing in any case, because Nurse O'Dowd had just left, taking with her the children of Senator and Mrs. Edward Kennedy—Teddy and Kara.

"Come along, children," she said. "It's time for your rest, now." They were not the type to plead for clemency or an extra minute. Caroline smiled and kept her place in her beginner's book. John began to round up the crayons from the floor. Miss Shaw had never had an occasion to feel sorry for them before. They were wealthy, they were handsome, they were "as good as gold," and their father was the President of the United States. Suddenly a sorrow welled in her heart as she watched them smile and hurry to obey.

She took Caroline's hand, and John danced on ahead, the little white shoes skipping in the dark corridor. In her room he received a little assistance in undressing, and he talked as volubly as ever, the flame-red little mouth busy with excitement. He loved helicopters and John was at his finest when he received permission to stand on the South Lawn and watch one come in or take off with his father and mother. Sometimes—on very rare occasions—his father permitted him to get on the helicopter and sit next to him and look out the window as the overhead blades slashed the air and the grass drifted away and the big White House grew smaller and smaller. When this happened, John squealed with delight and pressed his knees together. Caroline asked if she could rest "on top of the bedspread" and Miss Shaw said yes.

The nursemaid went to her small room, between those of the children, and waited for a second phone call which never came.

The clock hesitated between 1 P.M. and 1:05 P.M. as though, realizing the horror it had perpetrated, it desired to stand still so that it would not entertain fresh regret. The drag of time was so pronounced that, around the world, hundreds of millions of people heard the stunning news and consulted the time—for no purpose at all. Some would recall with clarity everything that was done or said at this moment; many who could not re-create the moment of marriage would recite this moment as though their powers of absorption had been speeded enormously and the second hand had begun to beat time in milliseconds.

A mile east of the White House, the Senate was in session under the big dome of the Capitol. The House of Representatives, except for two clerks studying their notes, was empty and dark. Senator Edward Kennedy had left the upper chamber. The Democratic leader, Senator Mike Mansfield of Montana, as thin-lipped as Kenny O'Donnell, contained his emotions and leveled the tone of his voice and asked that the august body of the United States Senate "recess at once, pending developments." There was no dissenting voice from the other side of the aisle. There was no voice anywhere. The gentlemen left their desks in twos, like aging schoolboys, whispering that there must be some mistake, a damned big mistake.

Nobody in an elightened century shoots at Presidents and Prime Ministers. Senator Wayne Morse, the mean, mustached maverick of the West, stared almost scornfully at the clock over the President's rostrum. "If ever there was an hour when all Americans should pray," he intoned, "this is the hour." It was a wry irony of politics that a body so powerful a moment ago could be reduced to the mystique of prayer as a means of sparing the life of one citizen.

In Wall Street, the Friday wave of selling was in full flower and the pale sun of a chilly day seeped to the street. Brokers in linen dusters crumpled bits of paper and dropped them to the floor. The bell clanged and there was a stunned silence, as though a hive of bees had been enclosed in a glass bell. The greatest tribute the Stock Exchange could ever accord to any man was to close. It closed.

The huge octagonal building on the Virginia side of the Potomac kept its hard face neutral, but, inside, men of rank were running. The vast Department of Defense was like a deadly snake touched. With the first news came reaction. The Army, the Navy, the Air Force reared back into a coiled position. No man knew whether this was an opening shot in a plot by a foreign power to assassinate the ranking ministers of the United States as a prelude to attack. It could hardly be accidental that, at this moment and this moment only, the President and the Vice-President were both out of Washington and the Secretary of State and other ranking dignitaries were on a plane westbound from Hawaii.

Who was left, not merely to direct the burden of defense but to grasp the reins of power? Who? Secretary of the Treasury Fowler? Who? Robert McNamara of Defense, who was barely back in his office from a trip to Honolulu? Who? House Speaker John McCormack, the old party wheelhorse who had devoted a lifetime to getting the proper legislation out of the proper committees to the floor for a vote?

The power was not in Washington, nor was there any man who could command it. If the President was injured, where was The Bagman? No one knew. Had anyone told Mr. Johnson that, should the wound render Mr. Kennedy unconscious, the frightful decision to launch a nuclear counterattack was now his? Had General Clifton told the Vice-President that it was now within his power—with that Bag—to dial any one of several types of attack? Did he know? Was he aware? Had anyone ever briefed this big, burly man in the matter of awesome and irrevocable decisions?

No. As the clock hung silent, the United States of America stood, for a little time, naked. As the radicals of the Republican Party had kept Abraham Lincoln from briefing his Vice-President, Andrew Johnson, on matters of war and peace, so, too, the men around Kennedy had kept the doctrines of power from Lyndon Johnson. He knew there was a Bag. He knew there was a man several booths away, standing with a Bag. But, if this shooting was a particle of a larger threat to the security of the United States, Mr. Johnson had neither the combination to The Bag, nor the exact knowledge of what to do with it.

McNamara ordered the Joint Chiefs of Staff to send a signal to all American military bases, domestic and foreign:

"1. Press reports President Kennedy and Governor Connally of Texas shot and critically injured. Both in hospital at Dallas, Texas. No official information yet, will keep you informed.

"2. This is the time to be especially on the alert.

"JCS."

In the archiepiscopal residence in Boston, the aged, asthmatic Richard Cardinal Cushing heard the news as a father might hear an ugly rumor about a promising son. To His Eminence, it was unthinkable. He had christened Kennedys; he had married them; he had buried them. To him they were not to be viewed as a rich or powerful Catholic clan; they were his children. At a dinner with them, he could raise his codfish voice in louder dissent than old Joe or young Bobby or laugh more heartily at the antics of the family than they could.

One, through some strange transmutation of baser metal, had turned to gleaming gold and was now leading the nation as the first Roman Catholic President. The Cardinal did not subscribe to all of "young Jack's" measures, but it was a benevolent blessing to have lived to see this boy run the country, not as a partisan Catholic, but as a patriotic President. The news that he had been shot and wounded was unfair and—please God—possibly untrue.

His Eminence, wearing the long black cassock which made him seem so much taller, led the nuns of his housekeeping staff into the little chapel. He was ready to sink his bone-weary frame onto a prie-dieu, when he called his secretary. He ordered the word to be sent out to all Catholic parishes in the New England states at once: "Pray, pray for the President."

For Mrs. Kennedy, the unbearable had to be borne. She sat. She stood. The pitifully whispered words of friends and strangers had to be acknowledged. She sat. She stood. At one time or other, the word death must have reminded her of Patrick Bouvier Kennedy. He had died last summer, thirty-nine hours old. His father, sleepless, had stared through thick glass into a pressure chamber as the infant fought valiantly against the fluids which seeped into his lungs. When the baby died, the young President pounded his fist against the metal chamber because he had not been able, in his strength, to breathe for his flesh and blood.

The impotence of his grief robbed Mr. Kennedy of his disciplined control. He broke down and cried. Alone he had knelt beside the small casket in the chapel of Cardinal Cushing, alone to pray for a son who, in the faith of his fathers, was in the serene company of heavenly hosts. The President had tried hard to reach that baby, to touch his hand. From his neck he had taken the gold St. Christopher Medal his wife had given him and thrust it inside the white casket beside the newborn.

A similar thought must have crossed Mrs. Kennedy's mind in the lonely cruelty of grief. She stepped back into Trauma One and walked around the contoured sheet and lifted it. She took his left hand and kissed it. Then she removed her wedding ring and put it on his dead finger. It could be worked up only to the second knuckle. She placed the hand back at his side and pulled the sheet down.

In the corridor, she saw Ken O'Donnell and told him what she had done. "Do you think it was right?" she said. "Now I have nothing left." The impassiveness of Mr. O'Donnell's face broke a little. "You leave it where it is." Silently he reminded himself to get that ring back later and return it to her.

The short trip was swift. The door to Trauma Two opened and the carriage came out, making the short turn on its casters, and the man under the sheet was whisked down the hall to the elevator, pushed by doctors who were taking him to the second floor, to Operating Room Five. The elevator was small. Some ran the stairs. Governor Connally's right lung had collapsed. In the operating room, the talented minds and hands and eyes of the doctors blended to their duties. At one time, twelve hands were over the patient. Orderly R. J. Jimison helped lift him from the carriage to the operating table and pushed the table outside, soiled with bloody sheets and the medical impedimenta of an emergency.

Dr. Robert Shaw established anesthesia and pushed an endotracheal tube into the patient to ensure positive pressure. The bullet, in traversing the downward plunge across the axis of the fifth rib, had lacerated the right lung and induced a pneumothora. Another pair of hands was busy shaving the chest and belly. The entrance wound in the area of the right shoulder was small and elliptical and looked like a black wart.

Dr. Gregory had the most difficult of the assignments. He was going to take a compound comminuted fracture of the right wrist and put all those small bones, and all the little pieces of them, back together again. The work of salvaging a life and restoring the full use of the body was under way by 1:05 P.M. Shaw was surprised to find that the intercostal muscle bundles, between ribs, appeared to be undamaged. Jagged ends of a fifth rib were cleaned with a rongeur. Two hundred cubic centimeters of blood and clot were pumped from the pleura; there was a tear in the right lung, but all the major blood vessels had escaped damage. Running sutures were employed and, on pressure from the anesthetic bag, the lobe of the lung expanded well with little peripheral leak.

The lower lobe sustained a large hematoma from a flying rib fragment. Bit by bit the repairs were made. The Governor's executive assistant, Bill Stinson, stood in surgical gown, watching. The patient had a strong, lean, well-nourished frame and, unconsciously, he was initiating a part of the fight to return to life on equal terms. Jane Carolyn Webster, a registered nurse, had her people ready with instruments and types of sutures before the doctors called for them. She had the Governor's clothing— all of it—placed in a bundle and put under the cart at the elevator. It was going to require a couple of hours of work before Connally would be ready for bed. When he was, Stinson asked that guards be posted and that the room next to the Governor's be reserved for the use of Nellie Connally.

The wound on the left thigh, where the bullet stopped after passing through the President and the Governor, was about the size of an eraser on the end of a pencil. As Doctor Gregory examined it, he surmised that the energy of the bullet was near exhaustion because the injury was barely surface deep. Dr. Shaw, assisted by Doctors Boland and Duke, spent considerable time on the exit wound in the chest. It was under the right nipple, about five centimeters in diameter, and the torn edges had to be snipped away. Dr. Giesecke, who monitored the anesthesia, had also worked on President Kennedy.

On the first floor of the big hospital, Trauma Two was being scrubbed by Audrey Bell. When she got to the nurses' working table, she found a group of bullet fragments and turned them over to police. Outside the door to the elevator, where the cart of Governor Connally was parked, a bullet bounced on the floor and was handed to Security Officer O. P. Wright. He put it in his pocket to be given later to the Secret Service or the Dallas Police Department.*

Admiral Burkley opened the door of Trauma One and edged inside. The place was clean and gleaming; the rows of instruments on sterile napkins sparkled. Only the contour of the sheet on the table spoke of another presence. The clay of John F. Kennedy was cooling. The admiral glanced across the floor and saw a wastebasket. In it were the bent and broken roses which the wife of the mayor had given to Mrs. Kennedy—how long ago? Eighty minutes ago at Love Field.

Two flowers had fallen out of the basket. The admiral-physician picked them up and placed them tenderly in his jacket pocket. Perhaps later, he would give them to Mrs. Kennedy. She might want to treasure the flowers on which the President had fallen and died. In the corridor, Mrs. Kennedy kept the vigil over the door to the room. Doris Nelson asked the Secret Service men what arrangements would be made for the body, and they told her that an undertaker and casket were en route to the hospital. She began to fill out the blanks in the death certificate. It would be signed by Dr. Kemp Clark, the neurosurgeon; the patient died of a brain injury.

Men began to do things by rote. Landrigan phoned Norris Uzee and asked him to lower the hospital flag to half-staff. It was done at once, but no one waiting outside noticed it. Dr.

---

* It has been suggested that, in a group of carts, this bullet may have fallen from Kennedy's. The President was still lying on his cart at this time, and, of those carts at the elevator, Connally's was the only one involving a bullet wound.

Clark gave the signed death certificate to Dr. Burkley, and the doctor tried to place it in the pocket with the roses. An FBI man grabbed hospital administrator Price by the arm and whispered: "Don't let anybody know what time the President died—security." Senator Ralph Yarborough, stunned by the situation, began to realize that a President had been assassinated and he moaned loudly and staggered to an upright column. He required treatment for hysteria and kept muttering: "Horror! Horror!"

In Washington, the tragic, secret word went from Jerry Behn's office to Secret Service Headquarters to Robert F. Kennedy. The phone rang. The voice of J. Edgar Hoover informed the Attorney General that his brother was "in critical condition." Robert Kennedy listened politely and said: "You may be interested to know that my brother is dead." Then he called his brother Ted and asked him to please break the news to "mother and our sisters." It could not be told to the President's father: Joseph P. Kennedy was convalescing from an extensive cerebral hemorrhage.

Mrs. Joseph Kennedy, as small as a vase of violets, took the news standing. "We'll be all right," she said. Then she put on her coat and walked out of the Kennedy compound on the Massachusetts shore, and paced the beach. The November winds were coming east and the breakers climbed up out of the green troughs white, falling in thunder on the sand. She walked, hands in pockets, the gusts tearing at her hair, with time to dwell on the hardships which can be imposed on a family by the will of God.

The city and the nation was in a daze. The President had been shot. It was not known that he was dead but the shock spread like mist along a shore. Lyndon Johnson was President but did not know it. To keep him secure in that little cubbyhole, Congressmen and Secret Service agents kept reminding the tall Texan that the assassination could well be part of a much bigger day of terror. Johnson began to believe it. Emory Roberts suggested that Johnson leave at once for *Air Force One*. Johnson said he would not leave, and would not board *AF-1* "without a suggestion or permission of the Kennedy staff." Roberts asked Kenny O'Donnell and he said: "Yes." Johnson refused to move. Roberts returned to O'Donnell and asked again: "Is it all right for Mr. Johnson to board *Air Force One* now?" "Yes" O'Donnell said, "Yes."

Mrs. Johnson asked if she could stop a moment and see Mrs. Kennedy again, and Mrs. Connally. Agents formed an advance guard for her. The new First Lady had a cast-iron gentility. She was opposed to violence of any kind, even in speech. She was surrounded by marching men, marching through corridors of silent men and, when the ranks broke, the young widow was standing before her. Mrs. Johnson's opinion of Mrs. Kennedy had been summed up in a sentence years before: "She was a girl who was born to wear white gloves." Mrs. Kennedy's opinion of Mrs. Johnson had also been summed up long ago: "If Lyndon asked, I think Lady Bird would walk down Pennsylvania Avenue naked." No one spoke. There was nothing worth saying. No miracle could repair the personal wound, the epicenter of which stood in silence, clasping and unclasping the bloody gloves.

Mrs. Johnson began to weep. She grabbed the young woman and said: "Jackie, I wish to God there was something I could do." The dazed expression was on Mrs. Kennedy's face. It was on Mrs. Johnson's face. In an hour, it would be on the face of the world. Lady Bird Johnson walked away, looking back and shaking her head and wiping her eyes. She went upstairs to Nellie Connally and the women hugged each other. Mrs. Johnson said: "Nellie, he is going to get well."

The dark, intelligent head of Malcolm Kilduff was also in the swirling fogs of bewilderment. He met Evelyn Lincoln, Mary Gallagher, and Pamela Turnure near the emergency area entrance. The face of Evelyn Lincoln was stricken. "Mac, how is he?" The three women stared at him, waiting. The assistant press secretary wanted to tell the truth, but the words hung in his throat. He couldn't even say them to himself. He waved his hands feebly and left them.

He walked dazedly in the opposite direction and met Kenneth O'Donnell. "Kenny," said Kilduff, "this is a terrible time to approach you on this, but the world has got to know that President Kennedy is dead." The presidential assistant looked surprised. "Well, don't they know it already?" To him, it was as it was to so many others: President Kennedy seemed to have died a long, long time ago. The horror-stricken mind, racing at top speed, seemed to have lived with this melancholy truth for a long time. The assistant press secretary was saying that the world did not know.

"Well, you are going to have to make the announcement." O'Donnell thought about it. He became conscious of a new order of things. "Go ahead, but you better check it with Mr. Johnson." The press man nodded, and shuffled off through the rabbit warren of passageways, wondering why he could not say the words: "President Kennedy is dead." As he approached Johnson's hideaway, Mac Kilduff found himself walking behind Mrs. Johnson.

The new President was sitting on an ambulance cart, his legs dangling. He nodded to Mrs. Johnson and returned to a moody look at the floor. Kilduff swallowed hard and said: "Mr. President . . ." Johnson brought his head up sharply; Mrs. Johnson turned as she was about to sit, and held a hand against her mouth. This was the first time Lyndon Baines Johnson had been so addressed; it was the first time he *knew* that he was the thirty-sixth President of the United States.

"Mr. President," the young man said, "I have to announce the death of President Kennedy to the press. Is it all right with you?" Johnson hopped off the cart and jiggled a hand in his trouser pocket. "No, Mac," he said. "I think we had better get out of here and get back to the plane before you announce it."

Kilduff had not thought of the assassination as anything more widespread than the death of Kennedy. "We don't know whether this is a worldwide conspiracy," Mr. Johnson said, quoting Emory Roberts and Clinton Hill, "whether they are after me as they were after President Kennedy, or whether they are after Speaker McCormack or Senator Hayden." He looked up and saw the fresh shock in Kilduff's eyes. "We just don't know," the President said.

Johnson looked at the Secret Service agents. "I think we had better wait a minute. Are they prepared to get me out of here?" Kilduff thanked the President and went back to discuss the matter with Roy Kellerman. The Secret Service began to lay its plans. If this was a plot, a conspiracy of some dimensions, Kellerman said he would feel better if they got Johnson back on the plane. Roberts and Youngblood wanted him to get aboard *AF-1* and fly at once to the White House. In that building, he could be given the utmost protection. Until then the craft was a sealed edifice with wings. It could be isolated from the rest of the field, from the world, and protected. It also had direct communication with Washington. *Air Force One*—or 26000—had brand-new, highly sophisticated equipment, some of which was directly related to The Bagman and his "football." The Vice-President's plane did not have this equipment; neither did the third presidential 707, which was en route home with Rusk aboard.

Kennedy's lieutenant, Kenneth O'Donnell, was not a man to quarrel with the political dice. He was sickened, but not to the point of misunderstanding the shift of power. He was a Kennedy man all the way, but there was no Kennedy. It seemed to him that at one minute he was sitting in a car tallying votes along the curb, and that of the next he was chief usher at an Irish wake. He went to visit Johnson.

The new President, the man who had assured Mr. Rubin, the restaurateur that his face would never be among the chased glass squares of the Chief Executives, was frightened. He had assimilated the doleful counsel around him, and believed it. In a trice, he became the only President who ever witnessed the assassination of a President, and it was too much for one set of shoulders to bear: at times his ideas had been treated with contempt by Kennedy's palace guard; now the palace guard attended him and called him "Mr. President." He thought of Harry Truman who, on an April day in 1945, chatted with old Senate colleagues and planned late-night poker games, who was asked to report to the White House and, in the blink of an eye, found that he was President of the United States. He thought of another Johnson named Andrew, who was Vice-President to a weary President named Abraham Lincoln, and who found himself at 7:22 of a Saturday morning the new Chief Executive. And Teddy Roosevelt, inaccessible in the Adironack Mountains when McKinley lay dying of a bullet wound; Calvin Coolidge being sworn in beside a kerosene lamp in his father's house in Northampton . . . Johnson knew American history.

The President asked O'Donnell if it might not be better to get to Carswell Air Force Base. It was military; security would be easy. No, it would not be better. Carswell was thirty-one miles away. No, Mr. President. The safest course would be to traverse those two miles from this hospital to that airport. Two miles. O'Donnell also pointed out that the short trip should be all the safer because it was not scheduled. No one knew about it.

Part of Johnson's political philosophy was to seek intelligent help with the utmost candor. He knew O'Donnell was a "take charge" man and the new President looked him in the eye. "I am in your hands now," he said. O'Donnell misunderstood. He thought that Johnson was asking for a pre-endorsement of his actions by the Kennedy group. To the contrary, Johnson was as dazed as any of the others and was in urgent need of good counsel.

"Well," Johnson said, "how about Mrs. Kennedy?" The small, thin smile adorned O'Donnell's face. "She will not leave the hospital," he said, "without the President." There was no doubt about which President. Mrs. Johnson nodded approvingly when her husband said that he would not go back without Mrs. Kennedy and the body of her husband. The smile disappeared and O'Donnell said that he still thought the best move would be for President Johnson and his "people" to get aboard that plane now. "I don't want to leave Mrs. Kennedy like this," Johnson said. Perhaps, he conceded, it would be just as well to wait for her on the plane.

Had O'Donnell been clearheaded, he would have recognized that, even though Johnson automatically assumed the burden of the Presidency the moment Kennedy was incapacitated by a rifle shot, he had none of the executive powers until he was sworn in. He was President but could not act as one until that oath had been taken. It was printed in almost all almanacs and could be administered by a notary public. This lapse cost the nation the services of a Chief Executive for two hours and five minutes. All Johnson had was the title.

Congressman Homer Thornberry of Texas came into the little room. The silence had thickened. Congressman Jack Brooks stepped inside and thought he was intruding. Someone took his arm and told him to stay. Johnson asked if he could see Mrs. Kennedy for a moment. Agent Clint Hill shook his head negatively. "You should not leave this room, Mr. President." Kenneth O'Donnell excused himself and left. He would like to get Mrs. Kennedy away from Trauma One before the casket arrived. He needed a good reason.

The Secret Service was, to a man, unsentimental. Their work consisted of protecting the life of the President. Officially they would not be involved in tracking an assassin. The agents were told that, if necessary, they were to place their bodies between a potential assassin and the President. They had lost one today. It was a dark and dismal thing for them to contemplate, and they were going to go "overboard" to protect the new one. They advised Johnson to get aboard *Air Force One* at once and to take off for Washington. Johnson was shocked. He asked where Mrs. Kennedy and the casket would go. *"Air Force Two,"* they said. Emory Roberts repeated this "suggestion."

Morally, Mr. and Mrs. Johnson could not consider the proposal. They would not fly back to the capital alone, with a dead President and a grieving widow on a following plane. Johnson said that he would agree to get aboard *Air Force One,* but he would wait "for President and Mrs. Kennedy." That settled it. Agent Youngblood filled a gap of conversational vacuum by announcing that the Secret Service had located one Johnson daughter, Lynda, in a Texas school and that she was now protected. The younger one had been found in a Washington, D.C., school and an agent was at her side.

The Johnsons, sickened and frightened, realized the country was certain to interpret a quick return to Washington as "fleeing" and leaving the widow alone with the body of her husband. The President solicited advice from everyone around him. He received none from his congressional confreres, plenty from the Secret Service, some from Cliff Carter, his assistant, but no one thought of the oath of office. If it occurred to the President, he did not mention it, for the same reason that he would not depart alone on *Air Force One*—it would look like a precipitous power grab. No one recited the substance of Article 2, Section I (7) of the Constitution of the United States, which is explicit: "Before he enter on the *execution* of his office, he shall take the following oath or affirmation . . ."

Legally Lyndon Johnson was no longer Vice-President and had none of the powers of that office; he was now President of the United States, with none of the powers of that office. He could not have protected the country if, as some surmised, the death of Kennedy was part of a much larger plot to bring the government to its knees.

• • •

One Secret Service agent returned to the scene of the crime. Forrest V. Sorrels of the Dallas office had an intuitive feeling that this case could be solved and closed out quickly if the police and sheriff's deputies sealed the Dealey Plaza area and endured the tedium of interrogating everyone. Someone, he was certain, saw something. That someone could be lost within minutes unless a clearing house for affidavits was set up. He had worked with Chief Curry and the Dallas police and with Sheriff Decker and the county officers. Sorrels, respecting both, feared a fatal division of authority or, worse, a conglomerate mass of law officers working without direction.

He was back in the Texas School Book Depository building within twenty-five minutes. Carefully, he had walked on the overpass, down through the parking lot and grassy knoll, and around the railroad yards behind the building. He saw a Negro standing at the loading platform and said: "Did you see anyone run out the back?" The man glanced up from his reverie and said: "No, sir." Without challenge, Sorrels went into the building at 12:55 and saw groups of officers sifting and questioning employees. The Secret Service agent, who had no police power in a Dallas County homicide, watched for a few minutes and then walked out the front entrance, still without being asked who he was or why he was there.

He saw more policemen on the lawn and more citizens babbling and pointing. Sorrels, raising his voice, said: "Did anyone here see anything?" A man pointed to another in a tin hat. The Secret Service agent was not in communication with Brennan, who had watched the assassin from the low wall in Dealey Plaza. This was an accident because Sorrels did not know where the shots came from and had walked through the Texas School Book Depository only because he thought that one of the employees may have seen gunfire.

Brennan glanced at the Secret Service identification which was flashed at him. "Did you see anything?" Sorrels said. The pipe fitter pointed to an upper window of the building. He began his story all over again. "I could see the man taking deliberate aim and saw him fire the third shot." Brennan said that the rifle was then pulled back into the window slowly, as though the rifleman was studying the effect of the shot at his leisure. Brennan pointed to the Negro boy Euins and said that he too had witnessed the shooting.

Sorrels began to feel a little better. He had leads. The police had the same ones, but this agent was going to ensure that these witnesses would be interrogated at length in an office, with a stenographer taking notes. Forrest Sorrels was sure that he wanted to question every employee of that building. As he crossed the square and walked into the sheriff's office, an officer pointed to a young couple, waiting patiently on a bench, who had also witnessed the shooting. Somewhere around was a man with pellet holes in his cheek, a man who stood in a direct line with a shot which ricocheted from the pavement beside the President's car. Mr. Sorrels began to feel encouraged.

The building was being shaken down for the second time. Policemen and deputies were on every floor, like armed treasure hunters, studying each step on the stairwells, examining the roof, on hands and knees in the small attic spaces above the seventh floor, shouting to each other across the dusty barn-like spaces, overturning cartons and standing on boxes to study the areas over the ceiling sprinkler system. Sheriff Decker was on the sloping lawn below the front windows, listening to Captain Will Fritz.

Some witnesses were escorted across the square to the sheriff's office; others were incoherent. Fresh groups, alerted by radio and television, were cluttering the square, listening to the police, offering suggestions, and conducting interrogations of their own. Motorcycles were on their sides in varying attitudes of disarray. Cops in helmets tried to maintain the flow of traffic on Elm, Main, and Houston. The squawk of police radios scattered metallic words across the lawn. An airliner, making the turn for final approach to Love Airport, emitted a subdued scream as the pigeons, in fatigue, gave up and stood along the roof edge watching policemen.

Luke Mooney, deputy sheriff, was on the sixth floor. He was one of many. In the southeast corner, he noticed that the boxes were piled higher than elsewhere and, to squeeze between them, he had to turn sideward and hold his breath. Once inside the little "fort," his breathing stopped automatically. At his feet, he saw three empty rifle shells. His eyes saw some low-lying boxes which could be used as a rifle rest. There was a diagonal crease in one which pointed out the window.

The first thing that Luke Mooney decided was not to touch anything. He leaned out the sixth-floor window and saw his superior, Sheriff Decker, and Fritz of Dallas Homicide, standing below. Mooney shouted, but no one heard him. He whistled between his teeth and shouted again. Both men looked up. "Get the crime lab officers," he shouted. "I got the location spotted."

Mooney kept the other policemen away from the area. In time, Fritz arrived. The Crime Laboratory, a mobile unit, had been summoned from headquarters on Main Street. The deputy sheriff was excited. Having made his find, he observed everything. The pile of boxes was high enough to serve as a private screen against prying eyes from anywhere on the sixth floor. The small boxes which had been placed inside, on the floor, were just high enough, with the window one-third open, to serve as an assassin's roost. A man could sit on the one nearest the heating pipes, while resting the gun on the one near the window, and looking diagonally down Elm Street toward the overpass. He would have an open, commanding view everywhere except as the motorcade passed the broad tree below. The only open space in the tree was furnished by the "V" of two main branches. Mooney was still dwelling on the subject when ranking officers and their entourages descended on him.

Channel One was busy with traffic about the "find," but Channel Two remained mystified. At 1:11 P.M. Assistant Chief Charles O. Batchelor, at the Trade Mart, asked once more: "Find out any further information at Parkland about the condition of the President, whether he can be here or not. Mr. Crull is standing by and needs to know immediately if you can find out so we can do something to these people out here." One minute later, Inspector Sawyer came on with a bit of misinformation: "On the *third* floor of this book company down here, we found empty rifle hulls and it looked like the man had been here for some time. We are checking it out now."

Lieutenant J. C. Day, with twenty-three years of police work behind him, was on the scene with his Crime Laboratory within a couple of minutes. He and his men had a Speed Graphic camera, dusting brushes for fingerprints, an array of technical equipment in the "bus," and the acumen to preserve a chain of evidence intact. The majority of witnesses to the assassination had insisted that there had been three shots. Day now had three empty shells to support this contention. Across the street in the sheriff's office, three Negro employees began to tell the story of how they had been watching the parade from the fifth floor, when they heard those shots and listened to the empty shells drop on the floor.

Will Fritz said that he was going back to headquarters to check up on a man named Lee Harvey Oswald. He stood in an area of the sixth floor, the cowboy hat back off his forehead, watching Lieutenant Day and his men photograph the empty shells, lift them by the ends, and dust them for fingerprints. There were none, and Day initialed the hulls and placed them in a container. The cartons around the window were examined, and palm prints were made.

Other policemen were working the sixth floor. The remains of a chicken sandwich had been found. Someone else found a roll of brown paper, fashioned as a long slender cone. It could have been used to hold a rifle or something like curtain rods.

Deputy Eugene Boone yelled: "Here is the gun!" The others ran to him. He was near the staircase leading down, farthest away from the window where the shells had been found. "Here is the gun!" When policemen reached Boone, some could not see the rifle. It was standing upright between two triple rows of cartons, squeezed tight.

Captain Fritz watched Day's men photograph it, then lift it from its position without marring the chance of obtaining fingerprints. Day knew, after a glance at the roughness of the wood in the stock of the cheap gun, that it would not hold fingerprints. However, the barrel might. It had a canvas sling on the underside and a cheap four-power Japanese scope on top of the barrel. When the Crime Laboratory finished its work, Fritz pointed the rifle at the ceiling, pulled the chamber open, and a fourth shell rattled to the floor. Some of the policemen, studying the contour of the gun, murmured: "Mauser." Boone nodded. "Mauser," he said.

Fritz was becoming increasingly interested in the "boy" who worked on this floor—the missing one. He wanted to start by checking the police records on Lee Harvey Oswald. Then he would send officers to Oswald's home at 2515 West Fifth Street, Irving. The captain could have lifted a phone and asked the police of Irving to pick up Oswald, provided, of course, that he was home waiting for police.

The assassin became aimless and languid. He walked south a few streets, then east a few, then south, then east. It was not a way to get back to downtown Dallas; it was not a way to get a bus to leave Dallas. Lee Harvey Oswald had passed Davis, the last street which might have brought him back to Marina at Irving, Texas. He was now in a rundown area of clapboard houses, faded roofs, weedy lawns, and used car lots.

The flagstone sidewalks were tilted and broken. The sun was high and hot and there were few people on the street. He walked down Crawford, turned left onto Tenth, and went toward Patton. Oswald was still on Tenth, crossing Patton, when he sensed a car in low gear behind him. Behind the wheel was Officer J. D. Tippit, the policeman who "moonlighted" two jobs. He was a dark-haired, good-looking man with a strong jaw and a slow, genial manner.

The cop was off his own beat—78—because Channel One had pulled so many Oak Cliff police cars into downtown Dallas. They were looking for a 25- to 30-year-old male wearing a work jacket and slacks, medium height, slender, black hair or brown. The nearest Tippit had seen to anyone of that description was this man walking ahead of him. Tippit had spent enough years on the force to know that the chances of nabbing the right man in a sprawling city of over one million persons will always be small. In truth, having stopped some persons in error, Tippit had long since become accustomed to opening such conversations politely. Had he been a trigger hysteric, he would have pulled to a stop behind his man and left the car with his gun drawn.

The right front window was open. Patrolman Tippit pulled up to the curb as his quarry completed crossing Patton. Whatever he said caused Lee Harvey Oswald to bend down on the curbside and respond to Tippit's questions. The conversation may have been unsatisfactory. Oswald seldom cared to converse with anyone. Tippit decided to get out of the car, possibly to ask additional questions and maybe to frisk the man. This one had a gun in his belt.

The few people abroad in the area became interested in the police car and the pedestrian. A hundred feet behind them, parked on Patton, William Scoggins, a taxi driver, watched them from across an empty lot. Helen Markham, on the opposite corner of Tenth and Patton, turned to watch. Domingo Benavides was driving his pickup truck in the opposite direction on Tenth and slowed it to a walk, twenty-five feet in front of the police car.

Tippit was in no hurry. He got out on the driver's side. Oswald watched from across the hood. J. D. Tippit made two steps. The assassin yanked the snub-nosed revolver from his belt and fired rapidly across the car. There were little flickers in the sunlight and a succession of explosive sounds rolled through the neighborhood like tumbling bowling pins. Patrolman Tippit was hit four times. He began to crumple slowly, and, as he fell on his belly beside the front wheel of the car, he managed to get his gun from its holster. He fell on it face down and his head hit the macadam. The uniform cap rolled off his head. He had about fifteen seconds to live and he spent it trying to say something. The mouth kept whispering to the pavement but nothing could be understood.

The cab driver was eating a sandwich behind the wheel. At once, Scoggins got out and crouched on the left side of the vehicle as he saw Oswald turn back across the open lot, coming his way. Mrs. Markham began to shriek and ran across the street to the policeman. The woman became rigid with hysteria. She held her fists against the sides of her face and roared: "He shot him! He is dead! Call the police!" There was a small frame apartment house facing the police car. Inside, two young sisters, married to brothers named Davis, had adjoining apartments. They were having a nap with two small children when the shots were heard.

They got up and stared in wonderment through a screen door. Lee Harvey Oswald was cutting across their lawn, going back toward Patton, holding a gun pointed upward and yanking empty shells from it. He dropped them in the weedy grass. He was reloading as he passed Scoggins, and the taxi driver heard him mutter: "Poor dumb cop." Oswald's marksmanship at close range was good. He had hit Tippit in the temple, in the middle of the forehead, drilled two shots into the chest, and missed with the fifth shot. One hit a uniform button and carried it inside the body. Vaguely Domingo Benavides remembered that he was out in his pickup truck to help a man whose car had a damaged carburetor. It seemed mad to be sitting in the truck trying to think of the type of carburetor when a policeman was dying in his own blood, a woman was shrieking, and people were running. He just couldn't remember the make of that carburetor.

Mrs. Markham, a waitress due to begin work at the Eat Well Restaurant at 2:30 P.M., found she could recall nothing about herself but everything about the shooting. She saw great gouts of dark blood pumping rhythmically from the policeman's forehead, but she couldn't hear her own screaming. A used car dealer and his helper watched Oswald loping down Patton toward them and they said: "What's going on?" The assassin kept trotting down the sidewalk toward Jefferson. He was not running, and his gun was no longer in his hands. One salesman, Ted Callaway, watched Oswald go by and he told another man, B. D. Searcy: "Keep an eye on that guy. Follow him." Searcy watched Oswald make a right turn on Jefferson, a main shopping street. "Follow him, hell," he said. "That man will kill you. He has a gun."

Callaway ran to the dead policeman, got the gun from under his body, and rode around Jefferson with Scoggins, the taxi driver, looking for the killer. Channel One, using the services of dispatchers Hulse and Jackson, was almost back to normal traffic when, at 1:16 P.M., they heard a voice say: "Hello, police operator . . ." Hulse said: "Go ahead, go ahead, citizen using the police . . ." Citizen: "We've had a shooting out here." Hulse said: "Where's it at?" There was no answer.

The citizen did not know how to use the police radio. Hulse called again: "The citizen using police radio . . ." Citizen: "On Tenth Street." Dispatcher: "What location on Tenth Street?" Citizen: "Between Marsalis and Beckley. It's a police officer. Somebody shot him." There were voices in the background. Citizen, correcting himself: "What's this? 404 Tenth Street." Dispatcher Jackson knew at once that it had to be J. D. Tippit. He had last heard from this man eight minutes ago, less than a half mile from the place of a shooting. Jackson: "Seventy-eight." This was Tippit's number. He did not reply.

Citizen: "You got that? It's in a police car numbered ten." Jackson: "Seventy-eight." The citizen was becoming hysterical: "Hello, police operator. Did you get that? A police officer, 510 East Jefferson." This was another incorrect address. Jackson: "Signal 19 (a shooting) involving a police officer, 510 East Jefferson." The citizen heard it. He said: "Thank you" and was advised to remain off the air. Within two minutes, the police had the correct location—Tenth Street and Patton—and squad cars roared into the area from all directions.

Officer Nick McDonald, a moon-faced man with dark skin and a high forehead was listening, almost absentmindedly, to the radio traffic at the School Book Depository building. He was working with his partner, T. R. Gregory, and they watched the Traffic division try to keep the curiosity seekers out of Dealey Plaza. The day now was summery and the patrolmen in the street were hotter than the weather, blowing whistles, diverting drivers, trying to prevent family cars from parking. McDonald had looked for an assignment at the School Book Depository, but the building was seething with cops.

The radio came on and Channel One announced, in the flat, toneless manner of police dispatchers, that word from Parkland was that President Kennedy had just expired. A moment later Nick McDonald heard the excited voice of a stranger announcing that a policeman had been shot. When he heard the area of the crime, he said to Gregory: "That's Tippit. We're not doing any good here. Let's go up to Tenth Street." On the way, they heard an additional report: that a suspect had been seen running into the basement of the public library at Marsalis and Jefferson. "Let's go to the library," said McDonald. In the back of the car, the two men had a loaded shotgun. They brought it up front.

• • •

Lyndon Johnson ordered the Secret Service to get him "and my people" to the plane. He still wanted endorsement for his actions, and he ordered Rufus Youngblood to go up the hall and ask Kenneth O'Donnell if he should use *Air Force One*. The agent returned and reported that "O'Donnell says yes."* The President suggested that the party leave in unmarked cars. Whether the assassination plot was large or small, he did not want to have his wife risk her life with him, so he ordered her to ride in another vehicle. The Secret Service got in touch with Chief Curry and asked for unmarked cars. Kilduff said: "After you leave, I'll make the announcement." Rufus Youngblood had sent Agent Lem Johns out front to requisition some automobiles. He said: "Mr. President, if we're leaving now, I wish you'd stick close to me." Johnson was pressed between Youngblood and Kilduff. He kept glancing over their heads to his petite wife to reassure her that it was going to be all right. Agent Youngblood also had asked Johnson to keep his head below window level when he got into the car.

The President said, "Let's go," and the party whirled out of the area at top walking speed. To keep up, Mrs. Johnson had to run between Secret Service agents. Out front, Agent Lem Johns had three unmarked cars and three drivers with rank: Police Chief Curry, Captain Lawrence, and Inspector Putnam. There is something profoundly humiliating to see a President of the United States emerge from a building in an American city running in fear. Some people, lounging at the bottom of the huge hospital building, became alert and shouted: "Tell us something!" "What the hell is going on?" "What happened?"

The party kept walking at top speed, the Secret Service agents fanning out ahead and some walking backward. The

---

* Mr. O'Donnell denied that he was asked about *Air Force One*. There is no doubt that Johnson, thinking ahead, wanted to show that, even in tragedy, the continuity of government would be smooth. Therefore, from the start, he wanted to be aboard 26000 with his dead chieftain and the widow.

President jumped into the back seat of Chief Curry's lead car and slouched as low as a big man can. Youngblood was beside him. Malcolm Kilduff hurried back to the emergency entrance to make arrangements for the death announcement. Congressman Thornberry jumped into the front seat beside Curry. Mrs. Johnson was shoved into the second car. Another group was in the third.

Lem Johns had not told the motorcycle cops whom they were going to escort or where. The cars started out, spinning stones behind them, and a male voice said: "Stop!" Youngblood ordered Curry not to stop. The President asked who it was. Someone said: "Congressman Albert Thomas." "Then stop," Johnson said, and the Congressman was literally hauled into the front seat, and Congressman Thornberry was dragged over the back of the first seat to a spot outboard of the President.

The ride amounted to flight. No one dared to trust anyone. Single-mount motorcycle policemen pulled ahead of the little caravan and asked, on Channel One, where they were going. Youngblood told the chief to tell them Love Field. The cars of the curious were parked askew all over the hospital grounds, and the three automobiles followed each other over curbstones, sidewalks, across open fields, to Harry Hines Boulevard. There the police escort started the sirens, and the President, with his face squeezed in the back seat between the arms of Thornberry and Youngblood, said: "Tell them to shut those sirens off."

Curry did it. Still the wailing shrieks could be heard for a mile. It required two or three requests before they shut down. Then the motorcade began to run a series of "pink lights." As they approached red ones, the phalanx of cycles had to ease out onto the intersection and wave motorists to a stop. Then the three cars, slowed for the moment, hit speed until the next "pink light." The last part of the run was made at dangerous speed. At the airport the cars skidded through a hole in the fence and ground to a halt.

The Boeing 707 never looked so big, so friendly and so impregnable. It sat on the apron, a proud blue and white bird whose home was not Dallas, but rather the blue vault beyond the runway. There was no time for a farewell to Dallas nor a wave of gratitude for the hospitality. People behind the fence saw some dignitaries get out of three cars and they cheered. The officials hurried to the ramps and ran up into the plane without looking back.

The most forlorn figure left on the concrete was Chief Jesse Curry. Once, a long, long time ago, he had been a truck driver. Then, by study and application, he had become a policeman, a good one who was rewarded by Dallas with promotion. Now he was the chief of police, with no ambition for higher office. His sole desire was to "keep his nose clean" and retire with honor.

He sat behind the wheel, a man alone. He had worked hard and earnestly with the Secret Service, preparing for this day. All of it, including his career, died in an instant at Dealey Plaza. Someone—he didn't know who—had disgraced Dallas in the eyes of the world. It didn't matter, Curry knew, whether they found the guilty man or punished him—Dallas was going to need a goat.

Curry picked up the telephone and called Channel Two.

In any situation, there are usually two ways to turn. Lee Harvey Oswald, dogtrotting down Patton, reached the main street of Jefferson and turned right. The police turned left. They had an urgent call from Officer C. T. Walker that a suspect fitting the description of the man who had shot Patrolman Tippit was seen entering the basement of the library at Jefferson and Marsalis. It didn't require much time to surround the building. Red-blinking squad cars winked all over the thoroughfare. Shotguns and revolvers were at the ready.

Oswald kept trotting along Jefferson in the opposite direction, toward Crawford, Storey, and Cumberland. He was in no panic. Pedestrians and car salesmen saw him run by, and he turned into a filling station, trotted into the parking lot, unzipped his white Eisenhower jacket, and tossed it under a car. Whatever description had been given of him, he was now a slightly different man. He wore a burgundy plaid shirt. To hide the gun in his belt, he pulled the tails out, came out of the filling station, and continued trotting on Jefferson.

Nick McDonald and the other policemen at the library ordered everybody in the library basement to come out with their hands up. The door opened, and a few frightened people came out. They came out slowly, including the young man in the white Eisenhower jacket—the suspect. It required only a minute or two to ascertain that this was the wrong young man. He had been spotted running at top speed into the basement of the library—true. But what had impelled him to do it was that he had just heard that President Kennedy had been shot in downtown Dallas, and his friends were in the library. He wanted to tell them. Also, he worked there.

The police scattered. A half dozen squad cars began to comb the side streets of Oak Cliff. How far can a man on foot run in three minutes? Four minutes? Five minutes? Six minutes? How far? Which way? Often a car swung into a little street and found another police group already prowling the sidewalks and alleys. C. T. Walker, who had seen the young man run into the library, was now cruising slowly up a narrow thoroughfare. Ahead he saw a man in a white shirt and long sleeves walking behind a low fence. Walker, who had a newspaper reporter in the car with him, could only see the man from the thighs up. He placed his revolver on his lap and approached slowly.

When he was within thirty feet of the man, Walker stopped the car. White Shirt kept approaching. Walker's nerves were taut. He fingered the revolver and said: "What's your name?" The man looked at the police officer and bent down behind the fence. Walker swung his revolver out the window. The man slowly raised up, with a small dog in his arms. "What did you say?" he asked.

In a radio shop on Jefferson, the loudspeaker was turned up loud and some shoppers stopped to listen. On NBC, Robert MacNeil in Dallas was speaking to Bill McGee in New York: "Last rites of the Roman Catholic Church have been administered to President Kennedy," he said. "This does not necessarily mean that his condition is fatal. Vice-President Lyndon B. Johnson walked into the hospital where the President is being treated. Mrs. Johnson said that her husband is all right. She did not want to say anything about the President; she is in a state of shock. A blood transfusion is being prepared for President Kennedy. . . ."

The press was told, in groups outside the emergency room, that a conference had been called. The White House chief of records, Mr. Wayne Hawks, asked them to go to a nursing classroom on the ground floor of an adjacent wing. Kilduff would make an announcement. "What announcement?" the reporters demanded. "Is he badly hurt?" "Is he dead?" "We have deadlines." "If I leave this spot, I lose the only telephone."

No one noticed the hospital flag at half-staff. Kilduff came out, tough and businesslike, but inwardly unstrung. He strode across empty lots and the journalists followed like disenchanted apostles. In the doorway, a nurse was sobbing. The assistant press secretary was looking for classrooms 101 and 102. Tom Wicker of *The New York Times* was loafing in the rear of the group when he heard the radio in the limousine which had been used by the Vice-President. "The President of the United States is dead," the voice said. "I repeat—it has just been announced that the President of the United States is dead."

It was untrue. It had not been announced. The death story had started when a reporter insisted that the two priests had said that Kennedy had died. Father Huber said Mr. Kennedy had been "unconscious" when the last rites of the church had been administered. Still, to Wicker, instinct counted for something. The announcement lacked authority, and yet it carried the same stinging reality as those loud cracking sounds in Dealey Plaza. Wicker hurried a little and caught up to Hugh Sidey, of *Time* magazine. "Hugh," he said, puffing, "the President is dead. Just announced on the radio. I don't know who announced it but it sounded official to me."

Sidey paused. He looked at Wicker and studied the ground under his feet. They went on. Something which "sounds official" meets none of the requirements of journalism. The press did not know the story. The nearest anyone had come to it was Smith's UPI phrase, "wounded, perhaps fatally . . ." In Washington each man had "sources" through which he might check a supposition. Seth Kantor was the only one with connections in Dallas and he had no more information than the others. All of them were first-rate reporters, men accustomed to the respect of the White House, men who not only recorded the news but who often tried to analyze it, shading the story a little this way or that, depending upon their inner beliefs and confidential opinions from their "sources."

They filed into the nurses' classroom, with its desk and chalkboard, shouting for the announcement. They wanted it now. Some were demanding telephones. Jack Gertz of American Telephone and Telegraph Company was installing instruments as fast as he could. A few of the writers interpreted this as bad news. Why would they require special installations unless they were going to be stationed at this hospital for some time? And why would that be necessary unless President Kennedy was dying—or dead?

Kilduff walked from the back of the room to the front and stood behind a clean greenish desk with the blackboard behind him. The reporters sat at desks or lounged against the walls. The folded sheaves of copy paper, the pencils and pens were ready. The assistant press secretary appeared to be flustered. His eyes were red. On his cheeks there was a hint of tears or sweat. Before him he had the sheet of paper with the precisely worded announcement.

He was going to say: "Well, this is really the first press conference on a road trip I have ever had to hold." What he heard himself say was: "Excuse me, let me catch my breath." He was rolling an unlighted cigarette in one hand. The faces confronting him were familiar to him; some were his friends. They waited patiently now. Some were afraid that Kilduff was going to faint. There was an uneasiness in the room. General Chester V. Clifton, the President's handsome military aide, took a silent stance near "Mac." Flashbulbs were going off; a camera crew was trying to plug in some "frezzy" lights. Kilduff lifted the piece of paper and spoke mechanically: "President John F. Kennedy died at approximately 1 P.M. Central Standard Time today here in Dallas. He died of a gunshot in the brain." A reporter roared: "Oh, God!" Some scrambled for corridors and telephones. One said: "Give us the details, Mac." Kilduff began to breathe heavily. "I don't have any other information," he said.

A cameraman glanced at his watch. The time was 1:33 P.M. A few of the writers did not move from the desks. They had just acquired the dazed, stunned expression which was spreading outward from Dallas. Good writers do not permit a story, not matter how heartbreaking, to touch them. Some of these men were re-creating the days of repartee with the President, the barb of Irish wit, the lucidity of his thoughts, the short era of youth which had permeated the White House with laughter, sweeping out the ghosts of solemn men of affairs—the Boston-accented words when the President said: "When the going gets tough, the tough get going."

This was a time for a clear, unsentimental head. Few of the writers could muster one. Any one of them could have thought of a dozen things he had postponed saying to President Kennedy, and time had run out. This morning there was nothing but time. Some of them, to keep the trip on the front pages, had to dig for the Yarborough intraparty fight. Until then, the journey through Texas had been small-town political huckstering on the fundamental level, praising each city, promising it more federal funds for more projects, endorsing its local Democrats, waving the flag of local patriotism, and closing with the hackneyed You-and-I-will-march-forward-together.

Each of these men knew, better than most, the permanence of the word "dead," and each riffled through his mental files for memories and unfinished business. The story was so monumental in size that they would be writing all day and half the night, trying to sew a literary crepe. A man in surgical white walked into the room. It was Bill Stinson, aide to Governor Connally, and he wanted to report on the condition of the Governor. Kilduff almost pinned him to the blackboard. "One o'clock, one o'clock," he whispered loudly. The Governor's public relations expert, Julian O. Reed, came into the room and, in answer to a reporter's question, began to draw on the blackboard the seating in the President's limousine. This became confused, corrected, and redrawn, until at last all hands agreed that this was where the Kennedys were sitting, and here, in the jump seats, the Connallys sat.

One of the writers started to ask a question and burst into tears. In the hospital hall, a woman married to a United Press International man dropped a dime into a public telephone. In one minute, teletype machines chattered everywhere:

FLASH
PRESIDENT KENNEDY DEAD
jt135pcs

"We must get her out of there," O'Donnell said. "If she sees that casket, it's going to be the final blow." Mr. David

Powers, small and bald and close to tears, said that he would give Mrs. Kennedy some kind of a story to get her away from Trauma One right now. It would help, O'Donnell said, if they could both get her into a room on a pretext that they had to talk over something confidential. Then let the Secret Service sneak the box into the room and it would spare her an additional shock.

It would have to be done quickly, because Oneal was expected at the emergency entrance. The plotters decided to do it together. They sauntered over to Trauma One and started the little whispers about the need for a private chat. At first, Mrs. Kennedy stared at her husband's dear friends, the mouth still half open with shock, the dark eyes pooled with grief. Suddenly she shook her head negatively. "No," she said firmly, "I want to watch it all." She spurned the easy way.

Vernon Oneal and two assistants rolled the four hundred-pound bronze casket down the long corridors, between the rows of faces awaiting medication or treatment and around the narrow corner to Trauma One. It was on a carriage and, as they passed Mrs. Kennedy, the three men glanced at her and mumbled their sympathy. Inside they turned it over to Nurse Hutton and offered their assistance. The casket, as Kenneth O'Donnell said, may have been the final blow to the widow, but she did not whimper. She saw the gleaming bronze sides and the silver handles and the huge convex lid, but she didn't flinch. She studied it.

Miss Hutton had a problem. The brains of the President were still oozing from the massive hole in his head. She had lifted the body by the neck, and wrapped four sheets around it, but it was still leaking through. Obviously it would stain the casket, with its white satin shirring. She asked Supervisor Doris Nelson for instructions. "Go up to Central Supply," she was told, "and get one of those plastic mattress covers."

The cover was placed in the casket, so that the edges hung

over the sides. Then the nude body of the President, covered with sheets, was lifted inside. The plastic was folded over him and the lid was closed. Nurse Bowron sighed. The British girl had never been to America before, but this was a day she would never forget.

The mayor's wife wished to be sympathetic. "Mrs. Kennedy," she said softly, "I am Elizabeth Cabell. I wish there was something—" "Yes," Mrs. Kennedy said, "I remember you gave me the roses." The tone was soft and musical, but the mind retreated from reality. "I would like a cigarette," Mrs. Kennedy said. Mrs. Cabell looked for her purse. When she glanced up again, the young widow had disappeared. She was in one of the trauma rooms. Mrs. Kennedy found her pocketbook on a carriage and dug into it looking for cigarettes. The mayor's wife said: "I have a cigarette for you." She held them out but Mrs. Kennedy did not see. When she found her own, she stuck one in her mouth and stared at Mrs. Cabell as though seeing her for the first time. "I don't have a match," she said.

They were back in the chairs, waiting for the casket to come out of Trauma One, when a priest came around the corner. He was the Catholic chaplain at the hospital, too late to be of any assistance, but not too early to irritate Mrs. Kennedy with pious platitudes and hand-patting. It is possible that she wanted to ask him where he was when her husband needed him so desperately; it is possible that the man was not in the hospital at the time, or, if he was, no one informed him that a Christian was expiring. It is even possible that his approach was too unctuously friendly. Mrs. Kennedy needed some assistance to break the conversation.

In the outer hall, a Negro preacher arrived and said that he had been called to comfort a dying President of the United States. No one asked him the name of his church or who called him. He was ushered out by the Secret Service, who assured him that the matter had been taken care of. Mrs. Kennedy, hostile to the living, stamped her cigarette out and went into Trauma One and sat in a chair with her head leaning against the cold side of the casket.

The forward door on *Air Force One* was closed by Clint Hill. He turned the handle inside and locked it. A Secret Service man was stationed there and another at the rear ramp. On the concrete below, Secret Service men stood quietly, facing away from the aircraft. City police details patrolled the airport, and detectives walked from counter to counter, looking over young men who were departing from Dallas. Uniformed policemen patrolled the fence and Gate 24.

When Lyndon Johnson got aboard, he ordered all the shades drawn. The interior was hot and stuffy. The air conditioning had been shut down when the engines stopped. Mr. Johnson and his party threaded the aisle through the communications shack, where sergeants with headsets crouched, looking up in wonderment as their new President passed. The group went through the galley and the crew's quarters, all forward of the wing, then into the staff and press area, where the seats faced the back of the plane. In the middle of the silvery wing was the door to the President's private stateroom. An attendant held the door open, and the Seal of the President shone in white.

The first sound inside was from the television set. Lyndon Johnson looked up to see the face of Walter Cronkite, in New York, discussing a dark deed in Dallas. The President shhh'd everyone, hoping to hear something new about the extent of the assassination plot. A commentator in Dallas told Cronkite that Mr. Kennedy had been pronounced dead; the shots came apparently from a school book building near the end of a lively motorcade; the police had clues and were looking for a suspect; Vice-President Johnson had left Parkland Memorial Hospital but no one knew his whereabouts.

The big stateroom with its wall-hugging couches and ornate desk and rug was just as John F. Kennedy left it, except that the Texas newspapers were now crumpled in a rack. Mrs. Johnson walked aft to the bedroom with tears in her eyes. She alone had noticed the hospital flag at half-staff and it had crushed her with its finality as the sight of the bronze casket had Mrs. Kennedy. The bedroom has a walkway on the port side of the plane. Outside the bedroom another Secret Service agent stood. In the tail of the plane was a small area near the ramp door for the President's staff and Secret Service men. There were two lavatories, a small galley, and a breakfast nook.

The President left the television set and walked toward the back of the plane. He instructed the stewards to hold the private bedroom for Mrs. Kennedy's use. However Mr. Johnson quickly discovered that there was no other place from which he could make a private phone call, so he removed his jacket, tossed it on a clothes tree, and signaled the communications crew that he would be using this phone for a while.

There were many phone calls; the shocked man had to know that, beyond this little hell of terror, there was a normal, sunny world which was still official and still functioning. One of the first calls was to Attorney General Robert F. Kennedy. This one required some thought. Words of sympathy sound superficial no matter how well intended. Johnson wanted to convey the depth of his personal loss as well as offering his hand to the Kennedy family; he also wanted to ask the Attorney General for a legal opinion on when to take the oath of office as President.

Neither of these was easy to say. Robert Kennedy, on the phone, was less emotional than the President. He had no report from the FBI or any other government agency that there was a broad plot against the leading officers of government; he knew that Governor Connally had been hit, but it could be an accident, because he was in the same car with Robert's brother. So far as the oath of office was concerned, he wasn't sure when it should be administered or by whom. He promised to have Assistant Attorney General Nicholas Katzenbach call back with the correct answers.*

Officials at the Pentagon were calling the White House switchboard at the Dallas-Sheraton Hotel asking who was now in command. An officer grabbed the phone and assured the Pentagon that Secretary of Defense Robert McNamara and the Joint Chiefs of Staff "are now the President." Somehow, in the flight from the hospital, the new President had overlooked The Bagman and Major General Chester V. Clifton, who understood the coded types of retaliation. If, at this time, the Soviet Union had launched a missile attack, referred to in the Department of Defense as a "Thirty-Minute War," it would have required a half hour for The Bagman and General Clifton to get to Johnson's side.

Jefferson Boulevard between Zangs and Bishop is a Friday night shopping area. The street is broad, and cars on both sides park in parallel rows. It is one of the brightest, busiest parts of Oak Cliff at night, but this was Friday at 1:30 P.M. The only parked cars were owned by store clerks. A couple of women chatted and studied the windows. A bus on its way out to Cockrell Hill took its time. One or two shop managers, with no customers inside, stood along the curb in shirtsleeves, absorbing warm sun.

At 1:15 P.M. the box office of the Texas Theatre had opened. Julia Postal, the long-time cashier, had a little radio on and she looked through the slotted window at fourteen customers lined

---

* This was a lapse of memory on all sides. Although the Constitution of the United States does not require a time element, the oath should be taken as quickly as possible to ensure smooth continuity of government in the executive branch.

up for the first show. The marquee, jutting out over the sidewalk, proclaimed:

CRY OF BATTLE
VAN HEFLIN
WAR IS HELL

She took ninety cents apiece from each of them. Mrs. Postal was not discouraged. There would be plenty of customers before the day was over. The Texas Theatre had a one-price policy: ninety cents no matter what time the customer arrived. The war pictures always attracted the men. The manager, Mr. John A. Callahan, was inside. He was excited about the shooting of the President. He was talking to Butch Burroughs, who handled the hot buttered popcorn and the candy inside the lobby. It was a small movie house, part of a chain, but Callahan and Mrs. Postal and Burroughs kept it clean.

On the same side of the street, Johnny Calvin Brewer managed Hardy's Shoe Shop. Mr. Brewer was only twenty-three years old, but he was ambitious and industrious. He had been entrusted with his own shop for fourteen months, and the big bosses did not regret it. At the moment, there wasn't a customer in the store, and Johnny Brewer, neatly dressed in a nice suit and tie, listened to a radio telling the awful events going on in downtown Dallas. A moment ago another flash had come on: some policeman—no name was given—had been shot at Tenth and Patton, right here in Oak Cliff.

The youthful manager wondered what the heck the world was coming to. He was listening and facing the open front door when he heard the shrill scream of a siren. It was approaching the store seemingly at top speed. Mr. Brewer was waiting to see which way it was going when a young man in a flappy shirt turned in toward the store. The windows were recessed from the sidewalk in a "V." The stranger appeared to be studying the shoes in one of the windows. The police car whizzed by and the stranger walked out on the sidewalk and continued on his way.

The manager thought that the man seemed suspicious. He couldn't say why, and perhaps if a customer had been in the store he might have paid no attention to the matter. But the store was empty, and the radio was full of flashes of terrible deeds, one of them only eight blocks away. Johnny Brewer stepped out on the sidewalk and shaded his eyes.

He looked toward the Texas Theatre and saw Julia Postal—now free of customers—out at the curb. Mr. Callahan was hopping into his car and she was talking to him. The stranger with the dirty-looking sports shirt and the slacks turned into the Texas Theatre, without buying a ticket, and disappeared. Callahan was telling Julia Postal that he was going to follow that police car to find out what the excitement was.

Johnny Brewer approached the cashier as she returned to her post, and he asked her if she had sold a ticket to a man "wearing a brown shirt." She said she couldn't remember one. Mr. Brewer, who is not easily dissuaded, said that a man had ducked into the movie while she had been out talking at the curb. The shoe store manager insisted that this was a most suspicious person because, as the police car approached his shop with the siren at its loudest, the man had pretended to look at shoes and then had walked on to the Texas Theatre and was now inside without purchasing a ticket.

She hadn't sold a ticket in the past ten minutes. The movie was just starting, so Brewer walked inside and asked Butch Burroughs if he had seen a man in a brown shirt passing through. No, the candy butcher said, he had been busy and he wanted to know why. "I think the guy looks suspicious, that's why."

It seemed like a lot of trouble for one ninety-cent gate crasher, but Brewer was going to follow his lead all the way. He reminded himself that the cashier's booth is flush with all the storefronts on the street and, if the man had stopped to buy a ticket, he would have been in plain view from the shoe store. Besides, Julia Postal wasn't in the booth. And another "besides"—why didn't the man look up to watch that shrieking police car go by? Who looks at shoes at a time like that? Brewer thought the stranger looked "messed up and scared."

The cashier was excitable, but she thought that Butch Burroughs was more excitable, and she warned Johnny Brewer not to look for the stranger but to check the exits to make certain that he was still in the theater. The exits were properly locked from the inside. Julia Postal could not contain herself any longer, so she dialed the operator and asked for the police.

They were busy with two homicides. Mrs. Postal told the officer that she thought "we have your man." He said, "Why do you think it's our man?" and the woman gave him a description of a floppy sports shirt and a young man of medium build. "All I know," she said, "is this man is running from them for some reason." The policeman asked why, and she said, "Every time the sirens go by he ducks." The policeman asked casually what kind of a complexion the man had, and Julia Postal said she had not really seen him but it was "ruddy." She heard "Thank you" and a dial tone.

Mrs. Postal then phoned up to the projectionist. He didn't understand the request, but the cashier asked him to look through his little peephole to see "if he could see anything." She said she had called the police. "Do you want me to stop the picture?" he said. He looked out at the screen. Audie Murphy, an American war hero, was explaining why "war is hell" as a prologue. "No," she said. "Let's wait until they get here." She didn't have long to wait. A moment after she hung up, police cars began to pile up in front of the theater in awkward parking postures, and men were running toward the lobby with guns drawn. Julia Postal pointed inside and said: "He's upstairs," although she was surmising.

Johnny Brewer had finished checking all the exits except one. That was a door behind the stage. He opened it slowly and found himself staring at a gun. A policeman said: "Who are you?" It was not a time to hesitate. Brewer said that he was the one who had spotted the suspect. "I'm the one who told the cashier to phone the police," he said. Four cops, including Nick McDonald, turned Brewer around and they went back into the theater. As they got onstage, in front of the screen, the house lights began to go on. They weren't bright. Policemen were in the balcony; others, with shotguns, sealed the aisles at the rear of the theater.

The customers, scattered thinly over the orchestra, began to look around in surprise. Nick McDonald heard young Brewer tell a policeman: "He's not in the balcony. There he is," and he pointed to a man sitting alone between aisles near the rear of the theater. McDonald took officer C. T. Walker offstage and up the left-hand aisle. The others—T. A. Hutson and Ray Hawkins—started up the right side.

McDonald was pretty sure that he saw the man he wanted. The officer ordered two customers down front to stand, and he frisked them as Walker stood behind him with his gun out. His eye was on the target, and he noticed that the eye of the target was on him. The stranger did not move. The house lights were up, but the projectionist forgot to shut the movie off, and the screen danced with pale figures. There was the crack of rifle fire and the whistle of bullets.

Hawkins and Hutson, working the other aisle, stood behind two seated customers and said: "On your feet." The men were frisked for weapons and told to sit and remain seated. Nick McDonald moved out of one row of seats to the right-hand aisle. His target was in the second seat off the edge toward center. The two men locked eyes for a moment and McDonald walked toward the rear at a leisurely gait. There was a man and a woman sitting behind the stranger, and McDonald kept looking at them, so that he could keep his quarry within the perimeter of his vision.

The police officer almost passed the target. He kept walking back and, at the last second, swung in quickly and shouted, "On your feet!" Lee Harvey Oswald stood, bringing both hands up and said: "It's all over." Nick McDonald reached from the row in front, to slide his hands down the sports shirt. Other policemen began to come in from both aisles, front and rear. It was at this moment that Lee Harvey Oswald had a change of heart. He had known, from the moment the house lights went up, that the Texas Theatre was full of policemen. There were sixteen—outnumbering the customers by two. There was no possibility of escape. If he had no plan to flee Dallas—and barely the means—this should have been an ideal way to achieve a public surrender. He did not know, of course, whether they were taking him in for the Kennedy murder or the Tippit, and this may have made a difference to him, although it is difficult to follow such a line of reasoning. Either one, on investigation, would lead to the other crime.

Suddenly he brought both hands down a little. With the left, he punched Officer McDonald and knocked his uniform cap off. The right went to his belt and he withdrew the Smith and Wesson revolver. The policemen began to react by instinct. All of them recognized the danger, and each knew that if this was the man who had killed Tippit, killing one or two more policemen would hardly alter the issue for him.

Some dove at him from behind. McDonald swung hard and punched Oswald over the eye. The other hand grabbed Oswald's right hand and both came up with the gun. The nose of it gouged Nick McDonald's cheek and he and other officers heard a click. There was no explosion. Oswald and McDonald fell down between the rows of seats. The cop yelled: "I've got him!" but he didn't. Hutson was directly behind Oswald and he caught the young man's neck in the elbow of his right arm and squeezed. C. T. Walker grabbed Oswald's left arm and Hawkins, on the opposite side, fell on the pile of writhing humans and kept pawing for the hand with the gun.

Detective Bob Carroll hurried into the aisle in time to see McDonald bring the revolver up by the butt. He grabbed for the wrist as two other cops, down in the pileup, tried to force the prisoner's hands behind him. In a moment there was a snap and one of Oswald's hands was handcuffed to a policeman's. The cop hollered that they had one wrong hand, and there was additional confusion as they tried to free the policeman and secure both of Oswald's hands behind his back. Carroll got the gun and put it in his pocket.

The prisoner was lifted up like a submerged object. His pouting mouth was framed in a painful "O" and he called the policemen "Sons of bitches!" and "Bastards!" A policeman brought his fist up hard and caught the defenseless prisoner in the head. McDonald, chubby and perspiring, was still down between the seats, looking for his cap and flashlight, both of which had rolled under the seats.

"Don't hit me anymore!" Oswald shouted as he was dragged out into the aisle. The customers down front turned to watch, but they remembered that they had been told to remain seated, so no one moved. The cops were not sympathetic. All of them had heard, on Channel One, that Officer 78 had been DOA at the hospital and had heard the dispatcher ask a sergeant to please stop at Tippit's house at once to break the news to Mrs. Tippit before she could hear it on radio.

"This is police brutality!" Oswald shouted as he was half dragged, half carried through the lobby. Butch Burroughs, nervous in normal situations, watched the big group go by and saw Oswald's hands being brought up high and tight against his spine. Oswald shouted "Ow!" and called upon the theater patrons to witness this violation of his rights. "Just get him out," said Sergeant Owens. As they passed the lobby clock, the hands pointed to 1:50.

Trade people and passing motorists had stopped to see the excitement and, as Oswald was shoved toward a police car at the curb, fifty or sixty men began to shout: "Kill him!" "String him up!" without bothering to find out the charge or the guilt or innocence of the prisoner. Sergeant Jerry Hill pointed to a sedan and said: "Put him in the back seat." Oswald, sensing alienation from the crowd, shouted: "I want a lawyer. I know my rights!" An excited middle-aged man in the crowd shouted: "That's the one. We ought to kill him." The prisoner was hustled across the sidewalk, protesting: "This is typical police brutality. Why are you doing this to me?" It amounted to more words than most acquaintances had heard from Lee Harvey Oswald at one time.

The car pulled away from the curb and Sergeant Hill got on Channel One and said that the suspect in the Tippit homicide had been arrested, after a struggle, in the Texas Theatre on West Jefferson. They were now bringing the prisoner to headquarters. Carroll, driving, got the revolver from his pocket and handed it to the sergeant. The car turned into Zangs Boulevard and moved at good speed across the Houston Street viaduct into downtown Dallas. As the sedan turned onto Elm Street, the School Book Depository flashed by on the left side. No one in the car gave it more than a glance.

"Why don't you see if he has any identification?" the sergeant asked Officer Paul Bentley. In the back, the policeman began to go through the pockets. "Yes," he said. "He has a billfold." Oswald, trying to bring his wrists down behind him into a more comfortable position, said: "I don't know why you are doing this to me. The only thing I have done is carry a pistol in a movie." Another policeman said: "You have done a lot more. You have killed a policeman." The net effect of this exchange was that Lee Oswald now knew which crime had led to his entrapment. "Well," Oswald said quietly, "you can fry for that."

The policemen thought that their man was in a talking mood, and they decided to take advantage of it. They didn't know that he had already taken advantage of them. "What's your name?" one asked. "That," said Oswald, "is for you to find out." Another cop said: "You'll fry." The prisoner shrugged. "They say it only takes a second." "Here's his name: it's Lee Oswald." The sergeant said: "You Lee Oswald?" The prisoner had lost interest. "No," the policeman said, "I have another card here. Are you Alex Hidell? Hidd-ell, or High-dell?" There was no response. The sergeant asked Carroll: "Is this gun yours?" "No," the driver said, jerking his head toward the rear of the car. "It's his."

Walker and Bentley, in the rear seat, tilted Oswald this way and that to get to his pockets. In one they found a handful of .38 cartridges. Sergeant Hill opened the chamber of the gun in his hand and found that one bullet had a dent in the back. The concern of the five policemen was to drive the sedan into the basement at police headquarters, get this man Hidell or Oswald up to Captain Fritz's Homicide division on the third floor, deliver him intact, and book him on suspicion of homicide—to wit, the slaying of Police Officer J. D. Tippit.

Back on Jefferson, young Johnny Brewer suddenly remembered his shoe store, open and unprotected. As he skipped along the sidewalk, he wondered how long he had been away. The total time was eight minutes.

The shock was now universal. It was as real among those who disliked or disagreed with John F. Kennedy as among his friends. Dark of night had descended on Munich, Germany, when the flash arrived at 7:44 P.M. At once, Radio Free Europe beamed the tragedy to Poland, Czechoslovakia, Hungary, Romania, and Bulgaria in several languages. At Parkland Hospital, the sun was still high and hot as an irritated student held a sign aloft which said: "Yankee Go Home."

Secret Service agents drove the President's car back to Love Field, and Dallas citizens who saw the bubbletop flash by paused to stare unbelievingly. *It could not have happened. It did not happen. It would all go away.* At a fashionable school in Dallas, a teacher sat forlornly at her desk, head down, hand clasping forehead. She knew that she could not teach these bright little faces anything more today. Weakly she said: "You may have the rest of the day off." The next few words were drowned in a mass cheer: "The President of the United States has been shot!"

A. C. Johnson and his wife were in their little restaurant at 1029 Young Street when a policeman friend phoned and said, "The President's been shot." They had no radio, so they went out in the parking lot and sat in their car listening to all the excitement. A. C. and his wife worked hard. They had the little sandwich place and they had bought a small house over at 1026 Beckley and rented rooms. One of the roomers was Mr. O. H. Lee.

The demise of the President had an effect everywhere, but not the same effect. Jack Ruby announced in his *Dallas News* advertising that he was keeping his two small nightclubs closed. He could, in a breath, phone his sister Eva and weep emotionally over the President; in the next breath he would ask a friend casually: "Well, what do you think? Will it have any effect on business in Dallas?" Secretary of State Dean Rusk, coming in to Hickam Air Force Base with his diplomats, had difficulty trying to convince himself of two items: one was the fact that President John F. Kennedy was indeed dead; two, there was a possibility that this was part of a plot by a foreign power. Neither of these appeared to be true. Sorrowing, he ordered the plane refueled and flown directly to Washington.

In Fort Worth, Marguerite Oswald had finished her lonely lunch. The mother of Lee sat on a couch and stared at her talkative friend, the television set. A commentator said that the President had died in Parkland Memorial Hospital, and Mrs. Oswald thought that she would like to continue to watch, but she had the three-to-eleven shift at a rest home, and she liked to be there early.

The Secretary General of the United Nations made the solemn announcement in New York and asked the General Assembly to stand and observe one minute of silence. The American Ambassador, Adlai Stevenson, said, "We will bear the grief of his death until the day of ours."

At Fort Myers in Virginia, Captain Richard C. Cloy had his state funeral section pretending to lift an American flag from a casket, folding it properly, and tendering it in triangular form to the next of kin. For a week, Cloy's spit-and-polish group had been rehearsing a state funeral with caisson and three teams of horses. Until now, the rehearsals had been for ex-President Herbert Hoover, who was reputedly near death.* Within an hour of the tragedy in Dallas, Cloy's ramrod-straight outfit was undergoing serious funeral rehearsals, this time for Kennedy.

There was a basket of baby clothes and they had to be hung. Marina Oswald was in the back yard at 2515 West Fifth Street, working with clothespins in her mouth. Mrs. Paine came out and used a second clothesline for her wash. "Kennedy is dead," she said. There was no response from Marina. Mrs. Paine said that a television commentator said that the shots came from the School Book Depository on Elm Street. Ruth said she hadn't known they had a place on Elm; she thought that Lee worked in the warehouse. Mrs. Oswald said nothing; she had been out in the garage and was satisfied that the rifle her husband owned was still inside the blanket. "It was not my crazy one."

The district attorney of Dallas County hurried back to his office. Henry Wade was a big tough Texan who was never known to miss much except his office spittoon. He sat at his disorderly desk, thumbing through law books. His graying hair was damp as he studied precedents in the murder of federal officers. He was shocked to learn that, while it was a federal offense to threaten a President of the United States, it was not a federal

---

* Cloy's alleged statement, "We were rehearsing for the funeral a week," led to the ugly rumor that Defense Secretary McNamara had the army practicing for the burial of John F. Kennedy before the trip to Dallas.

offense to kill one. And yet, in the hope of ridding himself of all the complexities of the case, Wade phoned his friend, United States Attorney H. Barefoot Sanders, and said solemnly: "It's your baby. . . ."

The flash had been filed. Now the press demanded the story. At Parkland Hospital they asked Wayne Hawkes and Steve Landrigan to summon the doctors who had attended the President. All over the United States and Europe, editors were calling airlines, booking seats for reporters and photographers who were headed for Dallas. Phone calls to the hospital, to police headquarters, to the White House, to District Attorney Henry Wade, to Love Field, to the Texas School Book Depository were coming in from all over the world, tumbling in their urgency so that operators from Australia were asking operators from Berlin to please get off the line. A communications mob scene began at 1:40 P.M. Central Standard Time and it would last for fifty hours.

On the stage in the nurses' classroom stood Dr. Kemp Clark, Dr. Malcolm Perry, Dr. Charles Baxter, and Dr. McClelland. One hundred journalists with deadlines and no time for tact began firing questions. It was agreed that they would be answered by Dr. Clark and Dr. Perry. Neither had experience in these matters; neither had turned the body of the President over to examine it for wounds; neither had autopsy experience. The questions flashed from scores of strange faces; sometimes two or three were in air as the doctors tried to respond to one.

Clark and Perry at once began to sound poorly equipped. "Where did the bullets come from, doc?" "We don't know." "How many bullets?" "It is possible that there were one or two, or more." "Could it have been one bullet?" "Yes, it's possible." The doctors began to talk about a wound in the neck and a massive one in the back of the skull. The reporters wanted to know how they could accommodate the one bullet theory if there was a hole in the front of the neck and one in the back of the skull? Well, a bullet could possibly have been fired through the front of the neck, hit the spinal column in back, and deviated—or caromed upward—through the back of the skull. This would make the neck wound an entrance wound, which would mean that at least one assassin was in front of the President. If Governor Connally, as Bill Hinson had explained, had been shot in the back, the hypothesis lent itself to two assassins—at least.

Some of the writers began to ignore the word "possible." Television blinded the doctors with big white lights. Tape recorders were thrust before the team of physicians at the nursing desk. The faces beyond the light beams appeared to be bathed in talcum. The questions were being fired like Roman candles; Clark and Perry were trying to respond to one, or two, when questions three and four and five were coming at them.*

Outside of Trauma One, Roy Kellerman waited for the death certificate. He wanted to get the body on *Air Force One* as quickly as possible. Mrs. Kennedy was still inside, leaning her face, or her hand, against the side of the bronze casket. A man of professional appearance approached Agent Kellerman. He introduced himself as Doctor Earl Rose. "There has been a homicide here," he said. "You won't be able to remove the body. We will take it down to the mortuary for an autopsy."

Kellerman, tall and dark and often humorless, looked down at the officious man and said: "No, we are not." The doctor acted as though he had expected this attitude. "We have a law

---

* The doctors complained bitterly that they were misquoted or quoted out of context. In truth, they were incompetent to discuss the wounds of the President because they had not examined the body. They had no knowledge of the crime, the scene of it, the trajectory of projectiles, or whether wounds could be called entrance or exit. They could have drawn up a preliminary draft of treatment and given it to Dr. Burkley, the President's physician. Instead they chose to lend themselves to a press conference. As a result, Burkley could not discuss the President's wounds with competence at the Bethesda autopsy.

here," he said, "and you have to comply with it." At this moment, Admiral Burkley approached, and Kellerman said: "Doctor, this man is from some health unit in town. He tells me we can't remove this body." Burkley, who had been losing patience with Dallas, became enraged and shouted: "We are removing it!"

Dr. Rose tried to say something, and Burkley, his voice rising, said: "This is the President of the United States; you can waive your local laws." Rose kept shaking his head negatively as he listened. "This happened in Dallas County," he said. "We have our laws, just as you have yours. Under the law, an autopsy must be performed." Kellerman, who felt himself to be in a position of strength with all his Secret Service agents up and down the corridors, said quietly: "Doc, you are going to have to come up with something a little stronger than yourself to give me the law that this body can't be removed."

The shouting started. Kellerman did not know that Burkley had the signed death certificate in his pocket. The strident voices could be heard in the halls. Any passing personnel whom Earl Rose called to bear witness that, in a death by violence, an autopsy is mandatory, said: "Yes, that's the law." O'Donnell, O'Brien, and Powers heard the call to battle and they tried to tell Rose that, no matter what the law, Mrs. Kennedy was not going to wait one more minute to claim her dead husband. It was as simple as that. Kellerman kept saying: "I'm going to need somebody bigger than you . . ."

Rose said he would get someone "bigger." He began to call justices of the peace. In some areas of the United States, a person with that title is an elected official of consequence. In others, it is a small and appointive post with few prerogatives beyond the power to conduct a marriage ceremony. A justice of the peace, in Dallas, is a fully empowered judge. The men of the President's entourage did not know this. Earl Rose left the area for whatever reinforcements he could find. General McHugh joined the Kellerman forces at a momentary reduction in rank.

Bill Greer, the driver, was walking back and forth with two bags laden with the clothing of the dead President. O'Donnell nodded his head. "Let's get out of here," he said.

The car turned in off Main Street, braking down to a walk, and then dove into the narrow ramp under the gray eminence of Dallas Police Headquarters. Sergeant Hill turned in the front seat and told Oswald that there would probably be some reporters and photographers waiting in the basement. "We can hold you so that they can't get a picture, if you want. Also, you don't have to answer any of these guys if you don't want to."

There was no response from Oswald. He was leaning forward to ease the ache of the handcuffs behind his back. It also gave him his first good look at the basement of Police Headquarters. The policemen got out of the car in the southeast corner and formed a wedge around the prisoner. The sergeant saw local reporters and photographers and he knew that the word was out that this man was being taken in for questioning in the Tippit murder.

"You can keep your head down," the sergeant said. If Hill only knew how long this unhappy young man had been forced to keep his head down . . . "Why should I hide my face?" he said loudly. "I haven't done anything to be ashamed of." Hill nodded for his men to proceed and they formed a wedge and almost half-ran their man across the basement and up the step* into the jail office.

---

* The spot where Oswald, who had shot two strangers, would be shot by one on Sunday morning.

The Kennedy people had marshaled their forces but no one moved. Dr. Earl Rose had returned with a justice of the peace. "This is Theron Ward," he said. "He is a judge here in Dallas." Roy Kellerman looked at a small, thin person who could not have been over twenty-three years of age. "Your honor," he said meekly, "we're asking for a waiver here because—" Rose snapped: "He will tell you whether you can remove this body or not." "It doesn't make any difference," Kellerman said wearily, "we are going to move it. Judge, do you know who I am?"

The Secret Service agent drew his I.D. card and displayed it. The young man nodded. "Yes," he said. "Yes." The O'Briens and Greers and Powers began to edge up belligerently, but Roy Kellerman waved them away. He had decided on an appeal to this boy's patriotism. "There must be something in your thinking that we don't have to go through with this agony," he said. "The family doesn't have to go through this. We will take care of the matter when we get back to Washington." Judge Theron Ward was wavering. "I know who you are," he said sadly. "I can't help you out."

"You can't break the chain of evidence," Dr. Rose said with finality. Kellerman suggested that perhaps Dr. Rose would like to come along to Washington, watching the casket all the way to make certain that the chain of evidence was not broken. "There is nothing," shouted Rose, "that would allow me to do it under our law. The autopsy will be performed here." "All right," said Kellerman, waving his hands for the argument to subside. "All right." Out of the corner of his eye, he had seen the door of

Trauma One open, and Vernon Oneal and his two assistants were pushing the casket into the corridor.

The medical examiner saw it. He jumped ahead of it, standing in a doorway, shouting: "We can't release it! A violent death requires a postmortem!" The tragic figure of Mrs. Kennedy began to emerge from Trauma One, and Dave Powers pushed her back inside. The cruelty inflicted on her passed all bounds. The early-morning statement: "Campaigning can be such fun when you're President!" had degenerated into sudden explosions which had taken his life and crushed hers; it was succeeded by a nightmare ride and shrieking sirens, a dead husband on her lap as pedestrians waved and shouted, "Hi, Jackie!"; the fruitless, witless separation of husband and wife at the hospital, where men of science in white masks went through the motions of preserving a life already gone; where women who had lost nothing wept and she did not; the vigil of congealed blood; the high-speed mental motion picture of marriage and children and the towering climb to the top of the world of society where, in the vacuum of a clear blue sky, came the triple clap of thunder and the end of the world.

It was a repeating motion picture, full of prayer and priests and the strong brown head broken on a bed of roses; the surprised "O" of the mouth; the fixed, open eyes rocking with the speed of the car; the officious nurses with no good word; the running doctors silent in sneakers; the lifeblood in floor puddles; the cooling skin under the sheet; the brown brimming eyes staring at what was left of love—and now an angry shouting man who held his hand up like a policeman stopping the traffic of caskets to say: "We have a law . . ."

She went back into the room. The casket moved and stopped. Moved a foot more, stopped. A policeman wearing a helmet and a revolver now stood at Dr. Rose's side. The youngster, Judge Theron Ward, decided not to buck the local establishment. "It's just another homicide case," he said. The Secret Service men began to form in front of the casket and down the sides. The men who were not permitted to protect the President's body with theirs in life stood as close to his clay as they pleased. The showdown began to approach physical violence, backed by guns.

The policeman assumed his high-noon expression. "These people say you can't go," he said. "One side," said Larry O'Brien. Ken O'Donnell said: "We're leaving." The ultimatum had been rejected. Greer, swinging the bags of laundry, stood in front of the casket and stepped off to walk through the cop and medical examiner. Theron Ward phoned District Attorney Wade. Dr. Rose stood his ground. So did the policeman. It seemed as though they would be run over by a casket.

Kellerman ran back and beckoned David Powers to bring Mrs. Kennedy out. She saw the casket ahead, watched it break bluntly through the blockers, and trotted to put her hand on the gleaming metal. She was on the left side, almost running down the corridor, hair bobbing in one eye, the fingertips of the right hand in contact with the lid. In the nurses' station, Judge Theron Ward was stunned to hear the deep, heavy voice of District Attorney Wade state that he had no objection whatever to the removal of the President's body. None at all.

Mr. Oswald looked offended. He got off the elevator on the third floor of police headquarters flanked by detectives in pale Texas cowboy hats. Ahead of him, reporters and photographers ran a few feet to turn and shoot. Microphones were held under his nose. He shouted that he had done nothing wrong except to carry a gun in a movie house. "I want a lawyer," he said to a television camera. The press asked him his name and he permitted himself to be led, silently, into the Robbery and Homicide office, a third of the way down the middle corridor from the press room.

Almost all of the offices were glassy rabbit warrens, poorly designed and leading through an endless series of doors from one office to another. A policeman motioned for Oswald to sit. From the hall he could be seen through the half walls, staring at his shoes. Detective Richard Stovall, who had been taking affidavits from Texas School Book Depository employees, asked Oswald his name. "Lee Oswald," he said, without looking up. A moment later, Stovall's partner, Guy Rose, asked Oswald his name. "Alex Hidell," the prisoner said. The detectives began to go through the billfold, saw the two names, and asked which one was bona fide. "You find out," he said. They asked his address. Oswald said: "You just find out."

Sergeant Jerry Hill stood guarding his man. Reporters in the hall asked to see the pistol which allegedly killed Officer Tippit. Hill held it up by the butt as photographic flashbulbs lit the scene. He asked Detective T. L. Baker if he wanted to take the gun; Baker said: "No. Hold onto it until later." Hill said that this was the suspect in the Tippit shooting and did Baker want Hill to make up the arrest sheet or would the Detective Bureau do it.

Captain Will Fritz walked in and said to Rose and Stovall: "Get a search warrant and go out to 2515 Fifth Street in Irving. Pick up a man named Lee Oswald." Sergeant Hill said: "Captain, why do you want him?" Fritz was a man of lean words and few interruptions. He stared at the sergeant through his glasses and said: "He's employed at the Book Depository and he was missing from a roll call of employees." Hill pointed to the prisoner: "We can save you a trip," he said. "There he sits."

Fritz looked at the quarry and was unimpressed. It required a moment to comprehend that this young fellow was a possibility as the man responsible for Tippit's murder and the President's, too. This was going to be an interesting afternoon. In another office Charles Givens, a Book Depository employee, was staring through the walls of glass and said: "Hey, there's Lee Oswald." A friendly policeman leaned out in the hall where reporters were waiting and said: "He's Lee Oswald, a suspect in the Tippit murder."

There were venetian blinds in the office of Captain Fritz. They were lowered. The big man glanced at the messages and reports on his desk. There was nothing that couldn't wait. He asked for the prisoner to be brought in. He ordered Detectives R. M. Sims and E. L. Boyd to remain with him until the Oswald matter was cleared up. Oswald came in and complained about the handcuffs. "Fix them in front of him," Captain Fritz said. Oswald sat on a chair at the corner of the desk, the shackled hands now on top.

Someone was drawing up a list of items found in the prisoner's pockets, so Captain Fritz took his time with the young man. When he got the list, his eyes ran down it slowly:

1. Membership card of the Fair Play for Cuba Committee, New Orleans, Louisiana, in the name of L. H. Oswald, issued June 15th, 1963, signed A. J. Hidell, chapter president.

2. Membership card of the Fair Play for Cuba Committee, 799 Broadway, New York 3, New York. Oregon 4–8295, in the name of Lee H. Oswald, issued May 28th, 1963, signed V. T. Lee, executive secretary.

3. Front and back of Certificate of Service, Armed Forces of the United States Marine Corps in name of Lee Harvey Oswald, 1653230.

4. Front and back of Department of Defense identification card #4,271,617 in the name of Lee H. Oswald, reflecting service status as MCR/inact, service 1653230, bearing photograph of Lee Harvey Oswald and signed LEE H. OSWALD, expiration date December 7, 1962.

5. Front and back of Dallas Public Library identification card in the name of Lee Harvey Oswald, 602 Elsbeth, Dallas.

6. Snapshot of Lee Harvey Oswald in Marine uniform.

7. Snapshot of small baby in white cap.

8. Social Security card #433-54-3937 in name of Lee Harvey Oswald.

9. U.S. Forces, Japan, identification card in name of Lee H. Oswald.

10. Photograph marked Mrs. Lee Harvey Oswald.

11. Street map of Dallas, compliments of Ga-Jo Enkanko Hotel.

12. Selective Service System card in name of Alek James Hidell which bears photograph of Lee Harvey Oswald and signature "Alex J. Hidell."

13. Certificate of Service, U.S. Marine Corps, in name of Alex Hidell.

14. Selective Service System notice of classification, in name of Lee Harvey Oswald, SSN 41-114-532, dated Feb. 2, 1960.

15. Selective Service registration certificate in name of Lee Harvey Oswald, bearing signature, Lee Harvey Oswald, Oct. 18, 1939.

16. Slip of paper marked Embassy USSR, 1609 Decatur St., N.W. Washington, D.C., Consular Pezhuyehko.

17. Slip of paper marked The Worker, 23 W. 26th Street, New York 10, New York.; The Worker, Box 28 Madison Square Station, New York 10, N.Y.

18. Snub-nosed Smith and Wesson Revolver, .38. Chambers loaded.

19. Two types ammunition, 6 cartridges, .38.

Fritz, like some policemen, can detect a complicated matter from a distance. The nineteen items added up to two persons who were possibly one, with overtones of communism or a communistic plot, in addition to evidence which pointed to this man as a possible double murderer. The captain did not know how smart the fellow was; he would know in a moment. He left his desk and asked a policeman to get a superior officer to start collecting eyewitnesses to the Tippit murder in headquarters. If any two of them could pick this man out in a lineup, it would be enough to hold him on suspicion of homicide and give everybody time to start unraveling the assassination from the murder.

At a door, a policeman whispered to Fritz: "I hear this Oswald has a furnished room on Beckley." Fritz blinked his owlish eyes and went into the room. He had noticed that the hall outside was beginning to look crowded with reporters and photographers. Policemen moving from one office to another had to step gingerly over long black cables which snaked the length of the hall. Technicians were bringing a heavy-duty cable up from a generator down in the street. It was being hauled up the outside wall.

The situation in Dallas began, at this point, to go out of control. Press relations was not a function of Captain Will Fritz, but he was not blind to the increasing number of reporters and photographers choking the hall of the third floor. The police department had a press relations man, Captain Glen D. King. He was a personable officer who realized that, while he had the law on his side, the press had the last loud word on theirs. His office adjoined Chief Curry's, and King was responsible to Curry. It was a pleasant situation, so long as the captain worked for a man who was afraid of the press and what it could do to him. King could go to Fritz or Jack Revill, the newly appointed lieutenant in charge of police intelligence, and get the latest information on any story and release what he thought was proper and discreet. Fritz remained aloof from it. He ran his Homicide and Robbery Division as though it was an entity unto itself.

Fritz decided that he did not want a stenographer to take notes while he questioned this prisoner. Nor did he desire a tape recorder. The first order of business would be to find out what kind of a fish he had in the net, who he was, where from, how clever and, most of all, how tough. There was a time, not too long ago, when the first order of business would have been for a couple of young detectives to rattle the prisoner around a locker room to still the stubbornness and develop some cooperation. Those days were past. Fritz had to match wits with his prisoners, frequently a tiresome game of listening to obvious lies hours on end.

He sat with a sigh. "All right, son," he said. "What's your name?"

The crew of *Air Force One* monitored the Secret Service. They had heard the walkie-talkies discuss an unannounced departure from Parkland Hospital. Colonel Swindal ordered the crew to unscrew two double seats adjacent to the rear entrance of the plane and to secure them in the tail. The box containing the President's remains could be carried up there and lashed to eyebolts in the floor. He called the tower and asked for taxi instructions. They gave him wind, barometric setting, clearance on Runway 31, and a handoff to Fort Worth FAA after passing the outer perimeter.

He didn't start an engine. A Secret Service agent came forward and told the crew that the Johnson party was already aboard and the Vice-President was waiting for Mrs. Kennedy. A generator truck stood below, whining at the nose of the air giant, but the crew engineer did not turn the air conditioning on. The captain of *Air Force Two* was told to open his hatches. Some luggage came from his plane and was carried by hand to *Air Force One*. Whoever was in charge was confused, because he began to take some of the Kennedy luggage from *Air Force One* and place it on *Two*. The pilot of the press plane asked for instructions and was told that no one knew the plans of the press. He was to remain on the stand for instructions.

Swindal was told by the tower that a small plane was en route from Austin with a "V.I.P." for Johnson. Could the private plane have clearance to pull up alongside *AF-1*? Swindal asked who was on the little plane. No one seemed to know. Then the plane, still en route, caught the question and said: "Bill Moyers." Permission was granted. Moyers was a young assistant to Johnson, an ordained Baptist minister who was not yet thirty years old and was associate director of the Peace Corps.

Richard Johnson, of the Secret Service, watched the casket roll up the hospital corridor like a box of laundry. He was approached by O. P. Wright, a hospital policeman, who handed an expended bullet to him. "The Secret Service may want this," said Mr. Wright. Johnson rolled it around in the palm of his hand. "Where'd you get it?" he said. Wright explained that there were some carts, used ones, standing between the restroom and the elevator. This one had rolled off a cart. "It may have come from the President's cart." It couldn't. Mr. Kennedy's cart remained in Trauma One until after his body had been prepared for the trip. The body had been lifted from that table to Oneal's funeral carriage. "I also found rubber gloves and a stethoscope on that cart." the cop said.

Johnson slipped the bullet into his pocket. He had been told to get on the follow-up car to the airport. There was no time for further conversation. He thanked the man and started out. Back in Trauma One, Miss Bowron studied the antiseptic brightness of the room—ready now for another effort to save another life—and she reached into her pocket. She had the President's watch. She ran through the corridors, holding it ahead, the gold band dull with dry blood, and gave it to a member of the party. At the same time, Doris Nelson found a blue coat. In it was an envelope marked "cash" and a card labeled "Clint Hill." They, too, were returned. It was the jacket Clint Hill had tossed over the dead President's face.

On the National Broadcasting Company network, Tom Whalen intoned: "The weapon which was used to kill the President and which wounded Governor Connally has been found in the Texas School Depository on the sixth floor—a British .303 rifle with a telescopic sight. Three empty cartridge cases were found beside the weapon. It appeared that whoever had occupied this sniper's nest had been there for some time."

The casket was moving fast now, except for the sharp turns leading back to the emergency entrance of the hospital. Vernon Oneal was pulling; two of his men pushed from behind. Jack Price, administrator of the hospital, ran ahead asking everyone to please step aside. Everyone did, except the priest, who had arrived late. He held up a hand to stop the casket and suggested that he say some prayers over it. "Not now!" an agent yelled hoarsely, and the body kept moving. Price touched the top of it. He didn't know why, but he had to do it. He told himself it was a sort of final salute.

An attendant held the door open and the gleaming casket emerged into the sunshine. A few uniformed nurses stood outside the door, under the eaves, weeping. A doctor at a third-floor window glanced once and turned his back to the scene. Tom Wicker of *The New York Times* stood near some parked cars watching. Larry O'Brien and Kenny O'Donnell, hanging onto silver handles, had their heads down. They were weeping. Some of the White House staff stood near the back of the white Oneal ambulance, mouths and eyes agape, not believing.

Mrs. Kennedy came out into the sunshine, a portrait of despair. The hat was gone. The pink suit was splashed with blood. The stockings, askew, stuck to the legs with dry blood and brain matter. The white gloves were darkened with deep stains. The face, the immaculate face, was almost wild-eyed, whether with fear that they might take him away again or with the crashing waves of reality which come in steady tandem to all who grieve or whether the emotions were cracking under the repeated cruelties—no one knows. The doctors had offered her sedation several times; one even offered to help her clean the blood from her person. She preferred to taste death at the side of her husband, and, at 2:05 P.M., her knees were beginning to knock without control, her fingers trembled, her brain might not endure one more brutality.

Nor was O'Donnell certain that they were going to be able to steal the body of the President of the United States from officious Dallas. He urged Vernon Oneal to hurry. The mortician asked the Secret Service if they were going to the mortuary. They said yes, yes. Suddenly, the flight from Parkland Hospital became more precipitous than that of Lyndon Johnson. Men were running for cars; motors were starting; police were trying to line up as escorts and were told to "get the hell out of the way."

No one paused to reason. Roy Kellerman ordered Agent Andy Berger to get behind the wheel of the ambulance and drive to Love Field. Mr. Oneal wanted to know why he was not permitted to drive his own hearse, and he was told to stand aside. Kellerman tried to get Mrs. Kennedy in beside the driver, but she insisted that she would sit in the back "beside my husband." Doctor Burkley got in the back and helped her up and in. The third person, Clint Hill, got inside and slammed the back doors.

Agent Stewart Stout got in front. Roy Kellerman ran around the ambulance to make certain that the right people were inside. Then he told Kenneth O'Donnell and Lawrence O'Brien to take the next car and, privately, head for Love Field and *Air Force One*. Audrey Riker, who worked for Oneal, ran up to the driver's side of the ambulance and said: "Meet you at the mortuary," and Berger nodded: "Yes, sir."

The ambulance left the parking area fast. It moved across the service road to Harry Hines Boulevard and Berger hit the siren to clear the traffic. Roy Kellerman was on the radio, telling the agents at Love Field to permit an ambulance and one following car through the fence. After that, shut it to everyone and seal it off. In the back, Clint Hill watched behind and saw Oneal and his assistants make the turn—the wrong way to get to the mortuary. They were following, but they were behind O'Brien and O'Donnell and the agents in the second car.

O'Donnell was radioing the same instructions from the second car. Let the ambulance and one other through the fence. Then lock the place up. Tell Colonel Swindal to get ready for takeoff at once. The party would be aboard in ten minutes.

The sun used *Air Force One* as an aluminum oven. It was unbearably hot inside, and yet people were running toward the front or the back with imperative instructions from someone else. Both entrances were sealed tight. President Johnson received a return phone call from Assistant Attorney General Nicholas Katzenbach with the precise wording of the oath of office. Johnson asked a secretary, Miss Marie Fehmer, to please take it down and type it. Mr. Johnson looked at the television set in time to hear a commentator say that the Dallas Police Department had just arrested a suspect in the assassination.

The President, on another phone, thought of Federal Judge Sarah Hughes, a Kennedy appointee. The communications people required a couple of minutes to find out her local office number. Then Johnson called, but her office said she was out. He asked that she be found at once and to call him through the White House number in Dallas. He paced the little bedroom, thought of other phone calls, and made them. Mrs. Johnson understood this reaction better than anyone else. Her husband was a man of action; inaction could kill him, but not work.

The day was so horrifying, so beyond belief, that she had to keep reminding herself that it had really happened; then when the reality crushed her, she tried to think of other things. Sometimes, she was seen with a fixed half smile on her face as though people were watching and she had to put up a front. She had been reassured that her daughters were now under the protection of the Secret Service; she wondered what they were thinking. She thought of the two little Kennedy children, and it was a thought impossible to sustain.

Judge Sarah Hughes phoned, and Johnson briefly explained the tragedy and asked if she could come right out to Love Field; he would send Secret Service agents to escort her. No, the judge said, she knew quicker ways of getting to the airport than the White House detail, and she would be there in ten minutes. That would bring her to *Air Force One* by 2:20. The President said to please hurry, that they desired to take off for Washington. He hung up and told Agent Youngblood to radio clearance for the judge. She would be along in a few minutes. "Check on the location of Mrs. Kennedy, too," Johnson said. "Let me know when she will arrive."

Fritz could hear the noise from the corridor, and he asked two of his detectives to tell the newsmen to hold the noise down.

An agent of the Federal Bureau of Investigation, James Book-hout, phoned his office in Dallas and said that a suspect in the Tippit killing had been picked up. His name was Lee Harvey Oswald. In the office, Agent James Hosty was still running through the old files, trying to dig up anyone who might be a suspect in the assassination, and, when he heard the name Oswald, he felt the chill that comes to all law officers who find themselves on the wrong end of a gigantic surprise.

At once Hosty reported to his boss, Gordon Shanklin, that he had been handling the case of a Lee Harvey Oswald, that he was a defector who had fled to Russia, returned to Texas with a Russian bride, left Texas for his native home in New Orleans, fled to Mexico recently in an attempt to get to Cuba, been turned down, and come home to Irving, Texas. Shanklin demanded a quick rundown, and Hosty said that Oswald could not, on his performance chart, be regarded as a potential cop killer. He wasn't even a member of the local Communist Party. Hosty knew, because the FBI had a man in it who kept him well-informed.

Shanklin got on the phone. He spoke to Will Fritz. Again he offered the services and facilities of the FBI—in Dallas and in Washington—to the Dallas Police Department. He also asked if he could send Agent Jim Hosty as the bureau representative to listen in on the interrogation of the prisoner. The FBI knew a lit-tle about Oswald. The police captain said to send Hosty on over.

The disparate work of the law enforcement agencies began to spread and dissolve, then congeal, only to dissolve again. Three detectives were en route to Oswald's home in Irving. Stovall phoned ahead to the Irving police and asked to have some local officers meet them at the city line. Their assignment was to find out who lived with the suspect and to search the premises and take with them anything which might be regarded as evidence.

At the Trade Mart, there were several anti-Kennedy pickets who had not heard the news. Lieutenant Jack Revill and his Intelligence unit arrested them to protect them from possible mob violence. At the Texas School Book Depository, policemen continued to work looking for an assassin. V. J. Brian and his officers had found some acoustical tile in the ceiling on the second floor. They were ripping it out because someone had suggested that an assassin could be hiding in the space above it.

Lee Harvey Oswald gave Captain Fritz his right name. The prisoner had a lump and a laceration over his right eye and another underneath the left eye. The captain sat at his desk, rolling a pen back and forth across the blotter, looking up at his man now and then. How, he asked, were the bruises acquired? Oswald said that he had punched a police officer and the cop had punched him back—"which was right and proper."

"Do you work for the Texas School Depository?" Fritz asked. He had a deep, deliberate tone, the manner of a man who is never in a hurry. An accent touched by the South and by the West. "Yes," said Oswald, and he too knew that now the forces had joined battle. It was important for him to know when to answer, when to lie, when to evade, when to lapse into sullen silence. He had to know these things, because it was a whole police department pitted against one man. He might tire; they wouldn't. Nor was he lulled by the soft, easy manner of the captain. That had its own built-in danger. Oswald's greatest asset was that he enjoyed this game; he knew that the innocuous questions could be answered glibly, as though he were an innocent person trying hard to cooperate; the difficult ones could be blocked by a display of anger or impenetrable silence.

"What floor do you work on?" "The second, usually, but my work takes me to all floors." "Where would you say you were when the President was shot?" "I was on the first, having my lunch." "Where were you when the police officer stopped you?" "On the second floor, having a Coke." "Tell me, why did you leave the building?" Oswald permitted himself a little smile. "There was so much excitement all around, I figured that there would be no more work. Mr. Truly isn't particular about the hours; we don't even punch a clock. I thought it would be all right to leave."

"Do you own a rifle?" "No, sir." "You don't own one?" "I saw one at the building a few days ago. Mr. Truly and some of the fellows were looking at it." "Where did you go when you left work?" "I have a room over on North Beckley." "Where on Beckley?" "1026." "North or South?" "North or South?" "Yes." "I couldn't say, but it is 1026 Beckley." "What's the area look like?" "Oh, a couple of streets come together at that point, and there is a filling station across from the boarding house—." The captain nodded. "That's North Beckley."

Fritz excused himself. He went out into the hall and told a couple of detectives to run over to 1026 North Beckley and search a room rented by Lee Harvey Oswald. "All I did," Oswald was saying, "was go over to Beckley and change my pants. I got my pistol and went to the pictures. That's all." He knew the next question, and he was prepared. "Why do you find it necessary to carry a gun?" The prisoner waved his manacles in an explanatory gesture. "You know how boys do when they have a gun." He shrugged. "They just carry it." In Texas, this is good rationale. Young men in large numbers carry pistols not to use them, but to establish manhood, perhaps even virility.

"Can I get a lawyer?" The question was not unusual nor unexpected. The captain nodded. "You can call one anytime you want," he said. Oswald said the one he wanted was in New York. His name was John Abt. Fritz shrugged. "You can have anyone you want." Oswald seemed to be calmer. "I don't know him personally," he said, "but that's the attorney I want." He remembered a case Mr. Abt had handled involving the Smith Act. "If I can't get him," he said, "then I may get the American Civil Liberties Union to get me an attorney."

Fritz said that Oswald would find a phone in the jail on the fifth floor. He would have to make the call himself, at his expense. The captain asked him why he lived on Beckley and his wife lived in Irving. Oswald appeared to be faintly amused, as though, a man among men, the incongruities of women were beyond understanding. He explained that Marina was living with a friend, Mrs. Ruth Paine. Marina was teaching Mrs. Paine to speak Russian, and in return Mrs. Paine gave Marina and the babies a room. Oswald suggested that it worked out all right for all parties; he stayed in town, except on weekends, because he worked in town.

The men in the room—Fritz, Detectives Boyd and Sims, FBI Agent Bookhout—blinked at him. "Why don't you stay out there with them?" Fritz said. "Well," Oswald said, the Paines didn't get along well. They were separated a lot of the time. "Don't you have a car?" "No, I don't. The Paines have two cars, but they don't use them."

"What kind of politics you believe in?" Fritz said. Oswald required no time for a response. "I don't have any," he said. "I am a member of the Fair Play for Cuba Committee. They have offices in New York, but I was a secretary of the New Orleans chapter when I lived there." Fritz leaned back in his broad straight chair. "I support the Castro revolution," Oswald said, without being asked. "Do you belong to the Communist Party?" "No, I never had a card. I belong to the American Civil Liberties Union and I paid five dollars dues."

"Why did you carry that pistol into the show?" "I told you why. I don't want to talk about it anymore. I bought it several months ago in Forth Worth and that's all." "You answer pretty quick. Ever been questioned before?" In a trice, the cipher felt himself become an intelligent man to be reckoned with. "Oh yes. I've been questioned by the FBI"—he nodded toward Bookhout—"for a long time. They use different methods. There is the hard way, the soft way, the buddy method—I'm familiar with all of them. Right now, I don't have to answer any questions until I speak to my attorney."

"Oswald, you can have one any time you want." "I don't have the money to call Mr. Abt." "Call him collect or you can have another lawyer if you want. You can arrange it upstairs." "Thanks." "Ever been arrested before?" Oswald nodded. "I was in a little trouble with the Fair Play for Cuba thing in New Orleans. I had a street fight with some anti-Castro Cubans. We had a debate on a New Orleans radio station."

"What do you think of President Kennedy and his family?" "I like the President's family very well. I have my own views about national politics." "How about clearing this thing up with a polygraph test?" "Oh, no. I turned one down with the FBI and I certainly won't take one now." The captain realized that this was not an easy subject. This man, whatever he was, communist, nut, socialist, malcontent, Marxist, was about to show off his knowledge of how to fence with the law without getting hurt. Fritz swung around to face the FBI man. "Any questions?" he said.

The men in Love Field tower had a respite. The empty runways, with one long diagonal, looked like a crooked capital H. *Air Force One* had twice asked for taxi instructions, but the generator was still standing under its nose, breathing power into the bird. There was time to stand up and stretch. The men walked around the glass enclosure, commenting on the Dallas catastrophe and listening to the local gossip. There was a story that a Secret Service man had been killed with President Kennedy. No one had heard about it officially, but the story went that the government was keeping it quiet and had carried the body away secretly because the agent was part of the plot to kill the President.

"Look," one of the tower men said. An ambulance with red blinker showing was coming off Mockingbird Lane into the airport. It was followed by two cars, all at high speed. Two Dallas officers and some Secret Service men ran to the fence and watched the small motorcade return John F. Kennedy to the place where he had shaken many hands. The ambulance made the turn through the fence. So did the second car. The third was stopped by the bodies of the lawmen. Vernon Oneal, middle-aged and proud, got out to protest.

That was his ambulance inside the fence. It was supposed to lead him to his own mortuary at 3206 Oaklawn. Something wrong was going on because the driver had come to the airport. It was his ambulance and he had a right to be inside the fence. The President was in an Oneal casket.* The Secret Service men glanced at him and walked away. They left him to the mercies of the local police. Mr. Oneal was told firmly that he could not get inside the fence. In time the ambulance would be returned. He could wait if he pleased.

The rear ramp of *AF-1* was opened briefly and a host of Secret Service men performed a final service for a dead chieftain. They carried the bronze casket up. It weighed 400 pounds. The body weighed 180. The men staggered and stepped forward and tilted the big box and, halfway up, appeared about to drop it. Two crewmen came down the steps and tried to wedge themselves along the sides. At last it got to the top, and a group of Dallas citizens stood behind the fence, unable to contain the tears. Clint Hill saw a photographer, "I'll get him" he said to Mrs. Kennedy. "No," she said. "I want them to see what they have done."

The door slammed shut. The casket was dragged across the floor. O'Brien noticed that a space had been made for the casket. He told the agents to secure it on the left side of the plane barely inside the rear door. Mrs. Kennedy dropped into a seat at the breakfast nook opposite. She appeared to be spent.

_____

* The cost of the casket was $3900. Oneal sent bills to Mrs. Kennedy for a year. He says that the family never paid for it. A government agency got in touch with him fourteen months later and said it would give him $3400—no more. He accepted. The check came from the government. Because he demanded proper payment, his business in Dallas fell off 50 percent.

The woman slumped as though lifeless. Kenneth O'Donnell motioned for the rear door to be secured and guarded. He requested that the ramp be pulled away.

At the moment, he was scared. O'Donnell was certain that official Dallas would protest the kidnapping of the President's body. If they rammed through an order forbidding *AF-1* to take off, the authorities could besiege the plane in their zeal to adhere to a local law. They could show up any moment in force and demand an autopsy. The President's trusted assistant and friend was determined that this was not going to happen. The only way to forestall it, he was sure, was to get this plane the hell out of Dallas before anyone realized what had happened.

He looked up, as the crew was tossing bracing straps over the casket, and saw General Godfrey McHugh. "Run forward and tell Colonel Swindal to get the plane out of here," he said. McHugh went through the corridor as fast as he could. Coming up the front ramp were two associates of Johnson: Jack Valenti, a Texas press relations expert, and Bill D. Moyers, the bespectacled preacher from the small plane. General McHugh passed everyone without a glance. He told Colonel Swindal to take off for Washington at once. "The President," he said, "is aboard."

The passengers were growing in number. There was no passenger manifest. Some, like Liz Carpenter, secretary to Mrs. Johnson, reported to *AF-2* and were told that the Johnsons were now on *AF-1*. The Kennedy people were aboard because this was the aircraft they had arrived on. Malcolm Kilduff, standing at the foot of the front ramp waiting for the newspaper pool car, was astonished to hear that President Johnson wanted to speak to him at once. The assistant press secretary did not know that Johnson was on this plane.

Few others knew about it. Larry O'Brien, still crouching over the casket, looked up to see the President and Mrs. Johnson coming down the aisle from the private stateroom. He was flabbergasted. The man was President. This was *Air Force One*.

He saw the Johnsons move silently over to the breakfast nook. Mrs. Kennedy looked up and emerged from her reverie. There can be no doubt that she was surprised to see them aboard this aircraft. It is understandable if she felt resentful, because the trip home to Washington would normally be a "wake," a private mourning.

It might even seem, to her shocked gaze, that the Johnsons were "taking over" abruptly. Until a few hours ago, they would not even be invited aboard *AF-1* because security dictated that the President and the Vice-President must fly on different planes. Normally, she was accustomed to seeing them perpetually at the foot of the ramp, welcoming the President and the First Lady of the land to each city. Sometimes there were three or four such welcomes in a day.

There were no welcomes now. The Johnsons were trying to find the proper words for a grief they felt but could not enunciate. The words came out soft and reassuring, but they were empty vessels. Mrs. Kennedy took Mrs. Johnson's hand in hers. "Oh, Lady Bird," she said. "It's good that we've always liked you two so much." It was a *non sequitur*, but this was a time for feelings, not the analysis of sentiment. Mrs. Johnson began to weep again. "Oh," said Mrs. Kennedy, "what if I had not been there? I'm so glad I was there."

The President stood big and helpless. Like many men, he quailed in the face of grief and could not cry. Mrs. Johnson kept thinking, in horror: "This immaculate woman . . . this immaculate woman caked with blood, her husband's blood . . . That right glove is caked. . . ." She suggested that she get someone to help her change. The iron returned to Mrs. Kennedy. "Oh, no," she said. "Perhaps later I'll ask Mary Gallagher. But not right now." She was determined to appear in civilized Washington in these clothes and show the world what Dallas had done. Her clothing, her forlorn expression, were more eloquent than the bronze casket.

"Oh, Mrs. Kennedy," said Lady Bird. "You know we never even wanted to be Vice-President and now, dear God, it's come to this." Mrs. Kennedy nodded. She was aware of the big fight in Texas between her husband and Lyndon Johnson—a dirty party fight in which both sides impugned the motives of the other—and of how, when Kennedy won his party's nomination for the presidency, Johnson didn't want the second position, even when Kennedy offered it. At this moment, the man who didn't want the vice-presidency ("Why should I trade a Senate vote for a position with no vote?") had backed into the leadership of his country because a nobody with a rifle saw an opportunity to become a somebody.

The disparity between the Kennedys and Johnsons was apparent to both. The Kennedys were effete Europeans, in manner and address; the Johnsons were earthy Americans. It was not a detriment to either family to be what it aspired to be, to nourish its own style of living and its culture. The subtle bon mot was an effervescent joy to John F. Kennedy and his Jacqueline; it was lost on the Johnsons. The beauty of the hill country of Texas was lost to the Kennedys; to the Johnsons, a frame farmhouse, hard furniture, and cattle silhouetted against a sunset were matters which brought serenity to the heart.

The latest book, the newest song, the gossip of high society, the galas at the watering places were daily food and drink to the Kennedys. To the contrary, it was said of Lyndon Johnson that he could ruin a good suit of clothes merely by putting it on; his humor was a rough Texas guffaw and his wife enjoyed buying dresses from a shop rack. Johnson was closer to the work-hard-and-fight-'em-all philosophy of old Joseph P. Kennedy than John Kennedy was. Mrs. Kennedy enjoyed her lack of knowledge of politics; Mrs. Johnson worked full-time as her husband's assistant from the time he left Texas to take a seat in the House of Representatives. Lady Bird also found time to take her inheritance and build it into a television and ranch fortune.

309

The meeting in the back of the plane was awkward for both sides. Suddenly the simple, blunt people were running the United States of America. The adroit, the charming, the sophisticated Kennedys were out. A single blow had reversed the roles, and no one was prepared for it. No one said: "Now the Kennedys must move out of the White House and the Johnsons will move in," but the shock wave of probabilities moved through *Air Force One* as the passengers sat in gloomy meditation. When General McHugh said: "The President is aboard," he assumed that there was only one. Many of the passengers could not acknowledge Johnson's supremacy, even to themselves.

The President and the First Lady retired from the aft compartment, and Johnson went into the private bedroom to make certain that Marie Fehmer had the oath typewritten correctly. He was barely in the chair when the door opened, and Mrs. Kennedy was in the doorway. She looked as though this was the final humiliation. The President jumped to his feet, asked Miss Fehmer to leave, and apologized to Mrs. Kennedy. He said he had been checking something—"there's a little privacy here"—and was leaving at once. He got out, and went into the main stateroom, the area of desks and couches and television sets, and Mrs. Kennedy disappeared into the lavatory.

The pool car arrived and Malcolm Kilduff waved to the men to get aboard at once: Merriman Smith of UPI and Charles Roberts of *Newsweek* magazine. A third man—Sid David of Westinghouse Broadcasting Company—was told that he could cover the swearing in, but could not return to Washington on *AF-1* because there wasn't sufficient room.

In the President's cabin, Johnson's intimates sat watching television. They understood little of what had happened, and, isolated in the plane, the only avenue of information was television or radio. David Brinkley, in Washington, was speaking:

"Senator Edward Kennedy of Massachusetts and Eunice

Kennedy Shriver arrived at the White House a few minutes ago to go to Andrews Air Force Base—perhaps to fly to Dallas. Robert Kennedy will fly to Texas. Congress has recessed, and several members of Congress have given their reaction to the President's death. Senator Mike Mansfield of Montana is 'shocked.' Senator Alan Bible of Nevada calls it 'one of the great tragedies of our lifetime.' Senator Harry Byrd of Virginia is 'deeply shocked.' Similar sentiments are being expressed by all members of Congress."

On the other side of the airport, Mayor Earle Cabell had a complement of Texas congressmen who desired to return to Washington. In the distance, the brilliant blue and white of *AF-1* appeared to be standing dead in the sun, but the mayor was afraid that it would take off at any moment, so he asked the nearby Southwest Airmotive Company to please contact the tower and ask permission to drive across the runways to the President's plane. The tower sent a police car to escort the distinguished men, who got aboard before Judge Sarah Hughes arrived.

Johnson knew that this inauguration would go down in history as one of the most somber. His impulse was to have it done quickly and secretly and to bring his dead predecessor back to Washington. He sent Youngblood for Kilduff. "Do we have to have the press in here?" he said. Kilduff had no doubt. "Yes, Mr. President. There should be press representation. Also Captain Stoughton should be here to make pictures of the scene." The President rubbed his big hand down the front of his face. "All right," he said. "O.K., Mac. If we must have them, then we might as well invite the other people to come in and witness the ceremony."

The President summoned O'Donnell and O'Brien. The Roman consuls left Caesar on his shield and sat with Johnson, listening. The new President admired these men. He wasn't certain that they were superior to his own team, but he knew that Ken and Larry had spent almost three years at the font of power in the White House, and they had an intimate working knowledge of the executive branch of government which he lacked. Johnson asked both men to remain in government. "I need you more than he did," he said, jabbing his finger toward the back of the plane. Both men glanced at each other and said they would give it some thought.

Of the two, O'Donnell was the more unyielding. For him, it was Kennedy or nobody. He drew a grim joy from his intense personal loyalty to Kennedy. It was not a transferable commodity. He'd think about it. So would O'Brien, but O'Brien was an intelligent redhead who, in spite of his deep affection for Kennedy, could envision a world without him. He would like to continue in government, especially in the daily political herding of votes in Congress and the brawling atmosphere of the quadrennial campaign for the White House. Yes, he would give it some thought.

O'Donnell didn't want to discuss this thing. His concern was to get this airplane out of Texas, and he hunched forward, listening to the President, hoping every moment he would hear the plane move off the blocks to waddle to the head of the runway. He was still convinced, even in the sanctity of United States government property, that official Dallas would be pounding at the doors any moment. He asked the President about taking off.

"I talked to Bobby," Johnson said. "They think I should be sworn in right here. Judge Hughes is on her way—should be here any minute." O'Donnell gave up all desire to burden the President with his personal problem. They sat watching Cecil Stoughton, the White House photographer, try to line up his cameras in a corner of the stateroom. "I would like you fellows to stay, to stand shoulder to shoulder with me," Johnson said. The Kennedy assistants did not commit themselves. They watched the photographer without seeing him. The loyalty of the OOP group—O'Donnell, O'Brien, and Powers—was innocent of patriotism; it was personal fealty to a man. The man was gone. They had no leader, no direction, no future.

A. C. Johnson came up North Beckley, driving carefully. He made his swing into the driveway of his rooming house. His wife noticed the strange cars in front. It didn't require much time to pinpoint the excitement. The Dallas detectives and two government men were in the living room, going over the roster of roomers with Earlene Roberts. The stout housekeeper said, "They're looking for a Lee Oswald, but we don't have anybody by that name."

Mr. Johnson, a tall lean man who earned his dollars the hard way at the little restaurant and in the rooming house, asked the men who they were. They identified themselves and said that they were looking for a man named Lee Harvey Oswald. Johnson shook his head. His wife said she usually remembered the names, but they never had an Oswald. One of the policemen said: "We came out without a search warrant." In a moment another one was phoning the nearest substation at 4020 West Illinois to get a warrant.

Earlene went back to the television set in the living room. She knew that there was no Oswald. The cops said he was young—under thirty—slender and brown-haired. Sometimes the Johnsons had as many as seventeen roomers, and some of the transients had rooms in the cellar. A policeman said he would like to go down there and look around. All of them went to the back of the house.

"Let 'em look," said Mr. Johnson. He sat on a couch watching the fascinating story of the shooting of a President. Earlene Roberts thought it was terrible and kept saying: "Oh, my. Oh, my," but her eyes remained fastened to the screen. They were still looking, ten minutes later, when the policeman returned with the search warrant. A. C. Johnson wasn't interested in studying it. He would take the policeman's word.

The camera moved from the emergency entrance at the hospital to the big silent bird standing on the airfield. The commentator spoke of the shooting of Officer Tippit, and the camera switched to the third floor of police headquarters and a bedlam of photographers, policemen, and reporters. In the middle was a suspect who was shouting for his rights. Earlene Roberts and Mr. A. C. Johnson studied the face, and both stiffened in their chairs.

"Hey!" yelled Johnson over his shoulder. "It's this fellow that lives in here!" He pointed to the little alcove bedroom. "That's O. H. Lee," Mrs. Roberts said. Mrs. Johnson, out back with the policemen, hurried into the house, but the television picture had changed to the empty tables at the Trade Mart. "Who?" one of the government men asked. "Who is it?" They said, "O. H. Lee," and Mrs. Johnson said: "Well, that's why we didn't know who you were looking for." She displayed the register. "Here he is. O. H. Lee." Mrs. Roberts was excited and she said that he had come home, right in the middle of the day, and he had gone into the little room and changed to a zipper jacket or something.

The police were in the small room, edging past each other to get around the bed. They appropriated almost everything Oswald had—his skimpy array of clothing, a wall map of Dallas with the Texas School Book Depository building marked off, a couple of books. They felt the walls, the surbases, rifled the little closet and turned the mattress and pillow over. They even took two pillowcases and a face cloth.

The picture of Oswald came on the television set and all hands shouted: "That's him! That's him! O. H. Lee!" Earlene shook her head. "I said to him, 'What's your hurry?' and he never said a word. Just skipped out fast. I even saw him standing down the street at the bus stop."

An apologetic breeze stirred at Love Field. Jesse Curry sat in his car, thinking his private thoughts, squinting up the road toward

Mockingbird Lane. He kept half an ear on Channel Two, but the air traffic was dying off except for reports from the Book Depository and one or two from a house at North Beckley. He saw one car careening down the airport road toward his position and the chief of police knew, without asking, that this was Judge Sarah Hughes.

He met her at the gate and ran with her across the concrete to the plane. The door at the top was opened, and he went inside with her. At the door to the President's stateroom, Mr. Johnson grasped the hand of Judge Hughes. He couldn't summon a smile. "Thank you," he said. "Thank you for coming, judge. We'll be ready in a minute."

It is difficult, even in the most adverse circumstances, to repress the surging joy which must be a concomitant to being sworn in as the Chief Executive of the most powerful nation on earth, but Johnson didn't smile. His eyes were rheumy and searching. He stared a moment at Charles Roberts of *Newsweek*, as though trying to place the face. "If there is anybody else aboard who wants to see this," the President said, "tell them to come in." On the flight deck, General Godfrey McHugh demanded to know why the plane had not taken off, as he had ordered. The pilot, Colonel Swindal, had received orders and counterorders; he had asked for taxi instructions several times and had not taxied anywhere. The general made it plain that, as a brigadier general, he ranked the colonel, and he demanded that *Air Force One* start at once. Malcolm Kilduff, passing the communications shack, heard the voices and told Swindal not to take off. The colonel hit the starter switch for the number three engine. It caught fire and emitted a dismal whine. With number three going, he could dismiss the generator truck below the nose and start the remaining three engines on power from number three. Then, if these government officials in the back could agree on one premise—to leave or not to leave—the colonel was prepared to obey.

Someone told Kilduff that the plane must leave at once. With extended patience he said, "Why?" He was told that Kenneth O'Donnell had ordered the plane to leave. Holding his temper in check, Kilduff said: "He may want to take off, but he isn't in charge anymore. Johnson is now President." The word filtered quickly to the aft section and it was interpreted as another indication of Lyndon Johnson's merciless grab for power. It was O'Donnell who kept goading McHugh to go forward and "get this plane out of here," although O'Donnell had heard from the President that he was going to be sworn in before takeoff.

The stateroom began to fill. Mr. Johnson told Lawrence O'Brien that someone should ask Mrs. Kennedy if she would stand beside him during the ceremony. He said he would like her to stand at his side and the oath-taking would be of short duration. The President said he would also need a Bible. There must be one somewhere on the plane. David Powers came up into the stateroom. So did Admiral Burkley and Major-General Clifton. They were followed by O'Brien and the busy Malcolm Kilduff. The Texas congressmen were already present. So were the three newspapermen.

O'Donnell came in. Photographer Stoughton was leaning against a bulkhead. "Mr. President," he said, "if you are squeezed any closer, I won't be able to make the picture." He tried one shot and the flash didn't go off. There was a second try, and the small room was struck by the silent lightning. Mrs. Johnson, still wearing the half-frozen smile of shock, looked small beside her husband. The president fidgeted with his shirt cuffs, and Judge Hughes, sixty-seven years old and cheerful, smiled patiently.

O'Brien found that Mrs. Kennedy was not in the breakfast nook beside the casket. He knocked on the bedroom door, and, getting no response, turned the knob and entered. The room was empty. He tried the knob to the lavatory and found it locked. Mrs. Kennedy was inside, alone. Whether she knew what was expected of her and was trying to avoid it, or whether the depression of spirit led to nausea, no one knows. O'Brien left and asked Evelyn Lincoln, Kennedy's personal secretary, to see if she could get Mrs. Kennedy's attention. She said she would try.

Looking around the room, O'Brien found a small gift box. Inside was what he thought was a Bible. It was a missal—the prayers of the Roman Catholic Mass in both Latin and English. He carried it out and gave it to the judge. The abnormal heat of the President's stateroom was worse. There were twenty-six humans jammed into a space no bigger than fifteen feet by seventeen. Each one's body heat generated the power of a one hundred-watt bulb. They waited for Mrs. Kennedy.

Kilduff couldn't find a tape recorder, so he used an electric dictating machine and put a cartridge in it. Then he placed the microphone between the judge and the President. Marie Fehmer handed the judge a sheet of *Air Force One* letterhead with the proper words typed. Mrs. Kennedy stepped timidly into the room. The President grasped both her hands in his and whispered, "Thank you." He nodded for the ceremony to start. Mrs. Johnson was on one side of the President; Mrs. Kennedy, still in bloody gloves and garments, the face still stunned and expressionless, was on the other. Witnesses, tiptoeing to see her, seemed to stop breathing. The overwhelming emotional bath, endured by all two hours ago, was renewed. Some averted their gaze.

Kilduff switched the Dictaphone set on. Judge Hughes held out the missal. The President looked down at his wife and placed his left hand on the book. The right hand moved up slowly, almost reluctantly. The twangy Texas voice of the jurist said: "Now repeat after me . . ." The words required but twenty-eight seconds. Johnson said loudly: " . . . so help me God." The thirty-sixth President, who now had the power to implement his decisions, turned to Lady Bird, grabbed her by both shoulders and kissed her. Then he turned to Mrs. Kennedy, put an arm around her, and pecked at her cheek.

Some rushed forward and tried to give him a hearty handshake and a congratulatory grin. President Johnson turned a stern expression on them, and the bud of conviviality was crushed. He was tall enough to look over the heads of the others, and his eyes sought Malcolm Kilduff. The press secretary was lifting the sound cartridge from the dictating machine. He gave it to Stoughton and told him to get off the plane and give the pictures to the press. The sound, too.

Mrs. Kennedy seemed unaware of what to do. She stood near the door with the President's seal emblazoned on it, and looked blankly ahead. Mrs. Johnson grasped her hand and said: "The whole nation mourns your husband." There was no response. Police Chief Curry tried to grasp Mrs. Kennedy's hand. His voice cracked with sobs. "We did our best," he croaked. "We tried hard, Mrs. Kennedy." She glanced at him, a small man with cross-hatched wrinkles on his chin, the spectacles gleaming in the dull cabin light. She nodded.

The chief shook his head. He took Judge Sarah Hughes by the arm. "God bless you, little lady," he said to Mrs. Kennedy, "but you ought to go back and lie down." Mrs. Kennedy summoned a smile. "No thanks," she said. "I'm fine." The President said: "Let's get airborne," and Chief Curry hurried forward with the judge to disembark. Sid Davis of the Westinghouse Broadcasting Company completed his succinct notes on the ceremony and left. A congressman got off. Valenti and Moyers, the O'Donnells of the new administration, sat slumped deep in seats.

The President phoned the White House.

The one "at-large" Secret Service man, Forrest V. Sorrels, left the Depository building and returned to police headquarters. It didn't require much acumen to find that the building seethed with excitement, that the Dallas Police Department felt that, in Lee Harvey Oswald, it had a good catch. Sorrels got to the third floor, determined to find out who the prisoner was and whether he had had the opportunity and motive to assassinate the President.

"Captain," he said to Fritz, "I would like to talk to this man when there is an opportunity." The police officer had completed his preliminary round of questions. "You can talk to him right now." The party moved to an enclosure adjacent to the captain's private office and when Sorrels opened by asking him who he was, Oswald switched to his arrogant mood. "I don't know who you fellows are—a bunch of cops."

The Secret Service agent decided on the open approach. "I'll tell you who I am," he said. He reached into his pocket and brought out an identification card. "My name is Sorrels and I am with the United States Secret Service." Oswald turned his face away. "I don't want to look at it." He glanced at the man from the Federal Bureau of Investigation, the Secret Service man, and the two Dallas detectives. "What am I going to be charged with?" he said. "Why am I being held here? Isn't someone supposed to tell me what my rights are?"

The interrogated had begun to interrogate the interrogators. "I will tell you what your rights are," Mr. Sorrels said. "Your rights are the same as any other American. You do not have to make a statement unless you want to. You have the right to get an attorney."

"Aren't you supposed to get me an attorney?" Sorrels shook his head negatively. "No, I am not supposed to get you an attorney." Oswald could not believe it. He seemed to feel that the state owed him the services of counsel. "Aren't you supposed to get me an attorney?" "No," said Sorrels, "I am not supposed to get you an attorney because if I got you an attorney they would say that I was probably getting a rake-off on the fee." Sorrels smiled at his man. He hoped the little joke would crack the stiffness. It didn't. The pale agate eyes stared. Lee Oswald was in a mood to confuse, confound, and bend the law to his wishes. With the press out in the hall, he knew that he could play the martyr plaintively. If, at his side, he could have smart counsel like John Abt of New York, it is possible that he could have turned the law off as one might shut a leaky faucet.

"You can have the telephone book and you can call anybody you want," Sorrels said. "I just want to ask some questions. I am in on this investigation and I just want to ask some questions." Oswald studied his adversary. He could have refused to make any statement, but the fact that he responded to some questions proves that he thought he could handle Mr. Sorrels.

The Secret Service man took him over the same route he had just traversed with Captain Fritz. Name, place of employment, domicile, reason for living apart from wife, the daily routine of working on many floors, travel in Europe, and the Soviet Union. All of these and a few more had been answered. Suddenly Oswald tired of the game and said: "I don't care to answer any more questions."

Captain Fritz walked into the room. He had been on the phone. His manner was still heavy-footed and bland. "That room on Beckley," he said. "Why were you registered under the name of O. H. Lee?" Oswald shrugged as though it was of no consequence. "The lady didn't understand my name. She put it down the wrong way." Fritz nodded and returned to his office. He might have asked the prisoner why he had signed the register "O. H. Lee."

All the engines were shrieking. Swindal's first officer filed a flight plan asking for 25,000 feet out of Love Field. The handoffs would occur at Fort Worth Center; Little Rock, Arkansas; Nashville, Tennessee; and Charleston, West Virginia. Estimated time of arrival at Andrews Air Force Base, Washington, D.C., would be 1803 local time. The plane received an "all clear" from

Gate 24, and Swindal moved the throttle settings up a notch. The big plane moved forward, rocking a little on the concrete.

She made the crossovers and moved up to the head of Runway 31. Her back was to Mockingbird Lane, the scene of the final triumph. A commercial airliner, about to turn in on final, was requested by Jack Jove to turn away on a 180-degree course and then make another one and come into the airport. Swindal was given approval for takeoff. He spent another minute, running down the check list with his first officer and flight engineer. At 2:47 P.M. the throttle settings were boosted and the four fan-jet engines howled their grief as 26000 rumbled down the strip, jostling the casket as Mrs. Kennedy watched it, shaking the shoulders of the new President as he returned to the phone, Roberts and Smith—the press pool—trying to find seats, Kenny O'Donnell sitting opposite Mrs. Kennedy, feeling relieved to shake the dust of Dallas, typists like Marie Fehmer afraid to touch a key for fear of shattering the sacred silence.

The first officer called V-1 and rotating speed. Swindal pulled the yoke back gently and lifted off. He was well out between Walnut Hills and Letot before permission was granted to turn the plane northeast for Washington. The ugly rubber legs were tucked up. The big ship climbed as it always did, without strain. The colonel was not content with 25,000 feet. He asked for 41,000. The weather chart showed a high-flying jet stream moving slowly northeast. It was close to the absolute ceiling for a Boeing 707. The picture on the television set began to scatter. The communications shack was in touch with Andrews, in touch with Dallas, arranging phone calls, picking up incoming messages. *AF-1* kept climbing for a half hour. The patchwork quilt of farms below assumed almost stationary figures. The sun on the portside became brazen, and the shades remained drawn. The color of the Texas sky changed from pale blue to baby blue to midnight blue. The sky became darker and darker and the plane seemed to be slower and slower. At 625 statute miles per hour, it looked like a piece of confetti pinned to the heavens.

One of the stewards thought: "How strange. For the first time in history, we have two Presidents aboard." On the radio, the first officer heard an announcement from a plane still at Love Field: "This is *Air Force Two*. Our designation has been changed to SAM nine seven zero. We will depart for Washington, under present instructions, at 1514 local time."

The "poor dumb cop" wore a tag on his big toe. Dr. Paul Moellenhoff of the Methodist Hospital took a bullet from the body and showed it to Patrolman R. A. Davenport, who guarded the dead officer. Davenport marked it and put it in his pocket. The body was about ready. Blood had been washed away from the wounds of J. D. Tippit. There was a sizeable hole in the forehead and a smaller one in the temple; one of the chest wounds was small. The other, the one which had hit a uniform button and had carried it into the chest, was big and ugly.

Dr. Moellenhoff could do no more. He looked at the nude, well-nourished male on the slab and shrugged his shoulders. The goods guys do not always win the gun battles. The solemn words "in line of duty" have never exhilarated a widow nor fed a child. The doctor covered him and called for an ambulance. "The body will be taken to Parkland Memorial for an autopsy," he said. "Doctor Earl Rose will be waiting." So would Mrs. Tippit.

The wellsprings of sorrow and guilt, shock and confusion, the aimless litany of repeating: "The President of the United States was shot in Dallas today" made November 22, 1963, seem to go on forever. Psychologically, the entire nation was trying to face it and admit it. But the brain teetered on the edge of truth and pulled back. In Dallas shops, clerks and customers discussing merchandise stared at each other once too often and burst into tears. An emotional cripple named Jack Ruby, who didn't really care at all, unexpectedly phoned a boyhood friend named Alex Gruber, who now lived in Los Angeles. Ruby prattled about a dog he had promised to send Alex, a car wash business, and the assassination. The nightclub owner lapsed into uncontrollable sobs and hung up.

Nor was it less emotional in the basement of the police department. In the vast underground parking lot, policemen got into cars and couldn't remember where they were to go or what they were to do. Others, arriving, stood beside police cars talking with other officers about the assassination, as though by conversational repetition it would become understandable. Lieutenant Jack Revill returned from the Depository with three policemen. He saw FBI Special Agent James Hosty and they discussed the case with disbelief.

Hosty ran over to Revill and exclaimed: "Jack, a Communist killed President Kennedy!" Revill, a tough face under a cowboy hat, said, "What?" "Lee Oswald killed President Kennedy." "Who is Lee Oswald?" "He is in our Communist file. We knew he was here in Dallas." They walked into the basement elevator with other police officers, and Hosty said he knew that Oswald was "possibly capable of this." Revill was excited. He felt that "the town died today." He shouted invective at Hosty, and they got off the elevator at the third floor, in the midst of mass media. Revill repeated the conversation to his boss, Captain W. P. Gannaway. He was ordered to make a written report on it at once, to be drawn to the attention of Chief Curry. Hosty had no idea that the use of the words "possibly" or "probably" could, in an angry police report, hang him. He had never believed that Oswald, the friendless pedant, would be capable of violence.

The Dallas Police Department, which had cooperated fully with the Secret Service, was in no way to blame for the tragedy. It had extended itself to the limit to protect the President. Still there was a feeling akin to guilt in the department. Men asked each other blankly: "What could we have done that we didn't do?" At 2:50 P.M. FBI Agent James Hosty became the accidental goat.

The blame was quickly nailed down:

Captain W. P. Gannaway
Special Service Bureau
Subject: Lee Harvey Oswald
605 Elsbeth Street

Sir:
On November 22, 1963, at approximately 2:50 P.M., the undersigned officer met Special Agent James Hosty of the Federal Bureau of Investigation in the basement of the City Hall.

At that time Special Agent Hosty related to this officer that the Subject was a member of the Communist Party, and that he was residing in Dallas.

The Subject was arrested for the murder of Officer J. D. Tippit and is a prime suspect in the assassination of President Kennedy.

The information regarding the Subject's affiliation with the Communist Party is the first information this officer has received from the Federal Bureau of Investigation regarding same.

Agent Hosty further stated that the Federal Bureau of Investigation was aware of the Subject and that they had information that this Subject was capable of committing the assassination of President Kennedy.

Respectfully submitted,
Jack Revill, Lieutenant
Criminal Intelligence Section

In the hysteria of the hour, it was an ideal report for taking the responsibility from the Dallas Police Department (which it didn't deserve) and tossing it into the lap of the Federal Bureau of Investigation, where it didn't belong. The long and cool shadow of time shows that if the motorcade was staged again with the same personalities, the same diligence to duty and the protection of the President, and the same knowledge of Lee Harvey Oswald, the President would be killed as before.

Hosty could mask his feelings, but not to himself. All great tragedies depend upon an assortment of miniscule events, each of which must be executed in an orderly manner. The chain is easily broken, as it was on the day Oswald told his wife he wanted to kill the Vice-President. She locked him in a bathroom. He pounded on the door and threatened to beat her. Within a short time, Oswald agreed to remain in the bathroom if she would give him something to read. The chain of events was broken. On the night Oswald tried to kill General Walker, the first miss with the rifle frightened him, and he ran, hiding the gun on a railroad embankment under gravel.

The FBI wasn't aware of these things. No government agency saw Lee Harvey Oswald as a danger. He was on the records of the State Department, the Central Intelligence Agency, and the Federal Bureau of Investigation as a Marxist defector. His disappointment in the Soviet Union was so profound that he had borrowed money from the U.S. Department of State to return to Texas—and had paid the money back, a little at a time.

On April 24th, 1963, Oswald left for New Orleans, and Special Agent Hosty transferred his authority over Oswald to the New Orleans office of the FBI. Oswald headed for Mexico on September 25, 1963 and rented a room at the YMCA. By November 1, 1963, Mrs. Paine had informed the FBI that Oswald was employed at the Texas School Depository Building and was rooming somewhere in Dallas. To any law enforcement agency,

Oswald would have been much more of a pest than a menace; much more of a petty disputant than a threat.

Captain Fritz waved Hosty into his office. The FBI man opened his commission card and displayed it to Oswald. There was no comment. Jim Hosty backed up and sat beside Bookhout. "Have you been in Russia?" Hosty said. The prisoner rested his manacled hands on the captain's desk. "Yes, I was in Russia three years." "Did you ever write to the Russian embassy?" Oswald gave it some thought. Then he said: "Yes, I wrote." "Have you ever been to Mexico City?" The sensitive nerve had been touched. The reaction was instantaneous. Oswald was not aware that anyone knew about his bus trip to Mexico. "No," he said loudly.

"When were you in Mexico City?" the FBI man said. Oswald stood. He sat. He pounded his shackles on top of the desk. "I know you!" he shouted. "I know you! You're the one who accosted my wife twice!" The captain grabbed the handcuffs. "Take it easy. Sit down." Oswald nodded venomously. "Oh, I know you." "What do you mean, he accosted your wife?" Fritz said. The mood of high indignation began to dissipate. "Well, he threatened her." "How?" "He practically told her she would have to go back to Russia." This, apparently, was not the kind of accosting the captain had in mind. "He accosted her on two different occasions," Oswald said.

The visits to the Paine house in Irving had been routine. They were "follow-up" pilgrimages, designed to learn Oswald's current status. Hosty wanted to know if Oswald was working; if so, where. The FBI was disinterested unless the man was working in a sensitive industry, such as the manufacture of bombers or missiles. The defector was fond of excusing his successive firings from jobs by blaming it on the FBI. He never worked as anything but a laborer or minor flunky. The only thing he ever learned in a job was, while working for a photocopying service, to superimpose photographs and to fake military service cards and I.D. cards.

On the visits when Oswald was not at home, Hosty sat in the living room with Marina Oswald and Ruth Paine and, notebook in hand, directed his questions in English to Mrs. Paine. Marina had glanced at Hosty's car at the curb and copied the license number. She, too, felt hostile to the FBI. In a free country, it didn't seem just to be harassed by secret police. Besides, if the agents realized how ineffectual her husband was in almost everything he attempted, they would not check on him. No one knew better than Marina that her husband was chronically unhappy no matter where he went or how hard he tried. There was reason to believe that he despised his mother; was frustrated in his desire to dominate his wife; fought the Navy Department for giving him a dishonorable discharge from the Marine Corps; despaired in his role as dollar-an-hour nonentity in the business world; was shocked that the Soviet Union had no use for him; couldn't believe that the Cuban consul in Mexico City would decline his offer to enlist as an officer in the Castro army; cringed when his wife made fun of him as a lover; became embittered when Russian expatriates gave his wife and babies gifts of garments; perhaps, subconsciously, hated the Lee Harvey Oswald he knew.

The people on the plane gravitated into two groups. The Johnson people sat forward, the Kennedys aft. The Johnsons pretended that the situation did not exist. The Kennedys—which is to say Mrs. Kennedy, O'Donnell, O'Brien, Powers, McHugh—sulked in the rear compartment as though Johnson had boorishly appropriated the President's stateroom, evicting them all. They were desirous of making the President look bad. Mrs. Kennedy, having surprised the President in *her* bedroom, sat in the tiny breakfast nook near the casket, trembling with the vibration of the tail section.

For two hours and twelve minutes, the two camps remained apart. They employed messengers to walk the corridor with whispered wishes. The alchemy of the hours had transmuted the grief of the Kennedy group to rancor; the assassination was a deep personal loss, but it was also a fall from power. The Ins were Out; the majestic were servile; the policy makers were beholden to a new man for a plane ride; a lucky shot had killed the President, but it had also paralyzed the Cabinet and the White House guard. Men who are appointed to high offices must please the man who appoints them. When he goes, they go; or they wait for the man they held in contempt to say: "I need you more than he did."

Mrs. Kennedy retreated from the private bedroom to the aft galley. There were only two seats in that part of the plane. She sat on one. Mr. O'Donnell sat on the other. Admiral Burkley stood, swaying with the turbulence in the deep blue sky. Lawrence O'Brien stood near the casket. To each of them, it seemed to grow in size as the trip progressed. General Godfrey McHugh stood. This morning he had had a career; this afternoon he saw sudden retirement. O'Brien, frustrated by his thoughts, tried to patch a silver handle on the casket which had been jammed against the door of the plane on the way in. The bolts hung loose. He was not handy.

The Gaelic antidote to grief is whiskey. O'Donnell stood: "I'm going to have a hell of a stiff drink," he said. "I think you should, too." Mrs. Kennedy said: "What will I have?" O'Donnell said he'd make her a Scotch. She thought about it. "I've never had a Scotch in my life." O'Donnell moved on to call a steward. The forlorn face looked up at General McHugh. "Now is as good a time to start as any," she said.

The moody passengers had listened to the subdued whine of the jets; the landscape, far below, held no checkerboard interest. Conversations began and stopped abruptly. The people turned to whiskey. In some, it loosened additional tears; in others, it shored the dam of emotions. The President supped two bowls of steaming vegetable soup. Mrs. Johnson saw the small packages of salted crackers and, knowing that her husband was on a salt-free diet, munched them herself.

The short fat glasses of scotch and rye and bourbon jiggled their ice as *Air Force One* swept northeast. The empty glasses were replenished. As the busy stewards swept by, the word became: "Do it again, please." It did not dull the shock and sorrow; alcohol made it bearable. Some had many drinks; many had a few; a few had none. Still, it could not heal a breach. There were two separate and distinct camps aboard because Mrs. Kennedy wished it so. At one time, she looked up at Clint Hill, "her" secret service agent, "what will happen to you, now?" she said, and burst into tears. He looked down, the law officer always in control of his emotions, and the tears came.

The Johnsons, anxious to show a smooth continuity in the transfer of government, desired the two families to appear as one. At least, the Johnsons felt, the former rapport between the two groups could be maintained. They were wrong. After the swearing in, Mrs. Kennedy did not return to the private stateroom of the First Lady. She returned to the casket, and those of the Kennedy camp who wished to sit the vigil, remained at her side.

No mean word was uttered; no gauntlet was thrown. Glancing at the bronze box, Mrs. Kennedy began to think of Abraham Lincoln. The buoyant, youthful, sophisticated John F. Kennedy became fused in the shadow of death with the weary, cavernous man who had sealed the fractures in the union with the blood of its best boys. He, too, had had his Johnson; he, too, had died on a Friday; he, too, had been sitting with his wife; he, too, had been shot in the back of the head; in death he, too, had turned over the affairs of the nation to a man who was earthy, a vindictive Southerner who was politically alienated from his area.

Mrs. Kennedy ordered another drink.

3 p.m.

Dallas lost its official mind. Two and a half hours after the event, the aura of fatalistic acceptance was shattered. The professional calm of the police department was replaced by shouting officers who elbowed their way in and out of headquarters; poorly thought-out orders were executed, amended, and sometimes revised by telephone. Chief Jesse Curry, who might have supervised the hunt for the assassin, spent time telling the widow that she should go back "and lie down." The district attorney, Henry Wade, was trying to make certain that the crime was either his "baby" or the "baby" of U.S. Attorney Barefoot Sanders. Cops were sent out to Beckley and to Irving to search and seize, but they had no warrants for such work, and neither they nor their superiors had thought of getting them.

A police dispatcher, speaking to Captain C. E. Talbert on Channel Two at 3:01 P.M., said: "A Mr. Bill Moyers is on his way in to swear in Mr. Johnson as President and he will need an escort, but we don't know when he is going to get here." At Love Field, Curry was telling Mayor Earle Cabell that his police department had a suspect in the killing of Tippit and Kennedy, but neither official hurried to headquarters to serve the cause of justice.

The district attorney, the competent Henry Wade, found out that the crime was not federal. It was his "baby," but he permitted the clerks in his office to go home. It was a Dallas County matter, as Dr. Earl Rose had insisted it was, but the authorities had only the most superficial pathological findings from the doctors to present at any criminal trial. Fritz questioned

Oswald, but employed neither stenographer nor tape recorder and would have to depend upon his memory if the prisoner disappeared, died, or was tried.

In the county building, Sheriff Decker's deputies were processing witnesses by the dozen, taking affidavits, having girls type them, asking witnesses to sign, permitting some to leave, asking others to remain. Three Dallas detectives were waiting on Fifth Street in Irving for a couple of county deputies from Decker's office. They would wait a half hour before the county men arrived so that they could search the home of Mrs. Ruth Paine and ask Mrs. Oswald if her husband had ever owned a rifle.

Portable television sets began to appear in the parking lot of Parkland Memorial Hospital. A police officer asked for reinforcements to get "these people out of here, because it's going to be worse when people start coming home from work." It was a macabre picnic. Off-duty policemen were being called in to headquarters, and they were falling over each other and the press. One lieutenant ran upstairs from the basement and had a good lead: he had just found out that someone named Oswald was missing from the School Book Depository and he might turn out to be the man they were all looking for.

There is a mass madness which begets madness. It is contagious: calm faces contort; mouths shout; impatience is paramount; ordinarily good minds become scrambled; feet run, hesitate, stop, and reverse themselves; dignity is discarded; and, when it is over, the memory of the victim is highly inaccurate. The third floor of police headquarters was, on the afternoon of November 22, 1963, the fulcrum of the madness of Dallas.

The city, sorry at first for Mrs. Kennedy, was suddenly sorry for itself. The vain community was crushed. Earle Cabell was ahead of his time that day when he moaned: "Not in Dallas!" Jack Ruby, the boss of the strippers, was not far off the mark when he asked what the assassination might do to business. The city whose sin was pride had blood on its hands. It had only the most perfunctory pity for John F. Kennedy. The tidal wave of reporters and photographers from all over the world, crashing in on the city all day and all night, the spotlight of the world focusing on Dallas brought the realization that with world attention could come condemnation. The community could be morally indicted by the nation and the world. The city which bought and sold money began to search frantically for its soul.

The third floor at headquarters was shaped like a crucifix. At the bottom was the press room, a dustbin of old stories and muted metallic voices coming out of a police radio. On the right side, coming up the green floor, was the Juvenile Bureau, the Forgery Bureau, an unmarked room, and a transcribing room for reducing taped confessions and affidavits to the written word. On the left side was "313, Auto Theft"; "316, Burglary and Theft"; "317, Homicide and Robbery"; a water cooler and a private elevator to the fifth-floor jail.

On the crossarm to the right were restrooms, elevators, and an office marked "Police Personnel." On the opposite crossarm was a curving stone stairway, a "Perk-O-Cup machine," and "Soda, fresh milk." Facing these was a cigarette machine and one which reduced dollar bills to coins. At the top of the cross, on the left, was a big square office for radio dispatchers. Straight ahead were the private offices of the police hierarchy—the chief, assistant chiefs, deputy chiefs, and inspectors.

The third floor was a madhouse. The press scrambled for advantage like ruffians. Thick black cables snaked up the outside walls and across the floor. Reporters invaded police bureaus and hid telephones in desk drawers and wastebaskets. Enormous television cameras on dollies stared myopically from above the crowd. Still photographers hung from the tops of glass partitions to get pictures. The local newspapermen were inundated by their alien cousins.

A police lieutenant said that there were a hundred people in the hall. A sergeant, who had charge of screening credentials at the elevators, estimated three hundred. The net effect was as though some giant crap game had been raided and there was no place to put the prisoners except in the corridor with policemen stationed at the elevators and stairways while everyone protested or demanded counsel. The gate had been opened by Captain Glen King, the police department press agent. He had worked with the press before; he had worked with as many as four or six at a time. His credo was: "If they have press credentials, admit them." They were in, and more were on the way. Six men were coming from *The New York Times* alone, to assist Tom Wicker.

The infestation of madness which had infected the police department now assailed the reporters and photographers. The shouted questions were incessant; the demands to see Lee Harvey Oswald became a chant; all hands called for a press conference with the prisoner. The structure of authority began to fall apart. Captain Glen King consulted Deputy Chief Ray Lunday about permission for television cables to come from outside, through King's office window, onto the third floor. Lunday felt that the cables were permissible but not the unwieldy television cameras which go with them.

The deputy chief was certain that King, working directly for Chief Curry, required no permission. The request was, in effect, a courtesy call. Deputy Chief George Lumpkin was in his office, but King did not consult him. "King was operating on his own," Lumpkin said. Curry had signed "General Order Number 81" long before it stressed cooperation with the press. He "saw no particular harm in allowing the media to observe the prisoner." A woman was sitting in a glassed-in office crying and wringing her hands and the press demanded to know who she was.

No one could identify her. A policeman had brought her in and, on receiving fresh orders, had left. Detective L. C. Graves walked across the hall and found out that she was Mrs. Helen

Markham, who had witnessed the shooting of Officer Tippit. The woman was hysterical and could barely speak. Graves tried to take a statement and said he would have to ask her to stay until they could stage a police "showup" or "lineup."

City Manager Elgin English Crull walked across from City Hall to headquarters and was appalled by the crush of human beings. He noticed particularly that those of the press who required electricity were using city power. Switchboxes had been opened and outlets tapped. He might have shouted for silence and asserted his authority, but Mr. Crull didn't. He seemed mollified because one of the local reporters said to him: "Please don't blame us for what is going on. We don't act this way."

The blinds were drawn across the office of Captain Fritz, and he tried to blot out the roar of sound. Oswald was saying: "I know your tactics. There is a similar agency in Russia. You are using the soft touch, and, of course, the procedure in Russia would be quite different." Unfortunately, the police department seemed to have lost control of the interrogation. The suspect did not appear to be frightened either by his arrest, the marshaling of damaging evidence, or the enormous amount of attention he was now getting from the world. If one could judge by appearances and responses, Lee Harvey Oswald felt that he was the trapper, not the trapped. He would answer the questions he could handle without risk; he would shout and snarl and lapse into silence when the interrogation touched a sensitive nerve.

He refused to discuss his military service record. He would not listen to questions about his handwriting as "Alex Hidell," nor about Hidell's renting a post office box in Dallas and purchasing a rifle with a telescopic sight or a snub-nosed revolver from mail order houses. He waved aside questions about his wife and children. He would respond, in the manner of a pedantic lecturer, to questions about the Soviet Union, but when asked if he had shot Officer Tippit he snapped: "No!" When asked: "Did you shoot President Kennedy?" he shouted: "No!"

With a shrug, he said he had once been to Tijuana. No one had asked. Fritz interrupted the interrogation by walking out into the bedlam of the hall to listen to reports from detectives and to hand out fresh assignments. Each time, before he left, he would glance through his bifocals at Bookhout and Hosty and say: "Any questions?"

The captain had no time for lunch, but he offered some to Oswald. The young man said he would like coffee and dough-nuts. A policeman went for them. The questions continued. Often they were the same questions. As Oswald sipped the steaming coffee, Fritz reminded him that he would be fed again "upstairs."

The panic which seized Dallas ran from its head through its nervous system. It did not show on the streets. The shops were open. Women feasted their eyes on expensive gowns at the air-conditioned Neiman Marcus store and, when the obsequi-ous clerks murmured: "Wasn't it awful . . . ?" the customers glanced up sharply and said: "Yes, it was awful. How much is this Hawaiian silk print?" Politicians teed off at the Dallas Ath-letic Club and remembered to keep their heads down. Lovers lounged in Turtle Creek Park, holding hands in the warm sun and dreaming the dreams they should.

As it was in Dallas, so it was in the capital of the nation. The buses ran their routes. Taxicabs with noisy transmissions whined through Rock Creek Park. The statue of Alexander Hamilton failed to lift a granite brow. On the edge of the Po-tomac, Negroes shucked bins of cherrystone clams, peeled the cellophane skin from pink curving shrimp, hacked the heads and tails from fat scaly bluefish, and gulls stood silently against the leaden sky waiting for the scraps to go overboard.

Washington did not panic outwardly. From the sky, scores of thousands of automobiles on the highways north and south and east and west picked up a ray of sunlight and bounced it briefly from windshields. The city went about its business. The shops, the vendors, the offices, the officious bureaus continued to function as though the body politic were not prostrate and numb. On this one afternoon, there was no government. The executive branch was momentarily headless; the legislative branch adjourned in grief and dissipated its august membership to the winds. The Supreme Court, which can only say "Do not," is not constituted to contribute a positive act to the well-being of the nation. Nine learned men in black cannot balance a casket nor alter the hysterical posture of government.

In code and in plain English, radio messages flashed across the skies of the world, assessing the assassination, asking directives, reassuring command posts in far-off places, creating false alerts, tensing military muscles, causing lights to burn in embassies and legations in many countries. A member of the Cabinet asked the rhetorical question: "Who has his finger on the missile button?" No one. And no one wanted to believe that no one did. An act as stunning in its magnitude as occurred at 12:30 P.M. in Dallas could not be accepted anywhere as the deed of a lonesome malcontent. Nor would Washington or Moscow or Peking or Paris be willing to truckle to the truth for many years to come. Who could accept the thesis that a meteor, racing across the heavens, could be brought down by an idiot with a cork gun?

It was this which shook the city of Washington internally and tied up the phones, the circuits, the switchboards, even the area code, so that paralysis muted a second great city. The world knew that aircraft 26000 was up there somewhere, returning a gallant young man to his fathers, but who the hell was Lyndon Johnson? A short time before this day, an amusing program called *Candid Camera* had scored a hearty hit on television by asking pedestrians in a remote city: "Who is Lyndon Johnson?" Some had said: "The name is familiar. Why don't you consult a phone book?" Others did not know. Whoever he was, he was also high in the sky with that metal casket and he was reaching for the reins of government, quickly and surely, acting patiently and almost obsequiously, which were not his characteristics. He was, basically, a boss man; a doer; a demander; a tall, awkward person blunt enough to think tact was something on which to hang a picture. When he was a young man, he and Tom Connally and Sam Rayburn and John Garner had agreed that there was no more lofty position in the world than being a senator from Texas. The vice-presidency was a step backward. On this afternoon, the tough man felt fear. Loneliness too. No one said: "Hey, Lyndon . . ." The form of address, even from old friends like Valenti and Thomas and Gonzalez and Moyers, was: "Yes, Mr. President."

Two men wearing small buttons in their lapels walked up the short marble staircase of the old Washington Hotel into the lobby and then to the elevator. They pressed the button marked "6." When the car stopped, they got off and walked down until they stood in front of the old rosewood door at the corner of the hotel. They knocked. The Speaker of the House of Representatives, Mr. John McCormack, had a lean face with a big nose and the nasal tone of the Boston politician.

They told him that they were Secret Service men. The Speaker was next in line for the presidency. If the plane crashed, killing Lyndon Johnson, this faithful old party warhorse would take the oath as President of the United States. John McCormack never tried to be what he wasn't. He was not a giant intellect nor a skilled debater. He understood his countrymen and their requirements; the old man with the white hair and the slight snarl had an instinct for national government. A strong President could set the policy of the administration and John McCormack would fight for it even though he might not subscribe to some of its measures.

The Speaker refused to admit the two men. He was brusque. It was not necessary for them to tell him he was the next man.

He and Mrs. McCormack were averse to altering their private lives in the shadow of the Secret Service. He would not have these men accompany him in an automobile or stand over Mrs. McCormack in the shops. "Please," he said as softly as he could, "get out of the hall."

A block away, Maude Shaw sat in her room on the second floor of the White House. The children still slept, but she knew that they would soon awake. The thoughts of the English nanny were gloomy. She could hear the President calling from his bedroom "John-John" and "Buttons." In her mind's eye she could see the delight in the faces of the youngsters as they ran down the hall to the bedroom with the open door. Inside, the young President of the United States braced his breakfast tray with both hands as the children leaped upon their father with the lavish love and wet kisses which are its concomitant.

She could, in this interval of lonely introspection, remember the time that John-John disappeared in his father's office. No one could find him and the President called his son's name with sharp petulance. Then, from the panel of the front of his father's desk, a door swung open and the little boy fell out, laughing uncontrollably. The desk had been presented to President McKinley by Her Majesty Victoria of Great Britain. No one knew that, behind the majestically carved Presidential Seal, there was a gateway inside the desk.

Robert Foster tapped on her door. He was a young Secret Service man and his assignment was to break the news to Miss Shaw. The facts did not lend themselves to tact or gentility. "The President is dead," he said, and the thin, middle-aged woman bowed her head. He nodded toward the children's rooms. "We have to get out of the White House by six o'clock," Foster told her. "Mrs. Kennedy is flying back and doesn't want the children around. Hurry, we haven't much time."

The suitcases were in Miss Shaw's room. Foster helped her to pack. "Where are we going?" she said. The phone beside her bed tinkled softly and the light flashed. It was Mrs. Robert Kennedy, wife of the Attorney General. "I think you had better take the children to meet their mother," she said. Ethel Kennedy's voice trembled. "She will be at Andrews Field at six." Miss Shaw did not know what to say. Somehow she felt that the suggestion was wrong. "Oh no," she said at last. "Surely not. I am sure Mrs. Kennedy would not want to see the children just now. Please don't ask me to do that."

Ethel Kennedy thought about it. "All right," she said. "Bring the children here. I can't think of anything else. Can you? Anyway, I'll leave it to you. You know best. . . ." The phones were hung up. Foster was in a hurry. His orders had been to get the Kennedy children out of the White House at once. To where? The Secret Service man was on the floor, jamming clothing into suitcases, looking at Miss Shaw pleadingly, and Maude Shaw thought of the proper retreat: their maternal grandmother's house in Georgetown.

Mrs. Hugh Auchincloss was ideal. She loved the children, and she was gifted with a sweet maternal manner. The children loved her and, in the late afternoons, their mother had taken them to "Grand-mère's" house many times. They would feel less "strange," less inclined to tension or alarm in her house than anywhere else. Maude Shaw phoned, and, when the two women said "hello," both burst into tears. Foster was hoping for a quick decision, but the sobs and the intermediate conversation delayed the decision. Mrs. Auchincloss was surprised that the children were not to be with their mother, but she kept saying: "Bring them over. Bring them over to me. This is the place for all of you. Come and stay here. . . ."

The grandmother fought for control of her emotions, then her voice softened. "Miss Shaw," she said, "there is something I would like you to do, and I know my daughter would too." "Of course," the nanny said. "Anything." "We feel that you should be the one to break the news to the children—at least to Caroline." Maude Shaw relapsed in shock. "Oh no," she said loudly. "Please don't ask me to do that."

"We feel that you should be the one . . ."

This made it the wish, the command, of the mother as well as the grandmother. "Please, Miss Shaw. It is for the best. They trust you. . . . I am asking you as a friend. . . . Please. . . ." The children's nurse was overwhelmed by a feeling of horror. She stood at the phone as Foster stared at her beseechingly. "All right," she murmured crisply. "I will tell Caroline when I put her to bed tonight."

The children were awakened. They were cheered to find that they were leaving at once for the Auchincloss home. Nighties and pajamas and spare dresses and slips and shoes and little suits and blouses were all tucked into the valises. Also a special toy or two, a doll. Foster led the little party out and down the broad dark corridor to the elevator. He had a car on the South Lawn and it had been waiting a long time.

Maude Shaw wondered, "Why me?" The family was full of intelligent people. There were cousins and uncles and aunts aplenty. Could not one of them sit in privacy with these babies and break the news gently? Could not someone explain that God often calls a soul suddenly, one that he wants in heaven at once? Could not someone have told them that death is nothing more than a postponed reunion? That their father would be as happy waiting for them to join him as they were sorrowful at his leave-taking?

Maude Shaw made half a promise. She would tell Caroline only.

The Cabinet plane, a third of the way back from Hawaii to California, was on an almost identical course with *Air Force One* and at practically the same speed, but they were several thousand miles apart. Dean Rusk's 707 begged for additional information, and it arrived, either on teletype or by phone, chopped in segments. The Secretary of State remained in the private cabin. The others wandered in the public area, brooding, trying to assimilate the fact that Kennedy was gone and trying to decipher what this would mean to each of the careers aboard this plane. The shock to the personal senses was the passing of Kennedy; the shock to the political senses was the accession of Johnson. No one doubted that the new man would achieve a keynote of continuity by announcing that he would adhere to the Kennedy policies; it would be a violation of historical precedent to do it with the same assortment of faces.

Someone suggested a poker game. A table was found and some chairs. There was nothing better to do. Cards would be preferable to thinking. The gentlemen placed money on the table—perhaps the money they had planned to use for shopping in Tokyo. Pierre Salinger, the cigar smoker, enjoyed the game immensely but seldom won. The players agreed on table stakes and, in a moment, the mourning period had been postponed and Kennedy's august appointees were saying: "Here's your ten and ten more." "Dealer draws three." "Smiling ladies, like the Andrews Sisters." "All pink."

Alone, except for a mess sergeant, Dean Rusk read the teletypes. At last a sketchy story on the suspect, Lee Harvey Oswald, began to click on the plane. It told about his defection to the Soviet Union, his life in Russia, and his membership in the Fair Play for Cuba group. This was difficult to believe, because most knowledgeable persons were certain that the assassin must have been an extremist-right-winger.

A Communist in Dallas, Texas? This would be as difficult to digest as a story that Josef Stalin had been killed by a Russian fascist. "If this is true," Rusk said to no one in particular, "it is going to have repercussions around the world for years to come." It is possible that he saw the news as the first evidence of a Communist conspiracy. This would have amused the loner who parried questions in far-off Dallas. When he worked in Minsk, the Russians couldn't even get him to attend party meetings in the factory.

The biographical material on Lee Harvey Oswald had also passed through the hands of Forrest V. Sorrels, the wandering Secret Service man. He asked Chief James Rowley in Washington whether PRS had been aware of Oswald's Marxist background. PRS—the alert file of persons dangerous to the President—did not have a listing under Oswald. Rowley, hearing of Oswald's defection to Russia, asked his superiors at the Treasury Department to contact the State Department to find out what they knew about the prisoner. Rowley would be interested to know why his agency had never been told about this defector.

Wheels were turning. They spun slowly at first but, with each passing minute, they accelerated. Files which were dusty with time were reopened, and cards of various colors withdrawn, scrutinized, and copied. The State Department had a dossier on one Lee Harvey Oswald. Several agencies became interested. Treasury wanted to relay all possible information on this man to the Secret Service, the agency primarily responsible. They wanted a digest of the dossier. The Federal Bureau of Investigation, which had a small file on the man, now wanted every shred of information. The Central Intelligence Agency, which sensed international complications, asked for copies. Aboard *Air Force One*, the news reached President Lyndon Johnson through Major-General Chester Clifton, who was sorting messages in the communications shack forward. The President asked for a quick check of the Oswald situation to find out if the State Department had erred in permitting this man to return to the United States.

There was blame to be spread, guilt to be impugned, punishment to be meted. No crime as monumental in size as this could be laid at the feet of a sullen ignoramus. It was a blessing that he was a Marxist, because, by negation, it absolved Dallas.

Of course Oswald's brand of Marxism was not related to the despotic socialism practiced in Russia, but the American mind lumps the two in political idealism, even though they are anathema to each other, and both have contempt for Bolsheviks, nihilists, and Mensheviks. It was sufficient to call Lee Harvey Oswald a Communist. The only other question to be resolved was to find out who was responsible for bringing him back from Moscow.

The poker players drank. They spoke in grunts, and Salinger won almost a thousand dollars. The drinks and the money were meaningless. For a time one player or another would break in with a fond recollection of Kennedy, but these sad pleasantries petered out. These were professionals with additional streams to cross and hills to climb, so they concentrated on Lyndon Johnson, wondering aloud what kind of a man he was. Everyone agreed that he was a master politician. Call him a horse trader, a locker-room Disraeli, a compromiser—Johnson was a winner. He was a doer; they knew that.

Some tried to speculate about the Johnson "team." The Cabinet was surprised at how little it knew. Lyndon Johnson had been Vice-President for almost three years and he worked across the street from the White House in an office in the old State Department Building, where there were baroque doors and high Victorian ceilings laced with heating pipes. They seemed sorry to admit that they had not cultivated him. It was recalled that Johnson had been the youngest majority leader in the United States Senate, but that showed legislative acumen, not judgment.

Secretary of Labor Willard Wirtz, the man with the pepper-and-salt hair, had campaigned for Adlai Stevenson in 1960. At the time, he had no appreciation for the upstart from Massachusetts. Now he said tactlessly: "I gather you don't think the world is at an end?" The poker game broke up. The gentlemen began to bicker. The plane hung between sky and sea and the men of power accused each other and used words like "rumormonger," "treachery," "hearsay." They were not sure that Johnson had not been shot, too. There was word that a Secret Service man and a Dallas policeman were dead—so the plot must be widespread.

They owed allegiance and each was eager to flex the knee in fealty, but to whom? Even Rusk, meditating in his private cabin, wasn't sure. The big Boeing shrieked through the sunny skies. There was no time for tears. The dry eye of power was focused on power.

The flow of information from Gordon Shanklin's office to FBI headquarters in Washington was steady. Except for a few lapses, the line might have remained open. Shortly after 3 P.M. the agent in charge spoke to Washington and said he had some news on the rifle found in the Texas School Book Depository. It was not a Mauser or a British weapon, but rather a cheap Italian military surplus rifle called a Mannlicher-Carcano. The caliber was 6.5 and the four-power scope was Japanese. The serial number, Shanklin said, was C2766.

His men had phoned local sports goods houses in Dallas and had learned that they had little call for Mannlicher-Carcanos. However, their catalogues showed that the importer was a firm named Crescent Firearms, Inc., of New York. The New York office should be alerted and start tracking that C2766 at once. There was similar information about the snub-nosed revolver carried by Oswald, but the FBI was much more interested in the genesis of the gun which they believed killed the President. It was almost closing time when the New York FBI descended on Crescent Firearms, but the records were brought out. C2766? That was part of a big shipment of rifles sent to Klein's Sporting Goods, Inc., of Chicago. The trail bent to the Midwest, and the FBI followed it.

Official Washington began to depart for Andrews Air Force Base, across the river. Second-echelon officials, diplomats from many countries, Supreme Court justices, bureau chiefs, wives, congressional leaders, all began the pilgrimage. The Attorney General was shocked. One of the many opinions of that day which cannot be rationalized was that Robert Kennedy seemed to look upon the homecoming as a private funeral. It was closer to being the return of Caesar to Rome, but Bobby thought of it as the return of his dead brother. He tried forbidding or dissuading some from going to Andrews. Then he heard that more and more dignitaries were already waiting there. Others were asking the military for helicopters. Black limousines were crossing the river to the air base with pale shocked faces silent in the back seats.

Others called at the White House first, although there was no one present to whom condolences could be offered, and some of these were put to work. Sargent Shriver, the handsome and articulate brother-in-law of John F. Kennedy, arrived and assumed the position of *chef de cortège*. He was director of the Peace Corps, but, even though the body was not back in Washington, the work of organizing funeral arrangements and—more important—the team of reliable men who would be needed night and day, every minute of every hour, to assist Sargent Shriver, had to begin now.

Two stenographers were in President Kennedy's office removing some of his keepsakes. The memento book of photos of his trip to the home of his Kennedy forebears at Wexford, Ireland, reposed on a table behind the desk and suddenly disappeared. A painting of a sailing ship followed. A mounted fish in an office across the hall came off the wall. The rocker with the U.S.S. *Kittyhawk* embroidered on it was placed on a dolly and wheeled into the hall. It was incredible that anyone could have issued such a callous order, but the mementos were being moved abruptly.

By hurrying them outside to be carried away to a private place, the press cameras could make the Kennedy bric-a-brac appear to be forlorn mementos, could make it seem as though the new man was in a hurry to take over the executive office. In time, the Queen Victoria desk would disappear, too, although it was the property of the United States.

Senator Hubert Humphrey of Minnesota phoned the White House and asked if he could pay his respects by waiting at the air base for the body of the President. He was told "No." "The hell with you," Humphrey said. "I'm going." The President's alter ego, Ted Sorensen, sat at his White House desk, his back to the fireplace mantle. He had researched and assisted Kennedy with a book called *Profiles in Courage*. Sorensen—an incisive phrase-maker—had written many of the speeches John F. Kennedy had delivered. The President had drunk deeply from the semantic genius of the quiet man, but there had been no public accolade for Sorensen. He was the Man in the Back Room, the man who could make mundane matters sound lofty and idealistic. All day long and far into the evening he had hand-polished words for his god. He sat in the office, thinking as the sun leaned west and bronzed the black oaks on the lawn. Ted Sorensen was asked if he would go to the airport. "If the others go," he said, "I will go."

The Pentagon monitored a Red Chinese radio at Hsinhua, which announced the sudden death of President Kennedy at 4:14 A.M. Hsinhua time. Radio Budapest played solemn organ music and commented that Cardinal Mindszenty, who had imprisoned himself years before in the American embassy in Hungary, would sing a memorial Mass for Kennedy. At 3:20 New York time, Martin Agronsky announced for NBC: "After the initial shock, President Kennedy's secretary began method-ically removing mementos from his desk—the family pictures, the PT-109 souvenir—making ready for the new President's ar-rival."

There was a world full of people who each would have liked to render some small service, and this applied also to men of position and power. The commanding officer at the U.S. Naval Hospital at Bethesda, Maryland, Captain R. O. Canada, Jr., ordered an ambulance to be dispatched to Andrews Air Force Base. So far as the captain knew, no one had asked for one. Aboard *AF-1*, two requests for an ambulance had been relayed to the capital city, but both had been refused on the grounds that the District of Columbia had a law prohibiting the transportation of the deceased in ambulances.

Canada, who sat at his television set, recalled that, eight years ago, Senator Lyndon B. Johnson had sustained a myocardial infarction. It had been moderately severe, and Johnson had been his patient. The crushing events of this day could—probably wouldn't—but could induce another heart attack. Captain Canada sent the ambulance and told the personnel to wait for *Air Force One*. Two attendants left at once, driving slowly through the rolling autumnal hills of Maryland, across the fresh-running streams, through the city, and down Route Five into the farm country where, almost a hundred years before, a man with a broken leg rode a horse to escape a shattered and disorganized government after the assassination of a President.

One of the first witnesses to get home was Howard Brennan. He carried his steel helmet up the walk at 6814 Woodward Street, Dallas, and his mind was troubled. For a long time he had been joining pipes along the railroad right of way at the overpass. Brennan had been content working in the open. He was hitting the middle years and he had learned the value of not becoming involved in anything but pleasantries. His daughter and a little grandson had returned to live again with him and with Mrs. Brennan, and Howard Brennan had accepted it, plodding on with his daily labor and taking his ease at home after supper.

He was an eyewitness. Sitting on a low wall, he had looked up and had seen a man in a window take careful aim and kill a President. Howard Brennan had submitted to the questioning of policemen and deputy sheriffs. He had stood in the county building with others milling about, shouting and declaring and denying, and his spirit began to shrivel. He was an eyewitness. Reluctantly he had read the statement and signed it, but Brennan had no heart for any of this.

"Will this be confidential?" he had asked, worried. The policemen said it was confidential. Brennan wished hard that he had not opened his mouth. He began to think that he was the only true eyewitness. Nobody knew who or what organization was behind the assassination. Somebody said a Secret Service man was already dead. Also a cop in Oak Clif. There may have been others. The Secret Service man who questioned him appeared to be excited.

Mr. Brennan began to feel a growing fear. They had asked him pointedly: "Do you think you could recognize the man with the gun if you saw him again?" Howard Brennan said: "Yes." Why had he said that? Why become involved? He was living in an era in which families prided themselves on not knowing who lived next door. No one wanted to be involved in any controversial situation—even a traffic accident.

In the house, he learned that his wife knew about the assassination. Everybody knew. There was nothing else on television or radio. Howard Brennan sat with his wife and told her quietly that he was an eyewitness. He had seen it. Actually watched it happen. The husband said that he had a growing fear that something bad could occur to the whole family. No one knew who or what was behind the assassination and Brennan did not want to be known as the man who could identify the rifleman.

He talked of moving out of Dallas secretly. They could take their daughter and grandson and get out. No one need know. It would mean giving up his job as a steam fitter. Mrs. Brennan was troubled because her husband was frightened. She listened to him patiently and said it wouldn't work. A person can't get away, she said. A person can't hide—no matter where he may go.

One of the surprising aspects of Lieutenant J. C. Day's work was that he couldn't find fingerprints. Normally there would be prints on the barrel of the rifle and the stock. There should be parts of prints on the empty shells from the gun. He and his Crime Laboratory left police headquarters and returned to the School Book Depository. Of course Day knew that, in his work, when a suspected lethal instrument was found to be free of fingerprints it was usually a sign that it had been carefully wiped by a person who might have a sense of guilt.

The entire sixth floor had been isolated by policemen. Day and his assistants went to work in that corner window where the empty shells had been found. They dusted the bricks on the ledge; they examined the heating pipes behind the assassin's seat on a cardboard box. The men moved about gingerly, disturbing nothing. They got nothing until they brushed the top of the box lying in front of the window. This, it was assumed, would be the low seat for the killer. On the front edge, facing the window, they saw a palm print come up clearly.

It was the first technological discovery, and yet it proved nothing. Anyone could have been sitting near that window, and anyone could have leaned on a box. The case against Oswald was to be built of chips and bits of evidence, the whole weighing more than the sum of the parts. The lieutenant backed his men away from the print, took strips of Scotch Tape and pressed it down on top of the white palm print. Then Day wrote on the box: "From top of box Oswald apparently sat on to fire gun. Lieut. J. C. Day." He tore the top off and took it back to headquarters.

The front door of the Carousel Club was open. The afternoon light bounced off the glassed-in photographs of strippers and shoveled a little radiance to the dusty interior. It was a place smelling of old cigar smoke and whiskey-stained tables. LaVerne Crafard kept the bar as clean as possible for his boss, Jack Ruby. Mr. Crafard was twenty-two; everybody called him "Smokey."

The chairs were still upside down on the tables. A machine behind the bar hummed as it made ice cubes.

Joy Dale arrived. She had an appointment to give a novice a dancing lesson. Miss Dale was an exotic dancer. Strip joints presented a difficult means of earning money. The pay was small. The customers demanded new bodies. The men were raucous and sometimes ungentlemanly in their comments. They tired of the same girls. A dancer with a good figure must, of necessity, develop new seductive routines and new and enticing ways of removing clothing.

Miss Dale had been to the hospital with her daughter. When she arrived at the Carousel, a light was over the bar. The rest was dim as such places usually are in daylight. The fat, rumpled figure of Jack Ruby emerged from his little office. "The club won't be open tonight and tomorrow night," he said briskly. "I don't know how long." He stared off for a moment, and then wagged his head slowly. "It's unbelievable!" he said. "How could a man shoot the President of our country?"

Joy Dale thought of it from a maternal side. "Can you possibly think how this woman feels?" she said. "She just lost her son, and now she's lost her husband." Ruby began to weep. His emotions, good and bad, were multitudinous and always close to the surface. Often they persuaded him to be unduly generous, or to fight, or to want to exterminate an anti-Semite, or to weep for someone he had never met.

"You shouldn't," he said, choking. "He should be killed." Jack Ruby called Crafard and ordered him to prepare a sign for the front door of the club, proclaiming that it would be closed. It was not to be posted until a late hour, because Ruby did not want his competitors to know until the last moment. It was, in a real sense, Ruby's contribution to the memory of a President. Business, on the other hand, had been poor. The sacrifice would not cost much.

His moods changed and blended as the colored prisms of light did on the stage. Jack Ruby found the world to be complex and confusing, and he simplified it to "creeps" and "good guys." The "good guys" sometimes became "regular guys," but the "creeps" were unalterable. Policemen were "regular guys" unless they gave Mr. Ruby a summons for a nightclub violation or a traffic citation, at which point the individual policeman became a "creep."

Ruby never charged a policeman for attendance at his nightclubs. He often took them into his office and pulled a bottle of whiskey from a drawer and set it on top of the desk. Some, who were more ambitious, were given personal introductions to certain strippers. "Be nice to this guy," Ruby would say. "He's a frenna mine." Ruby had felt his Jewishness from the ghetto days in Chicago, but he found no happiness or pride in it. Ruby was a defensive Jew who forbade his comedians to tell dialect jokes from the stage, who fought furiously with his fists any man who cast aspersions on Jews, but who was seldom seen in a synagogue.

This afternoon he thought seriously of going to Friday night services. It would be an additional mark of respect from Ruby to Kennedy. It was not something he would do for anybody, not even for himself. But closing the club and praying in a tallis in a temple had the mark of what Ruby referred to as "class." He would do it.

He watched Joy Dale giving the dance lesson to the student, without seeing any of it. Then he went back to his office and called his sister Eva. Mrs. Grant was tired of the phone calls. She tried to forestall her brother's nonstop dissertation by reminding him that they had no food for the weekend. He told her that he would stop in the Ritz delicatessen and pick up some cold cuts and salads. This, too, assumed the form of an accolade because Jack Ruby often bought pounds of salami and wurst and potato salad and cole slaw for policemen who worked late on a big case, firemen fighting a night battle with flames, radio commentators who talked through the late hours.

He told his sister that the policeman who was shot, Tippit, was a dear friend of his. This was an honest mistake. There was another Tippit on the police force. Ruby had never met the dead man. Mrs. Grant told him that the killer had been arrested. Someone named "Oswald." Her brother's reaction was typical: "He's a creep. He has no class." The grief, the welling tears were transmuted at once to roaring anger. He talked on and on about the dead President, his poor weeping wife, and the little kiddies who had no father. He was thinking, he said, of sending flowers to the place where the shots were fired. Thinking . . .

On the second floor at Parkland Hospital, Dr. Shaw was completing chest surgery on Governor John Connally. The patient responded well. He displayed good reserves. Shaw and his team were finishing the sutures and examining the drains when Dr. George T. Shires arrived. This man had been in Galveston when he heard the news and had managed to return to Dallas in incredibly fast time. Shires was professor of surgery at Parkland, and Chairman of the Department of Surgery at Southwestern Medical School.

He was senior to all the doctors present, but Shires did not impose this on them. He scrubbed and was gowned and, walking around the supine form of the Governor, he was given a whispered rundown of injuries and saw that the work remaining was the multiple fracture of the right wrist and the laceration of the left thigh. Without further conversation, Dr. Shires enlisted the assistance of Doctors McClelland, Baxter, and Patman, and they began the delicate job of putting a complex wrist back together.

On Oak Lawn, the ambulance of Vernon Oneal pulled into the back parking lot. He was irritated. The whole day had been frustrating. The phones inside jangled with calls from patients who, in the shadow of the catastrophic event, had heart attacks, fainting spells, and nausea. They wanted ambulances and they were wanted at once.

Oneal recalled that one ambulance, on an epileptic call, had been impounded at the emergency entrance to Parkland by the Secret Service. The ambulance and the personnel had nothing to do with the assassination but the men were forced to sit in the "bus" waiting for permission to leave. Then Oneal himself had responded to the crisis call for "the best casket you have," and that had resolved itself into a series of arguments between Dallas and the federal government over an autopsy.

They were forced to run the casket out of crowded hospital corridors and the Secret Service driver had agreed to meet Oneal at the mortuary but instead had headed at top speed for Love Field. Vernon Oneal had been warned by government men and police to remain outside the fence. They were not interested in the fact that the ambulance belonged to him—the casket in which the President reposed was also his.

He had been required to wait until that airplane was reduced to a small speck in the sky. Then the cops said: "Now get your ambulance out of here." Vernon Oneal was an old-fashioned man. He wondered whatever happened to the words, "Thank you."

As he parked the white ambulance, a black car was edging up the ramp of a C-130 plane at the airport. This was the President's Lincoln. It had been a triumphal vehicle on many occasions, a death trap once. All day long and through most of the night, evidence in the crime would be moving from Dallas to Washington. The seven-passenger car was the biggest item, and the Secret Service detail was sorry that hospital orderlies had sponged it out.

The slide rules and cellophane disks made the computations. Colonel Swindal leaned back in his seat on the flight deck. *Air Force One* was ticking off 625 statute miles per hour. To maintain that speed, the throttle handles were being yanked backward a little every fifteen minutes. The ship was burning seven tons of fuel per hour and, as the weight decreased, speed increased even though the throttle settings remained static. The plane was never permitted to go beyond .84 of the speed of sound.

The transponder was on permanently, so that the ground stations could track this aircraft easily. It showed up on radar screens as a larger, whiter blip, drifting like a slow rowboat across a large dark lake. Swindal yanked one earphone loose. Ahead and slightly to starboard an early moon was rising. Behind the thundering plane the sun had already changed from polished brass to dull bronze.

Estimated time of arrival at Andrews would be 2305 Zulu, or Greenwich, England, time. For two hours and fifteen minutes, the President had been snatched from earth. It was as though, subconsciously, those who realized that he must be given back to the earth permanently had taken him almost eight miles straight up to keep him, for an hour or two, from the destiny of all clay.

The air was smooth at this altitude. Now and then the flight deck shuddered, and the trim tab wheels spun forward and back like thinking things. Then the serenity of flight asserted itself again, and the subdued shriek became a hum, a one-note lullaby. There was time to think, time for coffee, time to listen to the latest information from the communications group sitting behind the flight deck.

Back in the ornate President's stateroom, activity was still the therapy of choice. The President refused to permit himself to sit and think. He had Liz Carpenter, Mrs. Johnson's secretary, working on a short statement to be read by Johnson after the body was removed from the plane. He had Valenti and Moyers as idea men. They sat near him, venturing thoughts that he should do this or not do that.

He was seldom more eloquent, or more helpless, than when he phoned Kennedy's mother. "I wish to God there was something I could do," he said. The tough fiber of the man softened. No words were adequate. And yet the impotence of the country was in the words. Rose Kennedy never lost her composure—had never lost it in the torrent of adversity which seemed, at times, to inundate the Kennedy family. She had her complete faith, her God, her Church. She thanked the President for his thoughtfulness in calling. She maintained her composure. Mrs. Johnson said: "Oh, Mrs. Kennedy, we must all realize how fortunate the country was to have your son as long as it did."

The conversation continued for a while, the proper words and the correct responses. Rose Kennedy did not ask Mrs. Johnson to switch her to Jacqueline Kennedy, who was sitting fifty feet behind the Johnsons. Nor did Mrs. John F. Kennedy phone her mother-in-law.* When the phone was put back on the cradle, the President asked for Mrs. John Connally at Parkland Hospital. When he learned that his political protégé was going to recover, the President's spirits seemed to lift. Nellie sounded bright and cheerful. She said that the Governor had sustained the surgery well and that he was having drains placed in his wrist.

Mr. Johnson noticed Charles Roberts of *Newsweek* and Merriman Smith of UPI writing the story of the plane trip on typewriters, one of which was borrowed. He stooped over both men and whispered that he wanted all of the Kennedy White House staff and all of the Kennedy Cabinet to remain on with the Johnson Administration. This, of course, was the first big pronouncement of the new administration. General Godfrey McHugh noticed the writers and reminded them that "throughout this trip I remained back there with the President." Admiral Burkley, Kennedy's presidential physician, wanted them to note "that I was with him when he died." In the gleaming, speeding

---

* Four months after the assassination, I sat with Rose and Joe Kennedy at their home in Palm Beach. Mrs. Kennedy said: "I have not heard from 'Mrs. Kennedy' since the funeral."

aluminum tube, each one knew that the only recorded history
would be what Roberts and Smith wrote, and each had a specific reason for wanting to be a part of it.

Johnson sat with Kilduff and made memoranda on sheets
of paper of personalities he should meet at once at Andrews,
of others who should be called to his office this evening at the
Executive Building office, of what time to have a critical Cabinet meeting in the morning. The more ground he covered, the
more there was left to cover. It seemed that he was phoning
McGeorge Bundy in the White House Situation Room every
few minutes. Bundy was in the basement amidst all of the "instantaneous" sources of information from around the world. He
was plotting the future. Upstairs, in the White House, Sargent
Shriver was mapping the past.

Still busy, President Johnson saw O'Brien walking by,
talking to a congressman. He called Kennedy's legislative assistant to his side. "Larry," he said, looking up earnestly from
the desk, "you have a blank check on handling this program.
Go ahead just as you would have under President Kennedy."
The redhead nodded and walked on. This early he could see an
obstacle ahead unseen by Johnson. Jack Kennedy did not live to
see his legislative program for his country enacted into law; the
Congress seemed to be disenchanted with the charmer. Now
that Kennedy was dead, if Johnson implemented the Kennedy
program into law, he would not be thanked by the Kennedy
group. In fact, this big, industrious man who took charge of the
nation so quickly and so firmly was going to meet rancor and
contempt for his work.*

The communications crew at the forward end of *AF-1* could
not handle the traffic. The outgoing calls were heavy enough,

---

* O'Brien and Bundy were the only Kennedy men who remained with Lyndon Johnson. Both, as noted by Charles Roberts in the *The Truth About the Assassination*, were
branded by the Clan Kennedy as "traitors." Bundy's response was that the presidency
is bigger than any man. O'Brien shrugged and said: "You do what needs to be done."

but too many officials in Washington wanted to speak to Johnson or to Mrs. Kennedy. She wasn't accepting any incoming calls unless they were from her brother-in-law Robert. Other calls were referred to her secretary, Pamela Turnure, the dark, attractive girl who matched Mrs. Kennedy in beauty. Texas politicians were phoning the new President. General Clifton asked Andrews to have a forklift ready to carry the casket down the rear exit; the loose handle would be dangerous if the casket was to be carried. He also phoned the Army's Walter Reed Hospital and said that the autopsy would be performed there.

The President was conscious that Mrs. Kennedy might, at this time, have composed herself and want to express her wishes. He sent Malcolm Kilduff aft, but the lady had no wishes. This was done several times, but there was nothing she wanted from the President. Kilduff felt the stiff politeness of the Clan Kennedy and recognized his role as the emissary under flag of truce. It was said that the trip to Texas was to be Kilduff's final assignment as assistant press secretary. Whether it was President Kennedy who was displeased with his work, or the hatchetman, Kenny O'Donnell, Kilduff was on the scene when the new President had no press expert.

Dr. Burkley, standing alone, noticed that Mrs. Kennedy was alone. He approached and, rather than bend down to speak to her, dropped to his knees. It was a comical attitude for the dignified admiral. He was at eye level with her and he said: "It's going to be necessary to take the President to a hospital before he goes to the White House." She was in a trance-like state, but the young lady came out of it quickly. "Why?" she said. The tone was sharp because she had had her fill of hospitals and their cast-iron rules.

Burkley looked like a supplicant at prayer. "The doctors must remove the bullet," he said. "The authorities must know the type. It becomes evidence." Mrs. Kennedy could understand the situation. The admiral did not use the word autopsy, which entails evisceration and the removal of the brain and other organs. She asked where the bullet could be removed. Burkley said he had no preference although he had. He was a United States Navy admiral, and Kennedy had been a Navy lieutenant. "For security reasons it should be a military hospital," he said.

Mrs. Kennedy was prompted to say the right word. "Bethesda," she said. The admiral was satisfied. He got off his knees and went forward to the communications shack to alert the naval hospital. The knees became a trend. General McHugh dropped to his knees to ask Mrs. Kennedy to "freshen up" before the plane landed. "No," she said adamantly. The words had become a set piece: "I want them to see what they've done."

David Powers replaced the general. The area where Mrs. Kennedy sat began to resemble an open confessional. Each man found it easier to converse with her by dropping to two knees or one. Mr. O'Donnell occupied the only other seat in the back of the plane and no man was prepared to challenge the high priest.

In the forward compartment, Liz Carpenter worked on a short statement to be delivered by the President at Andrews Air Force Base. She was block-printing it. Mrs. Carpenter would like to have used a typewriter but she reminded herself that "they are their typewriters. Besides, they make noise." She wished that the Kennedys would understand that the Johnsons had also lost a President. As she wrote, she remembered a ball in the East Room a month ago. Lyndon Johnson had danced with Mrs. Kennedy. He knew that the First Lady seldom accompanied her husband on trips. The Vice-President had put on his best smile, "Why not come to Texas with the President?" he had said. She wrinkled her nose. "You have never seen a real ranch," he said. She began to brighten. "A real Texas ranch. We're going to bring in some good Tennessee walking horses and have a ranch barbecue . . ." "I think I'll go," said Mrs. Kennedy. "It could be fun." Liz Carpenter, a woman thinking like a woman, wondered if Mrs. Kennedy felt any gratitude to Johnson for persuading her to go.

The deputy sheriffs arrived in Irving at 3:30 P.M., and they could see the Dallas group waiting at the corner. The newcomers wanted to know the story and Detective Guy Rose of Captain Fritz's staff said that Oswald lived in the middle of the block, the house at 2515. Rose was the senior in the group; Richard Stovall and Adamcik were present to assist him. The county men were invited because the Dallas detectives were outside their city line.

The men discussed the best way of finding evidence in the house and wanted to know what they might be looking for. The case was new and Detective Rose didn't have much information; headquarters said that Lee Harvey Oswald was probably a communist. He had spent years in Russia; he came back with a Russian wife. The thing was mixed up, confusing, but Oswald had a room on North Beckley, and his wife lived here with a family named Paine.

The neighborhood knew that something was amiss within a few minutes. The street was suburban, with cars parked between sidewalk and garage; bicycles and roller skates rested on lawns. Children home from school shouted to each other and saw the strange men and lapsed into silence. Mothers, spending 75 percent of their time with wash and vacuum cleaners and gas ranges, spent the other 25 percent glancing between curtains to see that the children had not wandered off. They, too, saw the newcomers, standing in a group and whispering. After that, no woman left her window.

Rose waved the men to follow him. The house at 2515 was ranch style, a young home which looked old. The gray paint was flat. The roof shingles were something between beet red and pink. A car stood in the short driveway before a closed overhang door. The hedges were of varying heights, depending upon the nourishment each found. The one strong healthy attribute was a sturdy oak which stood in the center of the front lawn, its gnarled limbs extending over the street and back across the edge of the one-story roof.

The front door was open. Rose and Stovall led the group. Two walked around behind the house, in case anyone inside tried to run. The television was on and Detective Rose could see two women sitting on a couch, their eyes on the TV screen. He had just reached the little porch when one of the women stood and smiled. "I've been expecting you all," she said.

Rose was astonished. He introduced himself and the others and the woman said she was Mrs. Ruth Paine. "I've been expecting you to come out," she said graciously. "Come right on in." They stepped inside, a bit cautiously, and Mrs. Paine introduced Mrs. Lee Harvey Oswald as a Russian lady who spoke no English. The policeman's impulse was to get on a phone and ask Captain Fritz what to do.

However Stovall began to move around the sitting room and the kitchen, and Adamcik came from somewhere and nodded to the ladies and walked into a bedroom. The search was on and Guy Rose wanted to ask questions, but he was confused, so he asked if he could use a phone. In headquarters Fritz took the call and said: "Well, ask her about her husband. Ask her if her husband has a rifle."

Mrs. Paine volunteered to translate, and Rose said: "Ask her if her husband had a rifle." Mrs. Paine said, "No" emphatically, but Marina Oswald, hugging the baby to her breast, said, "Yes" in Russian. The surprise on Ruth's face matched Rose's. "We have Lee Oswald in custody," Rose said diffidently. "He is charged with shooting an officer." Mrs. Paine translated the news to Mrs. Oswald, and she said something in Russian.

Rose was asked if he had a search warrant. He said no, "but I can get the sheriff out here with one if you want." The lady smiled as one does who has nothing to conceal. "No," she said, "that's all right. Be my guests." She was retranslating her opinions back to Russian for Marina's benefit, and it became obvious that Mrs. Oswald was not happy with her friend's show of initiative. Ruth Paine cheerfully answered what questions she could, without translating for Marina. Adamcik carefully examined the backyard, where the baby clothes swung from breeze-swept lines. Deputy Sheriffs Harry Weatherford and J. L. Oxford frisked the house, the eyes darting from end tables to couches, lifting cushions and ashtrays, opening drawers—a haphazard operation in which officers worked fast, repeating work already done.

Marina Oswald's expression changed to deep concern, perhaps fright. She pointed to the garage. In Russian she said: "He keeps a rifle in there." Mrs. Paine repeated the words in English. She thought it strange, maybe incredible, that anyone could conceal a rifle in her garage without her knowledge. She told Detective Rose: "He keeps a rifle in there." They went out into the garage.

The space was not used for a car. It was small and cheaply made, with four two-by-four beams overhead, a cluttered garage down a step from the kitchen. There were a heater, two old tires, a big band saw with sawdust underneath, some cinder blocks, several cardboard cartons used to store odds and ends, a box of tools and an electric light bulb hanging from the middle of the ceiling.

"Where?" said Rose. Mrs. Oswald led them into the garage. She pointed to an old blanket rolled into a conical shape. It was lying in the sawdust under the band saw. "I saw part of the rifle in that," she said. Mrs. Paine stepped on the edge of the blanket. "She says her husband kept a gun in here," she said. Officer Rose approached and Ruth stepped off. He stooped and placed his hand under the middle of the blanket. As he lifted it out from under the saw, Marina Oswald appeared to be stunned. The blanket hung lifelessly at both ends.

Mrs. Paine brought her hand to her mouth. She had thought that the police were merely investigating employees from the School Book Depository, the place from which shots had been fired. Lee Harvey Oswald was but one of many. Besides the policeman had said something about the shooting of a cop. She felt a grave realization overcome her as she thought about the missing rifle and the assassin in the window. She glanced at Marina. Her friend was not one to display emotion, but the pale complexion was white.

The policeman was also surprised. Before he lifted the blanket, he had been certain that a gun was inside. He thought he could detect the outline of it. There was a piece of white string around the narrow end of the blanket. They went back to the living room and Rose asked the women to sit there. He phoned Captain Fritz and told him about the empty blanket. Rose was ordered to bring the women to headquarters with whatever other pertinent material was found.

Another officer, out in the garage, emptied a cardboard box belonging to the Oswalds. It contained several hundred "Freedom for Cuba" leaflets. These were brought to the living room, too. Adamcik was prowling around the front of the house. A youngish woman approached him and introduced herself as Mrs. Linnie Mae Randall. She gave her address as 2439 East Fifth, and she pointed to it. Her brother, Wesley Frazier, drove Oswald to work this morning, she said. She and her mother had been listening to all the excitement on television and had heard Oswald's name mentioned. She said she wanted to report that she was looking out her window this morning and saw Lee put something long on the back seat of Wes's car. It was wrapped in paper or maybe a box.

Adamcik took out his little notebook and wrote some of it. "If you want to see my brother Wesley," Mrs. Randall said, "he's visiting my father right now at the Irving Professional Center." Yes, they would want to talk to Wesley. They would want to speak to her again. The policeman thanked her.

The ransacking of the house was haphazard. In the carton with the "Freedom for Cuba" leaflets were two photographs of Oswald holding his rifle. They were overlooked at the time. Stovall asked if Marina's husband had left a farewell note or said anything when he left home that morning. Mrs. Oswald shrugged. She had been half asleep when he dressed. A deputy went to her bedroom and glanced into a Russian teacup and came out of the room with Lee's wedding ring. It was of no great significance to the police, but it told the whole heartbreaking story to Marina. He had never removed his wedding ring. He had never returned it to her, even in the heat of arguments when he had beaten her with his fists. The ring in the teacup was a resignation from marriage. The end. In her heart, the young Russian pharmacist knew that, whatever the crime, "my crazy one" was in it.

The screen door swung open and a good-looking man walked in. He smiled at Ruth and said: "As soon as I heard about it I hurried over to see if I could help." The police asked who this was. It was Michael Paine, husband of Ruth, an aircraft executive. The marriage was a friendly estrangement, difficult to define. The police studied the man and wondered why he would "hurry over" as soon as he heard about it. Mr. Paine meant that the airwaves were laden with the name Lee Harvey Oswald and he knew that Oswald was a weekend boarder at his wife's house.

The cops found Russian books and, not knowing whether they were significant or not, stacked them in the pile with the wedding ring, the rifle blanket, Mrs. Oswald's passport, her birth certificate, immigration card, the birth certificates of the children—June and Rachel. There were some letters written in Russian from Marina's family in Minsk, a diploma, a few communist tracts. From Mrs. Paine's bedroom the officers removed a large assortment of vacation color slides, a Sears Tower slide projector, a metal filing box.

With no advance knowledge of what to take, they began to outdo each other in picking up material to be assessed at headquarters. Within a few minutes, they had a second slide projector, three boxes of high fidelity records, a telephone index book, even a wall bracket with instructions for mounting. A policeman riffled through a magazine, and, being mystified about its contents, tossed it onto the pile. It was *Simplicity*, a sewing periodical.

The chief walked into headquarters like an intruder. He came up out of an array of vehicles storming in and out of the basement, screeching brakes, and shouts, and Jesse Edward Curry took the elevator up to the third floor feeling that he had never known a day like this and wouldn't want to see another like it. As he stepped off at the third floor, he took a step and stopped. Ahead of him was a mass of male humanity jammed and murmuring. There were cameras, huge staring lights, and microphones.

Someone yelled: "Let the chief through!" and he bent his head and started toward his office, at the opposite end of the cross. Here and there he recognized a detective trying to buck the tide, working along the edge of glassy offices toward the elevator. Some men thrust microphones under the chief's chin and yelled: "Come on. Give us a statement. Did he kill the President? Did he?"

Curry kept moving slowly toward his office. He passed Fritz's Robbery and Homicide, but he couldn't see over the tops of all the heads. He remembered General Order 81, and he recalled how well Glen King had cooperated with the press. But what the hell was this! This was insanity, madness, bedlam. This was an aggressive group of strangers gone berserk. They were taking over headquarters.

He fought his way to his office. There wasn't time to look for messages. Cables were coming in thick and black over windowsills, curling across the floor and out into the hall. Curry saw Batchelor and Lumpkin, but they too were helpless. The chief learned that a policeman had had to be posted at Fritz's door to keep the press from crushing in on the interrogation itself.

Obviously police headquarters had been overrun by the press. The control points at elevators and staircase were worthless because the nation's reporters were descending on Dallas with credentials, and the European journalists were en route. It was a time to make a decision. The word must come from Curry. The situation was so bad that he had but two choices: either call his reserves and have the press dispossessed from the third floor en masse or permit them to remain there and hope that the situation would improve.

Curry was a "cooperation" chief. Editors and writers can be venomous. The story of the assassination was now bigger than anything Dallas could remember; it would go down in history as one of the major stories of the twentieth century, possibly the most dramatic. If Curry threw them out, there would be wails and protests and phone calls from men in high places. The chief and his department could assume a defensive posture in the assassination. It had happened in his city. Dallas, which had done its best, could be charged with according the President token protection.

It would be a lie. But it could be written. It could gain credence. These were not local reporters, men whom Curry knew by their first names. They were from big city dailies and wire services in San Francisco and Seattle and Salem and New York, Omaha, Philadelphia, Chicago, Atlanta, Miami, Washington, Baltimore . . .

If Curry weighed the matter, he confided in nobody. This was a day for people to be shocked and stunned for one reason or another, and it was his turn. He could not believe what he saw in the hall, but he would do nothing about it. A word from Curry and the harassed police would have been delighted to order everybody off the floor. A press room could have been set up on the first floor, where the courts and the traffic violations bureau were through with their work. Captain King could have reported to the press every half hour or every hour. He could have given them whatever Curry and Fritz wanted to tell. The reporters would have more space to set up their portable typewriters, their telephones. The photographers could have been allowed to see and photograph the suspect at the successive lineups which would be held several times for witnesses today.

There would be no scoops, no newsbeats. A sergeant and four officers could have maintained order on the first floor. But the word never came. The chief wanted to be a "good guy" and he was. But the press always had trouble spelling the word gratitude. It took what it wanted; it hanged whom it pleased; it unmanned officials who stood in its way.

Curry left his office and fought his way to the Homicide Bureau. The situation was intolerable. The chief couldn't understand men who were shouting in his ear. He got to Fritz's office and went by the cop at the door and inside to see what the captain was doing.

He looked at the prisoner, who looked up from his chair at the corner of the desk. Then he nodded to Fritz and the others. Curry's stomach began to sicken. This, too, was all wrong. This was no way to interrogate a prisoner. The proper way was to have him alone in a quiet room, with perhaps one other person—a witness or an interested party. He looked at the Secret Service men, including Inspector Thomas J. Kelley, the two men from the FBI, the Texas Rangers lounging against a wall, and two detectives from the Homicide Bureau. There was barely room to stand in the office. The air seemed to have left the place.

The chief of police looked at the young man with the bruise over one eye and a small laceration under the other. He didn't look like much, to have caused such a commotion. Curry left the office without asking a question.

● ● ●

The most positive person was Mrs. Marguerite Oswald. She had the righteously folded face of a woman who knows her rights. She arrived in Dallas dry-eyed. If there were tears for her son, she was saving them. Behind the glittering eye-glasses, her gaze was as steady as the flight of a bullet. She said she did not wish to speak to the police. She would not speak to them if she was brought into their presence. The custard jowls shimmered with determination. "I want to speak to the FBI," she said.

People by the thousands all over the world may have wept, may have wrung their hands at this time. Many who were totally unrelated to the crime were overcome by hysteria and could not continue their tasks. This woman dominated everything within her purview. All her life she had fought for a foot or two of living space, and the enormity of this crime, even the possibility that her son might be charged with it, would not alter her loud and indignant attitude toward the world.

She had dominated her husbands. She had dominated her sons. Marguerite was easily affronted and could nourish pain for a long time. The husbands, one by one, had died or divorced her. The oldest son, John Edward Pic, lived in Staten Island, New York, and did not communicate with her. The second one, Robert Edward Lee, Jr., lived in nearby Denton. She had not seen him in a year. The baby, Lee, had left her to run off to Russia. Marguerite had made trips to Washington, demanding to see highly placed officials, because her Lee had changed his mind and desired to come home. It was not a poor mother's duty to bring him back. That was the government's job. He had served his country as a United States marine. He had been overseas in the Far East. If, in Moscow, he had demanded the right to renounce his American citizenship, it had nothing to do with his present frame of mind, which was to come home with his Russian bride and their baby.

Marguerite Oswald was taken to a room where she was introduced to two men. They said they were FBI agents, although they showed no I.D. cards. Strangely they had the same name: Brown. Mrs. Oswald said she had something important to tell them—something they should know before this assassination investigation got out of hand. "I want to talk with you gentlemen," she said, sitting, "because I feel like my son is an agent of the government, and, for the security of my country, I don't want this to get out."

The men glanced at each other. They appeared to be shockproof. "I want to talk to FBI agents from Washington," she said. One of them nodded. "Mrs. Oswald, we are from Washington." The lady wasn't certain that these were the right men. "I understand you work *with* Washington," she said, "but I want officials *from* Washington." They told her that she had the right men. "I do not want local FBI men," she said. Her manner bespoke one who wants to reveal a hyper-secret which will climax the events of the day.

"Well," one of the Mr. Browns said, "we work *through* Washington." This did not mollify Mrs. Oswald. "I know you do," she said, pursing her lips and staring candidly at them. "I would like Washington men." The conversation ground to a geographical stalemate. They were not quite from Washington but they would not produce FBI agents who were.

She decided to tell them who she was. Mrs. Oswald, as was her right, always emphasized her mother role. Throughout her life, when minor debates appeared to be lost because of logic, she often said: "I am a mother. You do not know how a mother feels. . . ." The lady got a lot of mileage from pathos.

"For the security of the country," she said at last, "I want this kept perfectly quiet until you investigate." They nodded rapidly. "I happen to know that the State Department furnished the money for my son to return back to the United States, and I don't know, if that would be made public, what that would involve, and so please will you investigate this and keep this quiet." They looked at each other as though they weren't certain what weight to apply to the statement. The money of which she spoke had been lent to Oswald on his plea that he was broke. He had promised to repay the State Department, and he had, to the last penny.

"Congressman Jim Wright knows about this," Marguerite Oswald said. She also gave them the names of four officials of the State Department whom she had badgered to help her son. The two Browns left her. They thanked her and said they would contact Washington. She might have added that her son Lee, on his return from the Soviet Union, had protested a dishonorable discharge from the United States Marine Corps. He had addressed this demand for a hearing to the Secretary of the Navy, who was John Connally. Mrs. Oswald did not mention this, although it seemed to some outsiders that Oswald's protest had merit, inasmuch as he had served his hitch in the Marines honorably and had earned an honorable discharge. What he did afterwards as a defector to the Soviet Union occurred after his military service.

Marguerite Oswald had given the FBI something to think about. In a little while she left for Dallas Police Headquarters. She wanted to see her boy.

4 p.m.

The cold night wind swept the face of Europe. It came strong and steady out of the northwest, combing through the hedgerows of the Scottish moors, swinging the street lamps in Antwerp. There was a chill in it and pedestrians walked the Ring of Vienna with heads down and collars up to find that the opera had been canceled. The shops along the Champs-Elysées were bright with light, but the doors were locked. Under the Arc de Triomphe the eternal flame was whipped by a night wind which had no gusts but which pulled steadily at the crisp leaves along the Bois de Boulogne.

Radio Eireann canceled its programs as though anything more frivolous than Brahms would be sacrilegious. A Dublin commentator said: "It's as if there was a death in every family in Ireland." In the little Wexford town of New Ross, Andrew Minihan remembered that John F. Kennedy stood in the square and said: "This is not the land of my birth, but it is the land for which I have the greatest affection, and I certainly hope to be back again in the springtime."

Out in the hills beyond Dublin, Sean O'Casey, eighty-three and finished, the eyes dim beyond repair, sat under a bright light, the blue-brooked hands trembling, and wrote: "Peace, who was becoming bright-eyed, now sits in the shadow of death; her handsome champion has been killed. Her gallant boy is dead. We mourn here with you—poor sad American people." In London, the great bell of Westminster Abbey began the solemn bass tone which reverberated across the bridge and up toward The Strand. No one paid much attention until it passed the count of ten. It would toll for a solid hour, a tribute reserved for royal dead.

In Burdine's store in Miami, Mrs. Christine Margolis sobbed on the phone behind the cosmetics counter. Her daughter was trying to tell her what had happened, and Mrs. Margolis moaned: "Honey, don't cry. Don't cry." A Marine sergeant in Caracas, Venezuela, had an hour of daylight left. He strode smartly to the flagstaff in front of the United States embassy, saluted, and pulled the halyards until the banner was at half-staff. A Greek barber in New York said: "I cry."

Richard Nixon reached his home in New York and dialed J. Edgar Hoover. The FBI Director said that the Dallas police had picked up a suspect named Lee Harvey Oswald. He was a member of the Fair Play for Cuba Committee and a self-proclaimed Marxist, Mr. Hoover said. Nixon sat thinking of his Texas statement that Lyndon Johnson might be dropped from the Kennedy ticket. In the East, three race tracks closed—Aqueduct, Narragansett, and Pimlico—"in memory of President John F. Kennedy." Many of the bettors did not know that he was dead.

At Andrews Air Force Base the order went out to don "dress uniforms." The Air Force posted a ceremonial cordon of honor guards on the hard stand where *Air Force One* would stop. The Army sent three squads of men from Fort Myers, men properly drilled in a deathwatch. Helicopters coming in from the White House with distinguished mourners were requisitioned and ordered to stand by in case Mrs. Kennedy and the new President wanted to use them. The Marines, the Navy, the Coast Guard sent representatives. Someone suggested that it would be fitting if one or two men from each branch of service was used. They began to learn to drill together within the hour.

Nuns in convents all over the world knelt in dim chapels—no matter what the hour—intoning the rosary for the repose of the soul of a Roman Catholic chief of state. Dr. Russell Boles, Jr., was summoned from Boston to the side of

Joseph P. Kennedy at Hyannis Port to ascertain whether the father, convalescing from a cerebral hemorrhage, could withstand the shock of the news. At the United Nations in New York, the news was whispered to the pink bald head of General Dwight D. Eisenhower. For a moment, he showed outrage. "There is bound to be a psychotic sort of accident sometime," he said and put on his topcoat and left.

There was an astonishing river of flame in West Berlin. It started with students holding torches, parading toward the Rathaus. The parade picked up volunteers on each street. By the time it reached the big square with the rough-stoned buildings, 300,000 Germans were carrying torches, and a band began the slow sad strains of *Ich hatt' einen Kameraden*. In Bonn, Chancellor Erhard proclaimed a military alert; the German government feared a Soviet invasion.

An advice was received by the U.S. Naval Hospital at Bethesda, Maryland, that an autopsy would be performed there. Doctors were summoned to the administration office by flashing numbers in the corridors, and teams were made ready for the work. A suite of rooms on the seventeenth floor, including kitchen and sitting room, was prepared for the Kennedy family. The doctors at Bethesda were aware, from radio and television reports, that the dying President had been taken to a place called Parkland Memorial Hospital in Dallas. No Navy doctor thought of phoning Parkland to ask what procedures had been tried, what wounds had been treated, to ask to what surgical abuses the body had been submitted. Nor did it occur to Parkland, when the news was broadcast that the remains were headed for Bethesda, to phone with a summary report of Texas procedures.

Shortly after 4 P.M. in Dallas, Dr. James Carrico completed a two-page summary of medical findings and procedures at Parkland. It encompassed only the emergency work of Carrico and his confreres. No one had time to examine the President thoroughly before he died. Or after. In another office, Dr. William

Kemp Clark completed a two and three-quarter page medical summary in his own rapid scrawl. It might have helped Bethesda to know that the extruding hole in the President's neck had first been a small exit wound and that it had been enlarged surgically to permit a tube to be inserted into the bronchial area to assist breathing.

It was not a good day for professional thinking of any kind.

There is a penalty for being the so-called "good boy" in a family. Robert Oswald was the good boy. He wore the attributes of a responsible citizen when he was very young. His mother put him and his brothers into an orphanage. Robert understood unquestioning obedience, respect to elders, how to face misery, to live in hardship and poverty, and to protect a younger brother. Robert Oswald was born old. The only time he ever boasted, he said: "I do not go to pieces."

He was a medium-dark young man who married early and was a steady provider. He worked in a brick plant in Denton, Texas, and the company sent him to Arkansas for additional training. Life was exacting, but Robert and his wife knew that in time, they would own a little house and be able to stake out fifty feet of grass as their own. He kept away from his mother because she whined and had little tact. His older half-brother, John Pic, was in the same situation and managed to remain aloof from Marguerite except for the time, years ago, when she left Texas and tried to move in with the Pics in New York. Inevitably, there had been trouble between the women; inevitably, there had been maternal ultimata; the mother had taken the sullen little brother, Lee, and moved to another apartment. The boy was a truant, and the school authorities in New York had put him away for psychiatric evaluation. Marguerite managed to escape the courts of New York by running back to Texas with the little one.

Robert could not divest himself of the responsibility he felt for Lee. The publicity in the newspapers when Lee sailed for the Soviet Union fell on Robert. When Lee came back to Texas, Robert was at the airport to meet him. Lee said: "Hi!" and slapped his brother on the back. Then he looked around and said: "Where are the reporters?" There weren't any, and Robert hoped that the family name would not get into the newspapers again. He couldn't understand his younger brother's disappointment at not seeing newsmen at the plane.

Today, late in the afternoon, the name Oswald boomed from a small portable radio set in the plant. Robert had been at work and had heard about the assassination. Like his co-workers, he felt badly, doubly so because it had happened in Big D. It did not stop him from applying himself to his job. The radio could be heard as the men worked. There were bulletins about the hospital, about the death, about Johnson, the finding of a rifle. Much of it was sporadic and incoherent. Robert hoped that the police would get whoever was responsible.

Then the blow fell and Robert stood nodding dumbly. Men were around him saying that the police had arrested someone named Lee Harvey Oswald. They had him for shooting a policeman and the commentators said that maybe the same man shot the President. Robert nodded. That was his brother's name. Sure. Not just a relative; a kid brother. The men wanted to know what Robert was going to do. Young Mr. Oswald stood in the middle of the shop, thinking. What to do? The first thing to do was not to "go to pieces." The second thing would be to see the boss and ask for some time off to go to Dallas.

There was never any doubt about the second move. Didn't Robert always protect Lee from the big kids in the orphanage? Wasn't Robert the one who used his meager spending money to buy Lee a toy, a game, a ball to bounce? Robert could do anything except pry that kid loose from mother. Nor did he try. He knew, all along, and even when he was too young to know, that there was something wrong in an existence where a mother worked all day and ordered a little boy to play by himself in a room. There seemed to be something overly protective in having the little fellow sleep with mother until he was eleven. There was something wrong in that deep silent stare that the kid turned toward the world. When Lee returned from Russia, Robert and his wife tried to be friendly, but Lee talked about political doctrines which Robert could not comprehend. The two couples experimented with a friendship which died almost at once. The Russian bride could not be understood. Lee kept his family at Robert's house, but the younger brother immersed himself in books. He was not interested in old memories or discussions about mother. Robert said he might speak to some people and try to get Lee a job. His brother's eyes came up from a book and they were remote. Lee said he could handle his affairs.

The office phone was nearby. Robert Oswald called his wife. "Vada," he said, "you been listening to the television?" Yes, she said. The news was awful. Had she heard Lee's name mentioned? No, she hadn't. His name had been mentioned. He was arrested. "I'm leaving here for home," Robert said. It was said in a calm tone. As he hung up, the phone rang again. It was for Oswald. The credit manager, Mr. Dubose, was calling from the Forth Worth office. "Bob, brace yourself," he said. "Your brother has been arrested."

"Yes, Mr. Dubose. I know. I just heard." Robert Oswald felt a fear. He could steel himself against the words, "Your brother has been arrested," but he felt that he could not stand to hear such words as "for the assassination of the President of the United States." He took a deep breath, and Dubose said, "Your mother has been trying to reach you." Oswald said, "Thank you" and hung up.

In a moment, he was back on the phone. Oswald called William Darwin at the main office of the Acme Brick Company and asked permission to leave for Dallas at once. Mr. Darwin was sympathetic. "I know," he said. "I just heard. You go ahead and do whatever you have to do, Robert. Don't worry about the office." Oswald felt the security in the tone, and he was grateful. Had he studied American history he might have recollected that, when John Wilkes Booth shot President Lincoln, John's famous brother Edwin, the Shakespearean actor, was barred from stages all over America, even though he had no knowledge of his brother's deed. Disconsolately, Robert Oswald said that he thought the best thing to do would be to phone the Federal Bureau of Investigation first. They would know what he should do.

He was indeed panic-proof. At a time when competent minds were being swept up and fragmented in a vortex of fear, Robert Oswald disciplined himself to accept the deepest sorrow of his life.

The boy, a little too straight, his jaw a bit too grim, strode through the emergency area and, in spite of the arms which reached out to inquire where he was going, he kept in motion, got on the elevator, and went up to the second floor. He was seventeen, a time when the adolescent sees a man in a mirror, and he went past the guards, the Texas troopers in the corridor, as though he had no time for greetings.

He tried the most guarded door and it opened to his touch. Inside was Mrs. John Connally. He threw both arms around her and said: "It's going to be all right, mother." Mrs. Connally rocked in the embrace of her son John and, between sobs of joy, demanded to know how he could possibly have come all the way from Austin in so short a time. He wanted to know about his father. "I hitched a ride on an airplane," he said. He was man-like, offering courage and support to his mother and, in the same breath, insisting that he be permitted to see his father.

Mrs. Connally brought him through a door into the next room. There, the chief executive of the State of Texas reposed, looking like a scientific octopus. Plastic tubes ran into his body from overhead positions; others drained downward toward the floor. The fractured right wrist was suspended halfway over the bed. An oxygen mask was over his mouth, and the eyes turned to see the wonderment in the face of his son and the smile of pride shelving the tears on the face of his wife.

An old Navy roommate came barging through the door. This was the huge figure of Henry Wade, district attorney. Mr. Wade was most of all a practical man. He had seen death at close hand many times and he had no time for mourning or thumping his chest. Tonight he had a social appointment, and he planned to keep it. He stopped at Parkland to say hello to the living.

The Governor could not speak with the mask on his face, so Wade and young John wished him well. Mrs. Connally thought that this would be a good time to tell her husband some bad news. She was dwelling on it when Connally lifted the mask off with his left hand and said: "How is the President?" Nellie Connally said: "He died." The white head on the pillow nodded. "I knew," he said. "I knew." The mask snapped back on the mouth and nose.

They left as Doctor A. H. Giesecke, Jr., came into the room. The external signs, skin, lips, fingernails, had improved. The Governor emitted a groan with each breath and said he felt restless. His right shoulder was sore. It was impossible to get into a position of comfort. The doctor knew that all of this was unimportant. The Governor said that he felt a constant urge to urinate. Dr. Giesecke explained that a catheter was in him, and that this would cause urethral discomfort, but to bear with it for a while. Throughout the operation, the Governor had lost 1,296 cc. of blood and 450 cc. of urine.

He was dehydrated, even though whole blood had replaced what he had lost. As Doctor Giesecke concluded his examination, he noted minor signs of cyanosis, pink complexion, pulse 110, blood pressure 120 over 70, and the extremities were "warm and dry." It was too early to give the patient a sleeping potion. He would have to bear the pain until later in the evening.

• • • •

The first aid room was a quiet place. The woman moaned and sobbed without control, and then she stopped and talked rationally for a while. Detective James Leavelle reasoned with Mrs. Helen Markham, and she sat on the white enameled chair listening and nodding. Suddenly she would see again the young fellow in the jacket waiting for the policeman to come out of the car, and she would hear the shots and watch the cop fold toward the road in slow motion. She clenched her hands between her knees and rocked back and forth with uncontrolled hysteria. The screams were high-pitched and steady.

Leavelle and his partner, C. W. Brown, kept reasoning softly, and the screams diminished. Mrs. Markham required time to resume control. Then, when she had wiped her eyes with a handkerchief once more, she said she was afraid to look at the man who fired the shots. Leavelle reminded her that the Dallas police were not sure they had the right man; they thought they did. It would be up to her and the cab driver and the Davis sisters to identify the man. Mrs. Markham wasn't sure she could do it; the episode had made her hysterical with fear and she wasn't certain that she could stand in a lineup room and look at anyone.

Leavelle smiled. He picked up a phone on the other side of the basement room and called the third floor. To Captain Fritz, he said: "Mrs. Markham is ready." Word also went to the fifth-floor jail. Sergeant Duncan looked around to find young slender men who would approximate the build and age of the suspect. He ordered the jail clerk, Don Ables, to go to the basement showup room. Patrolman W. E. Perry was also young and slender. He reported to the basement. Richard Clark was another candidate. They passed muster as reasonable facsimiles of Lee Harvey Oswald, but he had something none of them had: a bruise over one eye and another under the other eye.

Fritz and his Homicide detectives brought Oswald from the office through the press mob. Questions were shouted; responses were mumbled while other questions were heard. Radio reporters with microphones were thrust away from the prisoner by detectives who wore their cowboy hats indoors. Leavelle watched Mrs. Markham. Her head moved birdlike from Detective Sims to the so-called prisoners, and the woman began to wring her hands. Ables was first, as Number Four, then Clark, Oswald as Number Two, and Perry as Number One. The moment Oswald began to climb the steps, Mrs. Markham began to weep. She held her hands to her mouth and said: "That's the man I saw." Lieutenant Leavelle, conducting the lineup, didn't ask her which man.

He ordered each of the four to turn profile, each side, then slowly completely around, and he asked questions like "What is your name?" "Where do you live?" "What is your occupation?" of each one except Oswald. The suspect stared toward the screen, then began to look upward as the others answered questions. He listened to the obviously spurious replies. Each man was referred to by his number and, when Leavelle completed the questions, he turned to Mrs. Markham and she tried to calm herself to say: "He is the one. Number Two is the one."

To make certain, she asked Leavelle to have the number two man turn sidewards again. "The second one," she said. "Which one?" the police said. "That second one—the one you called Number Two." Mrs. Markham began to feel weak. "Number Two from which end?" the police asked. Mrs. Markham pointed. Then she fell over in a faint.

The prisoner was returned by jail elevator to the third floor. Boyd and Sims cleared a path through the shouting press. Once Oswald tried to respond to a question, but the words were engulfed in sound. He was returned to Captain Fritz's office. In the hall, Ferd Kaufman, an Associated Press photographer, had to hold his camera high and aim slightly downward to catch the swift appearance and disappearance of the suspect.

Behind him someone said: "Eddie." It was not Mr. Kaufman's name, but he turned and a stout, middle-aged stranger said: "Excuse me. I thought you were Eddie Benedict." Kaufman knew that Benedict was a Dallas freelance photographer. He said: "That's all right," and the stranger gave him a card. "My name is Jack Ruby," he said. "I own the Carousel Club. This card will entitle you to be my guest at the club anytime." Kaufman nodded his thanks, watching to see if there was going to be any more immediate action with Oswald. Ruby broadened his smile, as though he didn't feel that he was making a sufficiently deep impression. "I'll be the only businessman in Dallas who will have an ad in the morning paper saying that his place will be closed for three days in memory of the assassination of the President."

Kaufman looked at his watch. "I don't know," he said. "There is still time to change an ad in the morning paper. Up to five o'clock," he said. They talked for a moment, and the photographer excused himself. He stuck the Carousel Club card in his pocket. Most newsmen would not wonder what a nightclub owner might be doing at police headquarters at a time such as this. To the contrary, one of the indexes to the truly big story is to count the number of outside individuals—politicians, department store executives, local characters afflicted with "copitis," social lions—who are impelled to be at the scene of a major crime or disaster.

Ruby had "copitis." Many times, he had bought the smiles of hardened police officers by bringing bags of cole slaw, cold cuts, seeded rye bread, rolls, and coffee. He had repeated his act at many big Dallas fires. He was never truly "in" with the officers of the law, but Ruby could get in and out of most places where the public might be excluded. A large number of cops had been his guests at his nightclubs and, though he might be a pest, no one wanted to order the man to leave.

• • •

Night fell quickly in Washington, as though for an event as solemn as this, darkness was *de rigueur*. The capital, always gifted with an acute sense of the proprieties, demanded the mood of the long shadows. Sometimes, in the bronze sunsets of autumn, the city clutches dusk to its bosom, but not on this day. The sun, so it seemed, had been there a moment ago, and now the great dome of the Capitol stood bright and pale against the black of the sky.*

Taxis returning to the city from across the Potomac spun the circle from behind the Lincoln Memorial and slowed to study the bronze face in meditation, the knuckles shiny on the arms of a granite chair. On the front porch of the White House, the great glass vial was alight, shedding its radiance between the great columns. Pedestrians peered between the tall iron pickets of the fence, sensing the historic majesty of the building which had lost a tenant. Inside, an usher carried the late newspapers to the table behind the President's desk. Among them was *The London Daily Express*. On its front page was a photo of Barry Goldwater. The caption read: "The Man Who Is Gunning for Kennedy."

Everywhere the dismay of the people was indoors. There, too, the lights were bright, but spirits were dark. The cocktail lounges were patronized. Office workers sipped drinks and talked of the event, both actions of therapeutic value. For some, it would be an excellent night to remove dinner from the list of events and place a bottle on the table.

There were a few happy people left. At 3044 O Street, in Georgetown, two of them scrambled upstairs and down, chattering about the things which please children. Neither John-John nor Caroline appeared to notice that *Grand-mère*, waiting inside the front door, stooped to give them an extra fervent kiss and a

---

* Throughout the book, all times given are Central Standard. At this time, it is 4:20 P.M. in Dallas, 5:20 P.M. in Washington.

381

long hug. She had dried her eyes and caressed the lids with a powder puff, and they had not noticed. Mrs. Hugh Auchincloss disciplined herself to face the children with a smile, but she could not bear to engage glances with an adult such as Maude Shaw, their nurse.

She wanted the children to eat early, so the dining room table was being set, but there had been no time to make preparations for an overnight stay. *Grand-mère*, a handsome woman who knew Washington as she knew the handrail on the staircase, led everyone up to a guest room. It was going to be awkward, she thought, because this room had twin beds. Miss Shaw could use one; Caroline was accustomed to a youth bed, but John-John needed a crib.

There was one in the attic, an old one with sideboards kicked by energetic feet, but it was in disrepair. Maude suggested that she could make it do. They went up to the attic, while the youngsters spun with the happiness of an unexpected visit to a loved one who granted them small liberties denied by their mother. For them, this was a good day. Caroline, who was usually sensitive to the moods and demeanor of others, did not even notice the shocked silence of the people at the South Entrance when she left the White House. They stared; some averted their eyes; one wept and turned to study a wallpaper of ancient ships.

The women brought the crib down. It was in sections, and something might have been done with it if some of the long screws and washers had not been mislaid. They sat on the floor of the guest room, trying to hold the pieces together. A maid brought some cord, and they tied the parts together. It was a sorry vehicle, but when the women shook it, the crib held.

Bedclothes and a mattress for the crib were found. It seemed good, in a way, to have many small things to do. When the guest room was ready, Mrs. Auchincloss led the way down to the dining room. There it was bright and cheerful, and the children consumed what was placed before them and kept up a running brook of comment. After a little play, they would be undressed and helped with their nightclothes. Mrs. Shaw kept telling herself that she couldn't do it; she just couldn't tell them.

The mood inside the White House was demanding and uncompromising. Men from bureaus and departments were impressed into service. Some, who did not have credentials, had to be endorsed by phone at the West Gate. They were planning a funeral for a President. He was still on his last flight and was still to be autopsied and embalmed, but the gentlemen acted as though there was not a moment to lose. They had yet to hear the wishes of the widow and those who were acquainted with Jacqueline Kennedy were aware that her wishes were adamant and positive.

Robert Sargent Shriver, Jr., handsome wavy-haired head of the Peace Corps, was the ranking officer. His wife was Eunice Kennedy, sister of John F. Kennedy. He had walked into the White House, selected a sizable office, and begun to recruit assistants. At the time, the government of the country was being directed by assistant secretaries and undersecretaries of the several ministries and it would limp along sans decisions until the new President reached the White House and began the true assumption of power. However, in the matter of a funeral, Sargent Shriver could and did make decisions all afternoon and evening.

The room in which he sat had the trappings of a dispatcher's office. Men came and went; no one seemed to stay for more than a moment. The commands were enunciated; the arguments were given brief ear; the alternatives were examined. At one time or another, Ted Sorensen, Kennedy's myopic speech writer, was present; McGeorge Bundy, who was handling the Situation Room in the basement, appeared and departed; Jerry Behn, Secret Service agent in charge of the White House, helped in whatever way he could; Ted Reardon, Kennedy politician, volunteered; so did Walter Jenkins, a Johnson assistant; old Averell Harriman, the elder ambassador; Dick Goodwin, writer; Captain Taz Shepard, the president's Navy aide; a priest from St. Matthew's Procathedral; numerous dragooned public relations men from several bureaus; and Lieutenant-Colonel Paul Miller, the U.S. Army's chief of ceremonial affairs. In addition, stenographers trotted in and out with phone calls to make, messages to type or deliver orally. The curving white hallway, with its Secret Service agents and White House police, was jammed with men who thought of this matter or that in connection with the impending funeral and who had to see Shriver about it at once.

Some men were there merely to keep other men out. The area generated an enthusiasm akin to a campaign. Words, ideas, suggestions, and questions flowed and ebbed through the room as though history was being made here and the participants wanted it to be flawless. It was possible that Mrs. Kennedy might desire to bury her husband in Brookline, Massachusetts, beside their infant son Patrick Bouvier Kennedy. This was the most likely choice. And yet Shriver thought of the National Cemetery at Arlington and phoned Superintendent John Metzler to resolve some of the difficulties attendant on such a matter.

Roman Catholics are admonished to be interred in consecrated ground. Shriver had to know whether Arlington could be considered such. Mr. Metzler said it could. How about children? Could they be buried with their parents? Yes. Would there be any objection to the interment of a President there? Had the matter ever come up before? No. Yes. Was suitable space available in case the family made a decision in favor of Arlington? Yes, there was. In fact, the men who administered the cemetery were prepared to offer a three-acre plot for John F. Kennedy. As a serviceman, even as a President, Kennedy was not entitled to such a large allotment of space, but this was not a time for anyone to be rational.

Colonel Miller told Shriver that a private funeral home would be necessary to prepare the remains. Everyone within earshot bristled. The colonel said that the military could not embalm and dress the body. It would have to be done by the family. This led to a discussion of Washington funeral homes. A firm decision was made to try Gawler's. Someone phoned Bethesda and the Navy hospital agreed to embalm Kennedy but could not dress him. Colonel Miller alerted Gawler's. A moment later, someone said that Bethesda could do the whole job. Shriver ordered the other arrangements canceled. All hands listened, and no one did it.

Miller also ordered a funeral detachment of ceremonial soldiers to report to Gawler's at once and commence rehearsing the Deathwatch. In the press room, second-stringers from the newspapers asked what was going on, and there was no one to keep them informed. Some were filing stories that the body was coming directly to the White House by helicopter; some followed the lead of the National Broadcasting Company, which announced solemnly that Kennedy would be buried in Massachusetts.

McGeorge Bundy ordered an assortment of top-priority filing cabinets emptied and locked. He sealed them in the name of President Lyndon Baines Johnson and thrust the keys into his pocket. The tall, spare figure of Angier Biddle Duke, chief of protocol, was almost grafted to a telephone as the embassies of many nations phoned and asked the proper means of expressing condolences, whether it was possible to pay respects at Andrews Air Force Base, should the minister wear mourning clothes, and who would greet him. It was necessary to find a large silver tray for the State Department, so that *cartes de visite* could be dropped.

The atmosphere was not mournful. The White House had no time for tears. Dozens of men of governmental second, third, and fourth rank hurried into the office with terse questions, listened to the battle orders, and rushed out to execute them. The chief had fallen on his shield. America was going to see a funeral designed to scar the conscience of every citizen.

In the Fish Room, a small radio was on and a few secretaries listened to NBC's McGee announce: "At no time has Mrs. Jacqueline Kennedy lost her composure, although her clothes were spattered with President Kennedy's blood."

The elevator was slow. It slid upward with Oswald and his guards as though their weight was overwhelming. The operator sat behind heavy wire mesh in a private prison. At the fifth floor, the door opened and the prisoner was taken to a counter at the left. A clerk and a policeman stood behind it. They were casual. Lee Harvey Oswald was asked to empty his pockets. A detective said that his pockets had been emptied twice.

The handcuffs were removed. The prisoner rubbed his wrists. He talked softly to one detective as though this man was a solitary exception to the enmity of the world. Oswald spoke briefly about his wife. He had two children; how many did the detective have? He did not feel abused because the policeman in the theatre punched him in the eye. After all, Oswald had lashed out first. He had no objection to anything except the damned questions about a dead policeman and President Kennedy. The suspect felt that the Dallas Police Department should start looking elsewhere for guilt. The FBI men didn't like him.

There was an exchange about life in Minsk, Russia. Oswald assured the officer that the Soviet Union, while severely disciplined, was better than most people would believe. The detective decided to capitalize on the friendly attitude. Did Oswald have a rifle in Russia? No, he didn't. There is a law in Russia that rifles and revolvers cannot be sold or bought. He had purchased a shotgun, which is permissible, but he found only small game in the forests around Minsk, and the weather was too cold for hunting.

The man behind the counter asked one of the detectives to go back down to Robbery and Homicide and fetch the items taken from the prisoner. They had to be placed in an envelope and signed for. Deputy Chief Lumpkin phoned to the fifth-floor jail and ordered Oswald placed in maximum security cell F-2 with two guards on duty twenty-four hours a day. A policeman frisked Oswald once more and said laconically: "No necktie. No shoelaces." The belt was taken from him.

The sergeant behind the counter glanced at Oswald. "Okay, son," he said. "Let's have the clothes." Oswald looked to his detective for help. "What clothes?" he said. The man leaned over the counter patiently. "Are you wearing underwear?" he said. Yes, he was. "Shorts; no undershirt." "O.K. Everything comes off but the shorts and socks." Lee Harvey Oswald began to retreat into one of his belligerent I-know-my-rights attitudes. "I don't have to undress."

The officer shrugged. "How is it going to be—the easy way or the hard way?" Oswald began to unbutton his shirt. "They told me I can use a phone," he said. The clerk pointed behind Oswald. "It's around that corner," he said. "Two men are using the phones now." Oswald removed his shoes and trousers. He could see two men behind glass, with the door locked, using the phones. One man had stretched the receiver so that he could lie on the floor and talk. A guard stood outside the booth with the key to the door. The Dallas Police Department had a sense of honor. It never tapped this line.

The clothing was shoved into a large paper bag. "When they want you downstairs again," the clerk said, "you just come out here and pick up your duds." Lee Harvey Oswald stood quietly in shorts and socks. The socks were falling down. The prison was warm, but the act of degradation was not lost on Lee Harvey Oswald. The brownish hair was wispy. The angry face tapered to a chinless point. The neck was long with a lumpy Adam's apple. The shoulders were polished knobs. The arms and wrists were well muscled. The belly was flat and hard. As a human he was, for a moment, as insignificant as John F. Kennedy under a hospital sheet.

The guard took him by the arm and led him back toward the elevator and past it and across a broken concrete floor. "I want a lawyer," the prisoner said. "I know my rights." The guard nodded. They walked through a heavily barred door. They were now in a narrow alley with a wall on the right and three prison cells on the left.

The first two were empty and open. In the third, a Negro was head down on a bunk. Oswald was herded into the second cell.* The door clanged behind him. He examined the square space which was now his, and the perpetual pout returned to his pursed mouth. There were four bunks, two on each side of the door. The lower ones were wooden and partly bare of paint. The upper ones had springs and skimpy mattresses. Across the top of the cell there were bars. Above them he could see the ceiling, lumpy with coats of pale paint.

Four light bulbs, screened by wire, lit the little alley. In the back of the cell, a porcelain sink with chipped sides entertained the steady drip of one faucet. Near it a sloping hole in the floor was ready for functions of the body. There was no flushing system, no bucket for water. Oswald looked around. His steady eyes fell on the Negro in the next cell. The head had not lifted. He must have heard the exchange of words as Oswald was brought into the maximum security area. And yet, whatever crime they had pinned on him, whatever despondency had been induced by stripping him to his underwear and putting him in here, he would not raise his head to look.

Oswald glanced out into the alley. At one end, an officer sat in a round-backed chair with his arms folded across his chest. His eyes were on the vertical prisoner. At the other end of the alley, only eight or nine steps away, another man sat. The eyes fastened themselves onto the prisoner's skin. "How about that phone?" Oswald said in his aggrieved tone. The man at the far end of the alley did not move. His lips said, "Soon as we get the

---

* After Oswald was shot to death by Jack Ruby on Sunday, November 24, 1963, the first cell was occupied by the nightclub owner.

word." "I want New York," the thin body said. The policeman seemed to have an allergy to conversation. He took his time. "In a little while," he said.

The dialogue ended. He knew and they knew that he would be brought downstairs for more interrogation. The next time he was guided through the press, he would have sense enough to make a *cause celebre* out of the matter of legal counsel. All he would have to do is to tell the world that he was being denied the right to an attorney of his choice. It would put the police department on the defensive. Not that a lawyer could alter matters for Oswald on this particular day; Texas law would not permit bail in a capital case. The lawyer would be of little assistance in the matter of interrogation because Oswald's attitude was that he could do well by himself. He responded to the questions which he felt were innocuous; he roared defiance and lapsed into sullen silence when the questions became dangerous.

A lawyer might press the district attorney's office either to file a charge against Oswald or to admit that it did not have sufficient evidence to hold him. He had not been booked for any crime, nor had he been charged with being an accomplice, a conspirator, or a material witness. He did not know that Helen Markham had been behind that screen a short time back. No one told him that, with her affidavit, they could hold him in the murder of Officer Tippit. The evidence was far from overwhelming but it was sufficient.

For Oswald, the real hurt was in being silenced. Here he could shout and no voice would respond. For a little while, he had been a celebrity, a curiosity worthy of the attention of ranking police officers, Federal Bureau of Investigation agents, and Secret Service men. His was a face on television. How could anyone look at Kennedy without also seeing the unknown who was charged with bringing him down? He was, in every sense, a figure of history. As an omnivorous reader, Lee Harvey Oswald was aware of this, and it may have been responsible for whatever exalted feelings he had. Guilty or not guilty, the name of Oswald would not die.

The two policemen had lazy eyes. They watched him flip himself onto the upper bunk. The hands were cupped behind his head. From the prison kitchen, he could hear the babble of voices. Prisoners were cutting chunks of meat to be cooked for supper. Potatoes were being peeled and halved for boiling. The friendly derisions of the damned flew back and forth. The giggles of the Negro prisoners lasted the longest. None of them had a radio or a newspaper. None of them knew who Lee Harvey Oswald was.

The cordial atmosphere in the Paine household dissipated into something akin to rancor. There had been a come-on-in-we've-been-expecting-you attitude when the police arrived, but it had not been reciprocated. The cops were still searching and piling their findings on the living room floor. In response to suggestions or questions, they grunted or did not reply. Out front, three automobiles were parked. This was sufficient to attract the neighbors who had been listening to television and who knew that a Russian woman named Oswald lived at 2515 Fifth.

They stood in gossipy groups on the sunny sidewalk, waiting, perhaps, to witness a mass arrest. The estranged husband, Michael Paine, was both gentlemanly and sophisticated and he could not understand why these men walked around his home for an hour with big cowboy hats on. His wife Ruth was a devotee of patience, but she was running out of it. The police were ransacking her bedroom. Her cheerful demurrers fell on blank stares. They told her to get ready to accompany them back to police headquarters in Dallas. She asked what she was going to do with her two children, and the police said they didn't have the faintest idea. Nor could they be convinced that Mrs. Paine had no connection with the case at all, except that she gave shelter to Marina Oswald and the babies and, with some reluctance, permitted Lee Harvey Oswald to board free on weekends and use her garage for the Oswald furnishings and bric-a-brac.

The phone calls to find a babysitter became desperate as the minutes ticked on. Michael Paine saw his file of musical recordings dropped on the pile of "evidence" and said: "Don't take that. It's just records." The policeman looked at him and went back into the bedroom. Paine, with some exasperation, said that anyone could see that the box contained recordings which could be purchased anywhere. The screen door was still ajar and the neighbors outside could hear the raising of voices.

Whenever anyone suggested that an item on the pile was insignificant and of no value to such an investigation, it was certain to remain on the pile to be taken to Dallas. The cops were opening bureau drawers, ransacking personal effects, returning aimlessly to the garage where the rifle blanket had been found, digging into cartons and boxes with no notion of what they were searching for. Ruth Paine asked if she could go to the home of her babysitter nearby. Permission was granted with police escort.

Marina sat with Rachel in her arms, watching the action but saying nothing. June still slept in the bedroom, although it would be difficult to understand why the sound of heavy feet and the deep tones of the strange men did not awaken her. Christopher Paine also slept. Lynn Paine sensed the excitement and ran to her father's arms. Mrs. Paine asked a neighbor, Mrs. Roberts, if she could stay with the children while the adults went to Dallas Police Headquarters, but Mrs. Roberts said she was sorry, she was on her way out.

The woman and the policeman walked to the next block, where a teenage girl worked part-time as a babysitter. Mrs. Paine watched the officer dogging her steps and she said: "Oh, you don't have to go with me." He said: "I'll be glad to." The mother of the babysitter felt that it would be all right if two of her daughters babysat but not one. They brought their school-books with them.

At the front of the house, Mrs. Paine, whose airy innocence seemed to unman the cops, saw the great assortment of material being carried into one of the automobiles. When she saw three cases of recordings, she smiled and shook her head negatively. "You don't need those," she said forgivingly to one of the policemen. "I want to use them on Thanksgiving weekend. I have promised to lead a folk dance conference. I will need all those records, and I doubt that you will get them back on time." She peeked inside the car. "That," she said pointing, "is a sixteen-millimeter projector. You don't want that."

The cop grabbed her arm. He too was running out of patience. "We'd better get down to the station," he said. "We've wasted too much time as it is." There was ominous authority in the tone. They stood on the curb a moment, and Mrs. Paine said: "Well, I want a list of everything you are taking, please." The police had no search warrant, nor bench warrants for arrest or detention, but they were tired of the hour-long dialogue. "We better get down to the station," they said.

She insisted on going back into her house. Mrs. Paine had already changed from slacks to a dress. Mrs. Oswald said she wanted to don a dress. This brought a flat no from the police. Acrimony was beginning to show. "She has a right to," Mrs. Paine said, voice rising. "She is a woman." The two babysitters watched openmouthed. In Russian, Ruth Paine told Marina to go into the bathroom and change. An officer barred the door and said, "No." Marina Oswald looked down at her checked slacks.

One of the policemen pointed to the children. "We'd better get this straight in a hurry, Mrs. Paine," he said, "or we'll just take them down and leave them with juvenile while we talk to you." The Quaker snapped at her daughter: "Lynn, you may come too." The threat backfired. The police took Michael Paine in one automobile, piled the Oswald and Paine effects in another, and put Mrs. Paine and Mrs. Oswald, in addition to Lynn, June, and Rachel, in a third. In the house, the two babysitters watched one child.

The women began to speak rapidly in Russian. A policeman in the front seat said he grew up understanding some Czech, but he couldn't decipher the conversation. The three cars rolled back down the highway toward Dallas. The Czech-descent cop said to Mrs. Paine: "Are you a communist?" "No," she said. "I am not, and I don't even feel the need of the Fifth Amendment."

Roanoke and Lynchburg stood still in the darkness below, embers in a dead fireplace. The big plane seemed to have no forward motion. Far below, the government trackers watched it and listened to Swindal and his first officer ask instructions for descent. The code names of beacon stations, the radio call wavelengths, the rate of descent were heard and repeated. *Air Force One* was two hundred miles out. The earth was black; the sky at forty thousand feet was still deep blue; the dying sun lingered behind, over Dallas. They had seen each other this day, that sun and this man, and he had gloried in the effusion of warmth and light. Behind the huge blue and white tail, the sun was still up. It had outlived him by four hours and more, but no one marked it or cared. Night was a fitting mask for faces.

He had wanted to say something while the sun was high. To him, San Antonio was romance; Austin was political friction; Houston was a muscular giant; Fort Worth was war planes, and Dallas was a snob. There was something to be said in each of those places and much of it was superficial and pedestrian, but he had polished the stone of the Dallas speech with his own hands. He had rubbed the words and refashioned the phrases. The knowledge that he needed Dallas but Dallas didn't need him raised the hackles on the back of his neck. It was necessary to bow deferentially to the self-sufficient when he would have preferred to use the words to whip these people.

Mr. Kennedy had no patience with those who could not see. To his way of thinking, they were mournful apostles trying to resurrect a world which died when the last cannon cooled in 1945. America could not shirk its responsibilities to a world which could look in but one of two directions for leadership. Never again could it withdraw in safety to its shores. The cost of leadership would have to be borne by the taxpayers who cried for relief. In peace the cost of the military machine became more expensive. Man's metal arced across the skies of space, and so did man.

To a young President with a lifted chin and one hand in his jacket pocket, it was a new world with new geopolitics, chronic tensions, and internal convulsions. The civil rights decision had been handed down by the United States Supreme Court in May, 1954, but it had waited for the Young Knight to implement it. To some he was too young, too swift; to others he was the sunny smile of tomorrow; to the politicians he was the leader of the liberal wing of his own party; to Dallas he was a radical.

Had he spoken, Kennedy would not have appeased Dallas. The speech, fired in ringing phrases over the heads of luncheoners at the Trade Mart, would have attempted to divorce the reactionary voters from their reactionary leaders.

Three paragraphs into the body of the speech, he wanted to say: "America's leadership must be guided by the lights of learning and reason—or else those who confuse rhetoric with reality and the plausible with the possible will gain the popular ascendancy with their seemingly swift and simple solutions to every world problem." He did not say who, but a discerning ear might conjure the voices of the Murchisons, the Hunts, the Algers, the Walkers, the Governor who sat beside him. "There will always be dissident voices in the land, expressing opposition without alternatives, finding fault but never favor, perceiving gloom on every side and seeking influence without responsibility. Those voices are inevitable. But today," he wanted to say, "today other voices are heard in the land—voices preaching doctrines which apparently assume that words will suffice without weapons, that vituperation is as good as victory, and that peace is a sign of weakness."

Would Dallas have understood the shaft or smelled the blood? It is doubtful. The words, high-flown and as incandescent as bubbles, might have drawn applause from those who were bleeding. "We cannot expect that everyone, to use the phrase of a decade ago, will 'talk sense to the American people.' But we can hope that fewer people will listen to nonsense. And the notion that this nation is headed for defeat through deficit, or that strength is but a matter of slogans, is nothing but just plain nonsense."

Deftly his oratory danced from subject to subject: military strength backed by national will; foreign aid; Polaris submarines; to paraphrase a paragraph, he desired to tell his audience that "our successful defense of freedom is due not to the words we use, but to the strength we stand ready to use on behalf of the principles we stand ready to defend."

The words hung dead like a clapper in a bell. No one heard them; no one clamored to hear them. Makeup editors of afternoon newspapers cursed their luck as they replated editions which said: "Today at the Trade Mart in Dallas, President John F. Kennedy said . . . Swindal eased the engines and the pitch subsided to a murmur. It meant nothing to the man of the many words. He lay wrapped in plastic, the face puffy and discolored.

It meant something to others. Stomachs had asserted themselves and the stewards brought soup and sandwiches, coffee, cheese, and liquor. Especially liquor. General McHugh had ordered the kitchen closed. Someone else had ordered it opened. The empty fifths of Scotch and bourbon tumbled into bags of refuse. A few passengers were sadly drunk and maudlin. They talked of better days and better times and remembered the day Jack Kennedy said . . . The electricity of shock was in others so deep that liquor sharpened the grief. Some passengers penciled notes as though time might prove to be anesthetic. Many, confined to the plane and the presence, regressed a little and, like children, wished it all away; it never happened.

The road to safety, for Lyndon Johnson, lay in immersing himself in work. He had spent a time of terror in that hospital, but it would not happen again. He had Valenti, Clifton, Kilduff, Moyers, and Marie Fehmer running. They manned phones, made calls. He made decisions and took the more important messages. The Kennedy minions asked that the press be barred from Andrews. Johnson said no. "It will look like we're in a panic," he said. He called the Situation Room and told McGeorge Bundy that he wanted to call a series of meetings tonight and tomorrow morning. "Bipartisan," he said.

In the back of the plane, the Irish had finished a few rounds of whiskey and the conversation became sporadic. Each was deep in thought for periods of time. O'Brien, Powers, and Burkley were fatigued from the long period of standing, but Godfrey McHugh and Kenny O'Donnell kept a wary eye on those who kept the faith. Somewhere in mid-flight, Mrs. Kennedy voiced a thought about the similarity of martyrdom between Lincoln and her husband. It may have been uttered to O'Donnell or Burkley. By the time *Air Force One* began to descend, Abraham Lincoln had become the theme, the motif of the three-day "wake" of John F. Kennedy. It was an understandable and exalted thought on the part of the widow, and, in the acute distress of sudden mourning, the others thought that Kennedy had been every bit as great as Lincoln.

Someone suggested that the Kennedys could avoid the glare of lights and cameras by debarking from the starboard side of the plane; a fork lift could be raised to the level of the galley entrance. The casket and Mrs. Kennedy would be in the shadow of the plane, and the sanctity of privacy could be maintained. Jacqueline Kennedy looked up from her empty glass in the breakfast nook. "We will go out the regular way," she said in that odd, litany-like manner. "I want them to see what they have done."

In the forward cabin, the President was revising a short statement written by Liz Carpenter. It had the correct note of humility without being slavish. He said to Merriman Smith and Charles Roberts—still writing their impressions of the trip for the pool reporters—"I'm going to make a short statement in a few minutes and give you copies of it. Then when I get on the ground, I'll do it over again."

The confinement of the Johnsons and the Kennedys in the plane for a period of one hundred and fifty minutes was sufficient to cleave the families in permanent schism. Johnson, the burly, earthy Texan, lost the battle for unity by succeeding to the presidency. He was not, and could not aspire to be, Kennedy people. He could be tolerated as a Vice-President because his loyalty to John F. Kennedy was complete and unquestioned. Within the family, only Bobby and Kenny O'Donnell could not abide him as Vice-President. To them, he was a rumpled wheeler-dealer, part Southerner, part Westerner with cowdung on his heels. He lacked what they might refer to as "class."

To those among the Kennedys who felt neutral, or apathetic, about Lyndon Johnson, his tragic ascendancy to the presidency tipped their opinions against him. He was not worthy to follow their fallen hero, and his every act lent itself to two interpretations so that, among themselves, they could make him look mean and avaricious. He did not belong on *Air Force One*, they felt. The least he could have done was to permit the widow and her dead husband the privacy of the plane for the trip home. He should not have burdened Bobby—even though he was the Attorney General—with a question about being sworn in as President. It was a crass grab for power.

On the plane—her plane—he violated her privacy by offering condolences, taking over the private bedroom, issuing statements, holding the plane until a judge swore him in, thus imperiling the remains of Kennedy which might have been impounded by Dallas County at any moment. He was a crude, impossible man. At a time of stunning shock, he had the nerve to call Kennedy's top lieutenants and offer to keep them on, to give them "blank checks" to carry on the Kennedy tradition.

Lyndon Johnson must be charged with a lack of understanding of the Kennedy mentality. They required a villain for their rancor. The world lay shattered in their hands and no one could put it together again. When their chief fell among the dead roses, the heart of their political cult stopped. They had no standing anymore, no prestige. Among the politically and socially dead were Bobby Kennedy, David Powers, Kenny O'Donnell, Larry O'Brien, General McHugh, McGeorge Bundy, Dean Rusk, Robert McNamara, Jacqueline Kennedy, Arthur Schlesinger, Ted Sorensen, Mrs. Lincoln, Pierre Salinger, Orville Freeman, Major-General Clifton—a host of men and women. They were dead and they were aware of it. Many of them held Johnson in such contempt that they could not endure his offer of resurrection.

In the first moments of Johnson's presidency, he did not feel strong enough to go alone. He needed these people. He was willing to bury his pride in the bottom of his pocket and tell them that he required their counsel, their guidance. In spite of his own considerable ego, Lyndon Johnson lacked the confidence of a John F. Kennedy. "When the going gets tough," Kennedy used to say, "the tough get going." Most of all, in the cold loneliness at the summit of power, Johnson needed a feeling of continuance of administration. And this is what the Kennedy clan would deny him.

As *Air Force One* began to surrender to the forces of gravity, the small group in the back of the plane began to plot ways and means of keeping the President of the United States out of the casket photos. The world would be watching, and the Kennedys did not want the Johnsons in their mourning pictures. At one point, when Major-General Ted Clifton went aft to ask a question, O'Donnell, sitting opposite the grieving lady, curled his lip and said: "Why don't you hurry back and serve your new boss?" It was less a question than a declaration of the honorable thing—that all must go down with the ship and the captain.*

The Secret Service suggested that the new President spend the night at the White House. There was lots of room without disturbing the Kennedy family. This was declined at once. Johnson was irritated by it. "We are going home to The Elms," he said. "That's where we live. If you can protect us at the White House, by God you can protect us at home too."

The long rays splashed red against the broken comb of downtown Dallas, and a timid westerly breeze swept the confetti of the parade against the curbs on Main Street. Lights went on at the burlesque house off Ackard, and homeward traffic bounded on the elevated highways. For Dallas, this was going to be a long night. The city, afflicted with a monumental ego, flinched. As Lieutenant Jack Revill had said: Big D died.

At the end of the maximum security alley, the last rays of sunlight splashed against the dirty opaque window. By glancing at it, Lee Harvey Oswald could detect the difference between daylight and dark but nothing else. He heard the clang of the security door and looked up to see jail guard Jim Poppelwell coming in. The guards on the chairs became alert. "All right," Poppelwell said, "he can make his phone call."

Oswald dropped down off the bunk. Poppelwell was turning a dime over in his hand. The cell door was opened, and the prisoner emerged in his shorts and socks. As Oswald was led out,

---

* A few months later, in Atlantic City, N.J., I saw O'Donnell holding a door open at the back of the limousine. Jokingly, I said: "Ah, you are now the Johnson door-opener." He grinned. "Yeah," he said. "I hold doors for him." Shortly after, he quit to run for office in Massachusetts and lost.

the Negro prisoner was awakened and told that he was being transferred to another wing. The three cells would be exclusive for Lee Harvey Oswald.

The two telephones in the glass booth were also exclusive. Guard Poppelwell placed his man inside and prepared to lock the door after handing the prisoner a dime. Oswald asked how he could phone New York with that. The guard shrugged. That was not his department. Oswald has thirteen dollars somewhere in this jail. Poppelwell explained that he had nothing to do with the matter except to follow orders—give Oswald a dime and permit him to make a phone call. It was a patent injustice to grant the right of a telephone and deny the prisoner the right to use his money to make it, but Oswald realized that a protest would be fruitless.

He deposited the dime and asked for long-distance information. The operator asked, "Where?" and he said: "New York." He was told that he could deposit his dime and dial 212-555-1212. He did. When he was asked whom he wanted, the young man spent considerable time saying and spelling John Abt. He might have asked for the law firm of Freedman & Unger, at 320 Broadway, but he didn't know. Besides, it was close to 6 P.M. in New York, and attorneys would be homeward bound with their briefcases.

The operator said she had a John J. Abt at 299 Broadway. He said that would do. The number was AC 2-4611. Oswald repeated it and hung up. Then he looked back at Poppelwell and asked him to open the door. He had forgotten the number. Could he have a piece of paper and a pencil? Jim Poppelwell considered the matter gravely. He could not think of a prison regulation against supplying a prisoner with a pencil and a bit of paper.

Oswald was locked in while the guard went for it. Poppelwell tore the corner from a telephone contact slip and handed it in with a pencil. The prisoner was locked in and the process started again. This time he wrote "John J. Abt, 299 Broadway, New York" and, underneath: "212 AC 2-4611." Then he redialed the long-distance operator and asked to place the call collect. In a moment, the operator was back with him. She wanted to know who was calling collect from Dallas, Texas. He told her "Lee Oswald." It turned out almost as he must have divined; the call was refused.

Could he call again later? Poppelwell took him out of the booth and said he would ask Captain Fritz. Oswald said that he would call home and ask his wife to call Abt. On the third floor, reporters were asking police officers if Oswald had asked for a lawyer. "He's phoning one now," they were told. None of the cops could recall the name of the lawyer, except that he was a New York man.

At Andrews, the drill team came to attention. The distinguished gentlemen of the United States, and the equally distinguished gentlemen of many foreign countries, were organized in disorganized knots on the concrete. Outside the fence, thousands of faces peered through the metal links, as a similar crowd had six hours ago in Dallas. The overhead lights in a big hangar tossed a pale carpet on the apron. Military officers in blue and in gold, bedecked with ribbons and fourragères, stood at ease, watching the television cameras being set up on wooden stands.

The company was impressive and strained. Conversation was whispered. Now and then a platoon of military in gleaming boots and steel hats would march out of the darkness into the light, the rows of feet lifting rhythmically and setting down hard on the strip, to come to a loud halt in the area of the lights. The White House helicopters, green bugs with pinwheel hats, sat on the edge of the night.

Someone said: "Here he comes." Eyes lifted to the night sky. The word was passed. "Here he comes." A youthful voice roared: "Ten-shun!" Fire trucks astride the runway turned their lights on. The revolving beacon on top of the control tower snapped green. A small brown staff car took off with a burning of rubber. The word having been passed, all eyes looked in varied directions. To the west, a yellow star above the horizon was the only thing that moved.

Air Force One was on its base leg. It could be seen, not heard. The small yellow light descended slowly. It moved toward the city of Washington. All aircraft aloft or on the strip at Washington International Airport remained in aeronautical limbo. The tower cleared them out with "V.I.P." warnings, but the captains and the first officers, usually unconcerned, searched the skies from their flight decks looking for the last moment of John F. Kennedy's last flight. Some were in a holding pattern as far off as Friendship Airport at Baltimore.

The big plane was low coming over the Potomac basin. The Pratt and Whitney fans were down to a whisper. Under the silver wings passed the pale vision of the White House, the Capitol dome, scenes of triumph. The big plane made the final turn. Those waiting at Andrews Air Force Base could see nothing for a moment. Then the plane's star-bright wing lights came on, and the vision looked like a steady yellow candle standing in the sky. It remained there, seeming not to move, then the engines could be heard and the plane dropped down and down, louder and louder, over the fence and holding its rubber feet a foot or two over the blue-ribboned runway, then it touched and the engines gathered their breath for one final shriek of protest.

Air Force One ran down the runway, the windows bright with lights, and came to a pause at the far end. A woman in the group of silent men dug into her purse for a kerchief. She caught her husband's stern glance and snapped it shut. The long night had begun.

Darkness was stronger than light. There was a faint roar to the city, a sound which seemed to emanate from within the ear. The pigeons were in repose on the roof. Their claws made scratching polyphonics, mixed with cooing and the reshuffling of feathers as jealous males met. They could hear the contrapuntal click of the Hertz clock snapping the minutes like matchsticks. There were whistles, too, but these were irregular sounds from the street. In front of the Texas School Book Depository, young men in metal helmets addressed themselves to pedestrians with whistles.

The confluence of Elm, Main, and Commerce at Dealey Plaza made a traffic sink. It was slow this evening. Dallas was fascinated by it. Some people jumped on the macadam and said, "This is where he was shot." Students romped on the grass. Elders looked up at the Depository windows and saw the lights wink off one at a time. A police car parked on the grass squawked the tinny dialogue of assassination.

Deputy Chief N. T. Fisher, was leaving Love Field for police headquarters. It had been a bad, bad day. "Has there been any developments that you can tell me on the suspect that shot the officer," he asked the dispatcher. "Was there any connections with the shooting of the President?" "At this time," the dispatcher said, "it is my understanding that he is the same person. He is in custody." Fisher said: "Ten four. Thank you." "That's not official," the dispatcher said. "That's just the rumor up here. . . ." The dispatcher returned in a moment. "Four . . . hold the presidential cars at the location. 508 is en route to print them." Fisher was sure that there would be no fingerprinting of the presidential cars. "As far as I know," he said, "these cars were loaded on an Army transport. I don't know whether they are still there or not. I'll check." It didn't take long. "For your information," Fisher said, "they have been loaded and left on the other transport."

A postal employee at the terminal annex was sorting mail, scanning and skipping it into bins. He kept muttering, "Oswald" to himself. He had seen the name or had heard it. "Oswald." It wasn't common. The envelopes flicked into air and managed to drop into the proper bins. He kept thinking about the name and the shocking murder and suddenly, without trying, he remembered.

He had rented a post office box to a man named Oswald. That was where he had heard the name. He got the list of owners of boxes. The card was found. Three weeks ago a man who had called himself Lee H. Oswald had rented Box 6225. The business of the applicant, as signed on the back of the card, was "Fair Play for Cuba Committee, Chairman." The post office clerk took the card to his supervisor, who called the postmaster. They examined 6225. There wasn't much in it, but what there was was suspicious: a Russian magazine addressed to Lee H. Oswald.

No one seemed to know who had charge of the assassination investigation, so the postmaster called the Secret Service, the Dallas police, the FBI, and the sheriff. Then they posted an unobtrusive guard over 6225 to see if an accomplice might come in and claim the magazine. The clerk returned to sorting letters, satisfied that he had a pretty good memory.

Seth Kantor had a superior memory. He was using it under adverse circumstances while trying to find a place on the third floor to stand still. Of the out-of-town reporters, Mr. Kantor was the one who knew Dallas, Fort Worth, and Lee Harvey Oswald. Well, not precisely. He had not met Oswald, but he knew him. Kantor had some of the jadedness of the effete big-time reporter, but it wasn't so long ago that he had been a young and energetic innocent on the *Fort Worth Press*.

He had read about a United States marine who had defected to Russia. One day—was it in 1960?—the boy's mother had come to the *Fort Worth Press*. The paper had agreed to pay for a phone call to her son in Moscow. Kent Biffle had arranged a three-way hookup between the son, the mother in her apartment, and himself at the city desk of the *Fort Worth Press*. Everyone had been disappointed. Biffle had trouble getting the overseas operator in New York. Then he had to go on to Europe and from there to Moscow. Operators were cutting in and out of the line, and the minutes dissolved into hours.

Seth Kantor had watched Biffle from the other side of the city desk. When the call had finally gone through, the *Press* operator rang Mrs. Oswald at home, and she picked up the phone and gushed her love at her wayward boy and told him how nice the *Fort Worth Press* people had been to arrange the three-way call. Oswald hung up. He wasted no time telling his mother that he loved her, or missed her, or would write to her. She had been willing to forget that he had obtained an early discharge from the Marine Corps as a "hardship" case because his mother was ill and he was her sole support. He had come home and left at once for Russia without telling her anything. In truth, the only contact he wanted with his mother was to remind her that she owed him some money.

As Kantor recalled, the call ended abruptly. As he leaned against a partition in the busy third-floor corridor, looking at the anxiety-swept faces of his confreres, he remembered this phone call. The reporter had gone on to a better position, and he had picked up a newspaper a year ago which stated that Lee Harvey Oswald was due home at Fort Worth with a Russian bride. Kantor worked in Washington. He had clipped that story and made a note that if Lee Harvy Oswald ever came to Washington Mr. Kantor would try to interview him.

Today the reporter had been part of the motorcade in his own Fort Worth and Dallas. It had been exciting, almost emotional seeing old friends and a best man at one's wedding at a crossroads. It was not a time, considering the familiar faces and streets, the places where a man had once found stories worth space in a newspaper, to think of a defector who had been home over a year. And now Lee Harvey Oswald was once more the story, a bigger and more catastrophic one than anyone might dream. For the first time, among the craning heads, the lights and the shouts, Seth Kantor saw the sullen face, the bruised eyes, the manacled wrists, and he wondered what warped mechanism ticked in that head.

The big-time reporters worked the running story hard. They took notes on everything, even journalistic rumors, and they fired questions at policemen all day and all night. They had "leg men" out in the city picking up material on the Texas School Book Depository; the scores of witnesses who had gone through Sheriff Decker's office; the Irving angle with Mrs. Oswald and Mrs. Paine; the bits and pieces from Jack Price at Parkland Hospital; Governor John Connally's condition; the reaction of Dallas. There were men in Washington covering parts of the story, and there were others at American embassies in Europe picking up bits and pieces to add to the whole. The men on the morning sheets had to keep writing fresh material marked "Add Kennedy." Those on the afternoon papers had time, and they inundated the file rooms and "morgues" of the *Dallas Times Herald* and *Dallas News* asking for the professional courtesy of a free look at old clippings and pictures on Lee Harvey Oswald.

The story had a stunning beginning but no end. The words were dropped into the huge maw of public curiosity, were masticated, and the maw opened for more. City editors were on the phones asking for copy. They suggested fresh angles, some of which were worthless. When they had exploited the death of John Fitzgerald Kennedy to the fullest, the world was aware of it, and the only way to keep the story alive was to keep fresh material coming in on Oswald. If Oswald was the assassin, then the story centered on the Who? What? When? Where? and Why? of the Book Depository order clerk. Reporters were urged to keep on the necks of the police. The cops must be reasonably certain that they had the right man—or didn't—and a good reporter would keep badgering the officers or else run the risk of having one of them "leak" the story to a local favorite.

The story had too many parts. It was impossible to fit together all the small pieces which make a large and dismal mosaic. At the moment the reporters were racing up and down the hall with a rumor that Oswald would soon be coming down from the jail for another session with Captain Will Fritz. At Love Field, no reporter saw an agent for the Federal Bureau of Investigation board an airliner for Washington. He carried a small box and put it on his lap as he fastened his seat belt. In it were two pieces of the President's skull found inside the car and one found beside the curb at Elm Street.

On the fifth floor, a guard came into the maximum security alley with a plate. He handed it to Oswald along with a cup. The plate had beans and stewed meat and boiled potatoes. On the side of the plate were two buttered pieces of bread. The prisoner took the cup of steaming coffee. He looked at the plate. The guard said it was dinner. There would be nothing more until morning. Two meals a day in jail. Oswald said he wasn't hungry.

The guards told him to finish the coffee. He was going to get his clothes back in a few minutes.

The plane was always awkward on the ground. It came back up the taxiway slowly, whining and rocking. President Johnson had read his short statement to Smith and Roberts and had put it into a pocket in his jacket. He had issued an order for a ramp to be brought to the plane. The order stated that the Secret Service men aboard would carry the body of President Kennedy down the ramp. The casket would be followed by Mrs. Kennedy on the arm of President Johnson.

The President looked around as the plane waddled toward the big circle of light and he wondered where everyone had gone. The cabin, except for a few of his staff, was empty. Mrs. Johnson sat gazing out the window at the darkness. In the back of the plane, Kenneth O'Donnell issued his orders. They too were explicit. As soon as the aircraft stopped, he wanted the Kennedy group to crowd the rear doorway. They and the Secret Service men would take the body out of this exit, down a forklift. President Johnson was not a party to this plan.

Mr. Johnson felt that the symbol of unity was important. As the new President, he should stand behind his fallen chieftain, and he should offer his widow the protection of his person. To the contrary, the Kennedy people felt that this was boorish and overbearing. The plane was still in motion when they formed an unbreakable clot at the rear exit. They knew what was expected of them. In the group were David Powers, Lawrence O'Brien, Ken O'Donnell, General McHugh, Mrs. Evelyn Lincoln, Mrs. Kennedy and her secretary, Pamela Turnure. Flanking them were the Secret Service men.

When the President came down the aisle, an engine was still idling and he found his progress blocked. A male voice from somewhere said: "It's all right. We'll take care of this end." He recognized the humiliation. The plane stopped and he walked back to the presidential cabin slowly, to join his wife. He was about to take the arm of Mrs. Johnson when he saw his Attorney General, Robert F. Kennedy, running from the front of the plane to the back. Sadly, the President stuck his hand out and said: "Bob!" The Attorney General ignored the hand and kept running toward the aft section.

It was evident that Kennedy understood the situation. He ran so that he would not have to pause and recognize the new President. He made it down the aisle of the front cabin, squirming past the people who stood in the aisles, opened the door to the private cabin, and ran straight through. At the human knot, people stepped aside so that Jacqueline could fall into Robert's arms.

The communications shack passed a final message to Roy Kellerman from Secret Service headquarters ordering him to accompany the dead President to Bethesda Hospital. Colonel Swindal looked down on the small dark pools of people. *Air Force One*, in the glare of lights, was a dead moth. An honor guard of service men followed the Attorney General up the front ramp. They were there to carry the casket. The silence at Andrews Air Force Base was so deep that, when Colonel Swindal and his first officer, Lieutenant Colonel Lewis Hanson, hurried down the ramp, the sound of their feet on the metal-tipped steps beat an irregular tattoo into the darkness outside the pool of light.

Merriman Smith and Charles Roberts, one with a typewriter under his arm, hurried off the plane and stood under the giant wing, looking aft. In the silence there was disbelief. The men of lofty station who were there; the television cameramen in the darkness; the friends, the officers, the strangers of rank, the ambassadors, could not believe that Caesar was dead. They had heard the radio; they had seen the story on television; the late afternoon papers carried mourning rules under the headline: KENNEDY ASSASSINATED. And yet the human mind rejects the image of catastrophe as the eyes hunt for irrevocable truth. The slender face of Secretary of Defense Robert J. McNamara stood almost alone, the eyes watching the rear hatch behind the frozen light caught by his glasses. The breeze caught the wavy hair of the Chief Justice of the United States, Earl Warren, and he studied the aircraft as though, legally, he entertained a reasonable doubt.

The forklift, on small yellow wheels, circumnavigated the plane and pulled up at the port hatch aft. The operator placed it snugly and then pulled the small elevator upward. It was at least three feet short. Inside someone was opening the hatch lug and the door swung backward and away. The unseen eyes from the darkness looked. They saw a group of people squeezed together in the doorway, and five Secret Service men, stooping and pushing, shoved the edge of the casket into the doorway. It was caught in the light, and everyone below knew that John F. Kennedy was truly dead. He was gone and they would never see him again, never see him step out of a crowd with hand outstretched, never hear the flat, twangy Boston wit, see the square pearly teeth as the head rocked back in laughter, never again see that stabbing left index finger as it punctuated his argumentative shouts above the roar of a crowd.

He was dead. He was gone. And there was something that each man had forgotten to say to him. It was not a moment for tears. There was a succession of swallowing and an unspoken accusation in three thousand hearts: "What did you expect?" Whatever the expectation of magic, it was over and the unseen eyes began to focus on the woman in the doorway with the slab of dark hair down over one eye. Her expression had the shock of a little girl who has just heard the colored balloon break. She still held the string.

A few men jumped down on the lift. They pulled on the forward handles. Others, at the rear, pushed. The honor guard found itself caught in the rites of the warrior. The Secret Service wanted to carry the man. So did Larry O'Brien and Ken O'Donnell and David Powers and General McHugh. Everybody could not find room around the casket. The men pushed each other. The heavy bronze instrument teetered off the edge of the plane and began to wobble in air. Out of the darkness the sibilant voices of the television commentators could be heard. The Attorney General watched the bronze rock in air, saw the men on the lift catch and steady it, and he dropped nimbly onto the platform with his brother. With arms outstretched, he reached up for his sister-in-law.

She crouched and dropped and Kennedy held her. The big iridescent lens of the cameras caught the scene, saw the pink burled suit, the stains of blood, the twisted right stocking. In a trice, the nation knew. The horrifying picture of the gleaming bronze casket and the handsome young widow and the blood was mirrored in seventy million homes. The guilt was upon them and their children. Dinner stopped. Plates were pushed away. The picture on the screen would never be scrubbed off. It was there and scores of millions of people would date their lives with this day. Things happened before this time or after.

"Will you come with us?" Mrs. Kennedy whispered to her brother-in-law. He nodded. She knew he would. Admiral Burkley was the last to jump on the small platform. It began the slow slide to the bottom. The Bethesda ambulance had backed up, and, before it left, Robert Kennedy wanted to hear the wishes of his sister-in-law in regards to the funeral, and he wanted to use a Secret Service beep phone to speak quickly to Sargent Shriver at the White House.

He walked her slowly toward the ambulance. He took her arm and bent and whispered and nodded at the responses. She was in the full glare of the lights and her head was down. Faces were in requiem all around her, but they knew that this was no time for greetings or even condolences. McNamara stood at attention. Acting Secretary of State George Ball followed her with his eyes. Postmaster John Gronouski could not bear to watch. Franklin Roosevelt, Jr., had witnessed similar scenes. Senator Hubert Humphrey may have been the only man who wept. His eyes were red holes. Senator Mike Mansfield kept his teeth clenched. Everett Dirksen hunched his great head into his topcoat collar and turned away. McGeorge Bundy, the bright mentality behind the self-effacing features, seemed nailed to a place on the concrete.

Under the wing, Swindal and Hanson stood at salute. The tall, slender chief of protocol, Mr. Angier Biddle Duke, approached Mrs. Kennedy. He coughed and she looked up. "How can I serve you?" he said. She knew. "Find out how Lincoln was buried," she said. Mr. Duke turned away. He had an assignment. He would require researchers and admission to the Library of Congress, which was closed, and a long night of labor to find out exactly how Lincoln was buried.

In silence, she walked to the ambulance and reached for a door handle. It was the wrong one. Mrs. Kennedy did not want to sit up front. She would remain at her husband's side. For a moment, she was indecisive. Then Robert Kennedy approached and opened the rear door. They stooped and climbed inside. At the rear, the honor guard, the Secret Service, and Kennedy friends carried the casket and slid it in. General Godfrey McHugh hopped in with it.

Roy Kellerman was in charge. He waved William Greer, who had driven the death car, into the driver's seat. Kellerman sat next to him. Agent Paul Landis squeezed in on the right side. Admiral Burkley said he had told Mrs. Kennedy that he would stay with the body until it was returned to the White House. Kellerman waved the doctor into the front seat and he sat on the lap of Landis. Standing beside the ambulance were the cardiologist, the nurse, and driver sent hours ago by Captain Canada of Bethesda in case President Johnson sustained a heart attack. They were told that there was no room for them.

The car started to move forward. Mrs. Kennedy and General McHugh sat on one side of the casket. Opposite, leaning across the bronze to speak, sat the Attorney General. The lights of the base flickered by, lighting up the mourners and plunging them back into darkness. Directly behind, Secret Service Agent Clint Hill had requisitioned an automobile. He was driving Dr. John Walsh, Mrs. Kennedy's physician. A third car carried the O'Donnell group. Behind that were some women in a fourth car and the forgotten man. This was George Thomas, the president's valet. For almost three years, he had worked hard on the third floor, pressing fresh suits most of the day for the President's several changes. He would not need that ironing board anymore nor the iron. George Thomas sat silently among the women, wondering what was to become of him.

America sat before the altar of the image, murmuring "*mea culpa, mea culpa.*" No event in the history of the country—perhaps in the history of the world—was tendered so quickly, with such exquisite agony, as the shooting of a man in Dallas. It was totally unexpected; it was darkly tragic; it was exploited by the television cameras to the fullest. The facets of the somber story were revealed at the instant they were discovered by the camera.

In New York, Martin Isaacs of the Department of Welfare watched with fascination and horror, a man mesmerized by the incredibility of the credible, numbed by a succession of shocks. The magnetism of the opaque screen ruled the land. He had seen the School Book Depository from which the shots were supposed to have been fired; he had heard that a young clerk named Lee Harvey Oswald had been arrested; he had seen the big bird come to a stop with two Presidents aboard. At times, the unfolding of the story became so emotional that people who had not voted for Kennedy and were alienated by his politics broke and cried.

Mr. Isaacs could not think of dinner. He sat. He heard the solemn announcement that the networks would preempt their regular television fare for the rest of this day, perhaps until after the funeral. On the screen came a short shot of the third floor of police headquarters in Dallas, and Lee Harvey Oswald held his manacled hands up complaining. The shock on the face of Mr. Isaacs deepened into a stunned, bloodless expression. "That," he said, "that's the man I helped get back to Texas last year."

It was. The State Department had granted Oswald's request for a loan to bring his family back to the United States. They had traveled by train across White Russia, through Poland, and on to Holland. They had boarded a Holland-America liner and Oswald had used the time negotiating the Atlantic Ocean to write a draft of his personal Marxist manifesto. They debarked at Hoboken and, after some travail, had appealed to the Department of Welfare to put them up overnight at a Times Square hotel and to see them on to Dallas by commercial aircraft. Martin Isaacs had been the official Good Samaritan.

The nation was becoming absorbed in the story. The cameras were again on *Air Force One*. It sat in its own halo. There was some movement in the front door and a commentator murmured, almost in an awesome whisper: "Here comes the new President of the United States, Lyndon Johnson." Mrs. Johnson and her husband emerged. He stood in the doorway, glancing around grimly, the mouth compressed. They started down the steps.

The statesmen, still in small dark groups, remained immobile. The eyes, scores of pairs, were jaded. They understood the Washington power structure, and this man, tall and twangy had this day risen from a position of sufferance to take it all. The gentlemen had involuntarily swapped a gifted idealist for a party wheelhorse; youth for middle age; grace for awkwardness; adroit phrasing for hyperbole; a commander for a committee chairman; a Galahad for a hillbilly.

They knew. The heavy-lidded eyes watched him reach the foot of the ramp and the reassessment began. One perhaps, Defense Secretary McNamara, may have been overcome by personal grief and not been able to equate the new chieftain with the pluses and minuses of practical politics. The others turned from their several positions around the area of light to face Lyndon Johnson. McNamara, close enough to do so, shook hands with him. "It's terrible. Terrible," the new President said. The rheumy brown eyes glanced around the circle of knots.

He nodded curtly to a few. Senate Majority Leader Mike Mansfield had an arm around his wife, who was sobbing. Chief

Justice Earl Warren reached out silently and shook Johnson's hand. Arthur Schlesinger, who had been retained by Kennedy as a day-to-day journalist-historian, a Boswell to Kennedy's Samuel Johnson, stepped forward to grasp Johnson's hand at the moment that the President saw the row of microphones in the powdery light. His momentum carried him through the handclasp. It is doubtful that he heard Schlesinger's offer of assistance.

With Mrs. Johnson, he stood before the microphones and rustled a piece of paper. The spectacles were adjusted. Mrs. Johnson seemed to look, not at her husband, but toward the darkness and the red-eyed cameras banked in wooden tiers. There were no loudspeakers at the base, so that only the television audience heard the hushed and halting words:

"This is a sad time for all people. We have suffered a loss that cannot be weighed. For me it is a deep personal tragedy. I know the world shares the sorrow that Mrs. Kennedy and her family bear. I will do my best. That is all I can do. I ask for your help—and God's."

It was the wrong voice; the tone was that of the supplicant when America hungered for a leader. The last few sentences made of Johnson an average man and those who feared for themselves did not want to hear "That is all I can do. I ask for your help—and God's." The irony of the moment is that Lyndon Johnson was not a humble man; politically and personally he was a boss and it suited his personality to arrogate to himself all decisions. Had he said: "Even in its grief, this nation, this leader of nations, must move forward and, toward that end, I have lifted the banner which fell in the streets of Dallas and will carry it honorably and forcefully, as he did . . ." he might have subdued the fears, infused confidence where confusion lay.

He took his wife by the arm and turned away. The big sleek head was bent forward. The stride was the step of the athlete, toes slightly turned in. He asked McNamara to come with him.

Then he saw the slight figure of McGeorge Bundy and nodded his head for him to follow. Acting Secretary of State George Ball watched, and the President asked him to join the group. Johnson shook hands with the congressional leadership of both parties. He had worked with these men well and truly, and they respected his judgment. The professional politicians had much more faith in Johnson than the electorate. To the people he was a twangy Texas braggart.

There were many men of rank who were not asked to join Mr. Johnson. They maintained their secure knots of topcoats and windswept hair in the night. Arthur Goldberg remained alone, the onetime labor lawyer cum statesman who was present, not to curry favor with the new source of power, but to say farewell to the old. Major General Ted Clifton, a mature mind with a sense of duty, stood at hand salute as his Commander-in-Chief walked by toward the idly slapping blades of a helicopter. He was no longer a Kennedy man; he had not been invited to join the Johnson Administration. He was alone in the light, the neutral soldier.

Another man stood alone. He wore a gray topcoat. The back of it was turned up against the sudden chill. James Rowley, Chief of the United States Secret Service, had lost a President. It did not matter that he had not been in Dallas; it was of no moment that, given a second chance, his men could not do more than they had done to protect the precious life; he had lost his man. He and Behn, Kellerman, Hill, Youngblood, Landis, Greer had failed one day out of a thousand. Mr. Rowley, the hair grayer, the eyes riveted on the man now walking out of the circle of light, had advised several Presidents and was accustomed to the abrupt and irritable "No!" Sometimes Rowley had felt like saying: "In that case, Mr. President, I cannot be responsible for you," but the words had remained in his throat because Rowley *was* responsible. He was paid to protect. The study of many assassinations in many countries had taught him new techniques;

from each he had learned a little more. Today a mental misfit in a high window had crumpled all the techniques of protection and tossed them away.

Mr. Rowley stood alone. If, as often happened, the Congressmen on the Hill held an investigation and looked for a goat, James Rowley would insist on donning the horns. He would not permit Jerry Behn, head of the White House detail, or Roy Kellerman, the agent-in-charge, to assume the burden. The chief himself would be responsible, and he would be the first to say that, even if John F. Kennedy had allowed the bubbletop to be placed on the limousine, "it was not bulletproof."

In the helicopter, the President sat wearily. The overhead light was on. He squinted up at it and flagged the others to be seated. Mrs. Johnson sat on the couch opposite. George Ball and McNamara and Bundy sat around the President. James Rowley stood inside the door. A crewman waited. The President said, "Go on" with his hand. The hatch was slammed and dogged. The President cinched his seat belt and leaned with elbows on knees. When tense, he made a chewing motion with his lips. He was doing it now.

The advisers leaned forward, as he was, and the fore and aft rotors sped until they fought gravity successfully and the people at Andrews Air Force Base saw the greenish craft lift straight up, bow its nose downward, and move forward in an awkward humpbacked manner. Johnson's hearing was never acute, and he leaned against the thwacking racket outside to hear better and to watch lips.

He spoke of Dallas as though he was talking to himself. The speech was disjointed. "It was an awful thing . . . horrible . . . that little woman was brave . . . who would have thought that this could happen . . . you fellows know I never aspired to this . . ." Below, the patchy darkness of Anacostia flats surrendered to the brilliant effulgence of Washington ahead. Mr. Johnson was going directly to the White House. "Kennedy could do things I know I couldn't. He gathered a fine team of men," he said, gesturing at the group. The deep-pooled brown eyes again moved from face to face as they had on *Air Force One:* "I need you. I need you more than he did."

The chopper was moving past the Lincoln Memorial, ablaze with light, as it began to blink its way down to the South Grounds. "Anything important pending?" he said. It was a congressional phrase, one he had used many times in his career. No matter what the agenda, he had to know the acute problems, the matters which required a presidential decision today. The three advisers studied their own faces. This was a thought, a question which had occurred to each of them this afternoon. Each man had gone over the affairs of state and, to put it bluntly, a transfer of power could hardly have occurred at a better time: there were no important decisions to be made; there was nothing of consequence for the new President to consider for the next couple of days.

The helicopter turned sideward in the chill breeze, and the pale whiteness of the Executive Mansion was visible to Johnson. This was now his. Not really his; it had not been Kennedy's or any man's. It was the office, the home, the museum of the current President. He would have sole and exclusive responsibility for the conduct of American affairs; the fourteen Cabinet departments; the one hundred fifty-four bureaus, administrations, and departments; the entire military establishment; the execution of the will of the Congress.

On the left, in the glare of the lights, were the playground swings and slides belonging to Caroline and John-John. The lights were on in the office John F. Kennedy had left two days ago. Inside, harassed workers were completing the change of rug and decor which Mrs. John F. Kennedy had ordered as a surprise for her husband. No one had told these men to stop working and so they continued with a grisly surprise.

Somewhere among all those lights, Sargent Shriver was working. The Secret Service told him that the new President was arriving, but no one moved to greet him because these men had much work to do with his predecessor. The director of the Peace Corps knew that he would receive many phone calls and orders tonight from Mrs. Kennedy, from Robert Kennedy, from both. The task was to translate the random wishes into action.

David Pearson, a minor public relations man, was among those who had been summoned and he was standing in the office watching the tides of ideas flow and ebb. Shriver heard the first word from Jacqueline Kennedy via Robert. He made some notes on a desk pad and turned that gray graven face to the young men, many of whom he had never seen before, to pass the order of battle: "I'd like you all to know," he said, "in a general way, what Mrs. Kennedy's and the family's wishes are. Mrs. Kennedy feels that, above all, these arrangements should be made to provide great dignity for the President. He should be buried as a President and a former naval officer rather than as a Kennedy."

There was nothing startling or new. Of course the family desired dignity in the obsequies. For Kennedy to be buried as a President of the United States, rather than as a private citizen, could also be anticipated. Those who listened asked no questions, but the matter of being buried as a naval officer probably meant that Mrs. Kennedy would have him dressed in the uniform of a senior lieutenant.

A priest from St. Matthew's Procathedral, a man who served as liaison between the Kennedy's and the Roman Catholic hierarchy—in this case Archbishop Patrick O'Boyle of Washington, D.C.—suggested to Sargent Shriver that the church celebrate a pontifical Mass of requiem. Shriver was shaking his head no, gently but firmly, as he listened to the priest. He glanced at Dr. Joseph English, a Catholic psychiatrist and friend of the family. "Let's take the low road, Sarge," the doctor said.

No one consciously offended the Church. The collective will of the Kennedys was iron. "Look," Shriver said, "he made it a point to attend a low Mass himself every Sunday. Why should we force a high Mass on him now?"* It would be a low Mass. Fresh problems accrued to Shriver. He and his capable young men thought of them, articulated them, cultivated the pros and cons, and made the decisions. The first tentative list of distinguished personages to be invited was already being drawn up. Angier Biddle Duke was on his way back from Andrews to request that the Library of Congress be reopened. Researchers would have to be recruited to sit inside the cavernous dark stone building studying every historic reference to the Lincoln funeral so that Mrs. Kennedy could decide which items would be appropriate to this one. It was already known that the catafalque on which the Great Emancipator's body had rested was in a dusty warehouse of White House treasures.

In a space of fifteen feet there was a soup bowl of faces. The word had been passed to the press that Oswald was on his way down from the jail. They crouched, stood, and, in the rear ranks, elevated themselves on chairs in the small space between the jail elevator and the office of Captain Will Fritz. In the Homicide office, an FBI agent studied a notebook to ascertain how much information the prisoner already had given to the law. The items:

He said his true name was Lee Harvey Oswald. His race was white; sex, male; date of birth, October 18, 1939; place of birth, New Orleans, Louisiana; height, 5 feet 9 inches; weight, 140; hair, medium brown, "needs haircut"; eyes, blue-gray; no tattoos or permanent scars; mother, Mrs. Marguerite Oswald, address, unknown; her occupation, practical nurse; father was Robert Lee Oswald, expired two months before birth of Lee Harvey.

---

* All final decisions were made by Mrs. Kennedy. Archbishop O'Boyle was the ranking Catholic churchman in Washington. Mrs. Kennedy did not want him. She substituted Richard Cardinal Cushing of Boston, a family friend. As a sop, O'Boyle's auxiliary bishop, Philip Hannan, was permitted to read excerpts from President Kennedy's speeches at the funeral Mass.

Refused to explain Selective Service card with name Alex James Hidell; denied shooting Officer Tippit; denied shooting John F. Kennedy; admits employment at Texas School Book Depository as order clerk; admits defecting to Soviet Union; says he has Russian wife, Marina Oswald, two small children, June and Rachel; lives apart from wife except for weekends. . . .

It wasn't much. The information did not tie the man to crime. The increasing certainty of the Dallas Police Department that they had the right man would be worthless in court. Fritz needed more eyewitness affidavits like that of Helen Markham. He sat behind his desk, facing the half-glass door, and a roar could be heard in the corridor. Oswald came off the elevator with his guards, blinking at the bank of hot lights, and squeezing his way through the bouquet of microphones held under his nose. Except for the bruise marks, which were more pronounced, he had the attitude of a man prepared to protest his innocence forever.

He heard the hoarsely shouted questions, but he was not in a mood, this time, to respond. His intelligence might have told him that he was now the center of focus; he was *the* story. The world had seen the gleaming bronze casket, the distraught widow. These were facts which time could not alter. The world, through the magnetism of its journals, its radios, its cameras, wanted to study the prisoner, perhaps to judge for itself whether a young man alone, without motive except for the notoriety involved, could perpetrate a crime so monumental and unexpected that hundreds of governments and billions of people paused in their tasks to dwell upon it. The almost instantaneous reaction was identical with that of Lyndon Johnson and the Secret Service: it was probably a broad plot involving another country.

Oswald was seated near the corner of the desk. Fritz nodded, but Oswald offered no greeting. The captain started by asking his target what he was doing when the motorcade passed the School Book Depository. Fritz was a low-key man and he asked the question softly, as though it had not been asked before. The suspect placed his manacled wrists on top of the desk, looked around at the FBI, the Secret Service, the two Texas troopers, and the Dallas detectives. Then he started his response as though he did not recall the three times the same question had been asked earlier.

He was having lunch with some employees. He was in the commissary on the second floor. When they heard the echo of the shots and the subsequent excitement, the others ran out. Lee Harvey Oswald remained, and put some coins in the soft drink machine as Mr. Truly and a policeman came up the stairway. The captain wanted to know why Oswald had left the building after the shooting. "I didn't think there would be any work done that afternoon," he said, "and we don't punch a clock and they don't keep very close time on our work and I just left."

"How did you get your job at the Texas School Book Depository?" Oswald said that a woman down the street from where his wife lived in Irving had a brother who worked there. They were looking for order clerks at a dollar an hour. Oswald had rented a room at North Beckley and had been looking for a job, tracing bus routes on a map he kept in his room, and, when he had been interviewed by Mr. Roy Truly, he got the job.

"What were you eating for lunch, Lee?" He said he ate a cheese sandwich, brought from home, and a Coca-Cola. Fritz was afraid to confront Oswald with evidence. He would not display to him the rifle and say: "Yours?" He would not draw to his attention the wealth of eyewitness statements now piling up in an office across the hall. Assuming that the prisoner was the person who shot Kennedy and Tippit and assuming that he was not more intelligent than the combination of hunters surrounding him, the interrogation was fruitless and repetitious. Oswald was not permitted to know what evidence the law had, and thus he was never forced into a defensive posture. He regulated all of these sessions and determined which questions he would answer and which would be met with silence. Had he known the amount of "ammunition" Dallas County had, it is almost certain that he would have tried to explode it to his advantage.

"Do you own a rifle?" No. "Did you ever own a rifle?" The prisoner said he enjoyed hunting and had owned one a good many years ago. "Did you own one in Russia?" "You know you can't own a rifle in Russia. I had a shotgun over there." Again he was asked if he had seen a rifle at the Texas School Book Depository. The answer was again yes, that Mr. Truly and some of the boys were looking at a rifle.

"What did you do after you left work?" He walked a couple of blocks to get away from the crowds and he caught a bus home. He changed his clothes and decided to go to an afternoon movie. He put his pistol in his belt and left. "Why did you take the pistol?" A small shrug: "Well, you know about a pistol. I just carried it." He was asked again if he had shot President Kennedy and he again denied it. Oswald began to develop an attitude of confidence. "Did you shoot Officer Tippit?" "No, I did not. The only law I violated was in the show. I hit the officer in the show. He hit me in the eye and I guess I deserved it."

The questions followed a sedentary pattern. The responses were the same, sometimes almost word for word with what he had said earlier. As it continued, officers appeared on the far side of the glass partition and Fritz stopped, sometimes turning to the federal men—"You got any questions to ask?"—and walking outside for a brief time to get an oral report on some phase of the investigation, at others to give fresh assignments to teams of detectives.

Fritz asked if Oswald was a member of the Communist Party. The answer was a shake of the head. He did not bring any package to work this morning. What had the police found on him when he was arrested? Oswald was enumerating the items when he said "bus transfer." His mind paused. It was as though he detected a chink in his own armor. Again he began the story of how he left work for his room, except that this time he told how he got on a bus, but the crowds were too dense for progress, so he walked until he found a city taxi. He remembered that, as he got in the front seat with the driver, "some lady looked in." She asked the driver to call a cab for her. He wasn't sure, but he may have made some remarks to the driver "just to pass the time of day."

He was in error in his recollection of the taxi fare. Oswald said it was "eighty-five cents." At times he sounded like an innocent eager to assist the police with all the miniscule details of the day. At others, he bluntly refused to answer any questions. The employees he had lunch with—the names escaped him; suddenly he remembered one was a Negro called "Junior." A lot of his personal belongings could be found in Mrs. Paine's garage, he said, but not a rifle. They were placed in her garage in September—two months ago—when he moved his family from New Orleans back to Dallas.

Where had he purchased the pistol? "I won't answer that one." "How did you feel about President Kennedy? "I have nothing against him personally, but, considering the charge I'm here on, I prefer not to discuss it." Would he take a polygraph test? No, he would not. It would be better for his counsel to decide, but "in the past I have refused to submit to those tests." "Do you know that Governor John Connally had been shot?" "No, I do not. This is news to me." "Do you deny that you shot him?" "Yes, I do." "Did you tell someone named Wesley Frazier that you would bring curtain rods in his car?" That, said Oswald savagely, was a lie. The truth was that he preferred not to be in the Paine house at any time. He went there last night because there was going to be a party this weekend for the Paine children. Oswald did not want to disturb the household. So, instead of asking Wesley for the usual lift on Friday, he asked on Thursday.

He was on ice too thin for skating. It could be proved that the party for the Paine occurred last week. At that time, Oswald's wife had asked him not to visit her in Irving. Fritz toyed with a sharp letter opener. He said that Wesley's married sister had seen him carrying a long package and watched him put it on the back seat of the car. The prisoner came to life. "She was mistaken!" he shouted. "That must have been some other time he picked me up." Perhaps it was. The captain replaced the letter opener on the desk.

Nothing had been opened.

There was a mounting tension inside of Mrs. Eva Grant. She was a small, bird-like creature and her posthospital convalescence was prejudiced by her brother's presence. She was emotional, but Jack Ruby was, at times, so unstable that she tried to have a calming influence on him. She was resting in her apartment, as her doctor had ordered, when he arrived for the second time. He was good-hearted; he intended well, but he swung wildly and suddenly between generosity and anger. He was the perpetually defensive Jew, and in another breath Ruby wept for the helpless of the world.

He flopped in a chair and said he heard that there were going to be services at the synagogue for President Kennedy. He ought to go. Eva thought that it would be a nice gesture. The services would start at eight-thirty. He slumped and made a futile gesture. Jack said he had not been to Friday night services in a long time. It would be good to go tonight, Eva said, somewhat encouraged. People show respect for a dead President. He got out of the chair and made a phone call. Then he sat again and asked her if she had eaten.

A little, she said. Her brother said she should take care of herself. He should eat a little himself, perhaps. Ruby got up and looked in the refrigerator. He made another phone call. Then he ate, talking all the time about Mrs. Kennedy and the children. They would have to return to Dallas for the trial, of course. He arose and made a short phone call. It would be unfair to force them to come back to Dallas, where Mrs. Kennedy's husband had been shot. Very unfair. The whole situation had no class.

He phoned Don Saffron, a columnist of the *Dallas Times Herald*, and asked him about the amusement page. Were the nightclubs closing in respect to Kennedy? He listened and hung up. "Eva," he said, underplaying the drama, "what shall we do?" They had two small nightclubs which were doing poor business. What shall we do? Eva Grant wanted to read her brother's turbulent mind and make the decision he would want her to make. "Jack, let's close the three days." He paused. "We don't have anything anyway," she said. "We owe to—" Tears began to well in her eyes.

Ruby picked up the phone. He called Don Saffron again. "Don, we decided to close Friday, Saturday, and Sunday." Saffron said: "Okay, Jack." The only reason Ruby phoned Saffron was to have the columnist publish it. The next phone call was to the *Dallas Morning News*. Ruby said he wanted to cancel the advertising for his nightclubs. He was told that it was too late, the pages were already locked up. He asked his copy be changed to read that his clubs would be closed for the complete weekend.

The short, taut conversations with his sister could not sustain him. Ruby cudgeled his mind to think of more people to phone. It was as though he felt an emotional burden on his heart. It might be dissipated by talking about it. The assassination had not meant a great deal to him when he had first heard the story at the *Dallas News* office. It was a thing remote from Jack Ruby's problems. But when he saw the city swept into an emotional vortex, when he saw friends shake their heads and saw a tear here and there, the nightclub owner began to feel himself swept up in the dust of the storm.

Anyone who spoke to him this day knew that, if Jack Ruby could, he would have restored John F. Kennedy to life. The impossibility of a personal resurrection left him to contend with matters as they were—to do something (go to a synagogue); to condemn the prisoner without a trial (the creep has no class). Ruby phoned Cecil Hamlin, an old friend. Jack found he could weep freely. His voice was choked with sobs. Hamlin listened as Ruby told how he had closed his clubs, both of them, for the whole weekend. He felt so sorry for Kennedy's "kids."

He phoned Temple Shearith Israel to ask about memorial services for the President. They would start at eight-thirty, he was told. A moment later, he called the synagogue again. Eight-thirty was the time. He hung up and told his sister: "It starts at eight-thirty. I'm going." Then he phoned Alice Nichols. The lady was surprised. Ruby, who had never married and who, at times, was suspected of latent homosexuality, had courted Miss Nichols for several years. The faith, the trust he could not grant to others, he reposed in her.

And yet he had allowed the old friendship to die. On this night, after a long, long time, he thought of phoning Alice Nichols. He told her how dreadful the assassination was, and she agreed. He said that he was going to attend temple services. She thought that was charitable. Miss Nichols may have expected a more personal message. Jack Ruby said good-bye and hung up. He looked at Eva sitting forlornly in a chair, and he said he was glad he had closed the clubs.

She thought he began to look old. There was a difference, she felt, between the face of her brother this morning and the same face now. He was older. In her mind, she used the word "broken." The face, which had never been handsome, was deeply etched with furrows and the eyes had retreated into dark and shallow wells. He phoned Larry Crafard at the Carousel. He seldom announced his identity. "Any messages for me?" he said. There were none. He turned back toward his sister.

"I never felt so bad in my whole life," Jack Ruby said. "Even when Ma and Pa died." Eva felt her nervousness increase. "Well," she said softly, "Pa was an old man. He was almost eighty-nine years."

427

• • •

The State Department of the United States has always been a separate church. It has its contemplative monks; parchment scriveners; mitered abbots; bishops who preach the gospel of Pan-Americanism to the heathens in the far corners of the world; a rota of cardinals who dwell within the sacred precincts to perfect a policy of no-decision; and, of course, a lower-case pope. Originally it was intended to be a foreign ministry, but early in the history of the republic the State Department achieved a status of apartheid, which was followed by an air of sanctification. In quarrels with other, lesser departments such as Treasury and Defense, the will of the State Department prevailed.

Some Presidents learned, to their chagrin, that State was above and hardly beholden to the Chief Executive. Other Presidents joined State and ran it like messiahs. In the century since Seward had been Secretary of State, it had grown in size, in power, in compartmentalized subdepartments, in numbers, and in piety. State had become an august church with many bonzes, and all the prayer wheels emitted dial tones.

The well-nourished George Ball, Acting Secretary of State until the return of Dean Rusk from Hawaii, received a phone call from the Federal Bureau of Investigation advising that the man arrested at the assassination was Lee Harvey Oswald, of Dallas, an expatriate who had fled to the Soviet Union in October 1959. He had tried to renounce his citizenship and, failing, returned to the United States with a wife and baby on or about June 13, 1962. Mr. Ball might want to check his department records.

President Lyndon Johnson had appointed the Federal Bureau of Investigation to take charge of the federal government's interest in the assassination. As the bits and pieces fall into place, J. Edgar Hoover, who had been Director of the FBI since 1924, kept his strong bulldog face behind his desk and surrounded himself with his best advisers so that the flow of information from Gordon Shanklin in Dallas to Washington was smooth, the delivery of evidence by air was swift, and the dissemination of information to responsible parties in Washington was almost instantaneous.

George Ball learned of the involvement of the State Department, and he decided that it could not be researched by a clerk. He phoned Abram Chayes, legal adviser to the State Department, at his home, 3520 Edmunds Street Northwest, and asked him to return to the department at once. The assignment was simple and formidable. Chayes was told that Dean Rusk would be in Washington by morning at the latest. His function would be to spend the night poring over State Department files to find any which might relate to one Lee Harvey Oswald of Dallas, Texas.

Mr. Chayes was a good lawyer. He recruited a couple of assistants and turned the lights on in a few dark departments. Some of the old files had red seals on them; some had other colors. It did not matter whether files were secret or not, locked or not, or hidden in vaults. Abram Chayes understood that a high crime had been committed, and his function was not to extricate the State Department from a possible cul-de-sac, but to dredge up anything bearing on the man who might have perpetrated the crime.

The subordinate sections of the State Department do not relish surrendering their sovereignity to anyone, including superiors, but Mr. Chayes knew that the press would be at the departmental gates by morning, so he had to work fast. The women who cleaned the floors and waste baskets that night saw a man possessed—working to learn all that he could about one man in one night. First he went to the passport file and found the records on Oswald. From there he hurried to the State Department security office to find out if there was any record that this man might be a risk to his country.

There were records, but Oswald was not regarded as dangerous. Chayes went to the SCS file (Special Consular Services) which covered the issuance of a visa to Mrs. Oswald and the loan of money to Lee Harvey Oswald to transport his family to the U.S. Locating them was not a simple matter in a vast bureau which prides itself on maintaining records. Secondly, the exhausted lawyer would have to start with the first item—an application for a passport in September 1959 by Lee Harvey Oswald in which he stated that he wanted to become a student at a college in Switzerland but wanted a visa for several other countries, including the Soviet Union.

The whole story would have to start there. Abram Chayes would then have to find the second bit of information, which was a secret report from a U.S. consular official in Moscow that one Lee Harvey Oswald had appeared and demanded the right to renounce his American citizenship. It was a long road of papers, documents, reports. Somewhere in the mountain of material was the final item, where Lee Harvey Oswald had concluded repayment of the loan. When the research was completed, at some small hour, Mr. Chayes would be expected to dictate or type a digest of all he had learned. Dean Rusk would have to be armed with more than press releases; President Johnson was expected to ask some questions, too.

Lyndon Johnson has two gaits. One, when he wants to talk, is slow, with his head cocked toward the listener, sometimes with one hand in a trouser pocket. The other stride is swift and reserved for when he is doing the listening. He walked with head up and arms swinging. He emerged from the South Lawn of the White House at the head of a small group of people, moving fast. The two at his side were Bundy and McNamara.

The President had some ideas of his own: 1. He was the only person who could not afford a display of maudlin grief. In tragedy, the people look for strength, not weakness. 2. He must persuade the Kennedy team to continue with him until he acquired full control and understanding of the reins of government. 3. He would need all the bipartisan congressional support he could get. To achieve this, he would be forced to trade on old friendships on both sides of the aisle, but he had to have it. 4. It would be necessary to be briefed at once on all executive matters to which a Vice-President is not privy.

Bundy was close to trotting. "There are two things I am assuming, Mr. President," he said. "One is that everything in locked files before 2 P.M. today belongs to the President's family, and the other is that Mrs. Kennedy will handle the funeral arrangements." Johnson didn't break stride. "That's correct," he said. It was a poor assumption because some things in the President's personal files could be related to official decisions, commitments, and policy. It might have been better for a person like Bundy, or Sorensen, to sift through those files at once and acquaint the new President with matters which might have a bearing on his official conduct. The President of the United States is in the position of a paid government employee whose acts and decisions are the property of the people.

Silently, Johnson passed the White House policemen who stood in the darkness along the walk. Throughout the White House, Secret Service agents heard the beep on their radios and the word that the President of the United States was about to enter the Executive Mansion. Lyndon Johnson had become the prisoner of protection, and would continue to be for a number of years. He would not make a move, from office to hallway, from bedroom to East Room, from ranch house to front lawn without knowing that the word was being passed. Later, when he moved into the White House, the mere removal of the newspapers from in front of his bedroom at 7 A.M. would cause the night man across the hall to open his microphone and whisper: "Volunteer One is awake."

Directly ahead he saw the lighted French windows of the Oval Office. It was now his, but he would not use it. The drapes were half drawn. The inside, in brand-new colors, was as serene and majestic as ever. He had known that office, with awe, from the days when his hero Franklin Delano Roosevelt sat behind the desk, the burnished eyeglasses winking with light, the cigarette holder tilted upward. There had been times when, in a conference with President Kennedy, a government problem had been recited and the handsome young man would turn his stern gaze through a semicircle of advisers, prepared to ask: "Well, what should we do?" and Lyndon Johnson always hoped that the President would not ask him first.

The door to Evelyn Lincoln's office was held open for him, and he walked through. Someone said that Johnson should use the office of the President, and the President said: "No. That would be presumptuous."

He passed it, went out into the curving white hall, kept walking down through the West Wing, out past the Press Room, into the night air again, and down the walkway to East Executive Avenue, then across the street into the dismal old structure which had once been the State Department, but which was now called the Executive Office Building or EOB. Any department which had no White House priority, from telephone operators to mail room to Vice-President, was jammed in among the old high ceilings and exposed heating pipes.

Johnson went upstairs to his office and said hello to his private secretary, Mrs. Juanita Roberts. They were always in excellent balance: he roared, she whispered. He looked around and learned that he had picked up some men on the way from the helipad. He went behind his desk, moved all the pending papers to one side to clear the blotter and looked up at the men who stood. He told Ted Reardon that he wanted a Cabinet meeting at 10 A.M. He was going to require a lot of service tonight and he wanted no excuses. Reardon left to begin phoning the Cabinet ministers—some of whom were on a plane coming in from Hawaii.

Johnson was not in doubt. He knew the necessary steps but to prod these people he put on his son-of-a-bitch face. Kilduff, who had worked so hard for the new President, was dressed down for not having the casket leave by the front ramp. The President didn't care for excuses; it would have been proper for him to leave the plane with Mrs. Kennedy and the body of John F. Kennedy. Who the hell's idea was it to get that forklift at the back of the plane?

A Secret Service man informed him that his home phone number at The Elms had been changed. It was now hooked into the White House. "Luci Johnson was picked up at school and is at the house. Lynda is at the home of Governor Connally with the Connally children." It eased the worriment in his mind to know that the girls were protected by agents; it made him feel better to know that Mrs. Johnson was on her way home. He knew that the scar of noon would never heal in his wife. The house would be a warm refuge for her.

All evening long highly placed men would come to this second-rate office to reassess the new Chief Executive and to be reassessed by him. To all, he enunciated the same battle order: "There must be no gap in government. We must go forward in unity." He sent for soup. He took phone calls. He made phone calls. At one point he was dictating a memo, and the President lapsed into reverie. His eyes stared at the far wall. "Rufe did a heroic thing today," he murmured, almost to himself. "He threw me down in that car and threw himself on top of me."

There was one facet of Johnson's character which few people knew. He was genuinely surprised when someone did something for him gratuitously.

A few automobiles were in the lot, parked against the wall of Holy Trinity Church. The big starry lights at the entrance were lighted, though few attended the first solemn high Mass for the repose of the soul of John F. Kennedy. It was fitting that it should be celebrated by the last priest to see him alive and the first to see him dead. Father Oscar Huber, looking smaller than usual in the enormous chasuble, holding the gold chalice under its cover, ascended the brilliantly lighted altar and felt a weariness in his body.

His acolytes held his garment up the three steps, then stepped back as Huber set the chalice down and touched the altar stone with both hands. The few communicants in the dimness of the pews arose with a shuffling of feet. Others, who had come only to say the Stations of the Cross, decided to stay. No one told them that this Mass was for a President. It had not been announced in the parish bulletin, and the few in the church assumed that it must be for the dead President.

It would have pleased the President. The Church, the Sacraments, Mass, religious love and fear were instilled in the Kennedy children early. Their mother retained the faith of a child. In Palm Beach, Jack Kennedy was an usher at the Catholic church and helped to take up the Sunday collection. When he went to Washington as President, a wry monsignor, leaning over the pulpit one Sunday, said: "And now let us say an Our Father and a Hail Mary for an usher who has left us. I'm not allowed to mention his name, but he got a new job in Washington."*

The Roman Catholic Church is always more of a joy to a sinner than a saint because within its gates lie forgiveness and love. The President, if his friends can be believed, was closer to being a merry sinner than a saint. He was attracted to the sins of the flesh and found them difficult to resist. In this, he joined hands with the average man everywhere. In his church, in his conscience he was a silent penitent. He seldom discussed religion

---

* At the White House a month before he died, President Kennedy told me that, as a matter of practice, he always said a night prayer but never a morning one. Lyndon Johnson, on the other hand, seldom uttered a formal prayer night or day, but inwardly said: "Thank God" whenever he heard good news about the welfare of his family or his nation.

and was never known to permit his Catholicism to influence his thinking as a statesman. Some thought that, as President, he was slightly antagonistic to his church.

Father Huber turned to the gospel and opened the big book, tilted on a stand, to the red ribbon which bisected a page. It was opened to the Twenty-fourth Sunday after Pentecost, Matthew 24, verses 15 to 35. " . . . for as the lightning comes forth from the east and shines even to the west, so also will the coming of the Son of Man be. Wherever the body is, there will the eagles be gathered together . . ."

On the third floor, Detective Guy F. Rose was busy working a neglected angle. A woman neighbor of the Oswalds had said her brother had driven Oswald to work with curtain rods. The cop flipped the pages of his notebook. That was Linnie Mae Randall; Linnie Mae Randall, who said her brother's name was what? Frazier. Wesley Frazier. She had volunteered that her brother was at some hospital if the police wanted him.

Rose wanted him. It had occurred to Will Fritz and his over-worked Homicide squad that Frazier might be a party to a plot to take the life of the President. Frazier worked with Oswald. Frazier and Oswald were buddies of a sort. Where did the Frazier boy go after the assassination? Where was he during the shooting?

Guy Rose phoned Parkland Hospital. The operator had no patient named Frazier. She would connect the policeman with the record room. He waited for a response. Rose could have taken Mrs. Linnie Mae Randall to headquarters with him while taking the Paines and Mrs. Oswald. Also, when she mentioned that her brother was at some hospital, Rose could have phoned Fritz and asked a detective to pick him up. They could be most important witnesses; anyone who had studied the Lincoln assassination might see a parallel between the Fraziers and the Surratts.

Parkland was sure that it had no record of a Frazier. Calls were placed to other hospitals, to sanitariums, to clinics in and around Dallas. The detective made a list of doctors in Irving Professional Center. He did not want to think that young Wesley Frazier had slipped through his fingers. It could be most important; it could be nothing at all. As he made the phone calls, Detective Rose watched Detective Senkel herding Marina Oswald and the Paines out of the Forgery office. He dialed the Irving Professional Center and identified himself. A nurse supervisor said yes, they had a Mr. Frazier senior as a patient.

Rose said, "Thank you," and hung up. He phoned Irving police headquarters. Rose said he was working on the Kennedy thing and he thought that there was a wanted man in the Irving Professional Center. He was put on the phone with Detective J. A. McCabe, who said he would go out to the clinic at once and try to pick up Wesley Frazier. "We understand that this boy brought our suspect to work this morning—drove him in," Rose said.

"Call me back in fifteen minutes," Detective McCabe said. Guy Rose agreed. "Just take him into custody," he said. "If you get him, we'll send a man out right away."

The ambulance and its entourage of cars moved out Massachusetts Avenue in the dark of night, but there was a feeling in Washington that the lights were on; the people were present, but there was a mood of solemn meditation. The ambulance drew abreast of slow-moving cars. Greer tapped the siren lightly, and the cars moved to the side. The movie marquees proclaimed their wares, but few Washingtonians looked for entertainment. The meat markets, the supermarkets were open with Friday night sales on hams and loins of pork. Few automobiles cooled in the parking lots.

The ambulance passed Woodley. In the dark hush, the tall and dignified stone of Washington Cathedral rolled by. Inside, Woodrow Wilson lay in a crypt. He, too, had had lofty ideals.

In death, two Democratic Presidents were close, and then the distance opened up. No bullet had cut Wilson down; the assassins of the United States Senate had sabotaged his dreams for a League of Nations and he had died slowly, defeated. The car followed Wisconsin Avenue and Route 240.

In the back, Robert Kennedy tried to console his sister-in-law. At one point, he had pulled back the glass partition which separated them from the driver's section and he asked Roy Kellerman if any of them knew that a suspect had been arrested in Dallas. Kellerman said no. "They think he's a communist," the Attorney General said. The widow was shocked. To be killed by a Red seemed, to her, to rob her husband's demise of significance. She thought he had been killed by a white supremacist; she had been sure he had given his life for civil rights. The martyr Abraham Lincoln had been cut down by a Southern sympathizer; the Negroes, free and slave, had wept. A warped, misguided communist could, in her mind, rob her husband's death of meaning.

The ambulance barely rocked. The highway was flat and smooth, and Chevy Chase slid by the big windows as a series of flashing lights through the curtains. The darkness of the hummocks of trees came again, and the two who loved this man so fervently stared at each other in the barely perceptible gloom. It was a macabre scene—morbid indeed—and yet they understood, without mentioning it, that they must be as close to Jack as possible; the hours were numbered. General McHugh sat with them, but he could not be a member of this triumvirate. Like a good soldier, he sat quietly, trying not to listen unless a remark was addressed to him. The Attorney General said that everything would be done as she wished it. He would help in every possible manner; right now Sarge was in the White House drawing up preliminary plans and he had a sizable team working with him.

She had time to tell her brother-in-law about the triumphal motorcade, the happy faces in bright channels of sunlight.

She had time, if she chose, to tell the brother of her cherished husband about the sharp, clear crack of the shots; the dreamy expression on Jack's face as he slowly leaned toward her; the spasm of the body as the back of his head flew off. There was time to tell the one man to whom she could bare her feelings. The interminable whipping speed of that car to Parkland; the bloody roses; the strange, cold faces of doctors and nurses and the long fight for something already lost.

That man, that execrable man who wanted to confiscate her Jack; the running flight to cars to the airport. The agony, the horrifying, lonely agony of it and then to find that the President had hardly died before the Johnsons were there in *Air Force One*—no privacy, no respect—waiting for a judge to swear him in and then asking her, actually asking her, to step forward to be photographed with him. There were things that Jacqueline Kennedy would never forget or, for that matter, understand. A kindred mind across the curving lid absorbed her words, her shock, her spite. Robert Kennedy, tense, taut, could sympathize with her position, feelings of grief, and rancor. He could husband a hate for a long time.

At Glenbrook the ambulance slowed. Three thousand people leaned on the double-railed fence around the huge skyscraper and adjacent hospital buildings. This was the Naval Medical Center at Bethesda, Maryland. The people were quiet. They saw the ambulance. There was no movement toward it. The faces watched the vehicle with that speculative look which said: "Up to now, I didn't believe it happened." Greer moved the vehicle toward the main entrance.

The commander of the institution, Captain Canada, was in uniform silhouetted against the lights of the main entrance. Admiral Calvin Galloway was at his side. For the United States Navy, the situation was sensitive. It would be dangerous to say or do the wrong thing. Canada had been wrong inadvertently all along, so far. He had sent the ambulance and cardiologist in case Lyndon Johnson had a heart attack. The bus was returning to the curving driveway with a dead President.

The captain had been advised that Kennedy would arrive by helicopter. He had placed an honor guard at the helipad; two helicopters arrived, but they carried the Andrews Air Force Base honor guard, which wanted to be on hand at Bethesda when the body arrived. Both honor guards were standing at attention at the pads. Canada had been told about the crowd collecting around the big institution. A short fence made of two rows of pipe would not keep them off the grounds. He had called General Philip Wehle of the Military District of Washington, asking for soldiers. The general was still at Andrews.

Smartly, Canada, accompanied by Admiral Galloway and junior officers, stepped down and approached the three cars. They helped the people to alight. Captain Canada had a chaplain with him in case Mrs. Kennedy needed comforting. She didn't. Her brother-in-law took her arm and led the group toward the towering entrance. They may have appeared to be reasonable people, but Canada soon learned that the Kennedys were in no mood to dicker.

General McHugh told Galloway that the Kennedys were here for an autopsy and embalming. The admiral said that Bethesda did not have the means of embalming a body. Godfrey was adamant. He demanded to know if the admiral was telling him that it was impossible. Galloway kept his temper. It was not impossible; it might be unsatisfactory work, which could be worse. McHugh called up his reserves. O'Donnell and O'Brien were flagged to his side. The general said that the Navy did not want to do the embalming—the admiral had recommended a funeral director. "You heard the general's decision," Kenny O'Donnell snapped. The admiral and the captain stood as O'Donnell and the general left for a sixteenth floor.

In a moment, McHugh was back. Most of the others, except for Roy Kellerman and his Secret Service group, had gone up in the elevators. The body was driven to a rear entrance. It was taken out of the ambulance and placed in an empty and well-lighted corridor. There it reposed. McHugh stood by it and wondered what had happened. Kellerman and Greer stood looking around. No one spoke. No one appeared. They waited.

The headlights of the solitary car lit the quiet street momentarily and the low branches of the trees looked greener. The car turned into the driveway of a Walnut Hills home, and Dr. Malcolm Perry turned the ignition off and went into his house. His day's work at Parkland Hospital was done. The thoughtful face, wiped clean of expression, brightened when he saw his daughter, Jolene, and his son, Malcolm.

Malcolm was three and chattered his joy at the sight of his father. Jolene held out her school work for approval. There were some papers covered with large printed letters. "Say," he said, "that's good work." Dr. Perry brought the school papers up for one more look, and his world shattered into fragments.

The papers dropped from his hand. "I'm tired," he said to Mrs. Perry. "I've never been this tired in all my life."

* * *

The
Evening
Hours

The coffee machine ran out. Policemen dropped coins in it, held paper cups under a spout barely dripping, and kicked the automatic vendor. This was supper time for the Dallas department and there would be no supper tonight. Almost all of the personnel had worked all day and only a few were permitted to leave at 4 P.M. when the next shift arrived. Two deputy chiefs sat in front offices handling phone calls from all over the world. There were newspaper editors, police officers, statesmen, diplomats, and the civic-minded.

Some wanted inside information about the crime and the suspect. Some asked for official statements. Many offered suggestions. One woman, excited, phoned and said: "Part of a chicken sandwich was found on the sixth floor—right? Well, all you have to do is pump Lee Harvey Oswald's stomach. If chicken comes up, he's your man." Some were disturbed. They had seen visions and could solve the crime at once if they could be flown to Dallas. A few excoriated the city and the police department. "If you had properly protected Mr. Kennedy, he wouldn't be dead." "You know who killed him—you did." "Dallas should hang its head in shame."

On the several floors of police headquarters, men worked harder than ever in the history of the department. Uniformed men and detectives breasted the crowds in the halls to run down a tip, find a witness, do an errand, pick up an item of evidence, or they were en route back into headquarters, bucking the same crowds, reporting in, and getting a fresh assignment which could not be delayed a minute. There was no time to buy a sandwich; it could have been chewed hurriedly if someone had brought it in. The Dallas Police Department was operating like a tentacled octopus with no body; all the legs were waving and threshing, but the effort appeared to be without direction.

The chief sat in his office at the head of the third-floor T. He was a man alone, as though he were too aloof to seek his men or they were too aloof to consult him. He "assumed" that Captain Talbert had the people in the building under control, but no one saw Talbert running to Curry's office with reports. Will Fritz handled the case as though Homicide were divorced from the rest of the department. Between the interrogations he did not report to the chief with progress or problems. On the fourth floor, Lieutenant Day of the Crime Laboratory had examined the rifle, the bullets, the revolver, the cardboard boxes found adjacent to the sixth-floor window, the blanket which had just arrived from Mrs. Paine's house, but Chief Curry had very little firsthand knowledge of the findings. Lieutenant Revill of Intelligence had handed in a quick report on FBI Agent Hosty, stating that Hosty admitted knowing about Lee Harvey Oswald as a communist and potential assassin, and Curry had read it through his rimless glasses in silence and locked it in his desk.

Curry left his office and walked back down the corridor to Homicide. The reporters, as always, pressed him for a statement. He offered none. It was barely possible that anyone who was watching all the angles of this case on television might know more about it than the chief. He inched through the taut, sweaty faces and the klieg lights and turned into the glassy section marked Homicide. He could see the prisoner, lips pursed, listening to the questions of Will Fritz, and he could see the other officers, some standing, some sitting, staring at the prisoner waiting for an answer.

Fritz came into the outer office to talk to the chief in private. Curry wanted to know how it was going. At times, the captain of Homicide would blink those eyes like an uncommunicative frog.

He was a big man, bigger in his Texas hat. It was apparent to Curry that he was not going to get a detailed report. The captain said he thought he had enough evidence on Oswald to "file on him for the murder of Tippit." Curry nodded. That would hold the man for a long time. There would be no bail in a first degree murder charge. This would give the Homicide division plenty of time to build a solid case. Fritz may have felt that he owed a little more to the chief. "I strongly suspect," he said, "that he was the assassin of the President." This was information which had been imparted to the world hours ago. The conference between the chief and the captain was brief and guarded, almost formal.

Across the hall in Forgery, Detective Rose was on the phone. He was talking to Detective McCabe of the Irving Police Department. "We have Wesley Frazier right here," said McCabe. "He was found at the Irving Professional Center visiting his father." Detective Rose was grateful. This kid could be a missing link in the case. McCabe said that Frazier had been arrested. The word hurt. Frazier should have been picked up, or apprehended, but hardly arrested. No one could think of a charge on which to hold him, unless the great big legal basket called "material witness" could be used.

"I'll leave here right away," Rose said. "I'm taking Detective Stovall with me." McCabe told him that the prisoner would be waiting. The two left their office, fighting their way toward the press room and the back elevator, as Curry stalked his way coldly in the opposite direction. In the county building, off Dealey Plaza, one of Sheriff Bill Decker's deputies noticed a Negro boy standing in the outer office. He asked Amos Lee Euins if he had signed an affidavit. The sixteen-year-old said yes. It was he who had watched the execution of President Kennedy from one of the vantage points in the plaza.

The deputy asked if anyone wanted this boy to remain in the sheriff's office. Deputies looked around and said no. Typists sat intently behind their machines, taking down the oral testimony of scores of witnesses. Some had seen something. Some thought they had seen something. Day had passed into night and a few refused to sign if a word or a phrase was not quite in context in the affidavit. Copies of the approved affidavits were being delivered to Captain Fritz regularly. His men sorted them and acquainted the captain with a digest of the important ones. Hour by hour, the case against Lee Harvey Oswald began to congeal. If Fritz did not share all of his findings with his superior officer, it is also true that he did not confront the prisoner with them. He might have dealt a more slashing attack on his man, causing him to retreat, to admit, to concede here and there, but the interrogation continued with repetitious questions and, whenever the prisoner felt sensitive to them, he refused to answer and sat staring at the hound dogs who stalked him.

The First Lady crouched in the back of the limousine. On the other side of the seat, silent, sat her secretary, Elizabeth Carpenter. Mrs. Johnson felt cold. The Secret Service agents up front—Knight and Rundle—turned on the heater, but Lady Bird Johnson felt spasms of shivering run through her arms and knees and her teeth chattered. She looked out of the window at the darkness impaled by street lights and all of the wealth of practical sense within her kept saying that this was a bad day; a tragic day; a stunning, horrifying day which she wished could be cast away into the blackness outside, never to return. She was going to have to live with this day, but it would take time. It was as though a blue bolt had fallen on a picnic, and everyone had been frozen into congenial attitudes in death. It was as though no one would ever smile again because, in the ghastly presence of this day, there would never be anything to bring a smile to a friendly face.

There were moments when it seemed not to have happened at all. The mind could not sustain the intensity of shock too long; it short-circuited and, for a brief time, everything was as it had been at noon. It is possible that, with the exception of Mrs. John F. Kennedy, no mind raced over as many despairing trails that night as that of Mrs. Johnson. She was wife and mother to her man, and she meant it when she said, over and over: "Dear God, not this!"

All her life she had been a Texas belle and proud to be one. She had worked as long and as arduously as her husband in the House of Representatives, but success, to her, meant a step closer to home. At any time she was asked about the most lofty post in the nation, Lady Bird Johnson always gave the same answer: "United States Senator from the State of Texas." In her mind, she could play back every campaign as though it were a motion picture in color; she knew the true friends from the false; she knew the right moves from the wrong. She husbanded her husband as one would a national resource; when he fell with a myocardial infarction, it was this small woman with the brown eyes who sat up nights at Bethesda Naval Hospital listening to the soft hiss of oxygen, watching the rise and fall of the massive chest.

The thing she had to offer, besides love, was scared courage. The brown eyes were frightened, but the brain forced her to remain cheerful, to keep him from being restless. When he left the hospital, she might have entertained a latent hope that the doctors would order him back to the ranch in Texas, to retire from politics. They didn't. His recovery was splendid. Lyndon Johnson stopped smoking; he cut his drinking down to a casual dinnertime cocktail; his workload increased—he was Democratic Majority Leader of the United States Senate.

The higher he went in life, the further the ranch faded behind him. To Mrs. Johnson, who was also the business head in the family, the ranch was all. The mountains of alien Washington had been climbed. They had the house, several thousand acres of land in the hill country, a television station, some good stocks, some loans owed to banks, and two daughters who would like to see their parents return to Texas permanently.

When the political drums began to beat a tattoo for Lyndon Johnson's nomination for the presidency, in 1959, Mrs. Johnson thought of it as the final accolade for her husband. So, too, did Lyndon Johnson.* He fought hard, at times bitterly, with the inner feeling that the Democratic Party would not offer the nomination to a Texan.

Kennedy won it and offered the vice-presidential nomination to Lyndon Johnson. There was a moment of hesitation on both sides. When Johnson accepted, Kennedy's palace guard—including his brother Robert—was enraged. The period November 1960 to November 1963 was to be the cruel and cutting part of Lyndon Johnson's life. President Kennedy used his Vice-President's services wisely, sending him on many missions to many parts of the world; asking his assistance in getting legislation through the Congress; keeping state leaders in the Kennedy corral. In return, the Kennedy group made Lyndon Johnson the butt of their jokes; they could make him look bad in clannish conferences; some even tried to sabotage his chances of running for Vice-President a second time. The onetime majority leader of the United States Senate was forced to truckle to the junior Senator from Massachusetts.

She had been through it all. Socially, the Johnsons lacked the glitter, the polish, the urbane wit of the Kennedys. At the White House receptions, none of the society columnists pressed to know what Lady Bird was wearing, but everyone was prepared to gasp with pleasure when Mrs. Kennedy appeared. The most protracted hurt of all was that Lyndon Johnson was not emotionally suited to be second man on anyone's team. Mrs. Johnson knew her husband.

---

* At the White House in June 1965, President Johnson told me that, even in the midst of the fight with Kennedy for the Democratic nomination, he was certain that he would not win it. When he was elected Vice-President, Johnson counted on four, or at most eight, years in the office and was certain that he would be able to retire to his ranch in January 1969 at the age of sixty.

Now what? She rode through this darkest of nights without elation. Her husband had become the President of the United States. He would start pulling all those people together; he would plead; he would give a little to get a little; he would work the late hours acquainting himself with every facet of this awesome post; Vice-Presidents are poorly informed; he would lead because he enjoyed being in front; but was any of it worth the LBJ Ranch? What good could possibly come of leading a nation in an era of chronic tension? What if it broke his health and he had another heart attack?

The car pulled into the drive at 4040 Fifty-Second Street Northwest. There was a crowd outside. A few trucks and cameras were there; these had attracted the neighbors. Mrs. Johnson felt small and alone in the back seat. She thought: "I love this house. I love it. Now we'll never live in it again." Under the dome light at the entrance, she saw the slender figure of Luci. Three Secret Service agents stood in the shadows. "Oh, mother!" Luci said. Mrs. Johnson pressed the younger daughter in her arms for a moment. "My school said prayers," the girl said.

The Elms seemed busy to Mrs. Johnson. She had expected to come home and undress and put on a robe and slippers. She had envisioned a quiet home with few lights on. The street outside was heavy with watchers. She stepped inside with Luci and was surprised to find people standing everywhere. They were personal friends, or co-workers, or people important to the administration. As she nodded and summoned her small smile, Mrs. Johnson realized that this was the way it was going to be. It would never be quiet and peaceful again. Even the ranch would be swarming with Secret Service men and political friends.

Luci was prattling, but her mother did not hear the words. Mrs. Johnson went upstairs with Liz Carpenter. Mrs. Carpenter could summon a moonlike smile at the most abysmal of moments, and this was one. Mrs. Johnson was rubbing her wrists. "How do you feel?" Liz said. Mrs. Johnson reached into a closet for a dressing robe and slippers. "I'm freezing," she said. "Please turn that set on. We can watch it up here."

She propped several pillows at the headboard. A great weariness overcame her. There was no sleep in it. Mrs. Johnson phoned Lynda at the University of Texas. She wondered what Luci had been talking about. The television set told the same story over and over. When there was nothing new to tell, the networks fell back on rerunning material already shown. She saw herself getting off the plane with her husband; she saw the motorcade of the morning; there were random shots of people running and falling on grass; the face of a sullen young man in handcuffs being led through a crowd of men in a hall; she heard her husband ask for "your help—and God's."

Lynda was saying: " . . . the first thing I did was to go to the Governor's Mansion to be with the Connally children." Mrs. Johnson nodded. "That was just right, darling." Inside the massive fatigue, the mother felt a lift. Her girls had thought of constructive things. One prayed; the other hurried to help Nellie Connally's little ones. It was good to know that both of her daughters were safe. She made a few calls to close friends. Mrs. Johnson glanced at Mrs. Carpenter. "I don't know when he'll be home. But he'll probably have people with him, and he hasn't eaten yet." The First Lady pulled a quilt over her and felt a spasm of shivering.

The Bethesda elevator brought the Kennedy party—the first section of it—to the seventeenth floor. Navy officers in dress blues escorted the group to a special suite of rooms. As in the Hotel Texas in Fort Worth, there was a short corridor. To the right was a well-lighted kitchen. On the left was a bedroom. Farther on was a living room with settees, mirrors, a fireplace, a few ornate chairs, and prints of paintings on the walls. Mrs. Kennedy examined each room. The television set was in the bedroom. Two Secret Service men stood outside the little hall.

The rest of the Clan Kennedy would be arriving in groups. There were facilities in the kitchen for making tea and toast; one could also phone the hospital kitchen. There was plenty of time to peel the bloody clothes from the body and to soak in a warm tub. Mrs. Kennedy kept them on, including the gloves.

Now that she was safe from the reporters and cameras, Mrs. Kennedy wanted to do what Mrs. Johnson had done—inquire about her children. She phoned her mother. "He didn't even have the satisfaction of being killed for civil rights," she said. "It had to be some silly little communist." Mrs. Hugh Auchincloss mentioned that the children were safe at her house. Her daughter was puzzled. She had sent no message to have the children taken there. "Mummy," said Mrs. Kennedy. "My God, those poor children. Their lives shouldn't be disrupted, now of all times." She thought about it. "Tell Maude Shaw to bring them back and put them to bed."

Grand-mère may have thought that Caroline and John were comfortable in her house. Still she would not dispute it. If Jacqueline wanted the babies back in the hurly-burly of the White House, then so be it. It seemed a pity to dress them again and send them back. The phones were replaced on their cradles. *Grand-mère* may have wondered why her daughter had not asked if the children *knew*. Perhaps she wanted to tell them herself. In the excitement messages became unreliable. If Jacqueline had not ordered Maude Shaw to bring the children to Georgetown, then maybe she didn't want the nanny to tell the youngsters about their father.

Mrs. Kennedy left the phone in a daze. Thinking about the children can be, at a time such as this, both a lovely and a heart-wrenching reality. To have them as a valentine from him is solace; to think of the innocents as not having a father, especially two who adored their father, is depressing. She walked into the living room and asked someone to phone Sargent Shriver at the White House. In the family sitting room on the second floor,

Mrs. Kennedy said, there was a large-size book on Lincoln. It held a lot of daguerreotypes and line drawings of the funeral of America's sixteenth President. Tell them, she said, to study those drawings and the lying-in-state in the East Room of the White House. She would like to have her husband's funeral correspond as closely as possible to Lincoln's.

On the ground floor, Navy doctors met the Secret Service and two FBI agents in the hall. The casket reposed on wheels. Enlisted personnel tried to move it. Major General Wehle, who had arrived from Bethesda, waved them away. Roy Kellerman and William Greer grabbed handles on opposite sides; General Godfrey McHugh stepped into position to help. So did General Wehle. Valiantly they tried to lift the casket up a short flight of railinged steps to the autopsy room. Collectively they were not strong enough. The enlisted men watched the box teeter from side to side. Silently they moved between the older men and grunted as the burden was lifted over the railing and set upon a trolly. Admiral Burkley and FBI Agents Francis O'Neill and James Sibert followed.

The body was wheeled into a large, square, bright room. It was tiled. Over a table in the center, adjustable lights diffused their beams. To one side, there was a place with eight ports for bodies to slide into a wall in drawers. As the casket stopped next to the table, Special Agent-in-charge Roy Kellerman took a census of personnel. The Federal Bureau of Investigation was also interested.

The senior officer was Admiral C. B. Holloway, commandant of the hospital, who stepped aside as a spectator. Admiral Burkley was the President's physician. Commander James J. Humes, chief pathologist of the hospital, said that he would conduct the autopsy. Captain James Stoner, Jr., chief of the Bethesda medical school, was an observer. One who would be present soon and who would participate in the work was Lieutenant Colonel Pierre Finck, of the Armed Forces Institute of

Pathology. Commander J. Thornton Boswell, would assist. A medical photographer who would take overhead photos, in color and black and white, was John T. Stringer, Jr. In addition, Lieutenant Commander Gregg Cross and Captain David Osborne, chief of surgery, could be expected to be in and out of the autopsy room, depending upon whether Commander Humes required assistance.

Major General Wehle announced that he was present only to ask when the body would be returned to the White House. He had no desire to remain. If someone could give him an approximate time . . . No one could. No one knew precisely what the injuries were, nor how much time the work would take. There would be photographs of the exterior of the body in different positions; there would be X-rays; there would be oral notations of the doctors of their observations. Then the autopsy would start. It would consist of tracking and diagramming the wounds; examining the wet plate X-rays against the observations of the human eye; the removal of the brain; removal of certain organs in the torso, examinations for grossness, pathology, weighing. There would have to be time for summary conclusions among the doctors.

Roy Kellerman guessed it would take a good part of the night. He also would guess that the body would be returned to the White House in the same ambulance. The family wanted the body to be embalmed here in the autopsy room. That would take additional time. General Wehle thought it best to post a guard of honor at the main entrance of the White House and wait. Kellerman polled eight additional technicians and enlisted personnel in the room; James Ebersole: Lloyd Raihe; J. G. Rudnicki; Paul K. O'Connor; J. C. Jenkins; Jerrol F. Chester; Edward Reed; and James Metzler. And, directly beside Kellerman on a bench, one additional man who was sinking uneasily, almost ill: General Godfrey McHugh. The total, including William Greer and Agent O'Leary, also FBI men O'Neill and Sibert, came to twenty-four. They would not all be present at any one time; this was the aggregate. President Kennedy was not counted.

Commander Humes took charge. He stood under the big lamps in his white coverall and drew on the long rubber gloves. He reminded the men around him that X-rays would be taken and that anyone not actively participating in the autopsy should sit in an adjacent room. Most of the observers could see everything from there without being exposed to dangerous rays. Some men moved. Some did not.

Commander Boswell signaled to the enlisted personnel to open the casket. The locks were unsnapped. The lid was raised. The men looked in. They saw a bloody mummy. The President was wrapped in plastic, in addition to a sheet. The awkward handling of the heavy casket had jogged the body inside. The enlisted men gathered around the casket, and tenderly they lifted the rigid form within the sheet. It was placed face up on the autopsy table. The doctors began to speak their observations; notes were taken.

The doctors began to peel the sheet and plastic away. It stuck against the throat and the back of the skull, and tenderly they lifted the head and cut the material away. Humes waved the enlisted men in, and they lifted the body again and yanked the loose material away. For the first time, they saw John F. Kennedy. He was nude, on his back. It was a lean, well-muscled body. The hair had remained combed, or dressed. There was a ragged-edged wound in the neck, obviously a tracheostomy. The face appeared to be fatter, or more bloated, than expected. The left eye was black and blue. A massive hole appeared in the right posterior of the skull and, without moving the body, it was apparent that some brain tissue was still emerging from the gaping wound.

There were three former Presidents alive: Herbert Clark Hoover, who lived in the Waldorf Towers in New York City; Harry S. Truman, of Independence, Missouri; Dwight D. Eisenhower, of Gettysburg, Pennsylvania. Hoover was convalescing from an illness. Lyndon Johnson phoned Truman. The men exchanged feelings of shock, and Truman promised the President all the support he might need. Truman was asked to come to Washington; he said he would arrive on Sunday. Johnson said he needed help.

The President saw Juanita Roberts stick her head inside the office. He raised his face inquiringly. She said that a group of Senators and Representatives were in the outer office. Johnson said to ask them to wait. He needed them, too; he needed these men most of all. They had power. Some political honeymoon would have to be hurriedly arranged between the Chief Executive and the Congress. The whole world was watching the United States, and there would have to be unity of purpose on display.

He asked Bill Moyers to handle the phone calls. The lean and taciturn Baptist minister, standing at the side of the big desk, started the next phone call as the President was speaking. Johnson was talking to Dwight D. Eisenhower, and the President offered to send a plane. Eisenhower was a Republican. He understood the importance of consensus and said he and Mrs. Eisenhower would arrive in Washington in the morning. The phone calls were brief and, as Johnson hung up the phone, Moyers handed another transmitter to him.

A call went in for J. Edgar Hoover. The Director of the Federal Bureau of Investigation was home watching television. President Johnson said that he wanted a complete investigation and a report on the assassination. He was appointing the FBI to take charge of it, and he was herewith giving Hoover whatever plenary powers he would require to see it through. Hoover called FBI headquarters and ordered twenty agents to fly to Dallas this night. He phoned Gordon Shanklin in Dallas—a man who had forty agents in the area—and informed him that the FBI was now in charge of the federal investigation.

Johnson arose from his desk and strode into the anteroom, where the Republican and Democratic leaders of the Congress waited. The President walked around the group, shaking hands, looking at each man eyeball to eyeball. Some were still stunned. They looked like distraught children, asking guidance from father. The tall Texan leaned against a table and said he was speaking, not as a President, but as "friend to friend." He needed their help. They knew he needed it.

In a manner of speaking, the tragedy had brought the nation to its knees. Strong leadership would be required, and he was prepared to render it, but without Congress he could do very little. If ever there was a time when they could forswear party labels and criticism, this was the time. He would have a message for the Congress in a couple of days—perhaps Monday or Tuesday—but Johnson wanted their support right now. The honored gentlemen murmured assent. He had nothing special to ask them now. The President returned to Room 274.

It was happening as Mrs. Johnson knew it would. He was "pulling all those people together, pleading; he would give a little to get a little. . . ."

The policeman's hand smoothed the bus transfer on his desk. Lieutenant Wells had finally traced that little piece of paper. He had known that each driver had his own punch, but the problem of finding out who used a punch shaped like an old-fashioned door key with one crooked tooth had not been easy. The lieutenant studied the piece of paper, and he called Dhority and Brown over. "This bus transfer was found on Oswald," he said. "It was punched today. The thing may be of no importance, and yet it may turn out to be something we should know about. The bus companies have gone over their records, and they say that this punch belongs to a man named Cecil McWatters.

"Drop over to Commerce and Harwood and wait for a Piedmont bus. They're sending the driver in on it. Pick him up, find out what you can about when he issued this transfer. See if he can remember who got it. Sometimes a driver doesn't punch more than one or two in a round trip. If he remembers, I want him to take a look at Oswald in a showup."

Someone said: "Oswald just came back up from a showup." Lieutenant Wells shook his head. "He's available. We may have a half a dozen before this night is over." Dhority and Brown wanted to know how they would recognize McWatters. "He'll get off the bus. He'll be looking for you."

Across the hall, Lee Harvey Oswald crossed his legs and said he was getting tired of answering questions. "I did not shoot the President," he said softly. "My wife and I like the President's family. They are interesting people. I have my own views on the President's national policy." He shrugged. "I have a right to express my views. . . ." Thomas J. Kelley, inspector of the Secret Service, was sitting in on the interrogation. He asked several questions. Oswald looked around the room, from face to face, the master rather than the slave, the feared rather than the fearful.

He said he would like to get one thing straight. Questions had been answered all day. He didn't think it was right to continue unless he had counsel. Captain Fritz reminded the prisoner that arrangements had been made for him to use the phone. Oswald gave him a little smile. He might have mentioned that his collect call to John Abt in New York had been declined. If the police department had returned his thirteen dollars, Oswald could have paid for the call. The cops had given him ten cents. The money was his; there was no reason to impound it. In a way, his rights were infringed. But Mr. Oswald did not mention these matters. He said he still hoped to retain the services of John Abt; failing that, he hoped that the Dallas Civil Liberties Union would help him to get a lawyer.

Captain Fritz sat listening. He could have said: "Use my phone." Some of the law officers present were qualified attorneys: they could have offered to make the call to Abt; they could have suggested that Oswald be permitted to use his own money to make the call. They could have called the Dallas Bar Association at once and asked for someone to advise Lee Harvey Oswald of his civil rights. No man present was prepared to buck Will Fritz. No one had a desire to help a loathsome creature who might have killed a President with no more motive than to become well known by doing it.

In spite of the recalcitrant attitude of the prisoner, the questioning went on. Whenever he balked, they gave him a rest for a few moments, talked about casual things—family, jobs, Russia—and returned to the crimes he denied as though, this time, the response might be different. No force was used or contemplated; a few officers believed that the deadly monotony of the questions would, in time, crack Oswald's resolve. To the contrary, it is probable that he enjoyed jousting with the police.

McWatters was brought in to an office on the other side of the hall. The reporters shouted questions at him, but he was as confused as they. Cecil McWatters was a veteran bus driver. All Dhority and Brown told him was that he was wanted in police headquarters to answer some questions. When he saw the crowd and realized that it had something to do with the assassination, he felt frightened.

The cops pointed to a chair and the driver sat. They said that they had a transfer, issued for Lamar Street and marked 1 P.M., and they would like to know if he recognized the paper. He took it in his hands, looked it over carefully, and held the punch hole up to a green-shaded light. "Yes," he said. "This is my punch." He dropped it back on the desk. The bus driver picked up a blank sheet of paper, took his metal punch, and squeezed. "See?" he said, holding it up. The marks were identical. No other driver would have a similar punch.

There was a roar of sound from the press in the hall. A detective and a Secret Service man had Howard Brennan by the arm. The man who had sat on the low wall in Dealey Plaza, his tin hat on the back of his head, and had seen the assassin in the window, was a frightened witness. As Mrs. Brennan had pointed out, there was no use running away; the Brennans would be found.

The pipe fitter was a strong man. The nervousness was plain. Forrest Sorrels of the Secret Service introduced him to Captain Fritz in an anteroom. Brennan said he could promise nothing. In fact, he was sure he would not recognize the man he saw with the rifle. He had seen the gun first; the man was dim in the background behind the partly opened window. Once or twice the gunman had leaned forward, so that he was clear in the sunlight, but it had happened so quickly that a man would have to be pretty sure before pointing a finger at anyone. Fritz said he understood. The police didn't want lies or exaggerations which would not stand up in court.

Brennan was taken to another office. He looked miserable. In the next glass cubicle, the bus driver sat waiting. Two offices away, Ted Callaway, an automobile salesman who had heard shots and had seen a young man running down Patton, also waited. A Negro porter, Sam Guinyard, was patient. He worked at a used car lot at Patton and Jefferson, and he watched a dog-trotting man stick a revolver in his belt and pull his sports shirt outside his trousers to cover the weapon.

Detectives herded the witnesses toward the elevators, through the dense crowd which shouted questions. McWatters told a policeman that the bus transfer was not on the Piedmont line, which he was working this evening. It was the Lakewood line, which he had worked earlier today. He was told that he could draw up an affidavit later, explaining everything.

In the showup room, the witnesses were sitting behind the screen. None knew any of the others, except for Callaway and Guinyard. They worked at the same car lot. A policeman turned lights on and the small space down front was flooded with brightness. Fritz and Sorrels came down with the prisoner and he looked at the same faces of policemen who would be shackled to him. Those who watched his face closely could detect no rancor, no fear. He must have known that, in the darkness out front, witnesses sat staring at him, perhaps ready to point a finger at him and tie him to the Kennedy and Tippit murders.

If this is so, Oswald acted as though he had no cares. He became the number two man in the lineup, and, this time, two unshackled plainclothesmen accompanied the four "prisoners" into the powdery light. Lieutenant Sims conducted the questioning. He asked each man to state his name, his age, his address, where he went to school, whether he drove a car, and, if so, which type. Each was asked to face front; to turn and stand left; turn and stand to the right.

The bus driver studied all six men, even though two were obviously policemen. He found them to be of different ages and sizes, but he recalled issuing a transfer to a woman who wanted to catch a train and a young fellow who left the bus at the same stop. Cecil McWatters was certain that, of all the passengers that day, he could not recall more than a woman and a young man leaving the bus a few blocks east of Dealey Plaza.

He looked surprised as he studied the men on the stage. He leaned toward a detective. "There's one feller up there is about the size and build of the man who got on my bus and then asked for a transfer and got off." "Yes?" the detective said. "Which one?" "That second one from this side." "The number two man?" "If you call it that. But I couldn't positively identify him. That's just the size and general complexion."

Brennan sat next to Captain Fritz. He looked startled. "That one," he whispered, pointing, "is the closest resemblance to the man in the window." Fritz was whispering. "You said you couldn't make a positive identification." "That second one is the closest," said Brennan. "Did you do that for security reasons personally, or couldn't you?" "I did it because I was afraid for my wife and family. Me and my family might not be safe."

Fritz nodded gravely. "He's not dressed in the same clothes,"

Brennan whispered. "In what way?" Fritz said. Brennan said he didn't know. The clothes were not the same as he had seen in the window. The pipe fitter was shocked because he knew that he could make a positive identification of Lee Harvey Oswald. He had sincerely believed that he would not be able to recall the fleetingly seen face in that sixth-floor window. He recognized the man and was willing to state that Oswald was "the closest resemblance to the man in the window," but it would take a long time before Brennan would be ready to admit that he had identified the man at first glance, positively and without doubt.

Officer James Leavelle leaned across Callaway and Sam Guinyard and said: "Take your time. See if you can make a positive identification." The car salesman smiled. "The short one," he said. Leavelle continued: "We want to wrap him up real tight on killing this officer. We think he is the same one that shot the President. But if we can wrap him up tight on killing this officer, we have got him." Callaway left his seat and went to the back of the room. He wanted to see all the men from the distance he recalled earlier in the day.

Coming back, he whispered: "The short one." "Which one is that?" "The short one—number two." He pointed to Lee Harvey Oswald. Guinyard nodded vigorously. "That's the one," he said. "Same one. I saw him running." Leavelle asked if he was positive. Both said yes. He asked them to return to his office and furnish a sworn statement. The showup room was dark again.

Six hours after the assassination, the Dallas Police Department had control of the case. The men had worked hard running down leads. There was an element of luck. It was the shoe store manager who told them where to find their man. Without his assistance, Oswald might have sat through the show—perhaps several of them. At night he could have emerged, but he had no place to go. No money to speak of, no refuge.

He could not return to his room. He knew that the police would check every employee of the School Book Depository. It was an automatic move. Ironically Oswald had made certain that he could not survive a police screening. He had spread the word that he was a "pure Marxist" and was eager to explain to anyone what it meant. He was in the newspaper files and the FBI and State Department records as a defector who fled to Russia and tried to renounce his American citizenship. To make himself conspicuously suspicious, he continued to carry his Fair Play for Cuba identification card on his person.

When he faced capture in the Texas Theatre, why did he shout: "It's all over!" What was all over? Why try to shoot a policeman with scores of them in the aisles ready to gun down such a gunman? Was this a quick exit he had designed for himself? He drew a gun and a hammer clicked. Is this the behavior of a man who is innocent? Is this the behavior of a man who is guilty but is certain that the state has no case against him? Or is this the philosophy of a man who feels that one more murder might bring merciful darkness to him and historic recognition?

The accidental discovery that the prisoner sitting in police headquarters was the missing employee from the Book Depository was of enormous help in redeeming the day for Dallas. For several hours, the police work seemed to be scattered and ineffectual. By 6:30 P.M. all of the loose pieces began to fall together. Two crimes overlapped their focus onto one face: Oswald's.

Superior officers were so busy that they didn't have time to assess the work of the men. In the murder of Tippit, they had affidavits from Helen Markham, Callaway, Scoggins the cab driver, Sam Guinyard, and the shoe store manager, Johnny Calvin Brewer. They also had a statement from stout Earlene Roberts, the rooming house housekeeper. They had others from witnesses at the scene of the shooting—sworn statements—which had not yet been typed.

In the Kennedy murder, they had affidavits from a dozen employees placing Oswald on the sixth floor; they had Brennan and Amos Euins pointing the finger at him with a rifle in the window; they had his rifle and his revolver and were tracing the ownership of both to him; they had a palm print taken from a cardboard box in the sixth-floor window; they had a cop who saw him in the second-floor commissary three minutes after the assassination; Roy Truly also swore to this. They had a woman clerk who spoke to him on the way out of the building; there was a bus driver who could identify a transfer; a cab driver who suffered a silent young man to sit up in front with him en route to North Beckley; Earlene Roberts in the matter of rushing in, changing clothes, and rushing out again. There was a statement from his wife that he owned a rifle; Wesley Frazier was ready to swear to a long, thin package called "curtain rods" carried to work that morning. Linnie Mae Randall saw the package, too. The pieces of bullets recovered—from the rifle as well as the revolver shots in Tippit's body—would be traced to Oswald's firearms. There were three employees who would testify that all three shots came from over their heads as they stood on the fifth floor, and they would swear that they heard the empty shells bounce on the floor above.

Considering the magnitude of the case and the weakness of motive, Captain Will Fritz and his men had achieved remarkable results within six hours.

A closed road separated the pristine luminosity of the White House from the dismal gray of the Executive Office Building. A bright light from an office in each of the buildings mingled on the old Belgian blocks in the middle of the street. Sargent Shriver sat in Ralph Dungan's office in the extreme west end of the White House, burying one administration. On the second floor across the street, the light from Johnson's office poured downward as he brought an administration to life.

In the outer office, Jack Valenti knew that the President was beginning to feel sure of himself, because the orders to "get me Averell Harriman"; "I want to talk to the ranch"; "Let me speak to Shriver" were enunciated patiently. The strident fever was gone from his voice. "Cliff," the President said to Mr. Carter, "go down the hall and you will find a White House secretary. Ask her for two sheets of White House letterhead and two envelopes."

He was going to pause in his labors to write personal notes to Caroline Kennedy and John Kennedy, Jr. The mute and welling grief which he had fought all day gained an ascendancy only when the work pace slowed. The new President would tell them how he felt about their father, how proud they should be of him. Lyndon Johnson did not expect that the notes would mean much to the children now; he was thinking that, when the children matured, they might like to know that his successor thought of the Kennedy children on the day that their father had been cut down.

Cliff Carter walked down the high-ceilinged hall. He found the office. A middle-aged woman was sitting behind a silent typewriter. He asked for the two letterheads and the envelopes. Her mouth became firm. "Who are they for?" she said. The Texan said: "President Johnson." The woman stared at him in disbelief. Then she opened a drawer and took the stationery out. "Goddamn that man!" she shouted. "The President isn't even cold in his grave yet and he wants to use White House stationery. Goddamn him!"

Carter was a big man with a colorless face. He was big enough to afford good manners in trying situations. He said thank you and departed. The sheets were handed to Mr. Johnson. Carter went on to the next duty without telling the President what had happened. Mr. Johnson wrote the notes and asked that they be delivered to the White House at once. It left him depressed. He sat behind the desk, staring at the blotter. The President was thinking of Mrs. Kennedy. He looked up at Moyers and Carter and shook his head negatively. "I wish," he said, "that I could reach up and bring down a handful of stars and give them to that woman."

• • •

The doorbell rang and Mrs. Grant opened it. Jack Ruby was back again. He was behind a huge grocery bag. "Here's twenty-two dollars in groceries," he said cheerfully. He put them in the kitchen. Eva may have wondered what she was going to do with all these cold cuts and delicatessen salads. Her brother bought cold cuts as some gallants buy flowers. She watched him pick up the phone to call Dr. Coleman Jacobson. Two old friends—Jacobson and Stanley Kauffman—often upbraided Ruby for not attending temple services on Friday nights. He wanted to tell Dr. Jacobson, and Mr. Kauffman, too, that tonight he would attend. He did not want to appear boastful, so his excuse for each phone call was to ask what time services would begin.

Eva, looking at the groceries, murmured: "I never thought in my lifetime I would ever hear of a President being assassinated." She said that barbarians were running around. She went out into the kitchen and made some scrambled eggs and salmon for her brother. Ruby got off the phone and went out into the kitchen and ate hurriedly and silently.

"Really," he said to his sister, "he was crazy." Then he rushed into the bathroom and was swept by waves of nausea. When he came out, Mrs. Grant said: "That lousy commie. Don't worry, the commie, we will get him." Her brother wiped his eyes. Eva Grant had watched the story of the assassination unfold on television. She knew more about it than her brother. "This guy could have been sent here to do this," Eva said. Her brother said: "What a creep!" He had to leave to go back to his apartment and change his clothes. The departure was as abrupt as the arrival. Eva Grant sat alone in her kitchen and finished eating the eggs.

The Secret Service man waited inside the door. "Miss Shaw," he had said, "I'm sorry but we have to go back to the White House immediately." The bags had not been unpacked. Mrs.

Auchincloss was distressed at parting with the children. The little ones, still in the dining room, were disappointed. "Children," Maude Shaw said. "Mummy wants us. Caroline, be my bestest friend and help John on with his coat." The little girl began to play mother. She got the coat and held it. "Come on, John-John," she said patiently. "Put your coat on. We're going home again."

In a few moments, the children punctuated their good-bye kisses with *Grand-mère* and were driven quickly from the elegant old streets of Georgetown to the broad boulevards of Washington. In the dark, they could see the crowds of people, like deeper shadows, clustered in Lafayette Park. They saw others like ink blots along the White House fence on Pennsylvania Avenue. White House police were asking the people, quietly, to keep moving. In the driveway, flashbulbs winked like giant fireflies, and Caroline said: "What are all these people for?"

The nurse was saddened. "To see you," she said. The car moved slowly around to the South Grounds. The little party alighted at the Diplomatic Entrance. White House police, in uniform, nodded at the children and Miss Shaw. A Secret Service agent on a portable phone announced that the Kennedy children and nurse were returning to the mansion.

The back of the building was bright with light. Inside, the ushers nodded gravely to the children and tried to look cheerful. In the downstairs lobby, Chief Usher Bernard West, who had served several Presidents, sat on a chair staring blankly at a wall. Men and women were trotting in opposite directions. There was a sound of many murmuring voices, as in a vaulted cathedral. For the first time, Caroline and John paused in their childish flight to look upward at the faces of adults. A secretary stared at them and burst into tears.

The little faces became grave. So much activity at night was unusual. This was their house, the only one they remembered. They had become accustomed to seeing many strange faces in it—walking, sitting, crouching to hug them, faces fine and faces fat and sweaty, some of which became familiar in time, others seen but once. It was an unusual house, but they knew no other existence and therefore the unusual was usual. This was a new experience. Some, whom they regarded as old friends, turned away. One or two wept. Others stared at them and shook their heads. Most people didn't want to see them.

The party was led across the corridor and up the elevator to the second floor. This, the private section of the mansion, the living quarters of Presidents, was full of people. The girl and the boy looked up at the faces, many of them dear friends, but a search showed no parents. Maude Shaw whispered to a few of these people, and they told her, "She's expected here soon."

In the small suite of rooms, Maude Shaw closed the door. The British woman felt more fatigued and more nervous than she remembered. The phone rang and she asked the children to be quiet. It was an usher. He said that Mrs. Kennedy might go to Bethesda first with the President and return later to the mansion. Maude Shaw, far from resenting it, appreciated it. She could get John-John to bed and then have a moment to speak to Caroline.

She undressed John and bathed him and put him in his nighties. The nurse kept reminding herself that the children were good, so good. The phone rang several times. Each message was different from the last. Mrs. Kennedy was on her way to the White House. Mrs. Kennedy expected to be home soon. Mrs. Kennedy might not get home until late. Mrs. Kennedy . . . Mrs. Kennedy . . . Mrs. Kennedy.

A depression engulfed Maude Shaw. She could no longer act the happy playlet of bedtime. There were no stories, no bright promises for tomorrow. As Caroline was bathing happily, the nurse returned to her small room and looked out the front window. The stately trees were still there, dark branches of arteries against the night sky. The people out front seemed to be gathering, as though grief were a vigil of many eyes. In the hall, she could hear running feet; sometimes there were muffled voices.

It was a night of running panic. A bad time. To her, John F. Kennedy was more than a President. He was a magnificent man, a powerful friend. More than anything else, he was a father. More a father, perhaps, than a husband. Maude Shaw knew these innocents as well as anyone. She could not imagine how anyone, in the kindest manner, could inflict such cruel news upon them. John-John might not comprehend, but he would miss the big, affectionate hero who took him on helicopter rides, who teased him and made small man jokes. Caroline would be conscious of the permanence of death. She would understand that the tall, loving man who studied her printed alphabet and who always said in exaggerated surprise: "Caroline! Did you do this?" would not be seen again. She could not rationalize death, but she could understand the permanence of forever.

As the nurse watched them, her spirit became oppressed. They noticed that she did not play with them, so they played with each other—John-John running big skidding circles around the room as Caroline removed the dolls from her pillows. Miss Shaw helped John-John recite his night prayers and tucked him in bed. He would squirm for a few moments and maybe call her on one pretext or another, but then he would fall asleep and, for one more day, he would not know.

She went across the little foyer to Caroline's room and sat on the edge of the bed as Caroline primly turned down the bedclothes. The big girl, at six, had a special privilege. Every night she was permitted to read a page or two from a child's book. Miss Shaw took the book from her and began to read. The sound of her voice was unreal, and the tears came.

Caroline looked up from her pillow. The shiny face frowned. The nurse could no longer see the words in the book. "What's the matter, Miss Shaw? Why are you crying?" The nurse leaned toward the child and placed both arms around the little body. "I can't help crying, Caroline," she said, "because I have very sad news to tell you." "What?" the child asked. Miss Shaw wiped her eyes and began the story of a terrible accident in Dallas. It could be minimized only to a point. When a small voice asks "How badly is he hurt?" the impasse is reached. There is only one way of saying "he died."

Caroline began to cry. Maude Shaw, having inflicted the involuntary cruelty, sat weeping and patted the child's hand. She held that hand and kept patting it until fatigue overwhelmed the little girl. She slept.

Halfway between the White House and the Capitol is a huge mocha-colored doughnut called the Department of Justice. The north wing, facing Pennsylvania Avenue, is headquarters of the Federal Bureau of Investigation. The lights were on. Agents in pairs entered the corner doors on the Ninth and Tenth Street sides. Twenty had already left for Dallas. The FBI was in charge of the federal investigation into the case. Gordon Shanklin, at the Dallas office, and his agents Vincent Drain and James Hosty had been working on it since 12:40 P.M. Shanklin had pulled in agents working on other cases and had put them in Sheriff Decker's office, in Captain Fritz's office, at Love Field, in Irving, Texas, and the School Book Depository building.

The Washington office required no special organization. It was ready. Alan Belmont, assistant to Director J. Edgar Hoover, was in command of all the skeins of investigation and evidence. Assistant Director Alex Rosen assigned the agents who would probe the mystery. One man, who worked as liaison in the exchange of common information between the Secret Service and the FBI, was in SS headquarters on M Street. Assistant Director William C. Sullivan was in charge of the internal security aspects—and background—of the suspect, Lee Harvey Oswald. Inspector James Malley took charge of all agents assigned to the case in Dallas.

The IT of the case was intelligence and tact. Except for presidential fiat, the FBI had no right to examine the prisoner, the background of the case, or the evidence. The assassination was not a federal crime. The agents would work gently and inoffensively with the Dallas Police Department. They would be in the same delicate position as the Secret Service men who sat with Captain Fritz, listening to the questions and responses, but seldom asking a question without permission.

Closed-circuit teletypes began to rap out information to field offices all over the United States. New York was listening. So were New Orleans, Chicago, Dallas, Denver, San Francisco, and others. Everything that was learned was put on teletype so that, if anyone in another office could offer assistance, it would be on a return teletype to headquarters in a few minutes. The offices in the long corridors of the building were well peopled that night, and men began the work of sorting information and misinformation.

In the Fire Arms Identification section, Robert Frazier cleared the decks for a long night. He was surprised when an agent walked in. The man was Elmer Todd, and he said he had a bullet given to him by a Secret Service man at Andrews Air Force Base. Frazier asked where it came from. "It fell off a stretcher in Parkland Hospital," Todd said. The bullet was almost pristine. Frazier smiled down at it rolling in his hand. "The first reports," he said, "claimed that the gun was a 7.35 Mauser. This is very interesting." He placed the bullet into a device. "Just as I thought," he said. "This is not 7.35. It is 6.5 millimeter. Did the Secret Service man know which stretcher it was on?"

"No," Todd said. "The man who found it thought it came off Governor Connally's stretcher." Frazier began his work of examining the bullet scientifically. "Can't tell who manufactured it," he said, "without a cartridge case." He held the back end of the bullet up below a microscope. "It's not foreign-made," he said. "This is American."

The FBI, knowing that it had a small file on Oswald—mainly spot checks in Dallas to make certain that the man who defected to Russia did not get work in a sensitive defense plant—examined the reports and turned them over to William Sullivan, who was building up a skeletonized background on Oswald. He contacted the Central Intelligence Agency to see if they had anything on the suspect. Another man was sent to the State Department, to check their records of Oswald's readmission to the United States. Abram Chayes was still in his office, trying to make a digest of this material for Dean Rusk.

The New York office came in on the teletype asking for more information on the ammunition and the gun. They had an idea that they could run down the principal manufacturers of cheap rifles quickly. New Orleans came in, offering a detailed report on Oswald's arrest for distributing Free Cuba leaflets on the streets; they also had a detailed report on his recent trip to Mexico. Two men, Francis X. O'Neill and James W. Sibert, were making notes at the autopsy.

"I'm going to stop for a minute," Henry Wade said, as he parked the car beside police headquarters. The district attorney was seldom seen at the municipal building more often than, say, once a year. He had a big air-conditioned office in the county building and a staff of investigators and assistant prosecutors who worked to smooth the wrinkles in criminal cases before Mr. Wade tried them. He was big and shaggy, a type-cast Texan with wild wavy hair with streaks of gray.

He left Mrs. Wade and a couple in the car and walked inside and took the elevator to the third floor. Police who saw him nodded, or smiled, or shook hands and said, "Hello, Mr. Wade. What brings you over here?" He got to the third floor and, big and broad, shouldered the press aside with bantering words. He passed the Homicide office and headed for the office of the chief. His big feet slammed tripods and skidded over black television cables. To all questions, he drawled: "Fellas, I don't know nuthin'."

When he achieved the sanctum of the superior officers, he turned right and found Curry sitting at his desk. "How is the case coming along?" the district attorney said. Curry began to speak. Wade listened and asked questions. The prosecutor was beset by an involuntary verdict: he doesn't know. The big man listened to the little one, but he wasn't getting the facts he wanted.

The chief opened a desk drawer and gave Wade a memorandum from Detective Jack Revill. It stated what Revill thought that FBI Agent James Hosty had admitted to him about Oswald. The prosecutor refolded it and gave it back to Curry. An old memory popped into Henry Wade's mind: he recalled that there was a woeful lack of communication between Jesse Curry and Will Fritz.

Over three hours ago, Sheriff Decker had told Wade that the Dallas Police Department had a "good suspect." If it were true, Henry Wade would have to prosecute the biggest criminal case of the twentieth century. He would like to know how good the case was. "What are you going to do with Revill's memorandum?" The chief looked up. "I don't know," he said. One thing is certain. Chief Curry thought it best not to draw the memo to the attention of the FBI; instead it might have more power if released to the press and television. Certainly it tended to show that the Federal Bureau of Investigation was aware of Oswald and felt that he had the potential of an assassin. As an incidental bonus, it would take the press off the back of the Dallas Police Department and point it toward the FBI.

Down the hall, Captain Fritz suspended the interrogation. Justice of the Peace David Johnston had arrived with a warrant for the arrest of Lee Harvey Oswald on a charge of first degree murder. At his side was a tough and coldly venomous prosecutor, Assistant DA William Alexander. Judge Johnston composed himself, unfolded a document, and read aloud to Oswald that he was being charged with the willful and deliberate murder of Police Officer J. D. Tippit. "I didn't shoot anybody," Oswald said. The tone was not belligerent; it was a flat declaration of innocence.

Johnston gave the document to the captain, who scrawled "Will Fritz" across the bottom line. "Didn't you also shoot President Kennedy?" Fritz said. His tone was a soft bass. Oswald shook his head. "I didn't shoot anybody." Mr. Alexander took the signed document for safekeeping at Wade's office. The justice of the peace said: "I remand you in the custody of the sheriff of Dallas County." This order should have been executed but wasn't.

The postal inspector, H. D. Holmes, said that he had a question to ask. Fritz nodded. Holmes reminded Oswald that the records he had in his hand indicated that Oswald had rented Post Office Box 30061 when he was in New Orleans. The prisoner saw no objection to this. He said he had rented the box. The inspector said that the application listed one Marina Oswald and one A. J. Hidell as the only persons, beside Lee Harvey Oswald, who were entitled to take mail from that box.

If the prisoner saw a trap, he pretended not to notice it. "Well," he said loudly, "so what? She's my wife, and I see nothing wrong with that, and it could very well be that I placed her name on the application." The postal inspector and every police officer in the room knew that one of the vital points in the interrogation was to prove that Lee Harvey Oswald and Alex J. Hidell were the same person. The name had appeared on the post office application in Dallas. "I know," said Holmes softly, "but what about this A. J. Hidell?" Oswald stared down at his handcuffs. He shrugged. "I don't recall anything about that," he said.

In the outer office, James Hosty marveled at the number of law officers who could be crowded into Captain Fritz's fishbowl.

It was impossible to count the people in the hall, but the FBI agent made an effort to tally the enforcement men inside. There were three or four Texas Rangers, five or six Secret Service men, four Federal Bureau of Investigation agents, two postal inspectors, six Dallas detectives, a deputy sheriff, and Captain Fritz. These were in adjoining offices measuring ten feet by fourteen. In this standing-room-only situation, Hosty sought Forrest Sorrels, Secret Service agent in charge of the Dallas office.

The FBI man wasn't aware that his agency had been appointed to press the federal investigation into the assassination, so he assumed that the Secret Service was the dominant body. Hosty said that there was additional information at FBI headquarters which could be furnished to the Secret Service. Sorrels asked, "What?" There were two items which Hosty had in mind, but he did not feel at liberty to reveal them. Liaison between the two organizations was good, so Hosty proposed that Sorrels advise his Washington office to ask for material on Lee Harvey Oswald.

Sorrels thanked him and said he would take care of it. The two items were the contacts Oswald had with the Soviet embassy in Mexico City two months ago and the several letters he had written to the Soviet embassy in Washington, D.C. Down the hall Hosty waited for a public telephone to inform Gordon Shanklin of developments in the case.

The ultimate indignity is not death but what men do to the dead. The man on his back under the lights was nude, defenseless, broken. The gentlemen of science, like white wraiths, moved about the body in the manner of whispering druids. They made small and sometimes indecipherable notes on pads. Commander J. J. Humes and Commander Thornton Boswell joined in a medical ritual of exactitude. There were prescribed steps to be taken in this profound abuse of the body and, if they were carried out precisely, the President of the United States would leave the room as a shell, and the physicians would be able to say with certainty that he had succumbed to a gunshot wound in the head.

Kennedy was measured. He was 72½ inches tall. He weighed: 170 pounds. They looked at his eyes: blue. His hair, they decided, was reddish brown. He was 46 years of age, a male, and the subject of autopsy number A63-272. "The body is that of a muscular, well-developed and well-nourished Caucasian. . . . There is beginning rigor mortis, minimal dependent livor mortis of the dorsum, and early algor mortis." The left eye was swollen and black and blue, obviously from the shot which hit the right rear of the head and pressed the brain violently forward toward the left optic.

Clotted blood was found on the external ears. The President's teeth were declared to be "in excellent repair" although there was some pallor of the oral mucous membrane. The doctors were observing. They moved about the body slowly, looking, pointing, noting. There was nothing they missed, from the midline of the head down to the squared toenails. He who would not appear in a country club locker room without a robe and towel was under the merciless eyes of a score of men.

The small diagonal scar in the lower right quadrant of the abdomen was noted. The fact that he had arrived at Bethesda without clothes was recorded. A ragged wound was noted near the base of the larynx. Gently the body was turned over. The posterior was examined. The head wound was gross and obvious. There was a small oval puncture wound between the spine and the right shoulder blade. An inserted probe was stopped by the strap muscles. A frown darkened the faces of the medical priests. How could a missile go in there and (1) not come out in front somewhere; (2) still be inside? The doctors had a mystery. There was a separate small hole in the back of the head.

There was a long vertical scar along the midline of the spine below the lumbar region. The man on the slab had felt that the

475

operation almost killed him. It had not even relieved the steady toothache-of-a-pain which wearied every waking moment; it was the excruciating lightning which he was fond of denying when a well-wisher shook hands and yanked the President toward him. It was the dull ache which rang an alarm bell every time he arose from a chair; this was the moment when he smiled with his teeth and died a little in the eyes.

Strong arms turned him face up again. The doctors had already noted recent violations of the skin. Near the nipples were two incisions, done at Dallas. No hemorrhage and no bruise mark showed; therefore he was probably dead when these were made. Two more were in the ankles. And one in the left mid-arm. Along the front of the right thigh was an old scar which no one but John F. Kennedy would remember.

Skin tone was good. Muscle tone was excellent. And yet one would always have to revert to that massive head wound. The doctors did not want to probe it yet. It was, they decided, best described as a "large irregular defect of the scalp and skull on the right involving chiefly the parietal bone but extending somewhat into the temporal and occipital regions." This involved the right top of the head and part of the side, almost up to the right temple.

Mrs. Kennedy had expressed it simply: "They shot his head off." As a result of the bullet which entered the neck, he was leaning forward, falling to his left. A bullet, flying at about two thousand feet per second, had then hit the skull in the right rear portion. This, the third bullet, had exploded the brain, and both brain and bullet had crashed upward through the skull, inducing pressure cracks in all directions. The cerebellum fits snugly inside its case and anything which displaces or engorges it—such as edema or hemorrhage—can cause extensive damage to the brain and, sometimes, death. The 6.5 bullet, once inside, was a tumbling, disruptive force which scattered the dura mater at high speed, giving it sufficient force to find the weakest part of the skull—the fissure cracks—and broke it open with flying bits of bone and hair, also providing an exit for most of the bullet.

The doctors measured the missing area of the skull. It came to a little more than five inches from back to front, about half the head. In addition, there were star-shaped fractures of the skull which radiated across what was left of the cranial bone and down the sides and back. The doctors completed their pre-autopsy examination and called for a radiologist to take X-rays.

Across the room, the men of the FBI—Sibert and O'Neill—made notes. William Greer, a hearty product of County Tyrone, Ireland, a man who always drove President Kennedy's car and who drew pride from it, felt his stomach sicken. When the shot had been fired, he had heard the sound, as though someone was snapping a dry twig against his ear, and he had heard the echoes carom around Dealey Square, but he had not been able to look back at his distinguished passenger. At Parkland he had attended the beautiful and stricken young lady outside Trauma One and inside, too. There everyone had been so busy, so desperately busy, that Bill Greer had had no time to look for damage to the man he admired.

Now he looked. Now he saw the tan nude body bereft of station or dignity. As he looked at the head, all Greer thought of was a hard-boiled egg with the top sliced off.

Detective Rose of Dallas got out of his car at 835 Irving Boulevard and, with Detective Adamcik, hurried into Irving police headquarters. Their man turned out to be a frightened boy. Detective McCabe had him on a bench in a corner. He was tall and slender, about seventeen years of age, a long-necked boy with a nervous Adam's apple, big feet and hands, and speech laden with the homely idioms of the clay country of Alabama. Guy Rose was an experienced man. He looked at the kid and felt disappointed. If the assassin had an accomplice, it could hardly be a scared boy.

The first interrogation was brief. The responses were forthright. He knew Lee; drove him home to his wife on weekends and drove him back to the School Book Depository on Monday mornings usually. Lee was a fellow who didn't talk much, didn't make friends, brought his own lunch, and never bought anything, not even a drink of coffee or a newspaper. It was funny how he suddenly wanted to come home yesterday, because yesterday was Thursday, see, and he always made it on Friday. He said he wanted to get some curtain rods for his room in Dallas. Oswald was not a fellow you asked many questions.

This morning he had the long package wrapped in brown paper. It was on the back seat of the car. Detective Rose said: "Oswald says that was his lunch." Well, it couldn't be his lunch because it was "this long," said Frazier, holding his hands out in a spread of two and a half to three feet. "This morning, he told me he was going to buy his lunch. I remember because I was surprised. That's one of the few times he was fixing to buy his lunch."

The detectives asked Wesley Frazier if he owned a rifle. "Yes, sir," he said. "I got me a rifle. A good hunting rifle." "Where is it?" "Home. I can show you if you want to see it." Where was his car? Detective McCabe said it was left at the Irving Professional Clinic, where Frazier had been picked up. "We brought him here in our car." Guy Rose said they would all go back and locate that car. He wanted to have a good look at it.

The car turned out to be a ten-year-old Chevrolet. It was well frisked by the police. They even pulled the mats up. The country boy watched. It was difficult for him to comprehend what had happened and what he might have done to bring the wrath of the law upon him. By nature he was obliging and polite. When the police finished their work, they took Mr. Frazier to his sister's house and went through it carefully.

Linnie Mae Randall helped in every way. The detectives found very little worth bringing to headquarters. Frazier pointed to his .303 rifle, with full clip, and part of a box of hunting ammunition. Rose asked him and his sister to accompany them back to Dallas police headquarters. Captain Fritz would want sworn statements from them regarding their knowledge of Lee Harvey Oswald.

Wesley Frazier shook his head. "The only time he smiled," he said, "was when I asked him did he have fun playing with the babies."

7 p.m.

A wave of nausea engulfed Big D. The city became physically ill. Hospitals and sanitariums were subject to an inordinate number of emergency calls. The Vern Oneal ambulance service had three vehicles racing through the city with red lights winking, sirens shrieking, to pick up citizens who thought they were sustaining heart attacks, strokes, fainting spells, and ulcer attacks. Parkland Hospital had two nurses standing inside the emergency room door to wheel the patients in for treatment. Doctors were making emergency house calls. A large part of it was hysterical reaction, but the community reflexes also turned in the opposite direction—some of the people were glad that Kennedy was dead, although regretful that it had happened in their city.

Gloom dominated joy. Dallas, Southern in emotion and Western in speech, was jealous of itself. The city was hypersensitive to criticism. It saw itself as rich and righteously Christian. No dust of civic scandal was permitted to cling to its boots. The world was going to give this story a lot of front-page attention and Dallas was being soiled by a lone communist who was not even a Dallasite—a native of New Orleans, perhaps, or Fort Worth. In Dallas he was a cheap laborer, a roomer in a boarding house. He had drawn a bloody scar across the beautiful face of the city and, even if it healed, the scar would show and people would always say: "Dallas—oh, yes. Where Kennedy was assassinated."

In a bare room at Parkland Hospital, Bardwell D. Odum sat watching. The work was almost complete. Doctor Earl Forrest

Rose was sewing together what was left of Officer J. D. Tippit. Cool and professional, he completed his work notes on the autopsy. The dead frame on the table was almost the same size and configuration as the President. The hair was dark brown and thick; the body was muscular and well nourished. The face was broad and serene. He was a fairly young Caucasian male who should have had productive years ahead of him. Life had fled in a barrage of bullets when least expected. The pain, the bitterness would be in the eyes of a widow and children.

Agent Odum of the FBI continued to watch and to make his personal notes. Dr. Rose completed his work, sighed, and peeled the rubber gloves from his hands. He studied his notes for a moment, added an observation or two, and nodded to an orderly. A sheet was tossed over the body of Tippit, and the stretcher was wheeled out. The cop could go home now.

Odum got to his feet. The doctor fished into a small stainless steel tray. There was a battered uniform button in it and some bullets. When lifted, they rolled around the tray and made a musical sound. The doctor took an instrument and made a small indentation on each object. "This is my mark," he said to the FBI man. "I will be able to identify it in court." Bardwell Odum opened an envelope. The doctor dropped three .38 bullets into it. They had been taken from Tippit's chest. So had the uniform button. There was one additional bullet which Doctor Paul Moellenhoff had found when Tippit arrived at the Methodist Hospital earlier.

Lieutenant Day would be interested in these slugs. He and his men were still working on the fourth floor at police headquarters. As evidence came in, they studied it, analyzed, made notes and photographs. The Dallas crime laboratory was doing well. Day and his assistants had found two smudgy fingerprints on the side of the rifle close to the trigger guard. A palm print was raised from the underside of the barrel. There was a good palm print from a packing case. In some instances, they photographed their finds without completing comparison tests so that they could work on new evidence.

On the third floor, the crush of journalists was worse. Newspapers had flown in extra men, and they were arriving in groups, demanding to be brought up to date on the status of the case. The crowd was so dense that movement became difficult. When someone like Captain Will Fritz uttered a few words publicly, only those in front could hear it, and they had to shout it over their shoulders until the quotation rippled the length of the long corridor.

Seth Kantor was certain that he had never been in a situation like this. He had covered many stories, big and small, but he had never been party to a stampede. In the long dull periods, reporters often stared at each other for periods of time, uttering solitary words periodically: "In-CRED-ible!" "UN-believ-able!" "IM-poss-ible!" The tide of writers engulfed the corridor and spilled over into police offices, where they appropriated typewriters and paper and used the telephones.

Detectives sought corners in which to question witnesses. Radio reporters, hungering for news, fell back on the cannibalistic practice of interviewing news reporters. Lack of information made the demands more strident. Getting to the men's room near the elevators was a time-consuming assignment. No man dared to leave for a snack unless he had a reporter substituting for him. An air of acrimony was detectable. Early deadlines for the morning newspapers were passing, and Fritz refused to reveal what Oswald had said or what evidence he had against the man.

In the Burglary and Theft office, a lieutenant and three detectives worked a telephone, running down assignments for Homicide. The glass door opened, and Detective A. M. Eberhardt looked up. Jack Ruby was smiling. "Hi, Mike," he said. Ruby shook hands. Eberhardt said: "What are you doing here, Jack?" The nightclub owner was in good spirits. He said he had brought some kosher sandwiches—"nice lean corned beef"—to the reporters.

Eberhardt, like most of the Dallas department, knew Ruby as a nightclub owner who would stake a cop to a drink or a free strip show. He was a "police buff," one who enjoyed standing on the fringe of excitement. The other policemen exchanged greetings, and Ruby displayed a notebook and a pencil. "What's that?" they asked. He said he was acting as translator, or interpreter, for the newspapers of Israel. Detective Eberhardt didn't know whether this was a joke or not. He was aware that Jack Ruby could speak Yiddish, but didn't the people of Israel speak Hebrew?

"Isn't it terrible, the assassination?" Ruby said. The men were back at work. They nodded. Two stenographers were checking statements before asking for attested signatures. "Mike," he said, using Eberhardt's middle name—"it is hard to realize that a complete nothing, a zero like that, could kill a man like President Kennedy." Ruby asked how the detective's family was. He pointed to his lapel and said: "I am here as a reporter." The police knew that this was not true. There were no reporters from the newspapers of Israel. The statement was a pleasant excuse in the event that a young policeman, not aware of Jack Ruby's generosity to the department, should ask questions.

He had an hour before going to temple services for the President. When a phone booth was empty, he phoned his nightclub. Larry Crafard answered. Ruby never identified himself on a telephone if he could avoid it. "Any messages for me?" he said. The handyman said there weren't. Ruby didn't say he was at police headquarters. He hung up.

Almost every community has an assortment of indefinable personages who are referred to as "characters." Most of them are neurotics who are anti-social. They strive for unknown goals, reach for an equanimity which is never grasped. The majority of them are sensitive and emotional, rising to hilarity or anger quickly and returning to a mildly depressive state within minutes.

Jacob Rubenstein was a character. Like Lee Harvey Oswald, his aspiration was to become well known. Oswald selected a swift and desperate path. Ruby tried to ingratiate himself with those whom he regarded as his betters. Above all, he hoped to be accorded respect. When he got it, he accorded it quickly. He found criticism to be insufferable; a challenge to his virility led to a fistfight. He paid stripteasers $110 a week and enjoyed posing with them in their brief costmes.

There was a hint of homosexuality in his belligerence. In most brawls, he made certain that his quarry was intoxicated or smaller than he. He was in trouble with the unions for underpayment of scale wages, for complaining about other nightclubs which featured amateur night stripteasers. As an automobile driver, he was arrested or subpoenaed twenty times. He blackjacked an employee, kicked customers in the groin, slapped girls, threatened to throw a customer down a flight of stairs, and asked a stripper named Jada to move into his apartment "platonically"—to prove his manhood.

Cash was his god. His two bank accounts seldom showed more than $200, but on his person and in the trunk of his car he often carried over $2000 in bills. At times he wrote invalid checks. The Internal Revenue Service had a claim of delinquent taxes in the amount of $44,000. Still he regarded himself as a strong patriotic American, a religious Jew, a gentleman of "class."

He was ingratiating himself with men he regarded as big-time reporters from New York, Chicago, and Washington when a phone rang in a nearby office. Lieutenant T. P. Wells picked up the receiver. A woman announced that she was Mrs. Barbara Jeanette Davis. He listened. She lived at 400 East Tenth Street. Her sister-in-law, Virginia Davis, had found an empty .38 shell on the lawn.

Wells knew that the address was close to the spot where

Tippit had been killed. He checked the name and address and asked if her sister-in-law would mind coming down to police headquarters to make a statement. Mrs. Davis said that she and her sister-in-law had seen the shooting of the policeman from their screen door. They saw the man walk off fast, holding his gun up in the air and emptying it.

"I'll be right there," Wells said. "We'll pick you up."

There was a desk outside the suite on the seventeenth floor. Behind it sat the handsome man with the wavy hair, Clint Hill. Although the Kennedy family would be at Bethesda only a matter of hours, the Secret Service man ordered direct lines to the White House. A naval officer came up from the main deck with a blank, ordering an autopsy on John Fitzgerald Kennedy. It was typed, and the name Mrs. John F. Kennedy was typed where her signature was to go. Hill had used good judgment all day. He was not going to ask Mrs. Kennedy to sign the order.

It was shown to Paul Landis, who stood leaning against the suite door. He was sent inside to ask the Attorney General to step out for a moment. Robert F. Kennedy left the group and came outside. He, too, realized that it would be distressing to ask Mrs. Kennedy to sign such an order, so he took his pen and scrawled "Robert F. Kennedy" on the left side.

It was an improper signature, but the United States Navy would not quarrel with the Kennedys. They had tried to convince the family that the medical and surgical complex did not perform embalming, but this had led to mass frowns and acidulous barks from O'Donnell and McHugh. The autopsy order was returned to Captain R. O. Canada, commanding officer. Robert Kennedy returned to the suite, where he crouched on the kitchen floor, chatting with Robert McNamara.

Grief, like ecstasy, is impossible to maintain at a high level for considerable periods of time. Tears, shock, hysterics are concomitants of grief, but they fall like thunderous waves on a beach, slide quietly up the sands of memory, and recede in ripples. There were tears in that sacrosanct suite. There was laughter too. There was a wistful penchant for "Remember-the-time-Jack-said . . ." There was speculation about Lyndon Johnson. His name, mentioned among the men, wrung no applause. It is doubtful that any successor to Kennedy could have won the endorsement of the people on the seventeenth floor. Johnson stood less of a chance because Robert Kennedy had never bothered to mask his animosity from the moment his brother had picked Johnson as his running mate.

Among the women, someone recalled that Caroline and John-John would have birthdays within a few days, but this was stifled. It is difficult to say whether the sight of the bloody Mrs. Kennedy was more of a shock to the women or the men. Now and then she repaired to the bedroom to watch the television set and, when she turned the corner to return to the living room, conversation sometimes died in mid-phrase. It seemed that everybody wanted to make a phone call.

This led to the disclosure by the Washingtonians that the District of Columbia lines had been so tied up that afternoon that panic ensued. Edward Kennedy, for example, jumped into his car and tried to use the phones of other residents because he could not get an outside line on his own phone. He did not realize, at first, that the situation was common everywhere. Hundreds of thousands of people were calling other hundreds of thousands with the shocking news, and these in turn were calling mothers, brothers, cousins, and uncles. Trunk lines were exhausted throughout the city and most of the Eastern Seaboard. Even emergency calls could not get through.

At dusk the situation had eased. The Kennedys and their friends had many calls to make. Until the tie lines with the White House were established, many calls were dialed to NA8-1414. Most of them were "idea" calls. Sargent Shriver, working in Dungan's office, took them one by one. An invitation list to the funeral was being drawn up, and names of the great and near-great, crowned heads and premiers, were being bandied as though the personages were divided into two sharp camps: *grata* and *non grata*. The funeral would be held Monday. This was a fixed point from which to work. There would be a mass, probably at St. Matthew's Procathedral.

How many would it hold? Who would be invited? How about diplomatic cables tonight? De Gaulle? Yes. Queen Elizabeth? Yes. Harold Wilson? Yes. Richard Cardinal Cushing? Oh yes. He would say the Mass; it didn't matter that Archbishop O'Boyle was the ranking Roman Catholic churchman in Washington. He would step aside for a family friend. Who else? Barry Goldwater? Who said that? Governors? Indeed. The Senate would send a delegation. So would the House. Did anyone know why the Church of Rome insisted on five conditional absolutions over the casket? No one knew. How about the apostolic delegate? What was his name? Something Italian.

The diplomatic corps? Well, not the whole group. Those ambassadors could fill a church. The Supreme Court? Now why didn't somebody think of them before? The Supreme Court, of course. Who from the United Nations? Who from civil rights? Did the State Department have a list of its own? It would have to pass family scrutiny. Was General Wehle taking care of the military side? There would be a big military side, with honor guard, caisson, muffled drums, a horse with reversed stirrups, representatives of each of the branches of the armed services. Musn't forget the Green Berets—they were Jack's favorites. How about the Joint Chiefs of Staff?

Casket opened or closed? Closed. How about the lying-in-state at the Capitol? Time had to be made for the people to file past the bier in the rotunda. Saturday or Sunday? Let Jackie make that decision. All over the White House men were thinking. And, because the death had come when death was least expected, no one was prepared. Notions, ideas, and suggestions were being tossed in air to be shot down or caught delicately and approved. For a proper funeral everyone was going to have to think at top speed because something, or someone, was certain to be overlooked.

Down in the big stainless steel kitchen of the White House, sad-eyed cooks and assistants, lugubrious in white puffy hats, made sandwiches by the score. A lot of distinguished ladies and gentlemen probably would be coming to the mansion tonight with the body, and sandwiches and rich hot coffee would be in order. The personnel, domestic and official, invented tasks. A Secret Service man in Evelyn Lincoln's office stood looking through an open door into the President's private office. The new rug, a surprise from a doting wife, was on the floor. The desk, a gift from Queen Victoria and fashioned from the ribs of a British polar icebreaker, gleamed in the half light of the empty office. The drapes were new and the office had an empty majesty. It was waiting for a man worthy of grandeur. A special man. He would not enter this office again, shoes gleaming, a hand jauntily in the pocket of the jacket, the back a little too straight, the face youthful and tilted, ready for the tasks as though he had been born for this particular office. As though all else had been orchestral overture, the difficult days, the indenture, behind him. No man had been this young in this place; no man had been as remorseless in purpose; no man, no matter how untried and naïve, felt more attuned to the problems—not merely of his nation—but of the world.

The Secret Service man guarded the emptiness of the place. The rocker between the couches was already gone. The book of mementos of the trip to Ireland had been taken from the table behind his desk. The American flag flanked the window, but no breeze stirred the folds. The silver stand-up appointment calendar said "White House." No name was on it, no time. All the people and all the minutes had run out in a shattering sound of shots.

On a lower floor, Chief James Rowley sat in the Navy mess with Secret Service agents who had flown in from Texas. The white hair, the scrubbed Irish face, was tilted toward the table. The questions were uttered softly, but they were endless. He wanted to know everything about everything. Step by step, he took them over the entire trip, the work of the advance men with the Dallas Police Department, the PRS file of dangerous persons in the area. He asked each man what he did, what hours he worked, where he was stationed. He wanted to know what orders Jerry Behn had issued as agent-in-charge of the White House. Could this thing have been prevented? Did any agency have any record on the suspect who was arrested?

The men knew how Rowley felt. His administration of the Secret Service had sustained the worst possible blow—it had lost a President of the United States. No matter how well the service had done its work, it had lost *the* man. Rowley clasped his fingers behind the coffee cup on the table. He realized that some of the politicians would demand his head for the deed. The defense, that every precaution had been taken, that no one but God could have foreseen and prevented it, would be discounted in some places. Someone would have to pay, and who is a better target than the chief of the Secret Service? He made no excuses to his men; he asked none. "I want every one of you to make a detailed report now—tonight—before you go home. I know you've all worked long hours, and I know every man feels depressed. But I want those reports on my desk tonight."

Upstairs on the main floor, artist William Walton consulted the book on the Lincoln funeral. It was replete with old-fashioned steel-point engravings. The catafalque looked bigger than it should in the East Room, but that was probably artistic license. In the White House warehouse, the dustbin of many administrations, the Lincoln catafalque had been found. Walton would have it set up in the East Room. The drapes, the wall candelabra, the huge center chandelier, he noted, had been draped in deepest black.

Well, that was a little too much black for this century and this man. Some small tiebacks of black could adorn the drapes, perhaps even the candelabra. But it would be gauche to blacken that crystal chandelier with dusty black. As he pored over the drawings and some old daguerreotypes of the Lincoln funeral train, Arthur Schlesinger, Jr., and Richard Goodwin passed by on their way to the Library of Congress. Someone had been found who had keys to the enormous collection of history. The place would be opened for the two men, who would seek the books on Lincoln and make notes on how the final rites had been conducted for America's sixteenth President and first martyr.

At Andrews Air Force Base, a huge C-130 plane stopped on the ramp and the pilot killed the big engines. The back of it opened as though it would lay an egg. Washington motorcycle police gathered expectantly. Slowly down the ramp came SS-100-X, the big Lincoln in which Kennedy had been killed. Behind it the heavy convertible follow-up car bumped easily to the concrete. Secret Service agents Hickey, Taylor, and Kinney started the motors.

The police escort asked the route. Both cars were going to the White House garage. Agents were waiting to cover both cars with plastic and to guard them. The motorcycles started from Andrews in an inverted V. By short wave, Secret Service headquarters knew that both cars were on their way. The presidential car was the one they wanted to examine carefully. Unknown to them, the FBI also wanted to take an unhurried look at that automobile tonight.

On the other side of the city a military guard of honor drew up before the pale brick façade of Gawler's Funeral Home. Someone in the White House had said that the President's body would be brought there. Joseph Gawler was surprised. So was his manager, Joseph Hagan. They had been contacted by the

White House and asked to handle the dressing of the body and to help in the selection of the casket, but no one had told them when or where or even who. There was a double line of servicemen in dress uniform, and perhaps they knew more than Gawler.

Clients were entering and leaving the funeral home, paying their respects to less noted dead. It was embarrassing to have that military braid outside the door. Along Wisconsin Avenue, automobiles began to slow down, some to stop. Neighborhood people collected. The word passed from lip to lip: "They're bringing Kennedy here. Let's wait awhile." It wasn't true. The military order was countermanded, and the men were told to board their waiting bus and return to the White House.

In far-off Ireland, the lateness of the hour found solemn men drinking and thinking. Television was rare in the southern tier of counties, but some had paid it "more mind" that night than before. John F. Kennedy was dead indeed but they had seen him on the opaque screen as he had visited the villages of Wexford and tipped a cup of tea with his cousins. Scores of millions of Americans had seen him on television, but none had thought to phrase it in the manner of an old man in Dublin: "Ah," he said, "it would make you lonesome to see him talking and him being dead."

Three automobiles paused at the head of the concrete ramp, then idled down to the basement of Dallas police headquarters. The lead driver waved the others into a corner. Marina Oswald had dark thoughts. "Isn't it true," she said in Russian to Ruth Paine, "that the penalty for shooting someone is the electric chair?" "Yes," Mrs. Paine said. "That is true." The answer wasn't sufficiently detailed. "Your Russian has suddenly become no good at all," she said.

The detectives herded the witnesses into a group. A detective told Michael R. Paine that he would be taken to a separate office for questioning. His wife, tall and stately, helped Mrs. Oswald to take care of the babies. A policeman gathered the evidence from the car. Marina pointed to Adamcik, the Czech-descent policeman. "Translate to English for him." Guy Rose phoned from the jail office to ask Captain Fritz where to take these people. Fritz said that they would have to find space in Forgery or elsewhere; the third floor was crawling with human-ity. He did not want Oswald to see his wife or the children. Rose was advised to detail one man to guard the evidence and to take it to Day on the fourth floor and remain with it.

Marina submitted to the ordeal with little grace. It is possi-ble that her chronically deprecating assessment of her husband was on her mind. In the Soviet Union, she had had a small and secure place in the society of Minsk. She was a qualified phar-macist and the niece of a man who was a colonel in the security police. No matter what one's private opinions might be, mem-bers of her family did not say or do things which would draw the attention of the police. To be picked up for questioning was debasing. There was a standing to be maintained in the com-munity. Had she been slightly more callous—or more practical, perhaps—she might have been relieved that her husband had been arrested, that he stood an excellent chance (if guilty) of dying in the electric chair and freeing her. There was always the odd chance that, in the process, the United States government might choose to deport her to the Soviet Union. If she had a poignant regret, it was embraced in the prospect of having to take the children and "go home." She had written the letters which Lee demanded that she pen to the Soviet embassy in Washington, asking to be repatriated. But Marina Oswald did this because of her European notion that the wife must always be subservient to the wishes of the husband.

A room in Forgery was cleared, and Mrs. Oswald, in plaid slacks and a head kerchief, sat with Rachel on her knee. Outside, reporters yelled: "Who are these people? Is this Oswald's family?

Which woman? How about an interview?" Ruth Paine sat on the opposite side of the desk, but she couldn't hold June. The little one kept breaking away and running back to her mother.

A dignified middle-aged Russian stepped into the room. He was Ilya A. Mamantov. He bowed, smiled, extended his hand. It would be impossible for Mrs. Paine to serve as interpreter for Mrs. Oswald, he explained, because Mrs. Paine was also a witness. Therefore he, a geologist living in Dallas with a Latvian wife and mother-in-law, had been summoned by friendly policemen with shrieking sirens and revolving red lights to attend Mrs. Oswald. He hoped the ladies would not be nervous—as he was. He assured Marina that he would do his best to translate her thoughts into impeccable English.

A stenographer was called and the interrogation started. Mrs. Oswald's life stood at its true crossroads in this room. She could, if she chose, protect Lee by lying, lying which would be difficult to disprove. She could say that her husband admired few politicians but that John F. Kennedy was one. She could say that, to her knowledge, he never owned a rifle. She could say that her Russian was misunderstood when she pointed to the blanket in the garage as the storage space of his rifle. She meant to say that the blanket "looked" like a rifle, had the conformation of a rifle. They had disagreed, yes, but they had made up this morning, and he had left $170 with her and had promised to return to her tonight.

The other road was to tell the truth as Marina saw it. It would help the police to hang her husband. If she chose this road, the marriage would die in this room at this hour. Her little girls would bear a stigma as daughters of an assassin all their lives. The American government might return them to Russia. She could hardly support the children in the United States even if the government was favorably inclined toward her. She could not work because her husband had prevented her from learning any English.

Mr. Mamantov listened to each response, stared at the ceiling in silence, and tried to think it out in precise English. The work was difficult because he also had to translate the questions of the police—sometimes spoken in idioms—to Russian. Captain Will Fritz stepped into the office and closed the door behind him. He told the detectives that Lieutenant Day was coming down with the rifle. Paine, he said, was being questioned across the hall in a room with Robert Oswald.

Day came down the corridor holding the rifle high with a finger inside the leather sling. The press was upon him, shooting pictures and demanding to know if this was the gun which had killed the President. Day shouted, "Out of my way!" He had been testing this weapon when Fritz asked for it to be brought down for identification. The lieutenant did not like to stop in the middle of his work, but the captain ranked him.

Did Marina Oswald recognize the weapon? She and Mrs. Paine studied it with curiosity. Even to those who neither understand nor appreciate rifles, this one would appear to be cheap. Black paint had worn off the fibers of the wood stock. The telescopic sight was twisted to the left side. The leather sling, with its pad of soft leather in the center for shoulder-carrying was dirty. Marina stood to examine it. She didn't touch it. Then she shrugged. In Russian she said it could be the rifle owned by her husband. One detective reminded her that she had said her husband owned a rifle in Russia. Could this be it? It could be, she said. Obviously, she did not know the difference between a rifle and a shotgun.

Mr. Mamantov felt obliged to volunteer the information that no citizen in Russia is permitted to own a rifle. A shotgun yes, but not a rifle. Mr. Mamantov subsided. He was not a witness; his function was to translate. "When did he buy his gun?" Mrs. Oswald shifted the squirming baby to the other knee. "I don't know. He always had guns. He always played with guns even in the Soviet Union. He had a gun and I don't know which gun was this." "Would you recognize his gun—do you know it by color?" "All guns are dark and black as far as I am concerned."

Fritz told his detectives to question her about the telescopic sight. A detective touched it. "Is this what you saw?" he asked. "No," she said. "No. I saw the gun. I saw *a* gun. All guns are the same to me—dark brown or black." He pointed to the sight again. "No," she said. "I have never seen a gun like that in his possession." She pointed dramatically to the sight. "This thing." The questions stopped. "No. I have only seen this part of the gun," pointing to the stock. "The end of the gun." They asked if she had seen it rolled up in that blanket. "Yes. Dark brown, black."

More questions were asked. The rifle went back to the laboratory. Fritz said, "Excuse me" and returned to his prisoner. Marina answered everything candidly but volunteered no additional information. She might have told them about her husband's confession to her that he had tried to kill Major General Edwin Walker at his home. She could, if she chose, have told them about the day he wanted to assassinate the Vice-President of the United States and of how she had locked him in the bathroom. She confined herself to the questions.

Mrs. Paine told of her relationship with the Oswalds and her trip to New Orleans in a station wagon to return Marina to Dallas. None of it was exciting material. Still it added bits and chips of information to the rapidly augmented pile. The affidavits were typed and ready for signature. Mrs. Paine read hers—it stated, among other things, that she heard Marina say, in the garage this afternoon, that her husband had kept a rifle in that blanket. She signed it. Marina's, written in English, had to be retranslated in Russian word by painful word. She said: "Da" and signed.

There was a commotion in the next office, and Marina looked up in time to see a stout middle-aged woman coming into the Forgery Bureau. She gave a cry and arose to hand Rachel to her paternal grandmother. Marguerite Oswald looked down at the tiny face in her arms. Tears glistened behind her glasses. The women fell into each other's arms, neither one able to communicate except by embraces and kisses. Marguerite was moaning: "I didn't know I was a grandmother again. Nobody told me."

Policemen glanced up from their work and returned to the study of affidavits. Ruth Paine stood and extended her hand. "Oh, Mrs. Oswald," she said. "I am so glad to meet you. Marina wanted to contact you, but Lee didn't want her to." The grandmother stopped weeping. She rocked the baby back and forth in her arms and turned the large eyes of the inquisitor upon Mrs. Paine. "You speak English," she said. "Why didn't you contact me?" Mrs. Paine felt embarrassed. "Well," she said, "because of the way they lived. He lived in Dallas and came home on weekends. I didn't want to interfere."

The grandmother began to dominate the scene. She told Mrs. Paine to tell her daughter-in-law that she had been on her way to work in her car when she heard on the radio that Lee had been arrested. The police had asked her about a rifle that Lee was supposed to have, but she as a mother knew of no such weapon. The inference could have been that she hoped that no one else would recall a rifle.

Then she heard the admissions in Marina Oswald's affidavit as Mr. Mamantov read them slowly. The grandmother may have felt that Marina did not understand the question. Besides, who would know what the young woman said when the police had their own interpreter? Ruth Paine wasn't paying much attention to Marguerite Oswald's debate with the police. She recalled that six weeks ago her friend Marina had said: "It is only proper to tell the woman of the coming baby." Her husband did not want Marina to contact his mother. He said he didn't even know her address. He ordered his wife to leave his mother out of family matters.

"I don't know what I'm going to do," Marguerite Oswald said, turning her gaze upon Mrs. Paine. "I want to stay in Dallas and be near Lee, so that I can help with this situation." Mrs. Paine gave the proper response. "You are welcome in my home," she said, "if you care to sleep on the sofa." The grandmother was grateful. "I'll sleep on the floor to be near Dallas," she said.

She asked the police if the ladies could leave. The men phoned Fritz. All had made statements and signed them. They could go. In the next office Michael Paine was signing his words. Robert Oswald had concluded his interrogation. Mrs. Oswald pulled her white uniform skirt out from her side. "It's all I brought," she said. She had demanded to see her son Lee, but this privilege had been denied "until later." He was thirty feet away.

A policeman escorted Robert Oswald to meet his family. His mood was depressed. His younger brother was involved in an infamous crime. The name Oswald, which Robert had carried with honor, would be anathema all over the world. The situation, beyond doubt, would affect Robert's life and his career. In addition, he did not get along well with his mother. She masterminded every family difficulty, and Robert felt that, in the main, she was concerned solely with herself. Then there was this Russian woman and a baby—who would take care of them? Robert?

He walked into the office and saw Marina with *two* babies. Robert was surprised. Neither could understand the other, so he nodded. A tall, slender woman stepped between them and said: "I'm Ruth Paine. I'm a friend of Marina and Lee. I'm here because I speak Russian, and I'm interpreting for Marina." Robert Oswald felt little interest. He had just met her husband in the next office, and, when they shook hands, Oswald felt an instant dislike for Michael Paine.

He was still trying to greet his sister-in-law when his mother said: "I would like to speak to you—alone" and took him into an empty office. The jowly face quivered, the eyes stared around the room, and Marguerite whispered: "This room is bugged. Be careful what you say." The young man thought: All my life I've been hearing her tell me about conspiracies, hidden motives, and malicious people.

"Listen," he said loudly. "I don't care whether the room is bugged or not. I'd be perfectly willing to say anything I've got to say right there in the doorway. If you know anything at all about what happened, I want to know it right now. I don't want to hear any whys, ifs, or wherefores."

Apparently this speech caused her to forget to whisper. She began to speak swiftly and dynamically. She wanted Robert to know that she was sure his baby brother Lee had been carrying out official orders, if he had done anything wrong. When he went to Russia, she said, she was convinced that he was a secret agent for the United States government. He could have been recruited by the Central Intelligence Agency while he was still in the Marine Corps. If so, and if he was involved in some act today, well, the boy was probably under orders.

Robert was half listening, half meditating. His mother had learned that her son was a possible assassin of a President of the United States with no sense of shock. He felt her reaction was that she was about to receive the kind of attention she had craved all her life. She would never again be regarded as an unknown woman among millions of them. And yet he knew he must stand here and listen. It was his mother. The chronic sorrow was that, of her children, he and his half brother John Pic were hardworking responsible citizens, whereas she and the little loner made harsh and inexorable demands of life. They wanted recognition. Now they had it. It was such a joy that neither mother nor gunman had time to devote a moment of grief for the man who had fallen among the roses.

The doctors stepped away from the body on the table. Alone, it had more dignity under the white light. The supple arms, the strong hands were at his sides. The radiologist, from outside the area, called out his orders. The enlisted men moved the body at his whim. The large X-ray plates numbered fourteen. Among them were a front shot of the head, a lateral shot, a posterior picture. The positions were repeated for the thoracic cavity and the abdomen. Several pictures were made of the complete head and torso.

Before the next phase of the autopsy began, the doctors waited for the wet plates to be developed. A Navy photographer used the pause to get on a ladder and make color photographs and black and white pictures of the body as it appeared—front, side, and back—when received at the hospital. The commanding officers explained that these negatives would not be developed or printed. It did not occur to them that there was no point in ordering photographs if they were not to be used to support and augment the findings of the autopsy itself. The cassettes of film would go to the Secret Service, which was expected to make them a shocking gift to the widow. The philosophy of the Navy, on the night of November 22, 1963, was to play it safe and survive. The death of a President is a sensitive political event.

The witnesses sat quietly. The doctors waited. They had schematic drawings of a male body, front and back; they had drawings of the human head looking down at the top of the skull. They made terse notes, early impressions, and dots to represent wound punctures. None of these were exact; they did not even agree precisely with each other. Neither Humes nor Boswell realized that, outside this room, in the world of darkness, were laymen writers who could and would distort the early misconceptions, the burning of erroneous notes, the underdeveloped photographs into a malicious and mysterious plot to deny the American people its right to know the truth about the wounds.

It is difficult to search for sympathy for the United States Navy, the most pontifical of the American forces, because the senior officers had decreed, without warrant, that the main focus of the night must be centered on secrecy. A Marine guard was posted outside the autopsy room as though some unauthorized citizen might try to force his way into a scene which no sane layman would want to witness. Heads had been counted; names had been recorded; the mutilation of the dead, which is a scientific concomitant of a search for truth, was to be executed with such aggressive secrecy that, for years to come, no outsider would be permitted to see this room, even when it was empty.

In addition, the United States Navy did not assign the best qualified physicians to conduct the autopsy because they were not available. Commander James Joseph Humes, a product of Villanova University and Jefferson Medical College, functioned as director of laboratories of the Naval Medical School. He was an administrator. Among his confreres, Humes was known to be an excellent pathologist, an expert in the nature and cause of disease and the changes it brings to the body. It could be regarded as an imposition to order him to autopsy a body and qualify as an expert in missile trajectories and damage.

To assist Humes, Commander J. Thornton Boswell had been ordered to report to the autopsy room. Among the physicians available at the center, Boswell enjoyed the confidence of the officers, but he, too, was hardly a monumental choice for an autopsy involving violent death. This is not to say that Humes and Boswell were unqualified to conduct an autopsy; it was not their specialty. Humes was so conscious of this that, when he was offered an opportunity to secure the services of a qualified second assistant, Lieutenant Colonel Pierre Finck, he agreed. The colonel was chief of the Wounds Ballistics Branch of the Armed Forces Institute of Pathology. He had a career of four hundred cases of bullet wounds and two hundred autopsies. Although Finck was the most experienced, Commander Humes was in charge of the work.

The preparatory examination and the photography were complete by the time Colonel Finck arrived. Quickly, he gowned and masked and came into the autopsy room just as the X-rays were being placed on big illuminated opaque screens. Dr. Humes invited the laymen witnesses to come over to the screens and listen to a dissertation on the initial findings. Kellerman, Greer, Dr. Burkley, Sibert, O'Neill, and the others looked at the big negatives. A radiologist with a pointer took the frames one by one. Where there was no pathological finding, as in the abdomen, it was so stated and the doctor moved on to the more dramatic studies.

Most puzzling was the wound in the right strap muscles. It was almost certainly a wound of entry, because the hole was small and round, slightly elliptical, with no ragged serration which usually attends a wound of exit. The exit is often larger than a wound of entry because the missile, after striking the resistance of a body, sometimes begins a tumbling effect in its progress.

Humes or Boswell, on the schematic drawing of the body, seem to have made the wound of entry slightly lower on the body than it was. On the X-rays, it appeared at the lower end of the sloping muscle branch which extends from the neck toward the shoulder. It was hardly possible that a metal missile, moving at close to a half mile per second, could pierce the fleshy muscles less than one inch and stop. In addition, there was the question of what had happened to such a bullet.

The FBI and Secret Service men listened and made notes. The driver of the car, Greer, asked if he had permission to speak. The doctors listened. He said that a bullet had been found on a stretcher—or rather as it fell from a stretcher—in Parkland Memorial Hospital and had been sent on to the FBI laboratory for examination. Could this be the bullet that went into the neck and, in the jostling of the President on the stretcher, fell out?

The doctors agreed that it was a possibility. They could hardly subscribe to a thesis which depicted a dying bullet which did not have the energy to go through boneless masses of flesh, but it must be considered. If some foreign object had slowed the bullet—like hitting the back of the car—it could not have made that clean 6-millimeter circle; it would have been tumbling end over end and a large ugly entrance would have resulted.

Greer's question should have taught caution to the doctors, because it pointed up what they did not know about the events in Dallas. The superior officers—Captain Canada and Admiral Galloway—medically oriented, could easily have recessed the proceedings for fifteen minutes or a half hour. All that would be required was to phone Parkland Memorial Hospital and ask for Dr. Clark or Dr. Carrico—whether at home or in the hospital. If not those, then any of the other attending physicians could have helped. Bethesda might have said: "You had the body of the President. What were your wound findings and what methods and treatment were employed?"

The news that an exit wound had been found in the lower front of the neck—one which frayed the back of the knot on the President's tie—would have settled, beyond doubt, that the bullet had gone through the back of the neck muscles and out the trachea. The Texas doctors could have stated that the exit wound had been enlarged to form a tracheostomy. The mystery could have been dissolved at once. No one pursued it.

Colonel Finck studied the X-rays of the head carefully. There was a hole about 6 millimeters in size in the lower right-hand section of the back of the head. It was round and consistent with an entry wound. If this portion of the head was hit by a 6.5-millimeter bullet, the hole in scalp and skull would shrink to 6 millimeters after the missile passed. Once inside the brain, it would bevel the inside of the skull, tumble, causing the massive hole in the upper right side as the wound of exit. If, after he had been hit in the neck, his head fell forward and the body tilted to the left, as witnesses swore, then the small hole in the skull would result in the big one.

The X-rays showed the metal fragments still in the head. There was a comparatively large piece of bullet—7 by 2 millimeters—

behind the right eye. There were a few grains inside the "cone effect" entrance wound. The remainder of this bullet, emerged with flying skull and hair, apparently broke into two sections, and they were found on the front seat of the car. The first bullet, which missed target and car, was torn to flying grains when it hit the roadway, nicked a curb, and peppered the face of Mr. Tague on Commerce Street.

Greer's thesis had a supporter. Roy Kellerman, Agent-in-Charge, said he remembered a Parkland doctor astride the chest of the dead President, applying artificial respiration. Kellerman, a solemn man and a deliberate one, thought the bullet in the back of the shoulder might have been squeezed out by manual pressure. If so, the man who found a bullet on a stretcher in the hall was mistaken in thinking it came from Governor Connally's cart.

Medical judgment was reserved. Colonel Finck was not convinced. In all of his experience with bullet wounds he could not recall a missile entering flesh and stopping short. The X-rays showed no bullet path to the throat because the shot, instead of tearing through the strap muscles, had separated two layers and furrowed between, leaving insignificant bruises on the under side of one and the upper side of another. It emerged from the throat with most of its speed intact to hit the Governor.

The prisoner was on the private elevator heading down for another lineup. The fawn-hatted detectives hopped in beside Lee Harvey Oswald and the door was closed. They could still hear the shouts of reporters in the corridor. "Why did you shoot the President?" "Bastard!" "Son of a bitch!" None of the older reporters could remember a story in which the journalists expressed personal venom. Reporters at Rheims who had witnessed the surrender of Germany expressed no hatred. Others, aboard the battleship Missouri in Tokyo Bay, had watched with equanimity as the Japanese signed the document of surrender. Some had put in considerable time at the Nuremberg trials without rancor. In this case, the police had nothing more than a suspect, but the press reacted toward him as the French underground had toward the Parisian women who had slept with German officers.

Oswald neither smiled nor frowned. The sullen mask had been pasted on tight and it showed nothing. The onset of personal hostility did not alarm Will Fritz. He assumed that Chief Jesse Curry knew what he was doing. Chief Curry assumed that Captain Talbert had charge of the security of police headquarters. The chief passed an order that, if the press was admitted to any of the lineups, they would have to stand behind the witnesses in silence. He did not want them to address any questions to the prisoner.

Off the edge of the lineup room, Oswald stood quietly as Don Ables, the jail clerk, nodded to him. Two of the men were new. Both were prisoners. Again Oswald was the number two man, shackled to three others. He was intelligent and he must have known that he would again be the only man in those lights with real bruises.

In front, behind the dark curtain, Troy Lee sat watching. He had come to police headquarters with his wife and his sister-in-law. One was the former Barbara Davis; the other was Virginia Davis. They were young girls, laden with the responsibility of marriage, babies, rent, and husbands. They told the police officers, again and again, that they had been trying to doze off after lunch when Officer Tippit had been shot.

They heard the explosions and jumped up and ran to the screen door. A policeman, on the outside of his car, had fallen beside the front wheel. Through the screen door they had seen a young man, not walking but not running either, cutting cater-cornered across their lawn. One hand was held high and he was pulling empty shells from a revolver. The sisters were nervous. Troy Lee told them to be calm and to tell the policemen if there was anybody up on that stage that they recognized.

The lights went on and four men, walking in profile, trooped onstage. The women's eyes flickered across the brightly lit faces. Virginia leaned across Barbara and whispered to the detective: "That's him." She nodded her head positively. "Which is him?" "That second boy." Barbara Davis agreed. "Yes," she said quietly. "That's the one. The second one counting from this side." They pointed and whispered and Troy Lee leaned across the policeman to listen.

The policeman made a few notes in a small loose-leaf. "What did you see him doing?" he said. "I have to write up a statement." The girls again told the story of the nap, the fusillade of shots, and the trotting "boy." Onstage, a lieutenant asked each of the four men to face front, to turn left, to turn completely to the right. The Davis girls were so excited that they were giving their statements simultaneously.

The eyes of Marguerite Oswald stared at the furnishings of the Paine house. She nuzzled little Rachel to her bosom and studied the living room. To her, life was a sequence of frustrations. On the way home, she had tried to communicate with her daughter-in-law, but the best she could do was to pat Marina's hand. There was so much that she wanted to say. She had never protested Lee's choice of a Russian girl as a wife. But truly the girl was a foreigner.

Marguerite saw the couch she would occupy, and she looked into the kitchen on the edge of the garage. Marina was in the bathroom composing herself. The Paine children and June Oswald frolicked on the floor. Someone would have to get something to eat for the little ones. It must be close to bedtime. Two men knocked on the front door and Ruth Paine admitted them with her friendly cry.

One was holding a camera. Marina emerged into the living room looking brighter. Mrs. Paine introduced Marina to the men, Allan Grant and Tommy Thompson of *Life* magazine.

Marguerite Oswald was not introduced. It was an affront—conscious or subconscious—which could not be dismissed. The bud of friendship between Marguerite and Ruth began to wither. Mrs. Paine dropped to the floor with the children, tucked her legs under her skirt, and said gaily: "Now, I hope you have good color film, because I want good pictures."

This somewhat obtusely, involved another facet of Marguerite Oswald's character. It concerned economics. She never lost sight of a dollar, real or fancied. It might be true that her son, her flesh and blood, was in dire peril on this night, but it could not be permitted to obscure the realities. She would be prepared, from time to time, to sell her story and her opinions; she would be ready to sell letters or mementos from her son, but she was not ready to give anything away.

Her glance became hard. Tommy Thompson was saying: "Tell me, are Marina and Lee separated, since Lee lives in Dallas?" Ruth Paine wore her holiday smile. "No," she said, speaking to millions of people beyond the pages of next week's *Life*, "they are a happy family. Lee lives in Dallas because of necessity. He works in Dallas, and this is Irving. He has no transportation and he comes to see his family every weekend." "What type of family man is he?" "A normal family man. He plays with his children. Last night he even fed June. . . ."

Marina did not understand the conversation. Marguerite Oswald didn't think it was Ruth Paine's story to tell. "Mrs. Paine," said Thompson, "can you tell me how Lee got the money to return to the United States?" "Oh, yes," Ruth said. "He saved the money to come back." Marguerite began to fume. This type of publicity was uncalled for. Her beloved son should be protected from outsiders. And yet she felt that she was on brittle footing because this was Mrs. Paine's home. Marguerite could be expelled.

A lady can remain silent only so long. "Now, Mrs. Paine," said Marguerite Oswald in the petulant, injured tone of her third son, "I am sorry." All the heads in the room came up. "I am in your home. And I appreciate the fact that I am a guest in your home. But I will not have you making statements that are incorrect." This was calculated to divert the newsmen to the place where the correct story reposed. "To begin with, I do not approve of this publicity. And if we are going to have the story with *Life* magazine, I would like to get paid."

There it was. The poor can afford to be tactless. Grant and Thompson glanced at each other. Marina realized that a new and jarring note had erased the smile from Ruth's face, but no one bothered to explain what grandmother was talking about. "Here is my daughter-in-law," said Mrs. Oswald, pointing dramatically with her free hand, "with two small children. And I myself am penniless, and if we are going to give this information, I believe we should get paid for it."

Lee Harvey Oswald was for sale. The type of story a writer would get would depend upon the source. Marguerite would defend her boy; Marina would give it a somber mood and gray skies; Ruth Paine could make it as cheerful as a Quaker picnic in the hills of Pennsylvania; Robert Oswald could analyze it back to the cradle; John Pic wanted to forget it.

The grandmother had a latent suspicion that Ruth Paine had engineered a secret deal with the men from *Life* and was being paid. The appearance of the two men at the front door, she was sure, was not accidental. It might even have been set up with Marina's assistance. Ruth spoke to Marina in Russian. This was an additional frustration because Marguerite could not present her motherly views to her daughter-in-law. Nor did she trust Mrs. Paine to translate her dicta accurately.

The conversation became a crossfire of two languages. One of the newsmen stood and, addressing Marguerite, said: "Mrs. Oswald, I will call my office and see what they think about an arrangement for your life story." This was even better. Mrs. Oswald felt that her life story, one of hardships and affronts to

Southern womanhood, had more drama and more appeal than Lee's. She had not seen much of her son since he enlisted in the Marine Corps on October 26, 1956. He had been away a long while, come home for a few days, gone to Russia, come back to Texas, and avoided her. There was a question of how much information she could supply about him.

The *Life* man went into another room and called his superior. Mrs. Oswald dandled Rachel and, beyond doubt, had visions of real money at last falling into her hands. Private nursing cases are drudgery. She carried bedpans, changed sheets and nightclothes, brought cool glasses of water with glass straws, listened to the feeble protests of the chronically ill, snatched a little television and a nap, and hurried home to spend her waking time alone. This could be the biggest thing in her life.

The man came out and said that *Life* would not pay money for Marguerite's story. He had a counteroffer. The magazine would pay hotel and food expenses for the group in Dallas. Marguerite was disappointed. She was hurt. However, she would think about it, she said. The men from *Life* did not leave. Allan Grant made photographs. The flash winked. Marguerite began to feel warm. She rolled her stockings below her knees and sat. The cameraman made a picture. "I am not having this invasion of privacy," she shouted. "I realize that I am in Mrs. Paine's home. But you are taking my picture without my consent—a picture that I certainly don't want made public."

The photographer followed Marina into the bedroom. The babies were going to bed. Marguerite hurried to follow and interposed herself between the family and the photographer. He continued to make pictures. "I've had it!" she said, waving her arms. "Find out what accommodations you can make for my daughter-in-law and I so we can be in Dallas to help Lee." Her tone brooked no argument. "Let me know in the morning!" The men left.

● ● ●

Time dragged. The night was long. The people on the streets shuffled aimlessly. The clocks on the banks flicked their lights to 7:55, and a man's mind would race with regret over broad spans of horror and, when he lifted his eyes eventually to another clock, the lights said 7:55. The eternity of time was the result of trying to turn it back to 12:29 P.M., when the roar of the downtown crowds assailed the ears of the pleased President. It did not matter, really, whether a man liked him or not; this was part of the fair and sunny world of democracy.

The dreadful thing happened and those who did not admire him mourned with those who did. A blackness had settled on the land and strangers in buses and elevators and planes and in shops said: "I was on my lunch hour . . ." "I called home to see how things were . . ." "It was Friday, I thought we'd go to a movie tonight . . ." "I slept late . . ." "I was in class . . ." "I rarely turn the car radio on . . ." "I was getting a roast for Sunday . . ." "We felt as though we knew him . . ." "We were going out to our country place for a last weekend . . ." "I saw this woman crying . . ."

The only thing which could reverse the clock was television. The commentators were solemn, the voices sometimes shaken. But there he was, alive again, the arm punctuating the words: "Let the word go forth from this place at this time . . ." "Ask not what your country can do for you . . ." " . . . she takes longer, but then when she appears, she looks better than we do . . ." Robert Oswald, the lonely man, walked seven blocks to his car and it seemed as though he arrived at the same moment he had left police headquarters. Time would not dissolve. Nor would it evaporate. This frightful night would go on and on, torturing the innocent and the helpless. How long ago was it that his mind had been set in flames in that shop in Fort Worth? It must have been a long, long time because the same thoughts had been slipping in and out of the revolving door of the mind.

He drove through the downtown area. Robert Oswald wasn't sure where he was going. He could go home, but there was nothing to say. A man does not like to plague his wife and family with this situation. He would be impelled to discuss it, but what would he say? What would he say Monday at the shop? Or, more important, what would the men say to him? Suppose it was true—suppose his brother had done this thing? What do you say: "My name is Robert Oswald—you know, the brother of the assassin."

The car moved slowly through the almost empty streets. It came to Dealey Plaza and rolled slowly down the incline past the Texas School Book Depository building. Two policemen stood in the middle of Elm with flashlights, hurrying the flagging motorists. Robert Oswald had no desire to stop. He didn't know where he was going, but he didn't want to look at that building. His wheels passed near the spot where life ended and eternity began for John F. Kennedy. At the underpass, he moved out across the viaduct and over the damp bed of the Trinity River and on out on Route Eighty.

Thinking would do no good, and yet thought imprisoned him. He drove slowly, carefully, doing the right things, going west on Arcadia, staring along through the windshield. He could not go to his mother because he lacked the faith. It would be almost as difficult to try to communicate with Marina. The night was cool; that was all a man could say for it. The false summer of the sun was gone. Vaguely, a man could see the bright cat's eyes of trailer trucks eastbound, making him squint, and then they were gone. He must have passed Cockrell Hill and Arcadia Park because they were behind the car. There were filling stations whizzing by and lights, a diner, a motel. Cars went by him showing broad braces of red lights in the back. He passed Arlington, and Robert Oswald asked himself where he was going.

Nowhere. He would not flee, even if he could. He would help Lee. Being a good brother carries a price. On this day, it was high. In random thought, he may have asked himself if there was a small key out of the past which might have triggered this deed. There were scores of scenes, unwholesome, unhealthy, which could be dredged from boyhood. To a child, a bad life is livable if he has seen no other. To an adult who has earned his own contentment, old memories can be a pit of vipers. It would seem to the Oswald boys that they never had a youth. They were always little men, doing as they were told by their mother; doing as they were told in orphanages; doing as they were told in school; eating when they were told; eating what they were told. There was a shy, timid joy when Marguerite married another man and, for a moment in time, they had a home and a bedroom and a few toys. Even that was a cruel come-on because the boys barely became acclimated to the joys of climbing a tree, throwing a ball, or breaking in a new pair of shiny shoes when it was gone.

He drove over Lancaster into Fort Worth and out near the Ridglea Golf Club and turned around. The car, like his thoughts, was on a carousel. Robert Oswald returned to Dallas. He did not know why. There was nothing he could do. And he had no place to go.

8 p.m.

A Secret Service man stood outside the seventeenth-floor suite with two cases. There was an overnight bag with fresh clothing for Mrs. John F. Kennedy. The small one was a makeup case. Both carried the monogram JBK. They had been packed by Providencia Parades, an attractive darkskinned maid from Santo Domingo. Miss Parades knew that Mrs. Kennedy required a change of clothing. The bags were taken into the suite. Mrs. Kennedy had them placed in the bedroom and left them unopened.

The guests tried to become accustomed to the blood and brains. It was impossible. The glances were masked. In spite of the several conversations going on in the sitting room, the kitchen and the bedroom, the sight of this remarkable young woman emerging from a room constricted throats and hurt eyes. It was as though they were looking at a murder. Part of the President of the United States was in the room. Secretary of Defense Robert McNamara sat on the kitchen floor, his back against a counter, as Mrs. Kennedy chatted.

The conversation drifted. He saw the "bloody skirt and blood all over her stockings" and he, the most composed of men, thought of it as "fantastic." "Where am I going to live?" she asked, at one time. This was specious and small-girlish. No one who lives in the White House has ever regarded it as a permanent residence. It was not as though she were being evicted, nor even as though she had no place to go. She had inherited a fine home at Hyannis Port; her mother had a big home in Georgetown; Mrs. Kennedy had riches. It is possible that she intended the question to mean "What place would be best for me?"

An hour had passed since Godfrey McHugh came upstairs with news about the President's remains. Mrs. Kennedy, alert with the energy which nature lends to those most deeply hurt, noticed that some of the women looked fatigued. She suggested that they all go home and "get some rest." The ladies declined. Some pointed out that she was the one who needed rest, that there would be much for her to do, many decisions to make, and that she should consider lying down. She too declined. The widow had promised herself that she would remain at her husband's side until she brought him "home." It was a sacrifice for Mrs. Kennedy to remain on the seventeenth floor while he was in the autopsy room.

Charles and Mary Bartlett arrived, and this brought a freshet of tears. Mr. Bartlett was a Washington columnist. Twelve years ago, when Jacqueline Bouvier had been an inquiring photographer for a local newspaper, the Bartletts had introduced her to the young and dashing Congressman from Massachusetts, Jack Kennedy. He was a ladies' man indeed, with an eccentricity, he seldom carried cash. Often, at a motion picture house, he fanned his pockets and had to borrow money from his dates. For a young man who was granted a trust fund of one million dollars before he earned one, it was embarrassing to watch the lady of his choice hold up a queue while she delved into her purse.

The tears came. The tears dried. Often, the mind of the widow regressed to the good days. She remembered and remembered and remembered. In some of the sad, sweet recollections, a joy suffused her wan face. The eyes became enormous pools of dusky light, the graceful hands augmented the stories, Dallas didn't happen. Her relatives and friends nodded and smiled and added some anecdotes of their own. Then she would speak of how she planned to conduct herself, and the doleful word "funeral" was uttered, and suddenly all the happy days lay shattered in the silence and on the stunned faces.

America would be watching this funeral; of that she was certain. And Mrs. Kennedy said that she was going to hold her head high. She would not break down because she would not permit it. Everything that she would do, or permit to be done, would have to conform with what she thought her husband would approve. She recalled that, as a former naval officer, he had looked forward with pleasure to the forthcoming Army-Navy game. That is why she had asked for a Navy ambulance and a naval hospital. The Navy, she felt, would remove the bullets from his body and dress it for burial. She was not told that the procedure involved a full autopsy.

Morgan Gies waved the car into the White House garage. He took it to the back and had the driver place it in a deep alcove. Hickey shut the ignition off and said: "Look at this." He pointed to a star-shaped crack on the windshield. Gies looked. He did not touch. Orders had come down from Chief Rowley to cover SS-100-X without touching it. Hickey said: "I noticed it coming in from Andrews. It isn't much, but it keeps spidering." The motion of the car seemed to spread it in radiation.

Gies looked across the hood of the car. "Whatever it is," he said, "the crack isn't on the outside. This side is smooth." Two agents were backing the other Secret Service car into a bin. The President's limousine had a huge plastic cover drawn over its length. Two men guarded it. They took their posts in the alcove. Deputy Chief Paul Paterni of the Secret Service and Floyd Boring, assistant Agent-in-Charge of the White House detail, were on the other side of the street in front of Blackie's Beef House, waiting for a traffic light before crossing. They wanted to see this automobile at once. With them were Chief Petty Officers William Martinell and Thomas Mills of the White House medical staff.

Paterni walked in and asked that the cover be removed. Additional lights were turned on. He and Boring walked slowly around opposite sides of the car, leaning forward to look in. The other agents stood back. No marks or scratches or indentations were on the front of the car. Both saw the small star-shaped crack on the inside of the windshield and assessed it as new. The deputy chief also noticed a dent in the chrome plating at the top of the windshield. This was old, but he did not know it. Suitable notes were made. Observations were made orally, so that everyone present could testify to them. The sides of the car were examined, and no marks were found. The outside of the trunk was scrutinized. There wasn't a discernible scratch to show that Clint Hill had climbed painfully up that back to push Mrs. Kennedy to safety as Greer jammed the accelerator to the floor.

They glanced inside the back. The first thing they saw were red and yellow petals. They were scattered across the black leather seat and on the rug. Governor Connally's jump seat was folded down. Mrs. Connally's was still up. Dry blood was everywhere. It had congealed on the seat and it shone on the rug. Great gouts of it must have pumped out of the President's head on the way to the hospital. Splashes of gray-white had dried on the upholstery. Where the brain tissue was absent, someone had sat. It was also heavy across the rear of the front seat. On the rug, they picked up a three-inch piece of skull and hair.

The men moved to opposite sides of the car and looked in the front seat. Paterni spotted the dull gleam of metal. He called attention to it and reached in at the spot where Roy Kellerman had been sitting. What he picked up was the rear half of a bullet. It was intact, and the lead core was exposed. A moment later, another was discovered on the driver's side. This was a good find. When Paterni held the two parts together, it was obvious that this constituted one bullet. The middle sections, where the bullet broke, matched pretty well. It was possible that, if the bullet found on Governor Connally's stretcher was the one which went through Mr. Kennedy's neck unimpeded except by muscle to hit the Governor and stop inside his trouser leg, then this one could be the shot which hit the President in the back of the head, exploded his brain, tumbled forward, hit the windshield, and broke into two fragments, both falling on the front seat. It was possible. No one would know whether the reasoning was accurate.

The hum of an air conditioner in the autopsy room breathed on the silence. Commander James J. Humes executed his tasks with professional detachment, but it was impossible for a man to forget that this assortment of organic clay had been, until a few hours ago, the most powerful man on earth. Humes made notes on pads; sometimes Boswell made notes on the same sheets; Colonel Finck, the most experienced, knew very little of the actual circumstances of the assassination, but he noticed many small things which had a cumulative effect on his knowledge.

For example, the entry hole in the back of the skull was one quarter of an inch across by five eighths of an inch vertically. The bullet can be assumed to be round. It was 6.5-millimeter. The 7-millimeter side of the measurement showed that the missile had gone through the bone and that the skull and scalp had puckered slightly after it passed inside. This was normal. The 15-millimeter up-and-down measurement of the same wound told Colonel Finck that the head had been bent forward when hit. The missile had struck the skull on a downward tangent, skidding a trifle before boring through the head and out the top.

Boswell made a schematic drawing of the back of a head on a pad and, at the entry point, added an arrow pointed to the left side of the head. The true path was to the right. The doctors, aware that they were not in possession of the findings already established in Dallas, assumed that they were working on impressions tonight, impressions which, when all the facts were ascertained from Parkland Hospital, would be corrected.

The men looked inside the cranial vault. It was easy to do because a third of the skull was missing, and at least 25 percent of the brain was gone. Tenderly, the head was turned so that full light shone inside. Two fragments of skull fell off and rolled on the table. The fissure fractures around the massive defect were brittle. To touch the hair along the edge courted more fragmentation. The X-rays of the skull were studied again and again. Gloved hands moved inside the brain and began to emerge with bits of metal. The pieces were small. The X-rays showed where they could be found. Humes traced as many bits as he could. After a half hour of work he had a total weight of 12 grains on the table. The original bullet weighed 158.6 grains; Humes had recovered 7.5 percent.*

At The Elms, Liz Carpenter wrung her hands as she answered the ringing telephone. She wished that the telephone linesmen would change the number quickly. People who knew the phone number at The Elms were calling to ask Mrs. Johnson, "When you movin' into the White House, honey?" It was sickening. The new First Lady asked Liz to please make excuses to the callers. Either they wanted to know the sickening details of the assassination, or they were anxious to know how quickly "Lyndon" would take over. Under normal circumstances, Mrs. Johnson was a creature of patience and tact. She huddled deeper into the bed, with the television loudly tolling the oratorio of the dead and the phone buzzing insistently.

Mrs. Carpenter hurried downstairs and back up. "The press is out front," she said cautiously. "They would like to have you say something, Mrs. Johnson. Anything." The First Lady stared

---

* As Dr. Humes expected, his call to Parkland and Dr. Perry the following morning altered his thinking. He was told that the tracheostomy in the neck had originally been a wound and had been expanded to improve breathing. A tube had been inserted. Cardiac arrest occurred and closed-chest massage was futile. Humes, Boswell, and Finck realized that the shot in the shoulder strap muscles had come from above and behind, emerging at the bottom of the throat. It was a wound from which the President would probably have recovered without incident.

at the ceiling. "It has all been a dreadful nightmare," she murmured. "Somehow we must have the courage to go on." She was talking to her secretary, her friend. But Mrs. Carpenter thought that the words covered the situation. She went back downstairs and, at the gate to The Elms, repeated them to the reporters.

When she returned to Mrs. Johnson, the woman was out of bed. It seemed painful to lie down, to stand, to sit, to watch that infernal machine repeat the horrifying story over and over. Mrs. Johnson reminded Mrs. Carpenter that Zephyr Wright, the family cook, was not at home. She suggested that they both go to the kitchen and make fried chicken.

"It will keep us busy," Mrs. Johnson said, "and he will probably bring some people in with him. Men forget to eat. Then when they come in they want to know what's ready now." A few minutes later, Ray Scherer was on the air announcing: "Acting Press Secretary Andrew Hatcher reported at a news conference that President Johnson met with leaders of Congress for forty-five minutes and asked for their support in this time of tragedy. The congressional leaders assured President Johnson of their bipartisan support . . . Mrs. Kennedy has left the White House. Her children are with her."

Shortly after, another commentator intoned: "President Johnson left about an hour ago for his home in Washington. . . ."

"You don't have to talk to those people," Captain Fritz said. Oswald sat, crossed his legs, and placed the handcuffed wrists on his thigh. "I know," he said. The reporters in the hall had stopped shouting. Fritz phoned Lieutenant J. C. Day again. When he arrived, he said: "We're going to make a few paraffin tests." Oswald nodded. He had no objection. Sergeant E. E. Barnes arrived, carrying portable equipment. The order to do this work in Fritz's office surprised the sergeant. Prisoners were usually taken to the fourth-floor laboratory.

Barnes nodded to Detectives Dhority and Leavelle. Oswald

watched as the equipment was unpacked. He seemed to be interested in the procedure. The paraffin was melted to a warm softness and the sergeant said: "I'm going to make a paraffin cast of your hand." Oswald shrugged. "It's okay with me." The handcuffs were removed. The prisoner said: "What are you trying to prove? That I fired a gun?"

The sergeant applied the stuff to one hand and kept firming it. "I am not trying to prove you fired a gun," he murmured. "We make the test. The chemical people at the laboratory will determine the rest of it."

The firing of a gun causes a small amount of recoil. From the ammunition, bits of nitrate are sometimes forced backward out of the chamber. In the case of a rifle, the specks sometimes lodge on the face and on the hands of the person firing the weapon. This test is so unreliable that laboratories have reported positive nitrate results from persons who have not fired a gun and negative results from hunters who have used guns all day. The Dallas Police Department used these tests and sent them to Parkland Hospital for evaluation.

Barnes dipped a brush into warm paraffin and painted the gluey hand of Oswald. It was done a layer at a time until a quarter of an inch of waxy substance had been built up. When it cooled, Barnes and Day wrapped the hand in cotton gauze and painted an additional layer of paraffin on top. When it hardened, it was cut off with scissors and marked "Right hand, Lee H. Oswald." The work began on the left hand.

Captain Fritz went to his outer office. He told Dhority: "When they finish, take him upstairs." The second hand was paraffined and peeled like a tight glove. Lieutenant Day said: "I have to make palm prints of your hand." Oswald was patient. He neither protested nor struggled. An additional test was made of the right cheek. The material went up to the laboratory. Officer J. B. Hicks assisted in making fingerprints and palm prints on an inkless pad.

When the work was concluded, Barnes presented the fingerprints to Oswald on a police sheet and asked him to sign his name across the bottom. This, Oswald thought, was carrying cooperation too far. "No," he said. "I'm not signing anything until I see a lawyer." A policeman snatched the card. "Makes no difference to me," he said. Fritz returned and told Oswald, in the toneless, almost sleepy, manner he had used all day, that the police had found a map of Dallas in his room.

Oswald was unmoved. He wore the expression of a man who senses neither surprise nor significance. "I marked that thing," he said, "for places to look for work." Fritz, still standing and waiting to talk to his detectives, said: "There was an X on the School Depository building." Oswald showed surprise. "My God," he said, "don't tell me there was a mark where this thing happened?" Fritz shook his head affirmatively. To ameliorate the effect of the one X, Oswald said: "What about the other marks on the map?" The captain didn't answer. Oswald was taken to the jail.

The FBI wing of the Justice Department building was ablaze with light. Authority had been delegated; compartmentalized investigation was under way. Off-duty agents had already reported for service. In Dallas, Gordon Shanklin was relaying evidence and bits of information as it came to him from agents at police headquarters. The New York office was already "in progress" on locating all mail order houses which sold rifles and revolvers. The background file on Lee Harvey Oswald began to build. An inspector reread the statute called "Assaulting a Federal Officer" and found that it did not include bodily harm to the President or the Vice-President. This placed the Federal Bureau of Investigation in the unwanted outside position with no legal jurisdiction in the assassination.

The Director, J. Edgar Hoover, had the express order of President Lyndon Johnson to take complete charge of the case but the Chief Executive, except for the majesty of his office, was powerless to do anything legally. The crime was against the peace of the state of Texas, county of Dallas. The man in charge would be Henry Wade, district attorney. His investigative body was the Dallas Police Department. This hamstrung the FBI, which, from 12:45 P.M. could do no more than pledge complete cooperation to Captain Fritz and Chief Curry. Across the inner courtyard of the Department of Justice, Assistant Attorney General Herbert J. Miller, Jr., called the FBI to advise that there was no federal statute applicable to the assassination of a President of the United States and that all legality in the case reverted to the state of Texas.

The situation was more delicate than dangerous. The FBI, with resources far superior to a city police department, phoned Chief Curry to offer the "assistance" of all their manpower in addition to the excellent facilities of their laboratory. Curry's attitude was detached and cool. In his desk he had a report from one of his officers that the FBI had a live file on Lee Harvey Oswald, a report which charged an FBI agent with stating that the bureau knew the potential danger of this man. It is possible that the chief saw this as an excellent document to release to the surging, shouting press—a document calculated to take his department off the hook of responsibility and put the FBI on it.

He said that it was his understanding that a couple of FBI men were already in Fritz's office. Everybody was cooperating with everybody else. The Secret Service had men in there, too. Texas Rangers were there, and the whole thing cluttered up the office and didn't give Fritz much of an opportunity to work. So far as the FBI laboratory was concerned, Dallas had one of its own on the fourth floor. It might not compare with the Bureau laboratory in Washington, but Curry would talk to Fritz and ask him about it. Possibly Fritz wanted all evidence to remain in Dallas.

The captain of Homicide found himself in agreement with his chief. He was unimpressed by the offer. "I need the evidence here," he said. "I need to get some people to identify the gun, to try to identify this pistol and these things. If it is in Washington, how can I do it?" Essentially, it was the same dispute which had occurred at 1:15 P.M. over the body. All the legalities favored Dallas; all the power was in Washington. There were other phone calls. Some came from officials in Dallas. They asked Chief Curry to complete his laboratory work and permit the FBI to take the evidence for a day or two and return it. "Who," said Chief Curry, "is making these calls from Washington?" The response was always the same. "Just say I got a call from Washington and they want this evidence up there."

Hoover phoned Rowley in Washington and offered assistance. The Chief of the Secret Service said that Shanklin had been cooperating with the Service. Alan Belmont, in charge of the FBI investigation, called Shanklin for progress reports and ordered him to funnel all information through the main office. Shanklin said he had men at the School Book Depository, the hospital, and police headquarters. Congressman Ed Edmondson called the FBI to remind the Bureau of something of which it was painfully aware—that Speaker John McCormack of Massachusetts, next in line of succession to the presidency, should have protection. Edmondson said that he and Representative Carl B. Albert had phoned Rowley and asked for guards, but that none had arrived at the Speaker's office.

The FBI and the Secret Service appreciated the crotchety Speaker's wishes, which were that he would not tolerate agents shadowing his gait. Edmondson said that the shooting of the President could be but the first act in an overall conspiracy to murder the heads of government. Cartha DeLoach, administrative assistant to the Director, phoned Dr. Martin Sweig of McCormack's office. The FBI had no jurisdiction in the field of personal protection, but DeLoach was glad he called, because Dr. Sweig said that McCormack wanted no protection and had ordered him to "remove" two Secret Service men waiting quietly in a room next to the Speaker's suite in the old Washington Hotel. Mr. McCormack, a stubborn second-generation Irishman, said that the city was full of fear and hysteria and he was not going to add to it.

In the file section, an FBI agent dug up James Hosty's reports on Lee Harvey Oswald. Another agent went through the identification files and found fingerprints on Lee Harvey Oswald made by the Marine Corps on October 24, 1956. The same record revealed that he had been honorably discharged on September 13, 1960, and had been arrested for disturbing the peace while distributing Free Cuba pamphlets on the streets of New Orleans on August 9, 1963.

Calls were coming in from all over the country. The FBI switchboards were alive with winking lights. Deputy Attorney General Nicholas Katzenbach phoned to ask that he be kept informed if anyone was arrested for the assassination. He was aware that Oswald had been charged with the murder of Tippit, but his interest centered on the President. Belmont had a man in the State Department with Abram Chayes, examining the Soviet side of Oswald. State asked the FBI to forward copies of anything it had in the files on the suspect. Norbert A. Schlei, another assistant attorney general, called the Bureau to ask what kind of people killed Kennedy. He was drafting a proclamation for President Johnson and the phrasing would depend upon whether "they" were madmen, office seekers, political malcontents, segregationalists, or so forth. Schlei was told that there was no definite information; the suspect in hand proclaimed himself a Marxist and had once sought citizenship in Russia, but no one could yet say that he had plotted the assassination or carried it out. Further, no one knew whether this man had acted alone or in concert with others.

Word reached Washington that a police official in Dallas claimed that there was "proof" that the FBI was aware of Oswald as a potential assassin. Washington correspondents phoned the FBI. The word they got was: "We are rendering every possible assistance in Dallas." Out of Washington went some highly placed phone calls demanding that Curry retract the statement and rephrase it properly: " . . . that the FBI had a file on Oswald as a defector but nothing that would point to him as a violent person." If Curry wanted to say that one of his lieutenants "claimed" that an FBI agent had stated otherwise, that would be all right provided that the Chief also announced that Agent James Hosty denied making the statement. The next phone call, from an embassy, announced that the Mexican government had closed its border to the United States and was screening all passengers at international airports. No plotter would get into Mexico.

Chayes was studying his State Department files and duplications of Central Intelligence Agency and FBI files on Oswald. The more he studied them—after the assassination the most innocuous reports appeared portentous—he wondered if his department had a "lookout" card on the accused. He got a man named Johnson, in charge of the Passport Office, to report in and open the door to the "lookout" section. This is an area of special files on persons who, for one reason or another, are, in the view of the United States government, "sensitive personages."

The room itself was so sensitive that the door had a combination safe lock on it. Chayes, with assistants and FBI men numbering five in all, admonished each man to recall each step he made in this room and to report it. Assistant Secretary of State Schwartz accompanied the party and Johnson went to the "O" section. There was no mention of Lee Harvey Oswald. "Why isn't there a card on this man?" said Chayes. No one knew. No one could say whose responsibility it would be to have a "lookout" card on a man who wanted to renounce his citizenship.

In the autopsy room at Bethesda Hospital, Secret Service Agent Roy Kellerman received a phone call and tiptoed out. It was Chief Rowley. "In Dallas someone found a bullet practically intact on a stretcher. It was flown up here and I ordered it turned over to the Federal Bureau of Investigation." "Yes, sir." "I don't know what the doctors are looking for, but they should be told that a pristine bullet has been found."

The disorder of the apartment seldom irritated Jack Ruby. He was dressed to go to services, but the spirit flagged. He phoned his friend Ralph Paul and said: "Meet me and we'll go to services together." Mr. Paul felt no inspiration. Ruby hung up. He was dressed in a gray single-breasted suit, a white shirt and a blue tie. He was ready, but he hesitated, as though whatever good would come from attending services for President Kennedy had already arrived on the wings of self-serving phone calls and speeches to his sister.

The single bed with the off-white headboard was rumpled. The sheets were gathered in the middle of the bed. The drapes behind the bed were pulled back and there was nothing outside that window except blackness. A late newspaper was on the floor. A pair of cheap bedroom slippers lay on their sides. A carton marked "Johnson's Baby Oil" occupied one night table. His private phone book and a telephone were on the other one. The dowdy broadloom was white with lint and bits of paper. The atmosphere was cheap and depressing.

It was time to leave for the temple, but Jack Ruby permitted the time to pass. Rather than listen, he preferred to be heard. He sat on the bed and phoned his brother Hyman in Chicago. Jack told how awful it was. A terrible thing. Such a fine man. "I'm thinking of selling the business and going back to Chicago," he said. In each phone call, he appeared to be anxious to establish his prior right to mourn because the tragedy had occurred in his town. Ruby could make it appear that he felt like a Kennedy. There was a slowly welling emotion which had to be wrung dry. He phoned his sister, Marion Carroll, also in Chicago. This was followed by one to another sister, Ann Volpert. He busied himself with the telephone.

This was, in truth, the busiest day in the history of the telephone. The attorney general of the state of Texas, Mr. Waggoner Carr, swore he had received a phone call from someone in the White House but could not recall who it was, although he called the party back. The Kennedy crowd would hardly call Mr. Carr, so the forgotten man was limited to such names as Johnson, Valenti, Moyers, and Carter. The mystery caller asked Waggoner Carr if he had heard a rumor that the Dallas County authorities were going to draw up an indictment alleging an "international conspiracy." The White House would be interested in having this eliminated unless there was proof of a conspiracy. Carr said he hadn't heard the rumor but he would phone Henry Wade and find out. The caller said that the White House would not want to influence Dallas County, but if they were thinking of dropping a charge like that loosely, then the White House would like to know about it.

Mr. Carr phoned Mr. Wade. The Dallas district attorney said he hadn't heard such a thing and wouldn't be a party to it unless there was some proof more tangible than high emotion. From what Wade had heard at police headquarters, the evidence appeared to be following a pattern which would implicate Lee Harvey Oswald and, so far, no one else. "It won't be in there unless it belongs in there," said Wade. A call came in for Wade from his old friend, Cliff Carter, who was now at the side of President Johnson. "Are they making any progress on the case?" Carter said. "I don't know," said the prosecutor. "I have heard they got some pretty good evidence." It pointed to Lee Harvey Oswald. Cliff thanked his friend.

At headquarters, Captain Will Fritz kept his men to the task of clearing the case up. Slowly, the captain was reaching an opinion: it was Lee Harvey Oswald and, quite possibly, nobody else. Mr. Fritz had a second opinion: this boy would never confess. He would play with the interrogations as a musical prodigy might with a piano. The Homicide Bureau was going to have to secure enough evidence to lock this case up without a confession. The boy talked quietly enough, mannerly at times. But he anticipated the meaningful questions and refused to answer them. Anything that would tend to clear the case up, or add to the evidence, was blocked or sidetracked.

The captain sat at his desk and wondered if Oswald had training in these matters. He had asked him once: "When you got to Minsk, what did you do, get some training, go to school?" The prisoner said: "No, I worked in a radio factory." The captain suspected that Oswald might have been trained in sabotage.

On the fifth floor, Oswald was complaining. He wanted to take a shower. "I have hygienic rights, too," he was shouting. The jailers paid no attention to him. They had word that his brother Robert had returned to headquarters and asked to see Lee. Captain Fritz had no objection, so it was being arranged in a room with a huge pane of glass bisecting the space. Robert Oswald sat in a miniature booth with a telephone on a shelf in front of him. His brother would come through a door on the other side of the glass and would pick up a phone and talk. It was the instrument of the day.

Robert watched that door through the dusty glass. He looked away, and, when his glance returned, the door was slamming closed, and Lee was walking toward him. Robert remarked to himself that the door was made of steel and so was its casing, but he had heard no sound. It seemed unreal that Lee strolled toward the other side of that glass so slowly, so carelessly, almost a lounging stride; he looked at his brother blankly, bereft of affection or rancor. Lee sat in the cubicle opposite Robert and motioned for the older one to pick up a phone.

The voice was flatly calm as he said: "This is taped." Up close, the bruise around the eye was plain. "Well," said Robert, "it may or may not be." Robert leaned forward. He could see iodine or Mercurochrome on the bruise. "What have they been doing to you?" he said. "They haven't bothered me. They're treating me all right." Robert, in his personal agony, hoped to get an unequivocal statement of innocence or guilt from his brother—if not by word, then by some sign.

To the contrary, Lee Harvey Oswald was relaxed and was willing to talk of other matters, other days, but not about the crime which had convulsed a nation. Robert thought that his brother acted as though all of the frenzy in Dallas and all over the United States swirled around his feet but did not touch him. Lee would neither deny nor confess. Robert no longer occupied the privileged position of confidant. The little boy had seldom wept his problems to the big boy.

The family situation had worsened. Robert, the only one to meet Lee and Marina at Love Field when they returned from Russia, was an outsider. Lee could be civil. And yet he appeared to be conscious that he, not Robert, was in the headlines of the world. They spoke a few minutes about nothing. Sadly, the thing they had in common this day was that both noticed that little June Oswald had a toe sticking out of one of her red sneakers.

Down the street, taverns blazed with light. The signs winked on and off. Inside, the jukeboxes blared their sad ballads of love. Men and women who had cheered and emitted rebel yells for the handsome man from Boston and his gorgeous wife now sat and sipped their drinks. In the half light of the bars, the dancers moved slowly in shades of charcoal; the glasses glistened; ice cubes collided in small clinks; beer rose to the top of steins, foamed, and slid like lace down the sides. Steer horns adorned the walls; cocktail waitresses in hiplength black stockings rotated their hips as they carried trays and wore revolvers in big belts. The conversation was not on Kennedy. It was on football and women and Indian summer.

The barflies brooded over the assassination but, when it was mentioned, someone said: "Yeah" or "Too bad" or "Why the hell did he have to come where he wasn't wanted?" The topic was delicate, not to be mixed with whiskey. A middle-aged man bent forward, his forehead glistening like his cowboy boots, dropped three dimes into the jukebox and played "The Yellow Rose of Texas." This changed the drinkers' moods momentarily. The customers sang loudly and poorly and fell back on that old Southern game of coy and meaningless flirtation.

Down Commerce or Main, between the tall office buildings, assortments of traffic lights flicked their colors against the few cars and buses which defied the gloom. A broad well-lighted window near the Statler Hilton Hotel proclaimed "Dallas Public Library." It was closed. Knowledge and culture could wait until morning. The city was in no mood for a book or a lecture. There were many more people indoors than out. From Richardson to Oak Cliff, from Mesquite to Irving, the enormity of the event began to seep darkly into the municipal conscience. The man who, after lunch, said: "It ain't gonna cause me to lose no sleep" would not repeat the remark at this hour.

Violent death was not as harsh an event in the West as in the East. The state of Texas averaged a thousand murders per year. The men were earthy and carried guns as phallic symbols of manhood. They talked tough and had a contempt for the soft emotions. Tears were for women. Little boys, at the grave sites of their mothers, were taught to stand tall and manly. Pride in being a Texan born and bred could be exceeded only by a municipal pride which pitted Dallas against Forth Worth; Houston against Galveston; San Antonio against Austin; Abilene against Wichita Falls.

Dallas was not callous. A shock wave went through the city at noon. Regret and sorrow were genuine. The metropolis would have been content to apprehend the guilty at once and hang him and be done with it. The grievous wound to the city was felt at once by Mayor Cabell and police officials and Jack Ruby and newspaper editors, but it was not until night that the people had time to absorb from their network television shows the nationwide indictment of Dallas as a "climate of extremists." The world was making much more of a fuss over the President's death than had been anticipated, and it was pointing a finger of scorn at Dallas.

The average householder did not worry about "what this will do to business in Dallas" because the average householder had no stake in business. He began to wonder, sickened, if his city was going to become a tourist attraction for those who desired to see the place where Kennedy was shot. In his heart, he knew that Dallas was bigger, better, and worthier than that, and, collectively, the Dallasite listened to the pundits in New York and Washington with sinking heart. "It could have happened anywhere" was his impotent response. It could, but it had not.

At one point in the interviews between Captain Will Fritz and his composed prisoner, the officer said: "You know you have killed the President and this is a very serious charge." "I did not," said Oswald. The captain unlaced his fingers and spread the palms apart. "He has been killed," he said. Oswald sat back. "The people will forget in a few days and there will be another President." In that statement, he spoke for Dallas at lunchtime but not for Dallas at 8:30 P.M. The city, helpless in its horror, began to realize that the world was saying that Dallas had shot a great man in the back.

What had to be done had been done. President Johnson stood behind his desk, a sign that he was ready to leave. He told Walter Jenkins, an overworked aide and friend, that Jenkins would be responsible for setting up the meeting of the cabinet at 10 A.M. "That plane from Hawaii is coming in sometime tonight," the President told Jenkins, "and I want someone out at Andrews to tell them that we'll meet in the Cabinet Room at 10. There is going to be no gap in this government." He stared through the gleaming spectacles at Jenkins. "No excuses either."

He would not permit himself to dwell on the assassination; he knew that if he opened that topic, the young men who stood around offering suggestions and executing orders would fall into melancholy soliloquies. The event had brought its own dark thoughts. Frozen in the little first aid cubicle at Parkland Memorial, Johnson had been convinced by the Secret Service that he too might be marked for death. It was shameful to think of a new President as a prisoner in a tiled room hardly bigger than a shower stall. The crouching run to get into an automobile, the race to Love Field and sanctuary were, at best, degrading images in a great democracy, even though the attitude of the Secret Service, "maximum possible danger as part of a massive plot," was the correct attitude.

"You know," the President said, "Rufe did a brave thing today." The man with the honey-sweet Georgia accent had folded the new President onto the seat and had exposed himself to rifle fire by sitting high on the shoulder of Mr. Johnson. The act was more than brave; it was dutiful. In a manner of speaking, Rufus Youngblood had taken charge of the President before Mr. Johnson had taken charge of the country. "You will follow me . . ." "We will stay right here . . ." "Our best bet is to get aboard *Air Force One* and get you back to the White House at once . . ." For an hour, the President who did not like to take orders accepted them from the agent who did not like to issue them.

The oaken door marked "The Vice-President" opened, and Lyndon Johnson stood in the light. "Come on, Jack," he said to Mr. Valenti. "You come home with me." He told Juanita Roberts and Marie Fehmer to finish up and go home for some rest, because tomorrow would be a big day. Emory Roberts and Rufus Youngblood fell into step ahead of the President. Behind him,

Cliff Carter, Bill Moyers, and Jack Valenti hurried to keep pace. This was the new O'Donnell, O'Brien, and Powers group. The corridor was empty. The office where a White House secretary had condemned Lyndon Johnson's soul was dark.

The party emerged from the side entrance. Rufus Youngblood was on the sidewalk first, waving the limousine into position and studying both sides of the closed street. Johnson walked around the car and occupied the right rear seat. Valenti came in the other door and sat with the President. The other two folded the jump seats back and sat sideward so that they could see the President. The car started and, on the right, the pale lighting of the White House brought it into view between the trees. Lyndon Johnson took a look and sat back. He had known this ghostly mansion for thirty-five years; it, too, was his. The government of the country was his, but, like the house, the tantalizing role of caretaker could be unrewarding. The government, the house, the fortunes of the country were more Johnson's responsibility than his proprietary right. For the next fifty weeks he must work to preserve them and enhance their value. Then the electorate would tell him whether he would be permitted to continue or send him away to let another man do the work.

The witnesses were in attitudes of fatigue. For them, too, time was slow. The brain examination was almost complete. The curiosity of science was almost satiated. The doctors stood behind the head, peering, whispering, making notes. The cerebellum was fixed with formaldehyde because the brain, in its common state within the skull, does not lend itself to adequate examination. Like an intact walnut, the brain forms two complete hemispheres. The flocculus cerebri—a tuft of wool-like fibers—had been smashed. More than half the right hemisphere was gone.

When the brain was removed, more photos were taken. The disruption of the tissue moved from a medium-low position in the back of the head to an increasingly shallow depth as it ran

toward the top of the head. The path of injury was parasag-
ittal about 2.5 centimeters to the right of the midline, which
connects the hemispheres. The parietal lobe was missing. Such
lacerations as could be traced had a center which was jagged
and irregular and which radiated into smaller lateral lacerations.
The corpus callosum—a thicket of fibers which connects the
hemispheres of the brain—was cut. By looking down, Humes,
Boswell, and Finck could see parts of the ventricular system,
where spinal fluid is normally stored. The smashing speed of
the bullet had jammed the front portion of the brain against the
right orb, causing a black eye.

The witnesses watched rheumy-eyed. The FBI men and
Kellerman continued to make notes, but, as none had been
medically trained or oriented, the notes amounted to personal
observations reinforced by whatever opinions they could hear
from the doctors. The rest sat stupefied. No one had a desire
to study this event, and yet the tan, lean body on the table was
compelling in its surrender. No one became ill. Men looked or
turned away. The doctors may have been a little more zealous
than usual, a little slower, but this was understandable.

Humes and Boswell cut the scalp down to both ears. Bits
of the skull continued to fall off, and fissure fractures ran like
tributaries to a deep lake on top. The doctors required little saw
work to remove the top of the skull. Studying the X-rays, they
were able to locate and lift two bullet fragments in the front of
the brain. As they worked, the doctors must have reasoned that
death from this wound would be practically instantaneous. On
the X-rays, they counted between thirty and forty bits of metallic "dust," too small for a search but large enough to be visible
on the plate.

Kellerman waited until the doctors appeared to be taking a
rest. Then he told them that an almost whole bullet had been
found. His chief, James Rowley, ordered him to report it to
them. One of the doctors made a note. After a brief respite, they

continued with their work. The government and the people had to know, but the work and the findings were dreary. The solemn thought which gripped the laymen sitting across the room was that early this morning this had been a working body and an intelligent brain, both buoyant and integrated, and on the table they lay destroyed. The driver, William Greer, who felt a personal affection for the President, was beset by a feeling that what was left on the table was no longer Kennedy.

Lee Harvey Oswald never had a friend. He had not been able to comprehend the concomitants of friendship—confidence, affection, and loyalty. He was born with the pout of the discontented. There was his mother—slavish, domineering, complaining—and there was nobody. A snapshot of him as a boy, smiling among his classmates, gives the impression that he was obedient to the whim of the photographer. His impatience with a world of three billion people who would not recognize his greatness was borne silently for years, eventually inducing an explosion of the brain second only to that which President Kennedy sustained.

People remembered him. They murmured: "Lee Oswald" and rubbed their chins or shook their heads. "A loner." "He could be polite, but he was far away." "Braggart . . . a liar." "Dreamer." "Didn't dance, date, play cards, never had a hobby. Never laughed." "You could say hello, but Lee wouldn't answer." People remembered him in Dallas and Irving and New Orleans and Fort Worth and El Toro, California. They were remembering tonight, as his face stared sullenly in their living rooms, but they were troubled about what to remember. Some said: "Yes, he would do a thing like that."

A few career sergeants of the Marine Corps snorted: "Ossie Rabbit? Not him." An intelligent liar is difficult to read. He is secretive and presents the face he wants one to see, whether it is a real face or a spurious one. To a few old buddies in the

Marine Corps, he was an okay guy who kept his gear clean, studied radarscopes for landing patterns, and enjoyed arguing politics. To a librarian, he was a bookworm. To a boss, lazy and disinterested. To a wife, weak on sex drive and strong on despotic domination. To a doting mother, an all-American boy. To God, an atheist. To the Russians, a potential suicide. To a New York psychologist, a neurotic with overtones of paranoia. To the women of an orphanage, a silent child who sat in one place with one toy all day.

He lived in a concrete cocoon. Lee Oswald was fifteen years old when he found room inside it for Karl Marx. He read *Das Kapital*, but he did not understand it. The boy enjoyed the dogmatic and doctrinaire phrases of communism. They had a ring of mystery; they were incontrovertible. They could be tossed into conversations and people would rage futilely against them without knowing, anymore than he, what they were fighting. The world struggle of the working classes was not appreciated in Texas or Louisiana so Lee Harvey Oswald could identify with it and feel himself part of a small secret band.

School was anathema because school was regimented knowledge. Lee was superior to this. He could attain passing grades, but he had to work hard to achieve them. A dispute with a teacher was a better thing than study. Someday, he said, he would go to Russia. The Soviet system would not permit his poor mother to suffer the way she had when jobs were difficult to find. The Soviets took care of everyone according to his needs. And yet he found it increasingly difficult to tolerate the presence of his mother. When he was thwarted, sometimes he would lash out with his fists.

He accepted the drudgery and degradation of being a "boot" in the Marine Corps. His marksmanship on the rifle range was not as good as his drill instructors expected. Some called his work "sloppy." When other recruits enjoyed liberty, Lee Oswald was in the rifle pits firing. "Come on, scum!" they shouted. "Drill that target." Little by little, he improved. In time, he was given a test "for the record" and he scored 212 points. The gradations were "Marksman," "Sharpshooter," "Expert." Lee Oswald became a Sharpshooter. He became proficient in the use of pistols. His battalion went to Japan, to the Philippines, but Lee Oswald did not feel captivated by alien culture or hospitable people. He called himself an instrument of imperialism.

Sometimes he hopped into his "rack" to read. At others, he might be persuaded to go to the nearest town with his comrades and drink beer. Once, he hit a sergeant on the head with a mug and was court-martialed. On other occasions, when the boys went to a house of prostitution, "Ossie Rabbit" waited outside. Some marines said Oswald hated the Corps. He argued that Nikita Khrushchev was a great man and stated that he would like to kill President Dwight D. Eisenhower.

Nobody knew the nobody. Except for peeps at the world, he lived inside the cocoon. Marine Peter Connor said: "When the fellows were heading out for a night on the town, Oswald would either remain behind or leave before they did. Nobody knew what he did." The cocoon, in time, becomes a warm womb. It is self-sustaining and requires no fuel. It resists the hand of friendship, the pressure of soft lips. Sentiment and affection are threats to the cocoon.

Many were pondering what they knew of this man. On Fifth Street, Ruth Paine was trying to unravel her skein of thoughts about Lee Oswald. She asked her husband to go out and get some hamburgers. No one was in a mood for cooking. Marina came out of the bathroom and said wistfully: "Last night Lee said he hoped we could get an apartment together soon." Another thought, another thread which matched none of the others. Mrs. Oswald was hurt, as though she wondered how he could have held out such a bright promise to her and the babies if he planned a dark deed. There was no proof that he planned anything. Casually, Ruth said: "Do you think Lee killed the President?" The Russian girl frowned. "I don't know," she said slowly. Marguerite Oswald swept out of the bathroom, proclaiming that if the Oswalds were prominent people there would be three lawyers down in police headquarters right now to defend that boy. "This is not a small case, Mrs. Oswald," Ruth said. "The authorities will give it careful attention—you'll see." Marguerite didn't see.

Mrs. Paine was perplexed. None are so blind as Good Samaritans. She, too, remembered last night. Lee had arrived before dark unexpectedly. To her, he was always a strange and aloof man. She had befriended this little family, had given it sustenance and a roof when it had neither, but Lee Oswald felt no gratitude. He had ordered the Russian community of Dallas to stop offering gifts to his children and his wife; Lee Oswald would provide. The dependency he required from his wife was so complete that he forbade her to learn to speak English.

Ruth had watched him at play with little June on the lawn. The little girl came the closest to teaching love to the man. He did not understand the emotion and ridiculed it in action, but the expression on his face approached pure joy when the child ran to his enfolding arms. Ruth had seen it. Marina had seen it. Wesley Frazier had heard about it. In innocence, the chubby little girl was tiptoeing on the cocoon and almost put one foot in it.

The thoughts of these people on this night were like recollections of the dead. They run at top speed and are not connective, darting in this direction and that, as though searching for a reason for a tragedy. Still the remembrances were dissimilar. At nine o'clock last night, Mrs. Paine was aware that Lee Oswald, crouching before the television set, had tried to reestablish himself with his wife and had failed. It was not a pleasant situation, but it had occurred under her roof and she could not turn her ears off. He had followed his wife back and forth as the babies had been fed, arguing huskily in atrocious Russian, but Marina had decided to keep him on parole. At times he seemed desperate to establish a rapport with his wife, as though it must be tonight or never. Marina rejected the ultimatum. By 9 P.M. Lee Oswald was in bed alone.

Michael would return soon with the hamburgers. Mrs. Paine recalled that, as Lee retired, she went out into the garage to set some toy blocks on the deep freeze for repainting. The naked bulb in the garage was lit. Irritably Ruth Paine wondered who had left it on. Bills for services were always high enough without wasting power. This was waste. Who would come out into the garage after dark and for what purpose?

She remembered, late this morning, that Marina said she had retired around 10 P.M. When she had crept into her side of the bed, Mrs. Oswald was aware that Lee was awake. She couldn't explain it; she knew. He said nothing. He did not try to touch her. She was certain that he was lying awake in the dark. It was a strange feeling. Sleep had never been a problem to Lee. Sometimes, she felt that she did not know this man. In her mind, he became "my crazy one." She did not mean it to indicate insanity; it was synonymous with unpredictability.

He had worked at the radio factory in Minsk, ignoring the summons of the commissar to attend party meetings and complaining that the doctrines were dull and repetitious. Lee Oswald was unteachable. The city was too big and too cold. He would prefer Moscow, but he could not get a permit to go there. His apartment, because he was a foreigner, was better than that of comparable craftsmen in the Soviet Union, and his earnings were augmented by a monthly check from an organization ironically called the "Russian Red Cross." He was a worker who was being "kept" until the foreign office could decide what to do with him.

Overnight he fell in love. Miss Ella German was dark and attractive. It is possible, when Mr. Oswald's restrictive emotions are understood, that he could not fall in love but rather that he condescended to offer himself to Miss German as a husband. The lady was unimpressed with the foreigner. She declined. Lee Oswald could not believe that, once he had decided to take a wife, that he could be spurned. He was an oddity, a rarity, an American. Ella German said no. He walked the deep snows of February 1961 pondering this crushing blow, barely seeing the stony bones of the great museums and libraries and, in venomous retaliation, proposed marriage to Marina Nikolaevna Prusakova.

The blonde pharmacist toyed with the notion and said yes. They married, moved to a larger apartment granted by the government, and, within a month, Lee Oswald was writing in his English diary:

"The transition of changing full love from Ella to Marina was very painful esp. as I saw Ella almost every day at the factory but as the days & weeks went by I adjusted more and more to my wife mentally . . . she is madly in love with me from the start." Lee Oswald was fascinated with the thought of owning a domesticated animal which would lick his hand. He could take, but he could not give. Oswald was making love through a pane of glass. He could see her, but he could feel no tenderness.

He switched his energies to getting out of Russia with his wife. He appealed to the Soviet government, which looked upon the rights of the individual as an oddity. Oswald wrote to the United States embassy. He appealed to Senator John Tower of Texas by mail: "I beseech you. Senator Tower, to rise the question of holding by the Soviet Union of a citizen of the U.S. against his will and expressed desires." This was a change in political direction from his fist-pounding rage of October 31, 1959, when he had shouted at the American consul in Moscow that no one could stop him from renouncing his American citizenship and becoming a Russian. The best the consul had been able to do was to postpone the renunciation, and Lee Oswald later changed his mind. He didn't want to become a Soviet citizen, but he didn't know whether charges had been filed against him in America. He wanted the State Department to lend him money to return to Texas, and he demanded a pledge that he would not be prosecuted for any fancied crime.

A letter went to John Connally, former Secretary of the Navy. Oswald was never known to flinch in the face of an outrageous lie; all he asked of himself was that it be palatable and close to being credible. The note to Connally protested that after he had been honorably discharged from the U.S. Marine Corps the record had been altered to "undesirable" as the result of a visit to the Soviet Union. As a fellow Texan, Oswald wanted Connally to know that he had been a good and true U.S. marine and he wanted that record to be changed back to "honorable."

The defector hoped that Connally had not seen the two interviews he gave to correspondents in Moscow, one in which he said he "hated" the United States, the other in which he said that the Soviets were about to accord citizenship to him. "In November 1959," he wrote, "an event was well-publicized in the Fort Worth newspapers concerning a person who had gone to the Soviet Union to reside for a short time. (much in the same way E. Hemingway resided in Paris.)" In the middle of the letter, he became a flag-waving Marine. "I shall employ all means to right this gross mistake or injustice to a bona fide U.S. citizen and ex-serviceman." Two years earlier, he had offered the Soviet government all the Marine Corps "secrets" about radar. He was crushed when the authorities displayed no interest.

Marina could remember a great deal. A woman, she felt, could work a little harder to make a success of marriage. She did not understand the Americans but, when she arrived in Texas, she began to appreciate what they had. For the first time, she saw supermarkets with small mountains of food. There was no rationing on merchandise or clothing or rooms. She saw sleek cars and buses which could go anywhere without permits. The government could be criticized; there were people in Dallas who spoke her native tongue and who offered hearts and help. Marina could not understand why her husband denounced the institutions of this country. Still, she worked to keep the marriage intact.

Six months ago Lee had come home pale. He admitted that he had tried to kill Major General Edwin Walker. In Russian her husband told her: "He is very bad man. A fascist." The girl was astounded. "This does not give you the right to kill him. . . ." He shook his head. "If someone had killed Hitler," he said, "millions of lives would have been saved." Earlier on that day in April 1963 he had left a note for her in Russian. It told her what to do in case he was suddenly taken from her.

One part said: "Send the information as to what has happened to me to the embassy and include newspaper clippings (should there be anything about me in the papers.)" The only embassy Marina knew was the Soviet embassy. Her head ached when she read: "If I am alive and taken prisoner . . ." Marina kept the document and threatened to show it to police if he attempted an assassination again. A short time later, he tried to leave the little apartment with a pistol in his belt to "kill the Vice-President." On this occasion, Marina had locked him in the bathroom until he promised to behave.

The cocoon became crowded. Lee Oswald was too big for it. The wild threshing inside his mind increased. Picayune jobs at a dollar an hour became impossible to hold. Marina was on a ration of ten dollars a week to maintain her little family. Her husband ran off to Mexico, ostensibly to head for Havana and join Fidel Castro. The defeat rankled deep when he learned that the socialist world had no place for him. The Cuban embassy refused a visa; the Soviet embassy said that such a document would require three months of waiting.

Oswald had slammed doors and had shouted insults. Again he had offered marriage to his love and had been spurned. No one appreciated his great worth. If they thought that Lee Harvey Oswald was a nothing—a cipher—he would have to do something to revise their estimates. Including Marina's.

The long black car came down the street slowly and turned in at The Elms. Hurriedly the television cameramen snapped their big lights on and a knot of neighbors set up a faint cheer. There was no acknowledgment from inside the car. The big man in back hunched forward to search among the faces, but the one he wore was long and dour. A few journalists shouted questions at the car, but it rolled through the gates, bumping a little on the hard stones.

Secret Service men were out front, looking over the heads of the curious, curious themselves. They had lost a big man today, and now they overreacted to situations. They pushed people back, and shouted harshly to step aside. Each of them knew that solitary lesson: in any crowd, one madman is enough. All their adult lives they were sifting people with their eyes, waiting for the suspicious reach into a pocket, the tossed bouquet, the rifle on a roof. Two agents at the wrought iron gates drew their revolvers. They wanted to get their man indoors, that's all. And that's what they did, with Youngblood and Roberts watching him come out of the car and them walking backward behind him.

Lyndon Johnson was surprised to see people in the living room. They were surprised to see him, and even though they were old friends and neighbors, no one said: "Hi, Lyndon!" He wore a new mantle. They knew it and they were abashed. Some said: "We must be going" and no one said: "Please stay." It had been a long day, a day of evil rather than triumph, and it had worn all the central characters down so that the spirit sank in fatigue but the body spurned sleep.

The President hung a gray hat on a rack. He looked up the staircase in time to see Mrs. Johnson coming down. She knew he was depressed, and she pasted a smile on her face. He didn't say hello. Johnson leaned down and wrapped his huge arms around her back. He held his cheek close to hers and she said that she had made a lot of nice chicken. Her man had simple tastes in food: chicken, beef, lamb, and pork, lots of it—and two helpings of tapioca pudding. He had the hearty appetite of the Texas rancher. Lyndon Johnson enjoyed eating for its own sake.

"I'm sorry," he said. "I should have phoned you, honey. I had a hamburger at the office." Mrs. Johnson said she was going to keep the chicken hot anyway; the men with him would want to pick at something. The President walked into the ground-floor den, idly waving his hand for Valenti, Moyers, and Carter to follow him. It was a small room with books and a desk, a cold fireplace, and the leathery atmosphere of a man's sanctum.

Someone made a brace of drinks. "All right," the President said. "I think I'll take a Scotch tonight. Put a lot of water in it." The drinks were made and he sat in a winged fabric chair, sagging in it like a man who has just walked offstage and doesn't have to pretend anymore. Jack Valenti knew that Johnson would not quit working, so he opened drawers until he found a pen and a white pad. Ideas would be tossed in air, some to be discarded, some to be acted upon, and Valenti wanted to record them all.

On the wall opposite the President's easy chair was an oil portrait of Sam Rayburn, master politician of Texas, the little bald man who had taken the freshman Congressman from Texas under his wing and taught him how to win, how to compromise, how to get bills through the House, how to lose. Mr. Johnson rotated the glass in his hand and heard the clink of ice. He was staring at the portrait of the dead Speaker of the House. Then he lifted the glass and said: "I wish to God you were here" and drank deeply.

"Who are these people?" the press asked. The detectives said: "See Captain King. We can't give statements." Rose and Stovall and Adamcik assisted Wesley Frazier into a glass cubicle with his minister, the Reverend Campble of the Irving Baptist Church. Mr. Frazier's sister, Linnie Mae Randall, was shown into another part of the office. The pile of sworn statements was thickening. There were so many that no enforcement officer had been able to read all of them. And yet each was a tiny piece of information ready to be set into place in an ugly puzzle.

Detective G. F. Rose could see that the big, skinny kid was nervous. "Relax," he said. "You have nothing to hide. Right?" Wesley Frazier sat and placed his elbows on his knees. "Right," he said in a small voice. In another part of the office, Mrs. Randall was reminding the people that it was she who had volunteered the information that her young brother drove Lee Oswald to work this morning with a long package on the back seat. She had been washing breakfast dishes at the window . . .

The innocent, in panic, make themselves suspect. Matters which appear to be bright and simple, on recollection, assume a darker hue. The young man began to answer questions, first about himself, then about his relationship to Lee Oswald. He recalled how he asked Lee why he was going to Irving on a Thursday instead of the usual Friday. Vaguely he remembered Lee saying something about missing a visit the previous week because the Paines had a birthday or something. Oswald wanted to pick up some curtain rods for his rented room.

Lee wasn't the kind of man a fellow asked questions. Sometimes he just kept looking out the windshield, saying nothing. Other times he might say something, but he wasn't a talker. This morning, when Wesley parked in the freight yard behind the School Book Depository, he wanted to rev his car engine a little because the battery was low. His friend Lee didn't even wait for him—just took his curtain rods and walked ahead. He didn't even wait and hold the plant door for Wesley.

The boy's minister told him to tell the whole truth and no harm would befall him. Frazier nodded. He understood. He had no desire to get into trouble of any kind; at work he was known to be amiable, obliging, a boy who seldom disagreed with any-body. At another desk, his sister watched the pen of a detective spell the words she spoke, about the Russian woman who lived up the street with the Paines, the surly face of Lee Harvey Os-wald getting in her brother's old car, getting out of it, waiting for Wesley in the driveway—a peculiar man who never smiled, sel-dom spoke, an assortment of features which formed themselves into an everlasting grudge.

Wesley was certain of his answers until the detectives asked them a second or third time; the more he meditated, the less sure he was. The curtain rods were "about this long" [twenty-six inches]. "How long?" "This long" [indicating about twenty-eight inches]. "Does he bring his lunch from home?" "All the time, ex-cept today." "Why except today?" "I don't know. Lee never buys lunch, but this morning he said he was going to buy his lunch and eat in the domino room." "Why do you call it the domino room?" "Some of the fellas, when they have lunch and there is still time, they play dominos." "Did Oswald ever talk politics?" "Not with me." "Did you ever see him with a rifle?" "No, sir." "Ever talk about one?" "Not to me." "Did he discuss the visit of President Kennedy?" "Not that I remember." "Ever see him with a pistol?" "No, sir." "Do you have any objections to having your fingerprints taken?" "No, sir." "Ever been arrested?" "Yes, sir. In Irving tonight." "Besides that?" "No, sir, never have. . . ."

On the fourth floor, Lieutenant Day looked at the side of the rifle, smiled, and murmured: "Yes, sir." It wasn't much of a print, and it was coming up slowly, but there it was as plain as a slap mark on a tender cheek. "The metal is rough," he said to an assistant. "If it was smooth, this print would be sharper." It was part of a palm of a hand, on the underside of the wood stock. The screws of the stock were loosened, and the print seemed clear. The police photographer took several closeup shots of it. Day took Scotch tape, carefully applied, and slowly lifted the print free. It was faint, but it was discernible.

He had a palm print on a carton taken from the sixth-floor window. If both were of the same hand and they matched, the lieutenant would put them on a projector beside some he had taken from Oswald. If all three matched, then Oswald handled this gun and also sat in that window. Vincent Drain of the FBI came up to the laboratory to see how the lieutenant was doing. Day showed him the material. The FBI man reminded Day that headquarters in Washington was prepared to lend any assistance required. The lieutenant said he appreciated it, but the chief would have to handle matters like that.

The tall, good-natured Drain left. The men on the fourth floor continued their work. They knew that the FBI wanted all this material. Day had orders to process it, and that's what he was doing. The room smelled of developer. Lights went on and off as negatives were fixed. The men worked in silence, at microscopes, cameras, acid baths, calipers, projectors, spectroscopes. There were hairs on the blanket which housed the rifle, but they were short and kinky. They were pubic. Someone had once slept in this thing nude.

A clear palm print was thrown up on the projector. The smudged print from the underside of the rifle went up beside it. The officers stopped work to look. The one from the rifle wasn't clear enough. Still, the swirls which could be defined appeared to match the ones taken from Oswald's left hand tonight. The photos were reversed, and the eyes of the men scanned them again. It wasn't the best of evidence, but both appeared to be made from the same hand.

"Be back in a minute," the lieutenant said. He ran down the stairs and into the chief's office. "I make a tentative identification from a palm print on the rifle which matches one I got from Oswald," he said. The chief smiled and looked up from his desk. "Good," he said. Day fought his way down the center hall into Fritz's outer office. He called the captain out. Fritz said that the prisoner was on his way down again. He was making a couple of phone calls. The lieutenant whispered the story of the print match. Fritz smiled a little. "Give me a report on it when you have it," he said. "We're moving along—a little at a time." "It's tentative," Day said. "It looks pretty good."

America was beset by an anguished desire to be punished for the crime. It was on family faces in Oklahoma and Oregon, in Salem, Sioux City, San Antonio, and San Francisco. The shock wave had leveled off into a mass what-did-we-do? guilt. Psychologically, the nation was on its knees. This morning it had been rich, powerful, and unafraid. Within nine hours, the self-appraisal had been revised downward. There was a subterranean violence in the national character, one which few suspected. Some remembered the abortive attack by Puerto Ricans on the life of President Harry Truman. Others, older, recalled that a madman named Zangara had fired at President-elect Franklin Delano Roosevelt in Bayfront Park, Miami, and had hit Mayor Charles Cermak of Chicago. Cermak had died. In time so had Zangara.

The sense of guilt was felt everywhere. A man named Oswald may have pulled the trigger, but what kind of a country had bred him? America was a sophisticated land where even the common union laborer owned a television set and a car and perhaps played golf on Sundays. Passions seldom rose high enough for the people to use any weapon other than a ballot or a fist to show displeasure with a politician. This image had been broken. The people looked into a mirror and the glass was intact but the face was cracked. By 9 P.M. Central Standard Time, millions of people were prepared to believe the worst about themselves. It was easy to blame the whole thing on a nut named Oswald. Too easy. The guilt could be shoved off on Dallas too. Or even the whole state of Texas. For once, something was too big for Texas.

The dark, lined face of Edwin Newman was picked up by the National Broadcasting Company and he was prepared to tell the people what they wanted to hear: "This event is unreal," he said, "absurd—one of the things we just don't let happen. But if one in one hundred ninety million wants to kill the President, he will. The unpleasant truth about America is that it is a country of violence." There, it had been said.

"Violence plays a part in our very lives—yet what we worry about is our image abroad. Today, America does not appear to be an adult country. Emotions run high—regional, religious, and economic. We must begin at the top, for the political climate is set by the President. In the days to come we will hear much of how we must stick together. It is within our power to take our public life more seriously than we have. Americans tonight are a grossly diminished people."

The camera left Newman and, in Washington, D.C., the lopsided, boyish face of David Brinkley came on: "If we have come to the point where a President cannot appear in public without fear of being shot," he said, "then we are less civilized than we think we are." As a mark of its sincere sorrow, the three major television networks canceled the sandwiches of advertising which are their meat more than the public's and fasted.

The city of Dallas was anxious to dust its municipal cloak of Lee Harvey Oswald at once. Mayor Earle Cabell did it before the cameras at eighteen minutes after 10 P.M.: " . . . It is hard to believe," he said, "but I don't believe this event will hurt Dallas as a city. This was the act of a maniac who could have lived anywhere—a man who belonged to no city." It probably did not occur to the excited mayor that a statement of that type hurt Oswald's chances of securing a fair trial in Dallas County.

Chief Petty Officers William Martinell and Thomas Mills watched as the Secret Service men completed the examination of the President's limousine. They knew why they had been brought from Admiral George Burkley's office. On the way in from Andrews Air Force Base, Special Agent Kinney thought he had seen some hair and skull under a jump seat. If it was true, the petty officers would be charged with taking it to Bethesda Hospital.

The examination was near its end when Deputy Chief Paterni of the Secret Service said softly: "Here it is." It was a three-inch triangular piece of skull and hair lying under what had been Mrs. Connally's seat. Martinell lifted it on a piece of paper and dropped it into an envelope. He asked if he could remove some of the whitish tissue from the back seat. The deputy chief and Floyd Boring had no objection. Martinell took his piece of paper, curved it into a small shovel, and removed chunks of tissue from the back of the seat and the area between the jump seats.

It was placed in a separate envelope. Mills, running his hand across the rug under the windshield, found a metallic fragment and turned it over to the Secret Service. The plastic cover was placed over the car. The petty officers signed receipts for the material and started out for the hospital. The three-inch piece of skull they carried, in addition to the piece found in the road on Elm Street, Dallas, represented about two thirds of the massive fracture in the President's head.

The doctors had completed that part of their work. A few of the witnesses were escorted to the naval commissary for food. The notes kept by the doctors were in scribbles, sometimes only a phrase to remind them of an entire event: "Blood & hair upper medias" "only a few in size 3–5 mm." "no missile in the wound" . . . Doctor Humes leaned across the chest and made a Y incision. It extended from both shoulders down to the center of the sternum, then cut straight down to the pelvis.

The rib cage was lifted open, exposing the thoracic cavity. The lower flaps were pulled back, opening the abdomen. The three doctors studied the torso. The final report would state: "*CAUSE OF DEATH:* Gunshot wound, head," but pathology decrees that an autopsy should be complete, even when the cause of death is obvious. The organs should be examined for grossness and disease.

On the seventeenth floor, the impatient young Attorney General was accelerating his pace. The phone was in constant use. Shriver told him that the President's office was now empty of Kennedy keepsakes. From the desk had been taken the coconut shell in which he had sent a message that a Japanese destroyer had sunk his PT 109. The phone calls continued and at last Robert Kennedy became irritated.

He asked the Secret Service where Dr. Burkley was. The rear admiral had made several trips up to the suite, advising that the Navy doctors would not require much more time to complete the autopsy. But the hours kept grinding onward. Kennedy looked at his watch. It was after 10 P.M. Eastern Standard Time. Clint Hill phoned the autopsy room and spoke to Kellerman. He spoke to Burkley, who said: "It's taking longer than they thought." The Attorney General hoped that the doctors would get on with it.

In the White House, Sargent Shriver completed a call to Richard Cardinal Cushing of Boston, hung up, and clapped a hand to his handsome forehead. "My God!" he said. "We forgot to invite Truman, Ike, and Hoover!" Ralph Dungan, whose office was being used for the massive funeral campaign, leafed through a copy of *State, Official and Special Funeral Policies and Plans.* Major General Ted Clifton kept calling Godfrey

McHugh at the hospital to ask, within reason, when the President was coming home. He had to know for several reasons: artist Walton had to complete the funeral atmosphere of the East Room. Clifton had displayed an etching of the Lincoln catafalque to the White House carpenters and demanded that they duplicate it at once (someone had neglected to pass the word that the original Lincoln catafalque had been located in the basement of the Capitol building).

General McHugh kept saying that no one knew when the doctors would be finished. An embalmer hadn't been summoned. Originally, without consultation, most of the autopsy observers had figured that the body would be in the East Room by midnight. Well, it would be later than that. One o'clock? said Clifton. No one could be sure. Maybe two. Maybe even later. Time had dragged all day. Now events were dragging.

At the Library of Congress, Arthur Schlesinger, Jr., and Dick Goodwin could not believe that, after authorities opened the place for them, they would have to hunt for Lincolniana up and down the musty aisles with flashlights. The interior lights of the library were on time clocks. The White House aides were guided in their literary voyages by two competent men: David C. Mearns and James I. Robertson. They picked up pertinent books and copies of old magazines and newspapers. No one would be able to read, absorb, and appropriate the ideas entombed within these publications, but the men thought that it would be better to bring back too much than too little.

A new officer sat listening to Fritz and Oswald. He was fiftyish, a man with blue eyes, a high, freckled forehead, and spectacles. He was Inspector Thomas Kelley of the Secret Service and he had been in Memphis, Tennessee, when the crime occurred. Rowley had called him and asked him to hurry to Dallas to supervise the Secret Service aspect of the investigation. Kelley was a low-key man. The first assignment he gave himself was to sit quietly and try to assess the prisoner and Captain Will Fritz. Earlier, in the outer office, Forrest V. Sorrels had given the inspector a summary of all that had happened up to 9 P.M. Kelley had asked numerous questions.

As he sat in the crowded office, he had a fairly good idea of the game. He felt that Fritz was on solid ground with this young man. He watched Oswald closely and he became impressed with the fact that, unless he was misjudging the gigantic ego of the prisoner, Lee Harvey Oswald was the type who wouldn't share the credit for the assassination with anyone.

Kelley was also impressed with the slow, deliberate manner of Will Fritz. He liked an officer who didn't press the quarry too closely or harshly. For seven hours in elapsed time these two men had fenced carefully, almost with respect. The captain's best weapon was that he knew how much evidence was already in, how many persons had already identified Oswald at the lineups. His questions led from strength. Oswald's weapon was that he could maintain a quiet conversation with the captain until a sensitive one surfaced; at that point he could refuse to respond.

The Secret Service inspector was satisfied that this case would be cleared up. He crossed his leg, placed his clasped hands on the top knee, and listened. A small smile puckered his eyes. He had the impression that Oswald was lying and knew that Fritz knew that he was lying. In this game, when two men are playing for one more life to follow the one destroyed, there is not time for bickering and charges of untruth. Both appeared to understand this, and Oswald followed his road of innocence, slightly hurt that Fritz could think a young fellow at a movie could be involved in a thing like murder.

"Now how about that Alex Hidell thing?" Fritz said patiently. "We find three cards on you with that name." All day long, the prisoner had maintained that he knew nothing about it and had refused to discuss it. His attitude shifted. He wanted to toss a small bone to his captor. "I picked that up in New Orleans," he said. "Where?" "It was part of the Free Cuba Committee." "Was it a name you used?" The tone switched to surly. "I've talked about that enough. I don't want to discuss it anymore." Fritz asked a few unimportant questions. Oswald recognized that they were on safe ground and began to prattle. Will Fritz permitted him to talk. The captain kept glancing at his desk blotter, turning a Ticonderoga pencil this way and that, listening until the prisoner was completely unwound.

He opened another field. Both men knew the delicate areas of interrogation. "Do you belong to any other organizations?" "American Civil Liberties Union." Fritz raised his brows. "How much dues do you pay?" "Five dollars." Fritz wanted to ask about the curtain rods. He kept trembling on the edge of the question, then pulling back. The captain's reasoning was that if there really were curtain rods in that package, he didn't want to stumble into being surprised by that fact. Fritz had heard about Wesley Frazier and the long package in the back of the car; he had also asked Mrs. Paine if Mrs. Oswald had sent her husband for curtain rods or needed curtain rods. The answer was "No." The men who had questioned Earlene Roberts at the North Beckley furnished room had been told bluntly that Oswald was not bringing curtain rods to his room and wouldn't be permitted to use them if he did. The owner, Mrs. Johnson, furnished every room with curtains and Earlene had never heard of a roomer bringing curtain rods or curtains.

Fritz decided to take the chance. He was as certain as he could be that there were no curtain rods, just a disassembled rifle. "Somebody told me you had a package in the back seat of the car," Fritz said. The prisoner sat a little more upright. "No package except my lunch," he said. The captain nodded. "Did you go toward that building carrying a long package?" Oswald shook his head vigorously. "No. I didn't carry anything but my lunch." "Where did you get that pistol?" Oswald relaxed. "I bought it about six or seven months ago." "Is that so? Where?" "In Fort Worth, I think." "Where in Fort Worth?" "That's all I'm going to say about that."

"Did you make your phone call?" "Yes, sir. Thank you." "Did you get your party?" "No."* "You can try again. You don't have to thank me. We do this for all prisoners." "I don't have any money." "You don't need it. Call collect." "That's right. I hadn't thought of that." "What kind of a lunch did you carry?" "Cheese sandwich and an apple." "What do you give Mrs. Paine for taking care of your family?" "Nothing. It comes to a good arrangement for her because my wife speaks Russian and Ruth Paine is a student of Russian."

"Did you keep a rifle in the garage there?" "No rifle. I have a couple of old seabags there, some kitchen articles in boxes— that's about it—no, a couple of suitcases with clothes." "Do you have a receipt for a rifle?" "I never owned a rifle." "Are you a member of the Communist Party?" Oswald smiled faintly. "I have never been a member of the Communist Party." "You know what a polygraph test is?" "Yes, but I wouldn't take one without the advice of counsel." "That Selective Service card with the name Alex James Hidell on it, isn't that your signature?" "I carried the card, that's all. I told you I don't want to discuss it." "I'm just trying to find out why you carried it." "I won't talk about it."

"Who were your friends over on North Beckley?" "Never had a visitor at North Beckley." "Can you tell me something about that address book we found on you?" "Yes. It contains names of Russian immigrants who live around Dallas. Sort of friends of ours."

---

* Attorney John Abt of New York was spending a weekend with his wife in a cabin they owned in a farm area of Connecticut. It was not equipped with television or radio. The next day, when reporters located him, Mr. Abt said that he doubted he could defend Lee Harvey Oswald because of his legal commitments. Failing to contact John Abt, Lee Harvey Oswald phoned his wife Saturday but found that the *Life* magazine men had taken her and his mother from the Paine household. He asked Mrs. Paine to contact John Abt in New York.

Inspector Kelley, who had been a party to such interrogations for twenty-five years, marveled at the interplay of the two men. He became convinced that Lee Harvey Oswald was engaged in a contest he relished. At one point, when Fritz and some of the other officers left the inner office for a moment, the inspector walked over to Oswald. Kelley wore his most engaging smile. "I just wanted to say," he said politely, "that I'm an inspector with the Secret Service. As you know, we're charged with the protection of the President. If you didn't do this thing, I wish you would tell us so we can find who did." Lee Harvey Oswald glanced up at a new and slightly more subtle antagonist. "I won't do it now," he said, "but I'll tell you what I'll do. After I see my lawyer, I'll talk to you."*

Kelley stuck a hand in his pocket and walked back to his chair. He was positive that this young man had an ardent desire to let the world know, in time, that he, Lee Harvey Oswald, had done the thing that no other man dared to do and had done it alone. To die, to serve a lifetime in prison, or even to go free without scratching his name on the granite face of history would be, the inspector felt, contrary to the aspirations of the ego and to the conceit and the superiority Lee Harvey Oswald felt in relation with the rest of mankind.

Too many people were talking, too many were foaming with too many notions, and all things had to be dealt with at once. The western edge of the White House had aspects of a solemn football game, with young men running in and out of a jammed doorway, older men walking, heads down, to an office with a

---

* The author wants to make it clear that no record was kept of any of the interrogations of Oswald. He makes no claim that the questions and answers are in proper sequence. From interviews and recollections of persons who were present, these scenes represent an approximation of what was said. There is no doubt that some of them are not properly juxtaposed, in spite of information from Dallas and Washington observers.

free telephone. Two would troop in with books and magazines relating to the funeral of Abraham Lincoln, while two others would debate the delicacy of inviting Senator George Smathers of Florida to the funeral. "He was a close friend of the boss." "I know, but he is also a Senator and he'll come in with the Senate group." An elder statesman, Averell Harriman, sat bowed like a scarecrow in a perverse wind. A young government press agent, David Pearson, asked: "Mr. Ambassador, aren't you a good friend of President Truman's?" The long, lined face lifted; the head nodded. "Would you please contact President Truman, President Eisenhower, and President Hoover and invite them to come tomorrow morning?"

Without a word, the old one got to his feet and walked to another office to make the calls, issue the invitations. It had already been done by President Lyndon Johnson, but those in Dungan's office were not aware of it. Eisenhower and Truman would be in Washington tomorrow, to pay their respects to Mr. Kennedy and to confer with Mr. Johnson. Mr. Hoover was too ill to attend. It would take two hours of Harriman's time to ascertain this.

The Kennedy group had swift and accurate reflexes, but the death of their leader thrust upon them an unexpected event of magnitude. His death undermined the power structure and, as it crashed in chaos this evening, they planned a funeral which only the most callous would forget. The sunburst vision of charisma which the young man had displayed in all his political battles must, somehow, be made to shine for three additional days, when the bright light would be extinguished forever. As they had planned the best, the biggest, the most dramatic battles in the political wars, so too the final homage to his remains must be enormously tragic. He was a lot more than Jack Kennedy, rich *bon vivant*; he was President of the United States.

Shriver cut off the phones for a moment to draw up a draft of a plan. "Let's call 10 A.M. tomorrow," he said, "the first hour." He wrote "Saturday 10 A.M." on foolscap and, beside it, "President's Family." It was a start. He wanted a priest there for prayers for the dead at that time. At 11 A.M.—who?—ex-Presidents, maybe the Supreme Court. Noon—noon, perhaps the diplomatic corps. One P.M.—the United States Senate. But what about burial?

Where? What city? What cemetery? Brookline, Massachusetts? Did anyone recall where the baby had been buried last summer—Patrick Bouvier Kennedy? Somewhere near Boston. A family plot probably. On the other hand, Mrs. Kennedy may have thought of a place—certainly she would express her wishes. Two of the volunteers recalled that in March 1963 the President had strolled from the Tomb of the Unknown Soldier down the green cascading hill of Arlington National Cemetery. It was a sparkling day. The capital, stretched below, was a geometry of broad boulevards and impressive buildings and monuments.

Three hundred feet below the Custis-Lee Mansion, he had paused to drink the exquisite view. He stood among the ranks of small white headstones, the military dead of several wars, and he said: "I could stay here forever." This, thought Sargent Shriver, might be the last opportunity to grant him a wish. At Bethesda Hospital, Secretary of Defense Robert McNamara had already remembered that day, that stroll, that wish. Mrs. Kennedy had greeted the recollection with an approximation of joy. It is rare to know a young man's last wish. The decision was made.

The hour was late, but the caretakers at Arlington were summoned. They consulted plans of the cemetery. They, too, were grief-stricken and, even though the average soldier-citizen gets no more than a four feet by eight feet section of the cemetery, they found an unused area on the spot where John F. Kennedy had stood and offered three acres of ground. Someone announced it to the press, and there was resentment among the people at anyone getting that much ground. President Kennedy could settle for less.

Shriver marked off Sunday afternoon for the lying-in-state in the Capitol of the United States. Here the people could form into long queues and file past the box. Monday, the funeral. That would be Monday morning. Probably 10 A.M. There would be a Mass of requiem in either the newly completed Cathedral of the Immaculate Conception or Saint Matthew's Procathedral. Mustn't forget a naval guard of honor. Lieutenant John Fitzgerald Kennedy was United States Naval Reserve. Mustn't forget many things and many people. How many chiefs of state would fly to Washington? The time was 3:30 A.M. Saturday in western Europe; there would be no point in using the so-called "hot lines" at this hour. Still a note could be made to start phoning at 5 A.M. Eastern Time.

A funeral doesn't accord time to its planners. No one could guess how many people could be accommodated in either of Washington's Roman Catholic cathedrals. Someone could call Bob Kennedy at once, at least, and find out what the family wishes were? Then, once the capacity of the pews was known, space could be reserved for the family, the personal friends, the chiefs of state, the diplomatic corps, the Senate and House committees, the Cabinet, the Joint Chiefs of Staff, the three pool men of the newspaper wire services.

At the hospital, the Attorney General placed an arm around his sister-in-law. He walked her away from the conversational groups, looking at the rug under his feet. Robert Kennedy knew how to "handle" Jacqueline Kennedy. She reposed great confidence in him. He knew that each decision she made tonight represented an additional wrench of the heart. But she had to keep making decisions. "We should get some clothes for Jack," he said softly. She had not thought of it. What kind? Mrs. Kennedy thought it over. She remembered that he had a dark blue pinstripe suit, a plain blue tie with a small pale figure in it, a white shirt, of course, and a pair of black shoes.

A Roman Catholic would want a rosary entwined in his

hands. The President had beads in his room. On the other hand, Prince Stanislaus Radziwill, married to Mrs. Kennedy's sister, offered his rosary. It was accepted. Mrs. Kennedy remembered that her husband cherished a solid gold Saint Christopher medal which she had given him. Mrs. Kennedy wanted that medal in the casket with Jack. It was a girlish sentiment. She had others: she planned to write a final note to her husband and to seal it in the casket with him.

The Attorney General walked out into the hall and asked Clint Hill to telephone the White House for the clothes and medal. The Secret Service agent phoned George Thomas, the valet, and listed each item. "Just get them together with underwear and give them to a driver on the South Grounds. Tell him to deliver them to the autopsy room at Bethesda."

The medal could not be found. It was in the President's wallet, twenty feet from his body. William Greer had it.

The two Secret Service agents had breakfast at 6:30 A.M. in Fort Worth. They had come a long hard way, so when an officer whispered: "I'll have a man take you up to the commissary," Roy Kellerman and William Greer looked guiltily at each other and said: "Thank you." It had been fifteen hours since they had coffee, and yet the requirements of the stomach seemed out of place in an autopsy room. An enlisted man took them from the room and, when they got to the restaurant, they asked what was ready to eat. "Chicken, rolls, and coffee," was the response.

"All right," said Kellerman. They sat, working their broad fingers on the formica tabletop. The stomachs were hungry, but the thoughts negated food. Kellerman and Greer had been with the "boss" from the start, and with other Presidents before him, but there was nothing to talk about. They glanced around the room, smelled the steam-heated chicken, and mouthed safe words about nothing in particular. Both were tough law enforcement officers. Since losing their man, it seemed heavy for the spirit to see and watch his autopsy, too.

Neither knew much about the afternoon and evening events in Dallas, except that Robert Kennedy had told Kellerman that a young self-professed communist had been arrested. The plates of steaming chicken arrived, and both men looked at them and decided to try the coffee. They sipped and stirred and ate two rolls. In fifteen minutes, they were back in the autopsy room.

Outside the room, Greer found two men in medical coats trying to get into the room. A check showed that they were newspaper reporters. They left without dispute. Inside, Sibert and O'Neill of the FBI were receipting a glass container with metal slivers taken from the brain. At the same time, Dr. Burkley's enlisted men were delivering a piece of skull. Burkley gave it to Dr. Humes, who made a sketch of it, examined it, and, with Boswell and Finck watching, found where it fitted.

Not far away, President Johnson sat in the dining room picking at chicken. It was something to do. This was going to be a long night for him—it already was—and he looked around at the young faces: Jack Valenti, Bill Moyers, Cliff Carter, and an old friend who had just arrived, chubby Horace Busby. One of the characteristics the President had managed to hide from most people—except his wife and daughters—was a deep-set loneliness which he denied. He felt it now. It came over him at night and he would call friends at unreasonable hours and say: "Aw, come on over and sit with us awhile." He noticed that no one this evening called him "Lyndon." He was "Mister President" to everybody except his wife.

It brought no pleasure. The feeling was one of remoteness imposed upon him. The man who enjoyed friendship and loyalty and the rough-and-tough game of party politics was on a solitary eminence in the dark. Briefly he chewed on the chicken and listened to the conversation and kept the rheumy brown eyes on the television set. The screen switched to Dallas, and there was John F. Kennedy, the graceful, grinning chief, strolling along the Love Field fence in a forest of arms. Mrs. Kennedy, in the nubby pink suit, was behind her husband, smiling graciously and being yanked almost off her feet by the hearty Texas handshakes.

"Shut it off!" Johnson snapped. Then, more softly, "I just can't take that." He was trying hard to think of things which would keep him from remembering. He wiped his hands carefully on a napkin and called Secret Service Chief James Rowley. "Rufe did a brave thing today," he said. "He jumped on me and kept me down. I want you to do whatever you can, the best that can be done, for that boy." He hung up. It had not occurred to him that Rowley, too, was lonely. If there was any blame, any official laxness, it didn't matter that the planning of the Texas trip had been in the capable hands of Floyd Boring; it meant nothing that Roy Kellerman was in charge, along with Emory Roberts; no one wanted to weigh the possibilities that, if a Secret Service man had been on the left rear bumper going down Elm Street, it would have been difficult to hit President Kennedy. All indictments filter upward, and Rowley was the man at the top of the Secret Service. He was pleased to hear a praiseworthy report about one of his men, but Rowley knew that he was going to be every critic's target.

The car was at the curb. The bulky figure came downstairs from the apartment. Under his arm he carried a small female dog. He got in on the driver's side and deposited the dog on the other side of the seat. It was 9:30 P.M.—late. Jack Ruby had said that he would attend services for the President and they would soon be concluded. He had bragged to everybody. Now he would drive to 9401 Douglas and try to make the boast good.

The car jerked down the road, the little dog braced against the turns. Ruby kept his eye on the road and switched the radio on. His favorite station was KLIF. They had Joe Long at police headquarters. He was giving a summary of all the information the police had of Oswald. In the car, Jack Ruby felt a psychological block against uttering the name "Lee Harvey Oswald." Although Oswald had not been indicted or tried, the nightclub owner began to think of him as "the person that committed the act." Jack Ruby frequently felt blocks in his mind against people and names. He saw himself as a good, wholesome person with strong religious beliefs, one who loved law and order and law enforcement officials—one who, perhaps, was guilty of a few minor things like traffic summonses and fist fights—but a true child of God withal.

Reprehensible people did not deserve to have names or faces. "The person who committed the act" was the newest on a long list. Ruby had no trouble detecting good men from bad; he could do it instantaneously. His keen intelligence told him, as a weather vane swings to the new wind. He knew that Joe Long of KLIF was a good man because Ruby recalled that Long had given his club free plugs on the air. Russ Knight was another good man. He was a disc jockey on KLIF, a man who could enunciate his thoughts in swift, sure words. Joe and Russ—good guys.

Cars were parked around Temple Shearith Israel. Ruby switched his lights off, reassured the little dog as he always did, and locked the car. He hurried inside in time for the conclusion of the ceremony. A bar mitzvah was announced for Saturday and the communicants were invited to remain, at the conclusion of tonight's services, for cake and coffee. The jowly face of Jack Ruby hunted among the features of dedicated Jews for friends. He envied these people; they were regulars. They felt their Jewishness as a freezing man feels the warmth of fire. They were here because the spirit told them to be here. It was not fellowship which brought them, but an atavistic emotion of belonging to the true faith of God. In the temple they heard the true words of all the prophets, the warnings and promises of God himself.

The nightclub owner felt cold. He saw no fire, felt no warmth. The temple and the replica of the Ark of the Covenant, the scrolls of the Talmud, all brought memories of an unhappy childhood. Among the *goyim* of the Chicago ghetto, Jack Ruby had been a Jew bastard, a Christ killer, before he understood the terms. His father's divorce from the family left Jack Ruby without an anchor. His mother's renunciation of reality placed her in a world apart from his, where the true test of strength was in hoodlum gangs and fists.

The temple was disturbing. All the bad memories of long ago swelled to flood in his mind. In the long ago, the beards, the curls, the black hats were a mockery to the unbeliever. How could a strong growing boy protect these ancients from attack among the tenements and still judge them as ridiculous? It was easy if the young man could subscribe to the thesis that the Jews were helpless, poor, and persecuted. Think of them, not as strong and righteous, a people with a heritage of culture and learning, but as the downtrodden. He could turn his electrifyingly swift mind in that direction, but then he could no longer admire them. No matter what the prophets said, this world could not be inherited by the meek. Never the meek. Looking at his parents, his temple, his heritage, Jack Ruby bent the golden rule so that it spake: *Do it to him before he does it to you.* A man can inflict this cruelty on his world if, in addition, he persuades himself that he is noble and generous and compassionate to the afflicted.

The service was over. The nightclub owner had endured fifteen minutes of it. The worshipers stood in groups, whispering, and some moved on to another room where the ladies cut plain cake and poured cups of steaming coffee and punch. Rabbi Hillel Silverman had more important things to do. His job was not only to preach the word of God but also to hold the people of his temple together in unity of purpose. The coffee and cake could wait. He walked back toward the exit, exchanging greetings, listening to questions, offering counsel, shaking hands, tapping the satin yarmulke on his head to make certain that the badge of the male had not slipped.

Ruby went to the refreshment room and had a glass of punch. Mrs. Leona Lane, with her mother and growing sons, paused to say hello. The woman reminded Jack that they had had Passover dinner together four years ago at Sam Ruby's house. He was vague. She said that the assassination was a terrible thing. "It's worse than that," he said. The rabbi knew Ruby as a man ruled by emotion rather than reason. Silverman recalled that Ruby had not attended services until 1958. The senior Rubenstein died, and Jack appeared in the temple, shaking and weeping. For eleven successive days he sought a minyan and recited kaddish. The display of the spirit died as quickly as it was born. To some, everything is "worse than that."

Only two months ago Ruby had returned to temple services for the high holy days. This time he burst into tears and told the rabbi that his sister Eva refused to sit with him. They had had a disagreement. Eva Grant disapproved of her brother's behavior in dating a girl "too young for him." Jack said that "at a time like this, families should be together." The rabbi was in an awkward position. He phoned Mrs. Grant and, by coaxing, arranged for her to have lunch with her brother.

Silverman made a mental note to exchange greetings with Ruby tonight. The rabbi's spirit was crushed, and he found it difficult to raise others. His President had been assassinated in this city of riches and splendor. The Jewish community, perhaps more sensitive to violence than others, was stunned. The dreadful thing which had happened to the most powerful could therefore happen to anyone. At the door, the Jews sought comfort from their teacher, an explanation of how such a thing could happen, but he had little to offer. He, too, was confused and depressed.

Jack Ruby went to the rabbi and shook hands. Silverman was certain that he was about to listen to an emotional dissertation on the assassination. Jack Ruby did not mention it. He sought the rabbi's sympathy through the illness of his sister. He told what she had gone through in the hospital and of how she was now at home trying to regain her strength. He thanked the rabbi for stopping in the hospital to see Eva. In a moment the night-club owner was gone.

The dark of the city matched Ruby's mood. It was a blackness which could be relieved by the miraculous flick of an electrical switch. Dark, bright, dark again. Big bare roads in the night chalked with warm light and, in the fields beyond them, nothing. Moods, too, had an electrical switch—high, low, on, off. Near the downtown area, the colored neons flirted with the mind. "Girls, Girls, Girls." "Strippers." "Sensational." "First time anywhere . . ." "Fresh from New York!" "Treasure Chest."

He kept the radio on. The Dallas Police Department was working overtime. The radio stations were working overtime. Civic, righteous, church-going Dallas would expunge this obscene graffiti from its conscience with swift, sure Texas justice. No stone would be left what? Bali-Hai was open. Jack Ruby, driving through the garish lights, took note that his clubs—Carousel and Vegas—were closed in memorial to the President of the United States, a mark of respect. But not Bali-Hai. The lights were on; the people were there. Ruby took note.

The Gay Nineties was closed. Note, closed. That is class. The trade would be forced to patronize the open places. The big sweaty men would sit at dark tables with their setups, their brown paper bags of liquor, and the more they drank the more courageous they would become, and they would stare with reptilian fascination at the soft white skin on the little box of talcum called a stage and chant hoarsely: "Take it off. Take it off." The memorial mood could not tolerate this. This was a night for putting it all back on, from neck to ankles; a night for men to weep for each other because of what had happened to *him*. The baggy pants comics would crack dirty jokes and the customers would laugh the louder, to prove that they understood. Who could laugh? Somewhere far off that poor woman huddled with those two babies, crying their innocent hearts out. Who could laugh tonight?

There was nothing else to do. The duties of motherhood were complete. The little ones slept; the house was quiet. Marina said in Russian: "May I borrow the hair dryer?" Ruth Paine got it. "I am not sleepy yet," Mrs. Oswald said. "A shower lifts my spirits. I will take a shower and set my hair." A vigorous shampoo would kill time. She wandered around the living room, carrying the dryer, and asking aloud what Lee could have "against" President Kennedy. Nothing.

That was the strange thing. He had read articles to her, translating into Russian as the avalanche of phrases and sentences tripped from the mind in English to the tongue in Russian, and she knew that he always injected his opinions. Marina tried to remember his rendition of these articles about John F. Kennedy. He had said nothing critical about the man. If he felt no hatred for the man, then he could not have killed him. The motive, the pressure, the compulsion were not present. Unless, of course, a man was willing to kill any great man who passed that close to the window that day—Khrushchev, Johnson, De Gaulle, Adenauer, Harold Wilson, Erhard, or Mao. Then it becomes motiveless, and the mind retreats from the field of reason.

Ruth Paine said good night. They might have felt a compulsive interest, not in dryers but in television. Both women were helpless, swinging in the orbital perimeter of the assassination, but they switched it off and another light was extinguished. Marguerite Oswald made a place for herself on the couch in the living room. There would be time to think in the morning, time to propagate a mother's views on what mysterious and secretive force had assigned her son to kill Kennedy. Her mind was made up, as it always was. All she had to do was to hammer the facts into shape to fit the jigsaw which told her that her poor son was under orders to do this thing. Assuming, of course, that he did it.

There were three women in this small house at 2515 Fifth Street in Irving. The night hours brought no tears, no beating of breasts. There was regret without understanding; curiosity without mourning; self-interest embodied in a hair dryer. Like anti-magnetic satellites, the closer these women drew the faster they would fly apart. For Marina, there was the Slavic sense of gloom encompassed by possible deportation. Whatever flashes of sorrow she felt for her husband were brief and bright. Whether he was guilty or not, he had hurt her world—hers and her babies'.

For Ruth Paine it was a new and exciting existence. Within the span of one afternoon, she had been whirled up and out of the drab life of dirty diapers and high chairs and the Book-of-the-Month Club and set onto the edges of the story of the century. She was a celebrity. Policemen, reporters, and photographers hung on her words, her opinions, and she, Ruth Paine, Quaker, was the umbilical cord between Marina Oswald and the U.S. government. Everything would turn out good in the end, because, to a dedicated Christian, it always does. To a woman hitherto stranded on the sandbars of marital discord, this was an exciting ride down the rapids.

Robert Oswald understood Marguerite. The dear, sweet, protective mother would fall asleep on time and wake up on time. Guilt or innocence could be secondary to the last, the final chance to stand before a derisive world and take a bow. All of the earlier bitterness, the lack of understanding by others, the snubs of the fine ladies in the shops where she worked, the marrying and losing of men, the teenage loss of three sons—all of it might be prelude leading to this great shining moment when once more she could storm into the capital city of Washington demanding to see the high and mighty and assert herself as a mother. If, in addition, there was a dollar to be earned, it might be a most attractive buck.

Marina felt pity for the grandmother tonight. The poor thing hadn't known that she was a grandmother again until she arrived at the police station. The first vague distrust of Ruth Paine was in Marina's mind. The young Mrs. Oswald, in her muteness, felt that she had been thrust aside. Ruth was too gay and buoyant a mouthpiece for Marina. Her personal opinions were strong and sometimes meticulous, as in her verdict that all guns were black or dark brown; they all looked alike in their deadly venom; one could not differentiate between one with a telescopic sight and one without—nor could any amount of interrogation alter that opinion.

There was female accommodation in the hair dryer, but, underlying it was Marguerite's determination that the euphoric Mrs. Paine would not be permitted to answer the questions and resolve the mystery. Grandma would cultivate the confidence of her daughter-in-law only so long as both of them agreed upon who was running the show. Disagreement, which was bound to occur, would cause Marguerite to revert to her original notion that Marina was nothing more than a "foreign person."

In a little while, all the lights would go out in that house. All of them.

There was irritation in Jesse Curry's telephone ear. For a couple of hours he had been receiving calls from men of importance in Dallas County which began: "I received a call from Washington . . ." or "I got a call from the White House . . ." The message was the same: please turn the evidence over to the Federal Bureau of Investigation for a day or two. The chief did not think he should. His captain of Homicide did not think he should. Who called from Washington? "Well, somebody pretty high up . . ." or "I'm not at liberty to disclose a name . . ."

What most people did not seem to understand, Curry kept saying with exaggerated patience, was that this was a Dallas County case. "It is a straight homicide as far as I'm concerned. Fritz says we need the evidence here. I have to back my men." The calls kept coming. They came to Captain Fritz, who deferred to his chief, and to Curry, who deferred to the captain. One call came in to Fritz from Henry Wade asking whether the prisoner was to be transferred to the county jail for security. Justice of the Peace Johnston had ordered him handed over to Sheriff Decker.

Fritz understood the order but had no intention of complying. "We don't want to transfer him yet," he said. "We want to talk to him some more." His opinion was that Oswald would neither crack nor confess. And yet, as the man in charge of the police work, he owed it to his department to keep tapping at what he could see of the iceberg in the other chair in hopes that one final rap would split it.

The district attorney backed off. He was pleased that the department had made so much progress in the face of official confusion. Wade was seldom sure who was in charge: Dallas Police Department, Secret Service, FBI, Texas Rangers, the United States Department of Justice. All of them were busy asking questions, exercising prerogatives, assuming responsibility. Each had an interest a priori which was not to be denied. There was no hostility between the groups, although the Dallas department felt frustrated and irritated with outside attention. Some, in ironic tones, asked if Governor Connally was strong enough to call out the National Guard.

Progress on the case was steady and inexorable, a masterful amount of minutiae which kept growing, kept pointing directly at Lee Harvey Oswald. Fritz didn't need help to pin this case on this man. He was convinced that, in a day or so, he would have sufficient evidence to convict him of either or both of the murders. The captain emitted an aura of shaggy modesty, but he would be foolish to want to share this one with any other agency. All he asked, at this hour, in addition to the evidence he had, was to connect the spent bullets to that rifle and that revolver, to prove that A. Hidell had bought both guns, and that Alex Hidell and Lee Harvey Oswald were the same person.

Fritz was an old-line, experienced officer. All the other homicides were preparatory to this one. He would like to have wrung a confession from Oswald, but it wouldn't be necessary, he thought, to the successful prosecution of the case. He had his man "coming and going." As soon as he could get a justice of the peace, Fritz would formally charge Oswald with the assassination. The captain had confidence in himself and in his men.

Additional help might have been welcome. A naïve police department, beset by sectional jealousy, would decline the services of the Federal Bureau of Investigation. Curry and Fritz were aware that, after they had catalogued and tested evidence, there would be no use for it in Dallas until the day of trial. They had no intention of lending it to any agency. They resisted the phone calls in concert, and this represented one of the times they were in agreement.

Fatigue was not obvious on the faces of the three doctors. The work was exacting, and weariness was there as they strode around the corpse, making their observations, nodding agreement, trying too strenuously not to overlook any aspect of the body and its wounds. The trio of white ghosts walked the post of the dead with such infinite care that their signed conclusions were predestined toward error. When mistakes are most costly, careful craftsmen are the first to pay.

With the chest and belly open, Humes, Boswell, and Finck examined the lining of the thoracic cavity and found it "unremarkable." The organs were removed one at a time to be washed, weighed, and examined for grossness. The man on the table, in spite of his chronic back pain, was a healthy human. The coronary arteries were smooth-walled and elastic. In the abdomen there was no increase in peritoneal fluid. An old appendectomy displayed a few minor adhesions between the cecum and the ventral abdominal wall.

Fresh bruises were found on the upper tip of the right pleural area near the bottom of the throat. There were also contusions in the lower neck. Humes called his doctors away from the table and asked the Navy photographer to shoot additional Kodachrome pictures. The lens picked up a bruise in the form of an inverted pyramid. It was a fraction short of two inches across the top, coming to a point at the bottom. A few of the contused neck muscles were removed for further examination.

The autopsy was complete. The men of medicine had been on their feet a long time. The outer covering of the body was sewed in place again. The Navy passed the polite word—this time to Admiral Burkley—that it had completed the autopsy and declined to do the embalming. The Navy photographer passed the cassettes of film to Roy Kellerman and waited for a receipt. A long sheet was floated across the body. Enlisted men began to untie the backs of medical gowns and the doctors peeled gossamer-thin gloves from their fingers.*

Witnesses stood. They stretched their limbs. The last act was over, but the spotlight remained focused on the long sheet. The long bony feet stuck out. Greer looked and remarked that they were amazingly white. The toes turned slightly outward. He wondered why the vision struck an echo in his mind. He had seen those feet looking like this at Parkland Memorial Hospital. Trauma One. A man could be forgiven for asking himself how long ago that might be. No one noticed the enlisted men hosing down the floor.

The dead man was scarred, but so were the living. He would not remember his scars, but they could not forget theirs. Greer, solid, strong, middle-aged, had years of dependable work in him but the thought had crossed his mind to get out of the Secret Service and spend more time with Mrs. Greer, who was not strong, and a growing son, who would appreciate male guidance. Kellerman was granite, but for years to come his mind would freeze in immobility when he thought of November 22, 1963. He could not force himself to discuss the day.

Some Federal officers would quit within the next month. Others would ask for other assignments. A few became embit-

---

* In the morning, Dr. Humes called Dallas and spoke to Dr. Malcolm Perry of Parkland Memorial Hospital. The Bethesda doctor identified himself and said that he could not discuss the findings of the necropsy, but he would like to know more about the tracheostomy. Perry said that when President Kennedy arrived at the hospital, the wound in the lower anterior portion of the throat was noticeable. Bloody bubbled air was standing in it. There was no time to dwell on it or even to measure it for size. Medically, breathing had to be restored at once if the patient was not already dead. The hole in the throat was enlarged and incised by Dr. Perry, who stuck a tube in it. What had appeared to be a surgical incision was proved to be an exit wound.

tered. Rowley, the Chief, would remain on to defend his men and expand the Secret Service, even though he had sustained the greatest loss—losing "the boss"—and the private knowledge that, at home, a most attractive daughter was losing her sight.

In the nation's capital, the day began its final hour by collecting the ticking seconds and forming them into sedate minutes. They were stacked neatly, one on top of the other, and they should have brought a resignation to the people, but the country remained in a continuing trance, shaken and disbelieving. In Lafayette Park a bronze general stared across the broad boulevard to the big light on the portico of the White House. Beneath him people stood in groups, staring as he stared.

It was now a blustery night. Gusts combed the crisp leaves from the branches. Behind the White House, a claret-colored light winked solemnly near the top of the Washington Monument. The people remained in Lafayette Park, and more passersby paused to look, to whisper questions, to wait. They knew, even if he came home now, they would see a vehicle, perhaps a box lifted from the back of it, but the people remained steadfast and cold, not in morbid curiosity, but to get a glimpse of the box and convince themselves that it had really happened. This would tell them that he was dead.

The salads glistened under glass. The ham, the corned beef, the imported salami reposed on wooden boards. In wells, the potato salad, the cole slaw, the tuna fish salad reposed. Behind the counter, John Frickstad waited for the customers who sit and talk and nibble. This, of all nights, was a time when Dallas could be expected to frequent the neighborhood restaurants—the pizza-pie pickups, the short-order beaneries, and the German-Jewish delicatessens. The housewives were too depressed for full-scale work in the kitchens.

A few people sat at tables, sipping coffee, chewing on Danish pastry, but most of the orders tonight came by phone. Phil's Delicatessen at 3531 Oaklawn was patronized by those who enjoyed a thick meaty sandwich on Jewish rye bread. The tables were peopled, but business was not overwhelming. The counterman glanced at a table where five young people sat idling the time away. They had a *Dallas News* and Robert Sindelar read parts of the big story aloud. He was a student at Southern Methodist University, and he and his friends, having no other place to go, had spent two hours dissecting the crime. Dennis Martin had an opinion; so did Rita Silberman, Bill Nikolis, and Marguerite Riegler. The repeated phrase—even after the dishes had been removed—was ". . . but here in Dallas!"

A stout, middle-aged man walked in, but they did not notice him. Frickstad did, because he had filled orders for Jack Ruby for two years. The nightclub owner appeared to be in a hurry. He walked over to the table where the five young people were in conversation and yanked the newspaper from Sindelar's hand. "Excuse me," the stranger said. "May I borrow the paper?" Young Mr. Sindelar was thinking of something to say when he saw the stranger riffle through the pages, study something, and set the *News* back on the table.

The five stopped talking. The man shoved a gray fedora back off his forehead and walked into a phone booth. A coin clinked and he sat with the door open, dialing. He asked someone a question, then said: "I'm at Phil's Delicatessen. If you need me, I'll be here a few minutes." He came out and turned to the counter. Johnny Frickstad said hello and asked if Mr. Ruby didn't think the assassination was a terrible thing. The nightclub owner was giving the counterman part of his attention. Yes, he said. It was terrible. Terrible.

Then he strode back into the phone booth and dialed RI 8–9711. The city hall operator came on and Ruby said: "Homicide and Robbery." The phone was picked up by Detective Richard Sims. "This is Jack Ruby," the voice said. "I know you guys are working late. I have some sandwiches for you." Sims said: "Thanks, Jack. We have been eating in relays, but we're wrapping it up now." "Oh." "Yeah, Jack. We won't need any sandwiches now." "All right," Ruby said and hung up.

The entrepreneur bounced out of the booth and said to Frickstad: "Give me eight corned beef sandwiches with mustard. Give me eight black cherries cold and two celery tonics. Also I want three cups of butter, a half loaf of Jewish rye, and some extra pickles." The counterman went to work. Pickles and potato salad were supplied free with each sandwich. Jack Ruby watched a moment as Frickstad began to slice the corned beef, the slices curling away from the electric knife onto a piece of paper. "I'm taking this to the disc jockeys at KLIF," Ruby said. "They're working late." The counterman kept working. "I still don't know how I'll get in; they lock the station up. But I'll get in with the sandwiches."

He went back to the phone and dialed someone and said: "If you want me, I'll be at KLIF. If anything should come up . . ." There was a pretension toward busyness, an important man with important connections. Mr. Ruby returned to the students' table and asked if he might see the paper again. Mutely it was given to him. The pages were flipped, and the stranger murmured: "I own the Carousel and Vegas clubs. I want to see if the ads appear as I ordered them." The students sat around the big table, looking up at the man. "My clubs," he said, "are the only two closed on account of the assassination."

He found the page, turned the paper inside out, and displayed the one-column advertisements enclosed with black borders, which said:

Closed
tonight and Sunday
CAROUSEL
Closed
Vegas Club

The customers took a look. "Nobody else closed," Ruby said. The tone was modestly triumphant. It might have meant more to Mr. Ruby than "class"; he had personally checked the competitive strip joints which were open. He had outmaneuvered his competitors. Ruby's mood was warm and friendly. He returned the newspaper for the second time and said: "Maybe I'll give you people free passes to my club." This generosity covered a small cover charge payable at the door. Bill Nikolis asked if the assassination would affect Dallas. "It will affect the convention business," the businessman said. "No doubt about it." Bill Nikolis said that he knew a girl who had entertained at the Vegas.

Ruby didn't ask for a name. He put on a paternal smile and said: "I don't think you people are old enough to go to my clubs." There were no free passes. He swept away from the table, adjusting his black-rimmed glasses and pulling the felt hat down. His finger beckoned to the counterman. Frickstad had the order in two big bags. "That will be $9.50 plus tax," he said. Ruby dug in a pocket and pulled out a roll of bills. He paid and took his change.

The counterman followed the customer out into the parking lot. A cringing dog sat on the front seat. The bags were placed in the trunk. Frickstad waited a moment, and Mr. Ruby took a card from his pocket and scribbled something. "Here," he said. "This card will admit you free to either of my clubs."

The White House rested, gathering strength for the trial to come. In the kitchen, a few waiters and chefs smoked and talked. The mounds of sandwiches had been prepared, and slightly damp napkins made them appear glacial. Someone had to stop the sandwich-making. The smell of rich Navy-style coffee permeated the kitchen. In the chief usher's office, between floors, a Negro with a neatly trimmed white mustache dozed on folded arms. The State Dining Room was dark. So was the Blue Room and the Red Room. On the second floor, the Yellow Oval Room was alight; it was assumed that the family would meet there.

On the walk around the White House, uniformed patrolmen made their rounds. The big fountain on the South Lawn still tossed colored prisms of water toward the sky. Secret Service men stood in the shadows. A man came out of the diplomatic reception room with a box. A carpool number was called hollowly on the loudspeaker and an automobile drew up and the man with the box got in. John F. Kennedy's clothing was en route to Bethesda.

The rheostat was turned down in the President's office, and the light was faint, almost dim, from the little pitching green which Dwight D. Eisenhower had left. On the second floor, Maude Shaw slept in the little room between the children's quarters. Near the elevator, a Secret Service man sat at a desk with a shaded lamp. The late papers had been stacked against the President's bedroom door. In the sitting room, the smiling, confident faces of the Kennedys gleamed from silver frames. On the left, the door to Mrs. Kennedy's bedroom was shut, but then so too were the doors on the other side which led to the intimate Kennedy dining room and the service kitchen.

The center of action remained where it had been—Ralph Dungan's office. There, in a blaze of light, the men chipped the details from the ugly rock of a funeral. There were responsibilities to being part of the Clan Kennedy: intense illogical loyalty; the swift flight of youth which owes nothing to its predecessors; a world of instant "Yes" or "No"; a private planet of "Us" versus "Them." There was a premium on intelligence, of knowing that all hands were "on the ball."

The clan never shone to better advantage. Within a few hours, a vast and nebulous situation was being resolved. Orders were shouted; suggestions were negated; the elder statesmen limped through the door to speak, to sit, to leave. The polished head of Adlai Stevenson came in, the pouched eyes defeated for the last time. The younger men swept around Sargent Shriver's desk with speed. Some said: "Hello, ambassador." Others said:

"Hello, governor." John K. Galbraith, tall and stooped, came in to finger the faces with his poetic eyes and leave.

The minority leader of the United States Senate, Everett Dirksen, walked into the office shaking his leonine head and, intoning like a church organ, said: "I still can't believe it has happened. I am stunned, shaken." There was no response. "Thank God there are those like you who are carrying the burden at this terrible time. Is there anything at all I can do to help?"

Nothing. He could leave. Stevenson worked his way toward the front of the couch and stood. He left. Dirksen left. In the outer office, four young women typed and retyped the decisions of the Clan Kennedy. The precise hours and days were resplit to accommodate every homage due a President. To remain on the list of those invited to the funeral became a matter of surviving a sifter. Some who were on the list were wiped off. Names were shouted and decisions made. Dungan read the list once more; if he heard the word "Yes," the name remained. If someone yelled "No" and gave a reason, or if Shriver said "No" and gave none, a line was drawn through it.

"Barney Ross?" This was a comrade from the President's PT 109 days. "Yes." "How about Billy Graham?" "Billy considers himself a close friend of the President." Dungan said: "No." Shriver remained silent. The well-nourished, jaded figure of John Bailey, chairman of the National Democratic Committee, a Connecticut politician, suggested names. They were old-line wheelhorses who had helped elect John F. Kennedy.

Each name in turn was shot down by the word "No." At last, Sargent Shriver's patience became lacerated. "John," he said, staring at Bailey, "we are not trying to return political favors here tonight. We are trying to ask only those people who we know were personal friends of the President." It was done in simple, cutting words. Ordinarily, no one would address a national chairman in any manner except the deferential. Without brutish practical politics, no dreamer can be elected. Without election, the noblest aspirations of the candidate lie in a dark part of a desk drawer. Bailey had helped elect Kennedy, but Bailey was one of "them." Listening to the words, David Pearson, a journalist from Florida, thought of Bailey: "A magical facility for saying wrong things at right times."

The names are barked without favor or sentiment and they fly circuitously around the office waiting for one of "us" to shoot it down. Each name must be prepared to submit to this screening and rescreening. Eight justices of the United States Supreme Court—all "them" are to receive a blanket invitation to arrive in a body; one, Byron White, a former football player (an "us"), will receive a personal invitation.

The car was on Stemmons westbound, and there was relief in the back seat. Wesley Frazier had answered all the questions of the police as honestly as he could, and he and Linnie Mae were homeward bound to Irving. The detectives sat in the front, and they were tired. Rose and Stovall had been working since 8 A.M., and it was becoming difficult to concentrate. The Reverend Campble knew that Frazier and Mrs. Randall had done their duty as good citizens, telling what they knew about Oswald, adding nothing.

The young man was happy to be free. He realized that just being friendly with a man like Oswald could lead to trouble. For a young man who had specialized in minding his own business, it was a frightening experience to be taken to headquarters, to be asked about curtain rods which might turn out to be a rifle, to be regarded as a buddy of Oswald's when the facts pointed to the neighborly Linnie Mae helping Oswald to get a job at the Texas School Book Depository. Wesley Frazier had added free rides on the weekend.

The car was approaching the Irving Boulevard exit when headquarters called. Detective G. F. Rose picked up the microphone and acknowledged call letters. Headquarters asked that the car turn around at once and bring Wesley Frazier and his sister back to Dallas. The driver slowed and made a U turn. No one asked why. Frazier and Linnie Mae couldn't think of any questions which the police might have forgotten to ask them. There was always a vague danger that Lee Oswald might have implicated Frazier in some way, but no one wanted to dwell on that. Wesley couldn't see how anyone could implicate him in anything, but Oswald was such a strange person—even more frightening now—that no one discounted the notion that he might try to drag his benefactor down with him.

Rose said he was sorry. It was an order; he didn't think it would amount to much. The car seemed to get back into the basement of City Hall much faster than it got out. The witnesses were taken up to the bedlam of the third floor, and detectives helped to pry a path for them. The two people sat with their Baptist minister. He, too, was trying to dissipate the gloom by reminding them that they had nothing to fear.

A detective came in, looked at Wesley Frazier, and said: "You got any objections to a polygraph test?" Another policeman explained that it was nothing; you sit in a chair with a blood pressure cuff on and they ask some questions. If you're telling the truth, the blood pressure remains pretty steady; if you're lying, it goes up. Frazier looked at his sister. He said he had nothing to hide.

"Good," Rose said. "It won't take long." They led the boy up the stairs to the Identification Bureau. Captain Dowdy said that the man who conducts the polygraph tests was at home. They might have to wait. He phoned Detective R. D. Lewis. The policeman was willing to come in, but it would take an hour to get back to headquarters. Dowdy told him to come in. A policeman was placed with Wesley Frazier. "Son," a cop said, "I think you're going to have to wait an hour. You might as well relax." "What's this test like?" Wesley Frazier said. Nothing to it, he was told. You just relax and tell the truth. He thought he already had told the truth. He couldn't imagine anything further he could tell the police, but it was obvious that whoever wanted him back here wouldn't want a truth test unless he was suspected of not telling it.

On the third floor, Frazier's friend was submitting to the last of the special interviews. Word had come to the Dallas office of the FBI that, so far, a biography and a physical description of Lee Harvey Oswald had been omitted. Washington would like to have this material. A twenty-three-year veteran, Manning C. Clements, asked Agent James Bookhout about the matter. Bookhout had been around all day, and he couldn't recall anyone asking for the vital statistics of Oswald's life or drawing up a physical description.

Clements, who lived in Dallas, asked Captain Fritz if he had any objection. The answer was no. Manning Clements introduced himself to Lee Harvey Oswald, and the prisoner had no objection. The interrogation about personal detail started, and it was obvious to anyone peeking through the glass partitions that the prisoner had outlasted the police. He was still calm, in control, and they were worn. At times, he seemed cordial. He was willing to assist Mr. Clements to draw up a family portrait with vital statistics and street addresses, cities and schools and jobs all tossed in.

The questioning went on for a half hour, when the prisoner was taken out to the restroom. When he left with two detectives, Manning Clements looked over the desk of Captain Fritz and saw a wallet. He flipped it open. It belonged to the prisoner. Mr. Clements removed the cards, one by one, and copied the information. By the time Oswald returned, the wallet was back on the desk. The prisoner had lost his desire to cooperate. He became peevish and argumentative. The FBI men remained polite, but the interview was over.

Down the street a few hundred yards, Robert Oswald emerged from the coffee shop of the Statler Hilton Hotel. He walked through the bright lobby, vaguely seeing the small group of guests clustered in front of a television set. There was a newsstand which featured black headlines. Some newspapers carried black-bordered photos of the President and, nearby, a photo of Lee with a bruise under his eye and his mouth wide open. There were souvenir shops, too. Perhaps a hundred years from now, some such shop might feature a miniature rifle with the words "Souvenir of Dallas."

Robert didn't feel that he was Abel to Lee's Cain. The thought had not occurred to him because it wasn't in his nature to equate matters that way. He decided to remain in Dallas tonight. There was a sickening feeling in his belly that his brother was in deep trouble, deep trouble. An older and protective brother can afford to address justice in a demanding tone when he is certain that a mistake has been made, but Robert had left Fort Worth this afternoon whimpering and afraid.

Nothing that had happened since ameliorated that sensation. In that glassy room, Lee's sauntering, carefree attitude worried him. His brother was enjoying his predicament. Lee could have asked for help when they spoke. He had not. Even the civility was cool, as though Lee felt that he was in a different sphere from the rest of the family and might be pleased if he drew no further attention from them.

Robert Oswald strode to the desk. He asked for a single room. The clerk swung the register around. The pen was ready. A man with no baggage, a man who will be asked to pay for a room in advance, such a man can afford, for a moment, to seek respite in signing "John Smith" on the register or "James Jones." Robert Oswald decided not to retreat. He signed Robert Oswald, placed his home address underneath, and took his room key. He didn't even bother to examine the expression on the clerk's face when the man looked at the signature and said: "Good night, Mr. Oswald."

One question remained: the wound in the back of the neck. It could not be resolved now. Humes knew this, and he was in no hurry. The hour was late and he was half-persuaded that a bullet, reported found on a stretcher in Dallas, could be the one which had inflicted this wound and, that when manual respiration of the chest had been instituted, the pellet had fallen out. It was a possibility. Neither Humes nor Boswell nor Finck could be sure tonight what had happened.

They had used a lot of time making certain of their findings. They had studied that body with great and minute care. The X-rays were more than would normally be taken; the color photographs; the black and white photographs; each doctor had placed a finger into that small hole at the base of the neck; resistance was felt between the first and second knuckle. The FBI men, Sibert and O'Neill, had been ordered to draw up a summary of their observations and, even though they had no medical qualifications, they could not wait for word from Parkland Hospital.

Their report would state: "This opening was probed by Dr. Humes with the finger, at which time it was determined that the trajectory of the missile entering at this point had entered at a downward position of 45 to 60 degrees. Further probing determined that the distance traveled by this missile was a short distance inasmuch as the end of the opening could be felt with the finger." The use of the phrase "end of the opening" was a conclusion. No one had called it "the end of an opening."

It is one thing to draw attention to a mystery; it is another to resolve the mystery without qualification. Secret Service Agent Roy Kellerman followed the agents to a similar conclusion as a result of the superficial findings of the physicians. "There were three gentlemen who were performing the autopsy," he wrote. "A Colonel Finck—during the examination of the President, from the hole that was in his shoulder, and with a probe, and we were standing right alongside of him, he is probing inside the shoulder with his instrument, and I said: 'Colonel, where did it go?' He said: "There are no lanes for an outlet of this entry in this man's shoulder.' "

Doctor Humes, in his preliminary notes, courted the same easy conclusion: "The pattern was clear," he stated. "One bullet had entered the President's back and had worked its way out of the body during external cardiac massage, and a second high-velocity bullet had entered the rear of the skull and had fragmentized prior to exit through the top of the skull." By the time Humes was ready to write his official findings, to be signed by Boswell and Finck as well, his opinion of that neck wound had been reversed by information from Parkland Hospital:

". . . The missile contused the strap muscles of the right side of the neck, damaged the trachea, and made its exit through the anterior surface of the neck. As far as can be ascertained this missile struck no bony structures in its path through the body." The important phrase, this time, is *through the body.* It is to be doubted that any physician, encountering strap muscles which had reclosed the lane after opening it for the neck bullet, could have divined that the tracheostomy, so plainly surgical on the front of the neck, could have started out as a small exit wound. But then it is doubtful that many physicians would have permitted themselves to be badgered into a summary opinion.

The event is rare, but it sometimes happens that animals can, if they persevere, overcome their keepers. The reporters had been pressed back against the corridor walls to form an open lane, and they had closed it at once. The situation had an element of danger. The writers were being pressed hard by editors for a story on Lee Harvey Oswald. The police knew that it would be poor tactics to reveal the case they had against Oswald; they had no right to try the case in the newspapers. The reporters pressed the police with louder and louder demands. The target of their venom had been the prisoner; now it was the police.

No detective shouldered his way down that hall without being pelted with a hail of questions. Captain Glen King, in charge of security at police headquarters, realized that cooperation had been too complete; Assistant Chief Batchelor had ordered police at the foot of the elevator to check newspaper identification cards and to issue Dallas cards because, in the morning, a man without a Dallas card would not be admitted. Curry had bucked the line several times and been mauled orally by the tigers of the press. A local reporter apologized to the city manager: "It isn't us; it's the out-of-town press." Forrest V. Sorrels of the Secret Service had the sober impression that the press had taken over police headquarters.

The editors of the morning newspapers were on solid ground. They could assume that all readers were aware of the death of the President. In the parlance of the trade, he had died "for the afternoon papers." The morning newspapers could not headline: "KENNEDY ASSASSINATED!" They required an overnight lead, something akin to:

DALLAS COMMUNIST ARRESTED
FOR MURDER OF PRESIDENT!

The small bits and pieces of material filtering into newspaper offices all over the world supported this story, but the reporters could not find enough quotable material to support it. Many did not know that Lee Harvey Oswald already had been arrested for the killing of Officer J. D. Tippit. Nor were they interested. The killing of a Dallas policeman was a local story; Kennedy and Oswald were the big news and, as the hour passed 10 P.M. and morning editions went to press, the reporters became louder, more unruly. Either the police had a case against Oswald or they didn't. They demanded a statement. They demanded to know if he was charged with the assassination—yes or no. The reporters were caught between the inexorable pressure from the city desk and the immovable Will Fritz.

The Dallas Police Department, especially the upper echelons, was imbued with an ardent desire to cooperate with the press. These officers knew how succinctly the man behind the typewriter could make a law enforcement body appear ridiculous. Curry, King, and a privately hired press agent labored to divert the animals by feeding them scraps. It didn't work. Henry Wade returned to headquarters after dinner to make certain that no one would charge "international conspiracy" unless there was one. Wade could not believe that a man as big as he was would struggle to walk fifty feet. He had another fear: that the Dallas police, in their anxiety to remove this horrifying crime from its shoulders, would file a murder charge against Lee Harvey Oswald without having sufficient evidence for Wade to prosecute successfully.

Wade got to Fritz and brought him into an anteroom. The district attorney towered over the chief of Homicide. Fritz began to recite the case he had on the Tippit murder and Henry Wade said he wasn't interested. If Fritz was going to file a charge in the Kennedy case, Wade wanted to know just what evidence was in the hopper. The captain would have been in an excellent position if he had ordered someone like Jack Revill to summarize the material in a written report, but the best he could do was to rely on his memory. He explained the curtain rod story; the affidavit of Mrs. Oswald that her husband kept a rifle in the garage; three witnesses on the fifth floor who heard the rifle blasts overhead; the cop who caught Oswald in the commissary; Brennan and Euins and others who had identified Oswald as the man in the sixth-floor window; the testimony of Earlene Roberts that he had hurried to his room to change his jacket; the revolver; the shooting of Tippit; the capture in the Texas Theatre; the finding of the rifle on the sixth floor, the— "All right," said Wade. "Do you have anything that hooks him up to anybody else?"

No, the captain said. They had a neighbor up on the fourth floor, waiting to take a polygraph test, but Fritz didn't feel that this boy was part of a conspiracy; he just wanted to make sure he was telling the truth. Was it possible that Oswald was a member of a Dallas Communist Party cell? No, the FBI had been cooperating on the case, and it was possible they had a plant in that cell, and the word was that Oswald never joined anything. He was a little bit unbalanced on Fidel Castro, and Fritz believed that Oswald may have organized his own unit of a Fair Play for Cuba Committee; he may even have signed the membership cards with the name Alex Hidell.

The district attorney was satisfied. The Dallas Police Department was not dumping a weak case into his big lap. He conferred with his assistant district attorney, William Alexander. Wade heard that there was a map, found in Oswald's furnished room, depicting the parade route in relation to an X marked at the School Books Depository building.* The cops had the blanket in which the rifle had been wrapped. They also had a palm print on the rifle which seemed to match the grooves and loops of Oswald's right hand.

At the elevator a reporter stalked a detective for information. The cop said he had spent the day running from Sheriff Decker's office back to headquarters with signed affidavits. The writer said that a little information had been dribbling from Glen King's office but it was not the stuff about which leads are written. The assassination had, in effect, brought all the nuts out of the woodwork. They phoned by the score from all over the country, and everyone had a means of finding out if Oswald had really done it. The funniest was the old lady who reminded the police that a partly eaten chicken sandwich had been found on the windowsill of the sixth floor. The sugges-

---

* Untrue. Oswald appears to have spoken the truth when he said that the map was used in conjunction with job opportunities published in the newspapers. His habit, before he secured work at the Depository, was to mark off several places where a job might be obtained and then try to figure the cheapest way to get to them by using a bus and free transfers. The X at the Depository building turned out to be the final place he looked for work. There were no street markings showing the route of the President's motorcade.

tion was to examine Oswald's stool for the next few days and, if chemical analysis detected chicken, they could be sure they had the right man. The policeman who took the call said that this would make Oswald the chicken-shit assassin.

Curry learned that Wade was conferring with Fritz. He left his office and lunged through the hall. The chief was not a big man, but he had a big voice. "When are we going to see Oswald?" one man shouted, backing in front of the officer. "When are you going to let us talk to him?" someone else shouted. "Has he admitted anything yet?" "Come on, chief. We don't have all night. What's the story?" They retreated before him, the microphones nodding like holy water censers. This time, he offered no information. He was on his way to discuss the press and publicity.

The exchange of ideas did not require much time. The men in the hall could see the three, inside Fritz's anteroom, debating the matter. Fritz entertained the least interest in the plight of the press. He and his men were plodding with secrecy and luck up and down this case, and they desired to continue to walk alone. An observer might surmise that they would prefer to dispense with all assistance—from Sheriff Decker to Secret Service and FBI. They ignored their own chief of police to a marked degree. Fritz was in favor of sitting tight on the whole case. The chief felt that he had to "live" with the press and it was up to him to decide how much information could be given out at this hour without prejudicing the case. Something would have to be done. The hour was late.

The district attorney was in favor of making some sort of statement to keep the "boys" happy. He was not inclined to publicize the items of evidence but, if Fritz was about to file against Oswald in the assassination, there could be no harm in announcing it.

The three walked into the hall together. The reporters in front began to shout to those far in the rear. Floodlights switched on. Sound gear began to whir. Humans began to press down from both ends of the hall toward the center. Thayer Waldo, a veteran reporter of the *Fort Worth Star-Telegram,* glanced around and assured himself that there must be two hundred and fifty men crushed in this space.

Fritz, laconic, blinking behind those trademark glasses, began to speak in a deep whisper. A wave of shouts swept backward: "We can't hear you! We can't hear you!" The captain glanced at the district attorney. Henry Wade, who had a booming courtroom manner, announced that Lee Harvey Oswald had been formally charged with the assassination of President Kennedy. The news was electrifying, and some of the stringers left at once to file the flash. Others shouted: "Henry, we can't hear you! We can't hear you! Can't we hold this someplace else?"

There was a whispered conference. The chief faced the microphones and the sea of heads peering over a sea of heads. Someone asked: "Has he confessed, sir? Has he made a statement?" The chief, who had planned to offer a sop to the press by suggesting the police assembly room as a spacious place in which to make the announcement, found himself responding to questions. He listened to each one, looked down to think, then squinted into the lights with an answer. Curry was caught in a dress rehearsal to a press conference. Although Fritz had maintained his independence and had worked the case alone, it was the chief who was now seizing public relations by the horns, and it was Curry whose face would adorn millions of television sets tonight. "He has not confessed. He has made no statement. Charges of murder have been accepted against him."

The voices, of different pitch, different intent, piped up from all over the hall. There were polite questions, incisive ones, sarcastically framed queries, inane enigmas. "Any particular thing that he said that caused you to file the charges regarding the President's death against him?" "No, sir," said the chief. "Physical evidence is the main thing that we are relying upon." "Can you name that physical evidence?" Before Curry could respond, another question was flying his way. "When will he appear before the grand jury, sir?" "I don't know."

"Is that the next step?" Curry nodded. "The next step would be that." Henry Wade stepped back to listen, to stand by in case Curry began to impinge on Oswald's right to a fair trial. Fritz looked at all the faces as though he had not seen such a collection before. The chief was at stage center. An old reportorial axiom is that if you keep asking questions long enough, the victim will respond to an explosive one. The legal watchdog, Henry Wade, was sucked into the oral vortex and disappeared without a trace.

"Do you think you have a good case?" Wade brazened the big lights and said: "I figure we have sufficient evidence to convict him." In a community where the utterances of the district attorney are accorded more respect than the denials of the prisoner, this could be considered prejudicial. "Was this, was there any indication that this was an organized plot or was there just one man?" The questions were being fired like rocks, and they hit just as hard. "We believe," said Wade, "there is no one else but him."

The caged animals were devouring the keepers. "Do you know whether he will be tried in federal court, county court, or where he will be tried because this was a presidential murder? Do you care to comment on the jurisdictional dispute which has been arising?" No jurisdictional dispute arose. Each agency which checked the statutes arrived at the same quotient: the case belonged to Dallas County and no one else. Wade set them straight: "He has been charged in the State Court with murder with malice. The charge carries the death penalty, which my office will ask in both cases."

The journalists realized that the long-awaited morning story was coming rich and pure. "Mr. Wade, within forty-eight hours do you think he might be before the jury?" The chief growled: "Let Mr. Wade make a statement." The district attorney mulled it for a moment. "There are still some more ends that we're working on. This will be presented to the grand jury just as soon as some of the evidence is examined. It will be examined today, tonight, and tomorrow. He has been filed before, filed in Judge David Johnston's, justice of the peace, Precinct Two of Dallas, and been held without bond on this case and the other case too.* It will probably be the middle of next week before it goes to the grand jury because of some more evidence that has to be examined by the laboratory."

Officials are swept through a press conference by exuberance. A difficult case well handled, on the way to solution within a few hours, sometimes establishes a rapport between the hunter and the hounds of the press. It sometimes results in an editorial lynching. The law enforcement element joins forces with the press to overstate the case. "Mr. Wade, can you tell us if he has engaged a lawyer?" The chief decided to field that question. "We don't know that," he said. "His people have been here but we don't—"

It was a snide response because Fritz—listening—knew that Oswald had not been able to contact the counsel of his choice. The captain did not correct the chief; Curry should have known whether Oswald had a lawyer. It would be Wade's business to know the name of his adversary. "His people have been here" constitutes verbiage which could give the impression that the prisoner had a lawyer; it could also be interpreted as meaning that his family had been with Oswald and that they were taking good care of the matter. The questions began to come faster; they tumbled over each other so that a man might hear three before he could frame words to answer one. "Are there any fingerprints on the gun?" "Mr. Wade, can we get a picture of him?" "Are you going to bring him out?"

The district attorney said "I . . ." "Could we get a room where we could get a picture of him?" "Can we get a press conference

---

* Untrue. Johnston had yet to read the assassination charge to Oswald.

where he could stand against a wall and we could talk to him?" "Has where he will be tried been determined yet?" Wade was beginning to weary. "It will be in the Dallas County Grand Jury," he said. "Where did he say he was when the President was killed?"

The interrogators, under interrogation, became confused. Fritz, Wade, and Curry turned their faces inward and whispered. The situation in the corridor was out of control. "Wade! Henry!" "Captain Fritz, can we go to the assembly room, sir?" The district attorney stood tall and ran his hand through the wavy gray hair. "We will get in a larger room here," he shouted. "That's what we're talking about." The three men returned to the *sotto voce* conference.

"What about the assembly room?" Wade looked at the mass of crushed faces. "Is that all right?" Captain Fritz said: "That's . . ." "Let's go down there," Wade said. This opened the door to new questions. "Will there be a way to make any pictures?" "—make pictures right then and there?" Wade became helpless again. "I don't know," he said. "I don't even know where it is." Someone asked if pictures could be made of Lee Harvey Oswald. This triggered an uproar of shouts. "I don't see any reason to take any picture of him," the district attorney said.

"Of Lee?" a television man said, incredulity dripping in the tone. "Yes," said Wade. The newsman looked behind him for confirmation. "Well," he said, "the whole world's only waiting to see what he looks like." The district attorney began to lose patience. "Oh," he snapped, "is that all? The whole world?" "That's all," the reporter said. "Just the world." An edge of acrimony set in. The chief, perhaps to continue an amiable relationship with the press, held up both hands and announced: "All right. We'll set it up in the police assembly hall in the basement for Mr. Wade to make his announcement if that's what you want." The three men returned to their whispering attitude, and Curry broke away to say: "And I'll have the prisoner brought down for you, too, if you like."

The television cameramen shouted: "Not right away! We have to get these cameras downstairs!" Others pleaded for additional time. In the rear some began an exodus by stairwell and elevator to stake out the best positions in the assembly room. Curry shouted over the bedlam: "We can do it in about twenty minutes!" It was too important to be rushed. Some shouted back: "We need more time. Don't rush it."

No one asked the prisoner if he was willing to participate. He had been told by Will Fritz that he need not respond to questions from the press nor pose for pictures. Whether Fritz was overruled by Curry or agreed to the press conference, the fact was that Lee Harvey Oswald had been promised to the press. He would be delivered when the newspapermen were ready. If the thought had occurred to the conferees, they might have guessed that any restrictions placed upon the press would be violated, once Oswald got into that room. It bordered on the ridiculous to ask the press to stand mutely and look at the prisoner. Conversely the prisoner could be expected to use the press conference as a propaganda platform for himself and his political notions. It is possible that Oswald looked forward to this confrontation with satisfaction. It was the keyhole through which he could squeeze from jail to the outside world.

The press scattered like the pigeons on the roof of the School Book Depository and, for a time, it became possible to walk through the third-floor corridor. The elevator door near the press room opened and three men stepped off. The flankers wore big red press badges on their lapels: "President's Visit to Dallas—PRESS." The man between them had none. He crouched low, pencil applied to pad, writing industriously as they strode past the policeman who was asking for credentials. They were not stopped, and Jack Ruby, the harmless aficionado of law and order, was once again on the edge of excitement. In the corridor he was not challenged. Now and then he heard a friendly call: "Hey, Jack. What are you doing here?" The rumpled figure smiled apologetically and pointed to the notes and said he was translating for the press of Israel.

No one believed it. He inspired no rancor. To the police he was Jack Ruby, the suppliant Jew who bought police goodwill with free passes, free drinks, good sandwiches, and hot coffee. Sometimes, when a policeman sat in the charcoal darkness of the musty nightclub, Ruby would try to set him up with one of the strippers. Like the sandwiches, it was a service.

The friends in the living room began to melt off. No one suggested that they leave. The clock on the mantle had two gold hands which began the slow climb toward twelve. The people said it was late. If they thought that the President and Mrs. Johnson might sit among them and discuss the grave events, they misjudged their hosts. Mrs. Johnson, in nightgown and dressing robe, slid into the big bed in the master bedroom on the second floor. With the blankets over the small body, Liz Carpenter could see the intermittent spasms of shivering. They hit and went away. They hit again.

Mrs. Carpenter had hurried home to get some nightclothes in case she was asked to stay at The Elms. The new First Lady didn't think it was necessary. She was "freezing," but she would be all right. She wished that Lyndon would get some rest, but she knew he wouldn't. Across the room, the television set repeated the assassination over and over, as though, by infantile logic, it might become more understandable or come out differently. Every few minutes the bright, good voice of John F. Kennedy could be heard, telling the crowd that no one cared how he and Lyndon dressed but that, when Mrs. Kennedy arrived late, she looked better than "we do." Mrs. Johnson pulled the quilted blanket over her head. Liz Carpenter departed.

The President stepped away from the dining room table, and his young friends came to their feet. There was a sharp thud outside the window. Mr. Johnson stopped. "What was that?"

he said. From the doorway, Rufus Youngblood said: "They're out there changing the house lines over to the White House number. I expect they're having a little trouble." The tall Texan shook his head. He could not believe that he was living in an era when someone would have to explain a noise in the dark. He had not time to think that he was the only President ever to have witnessed an assassination, and, no matter how high his courageous resolves, an unexpected noise from any quarter would trigger tension the rest of his life.

There were other factors. In the United States, a Vice-President lives in the silent womb of history. He is not expected to hear or feel or understand—only to grow. A sudden propulsion from darkness to light, from sufferance to power, from ridicule to majesty is too much for the intellect and the neurological aspects of a man. "We really have a big job to do now," Johnson said. He started upstairs and waved Valenti and Moyers and Carter to follow. The soft protests that they had no clothes, not a toothbrush between them, did not impress the President. His response to all such excuses remained the same: "We can talk about that tomorrow."

They were led into the bedroom, hearing the voice on television. The President began to remove his jacket and tie. He hung them on a wooden clothes tree in the center of the room. The three men noticed a small mass huddled on the right side of the big bed. They began to retreat. Mr. Johnson called them back. "We won't bother Mrs. Johnson," he said. His tone was pleading. "There is so much to do, so much to talk about. Sit with me for awhile."

The trousers were draped over the valet. The big form disappeared into the bathroom. The waiting men were apprehensive. They watched the carousel of catastrophe on the television set. The President came out in loose striped pajamas and slippers. "Let's get one thing settled," he said. At the door he pointed. "Bill," he said to Moyers, "you may use the bedroom on the third floor. Cliff, there's a second bedroom down this hall. It's Lynda's room." This brought a smile to Moyers and Valenti, who envisioned this huge Texan reclining under a frilly canopy, with pandas and dolls on the counterpane and Carter's ham-sized feet hanging in midair at the foot of the bed. "Jack, you take the bedroom next to mine."

Having secured the night arrangements, the President slipped into the left side of the bed. Adjacent to it was a night table with a lamp and a telephone. Valenti found a chair and placed it near the phone. He sat, continuing his notes as the others talked. Bill Moyers stood. Cliff Carter was on the edge of the right side of the bed at the foot. The President placed a pillow halfway up the head of the bed and composed himself with both hands clasped behind his head. Sometimes he listened to the television; at others he exchanged ideas and refined them. He asked Jack to please make a note to remind him tomorrow to get in touch with the Governors of the states, perhaps suggest a conference in Washington.

The address to the Congress would be important. Someone suggested that the emergency had such anarchistic possibilities that perhaps Mr. Johnson should make the talk before the Kennedy funeral. This, he thought, could be interpreted as unseemly haste, even panic. However, it was another note for Valenti's pages to be cleared with congressional leaders in the morning. He would accede to their decision on this.

None of the Johnson group was sure when the funeral would be held, even though Johnson had phoned his condolences to Sargent Shriver. Moyers had been Johnson's liaison all evening between the Executive Office Building and the White House, and it was Moyers' impression that the services would be held on Monday. If that were so, the new President could not address both houses of Congress until Tuesday or Wednesday, a long time to keep the nation waiting. Moyers thought that a press conference tomorrow, or a presidential statement, might cover the intervening days and build confidence among the people.

The bedclothes on the right side were turned back and Mrs. Johnson, still huddled in robes, stood with a pillow in her hand and murmured: "Good night, all" and left the room. Before she did, the President leaned across the bed for his nightly kiss, murmuring: "God bless you, honey" and returned to the conversation. "The important things now," he said, emphasizing with a finger hitting an opposite palm, "are a Cabinet meeting, a Security Council meeting, a White House staff meeting—maybe we ought to call those boys together at nine tomorrow morning, before the Cabinet meeting—and the address to both Houses."

" . . . and now," the voice of television said, "we bring you some biographical film clips of the new President of the United States—Lyndon Baines Johnson." Conversation stopped. The men studied the film material, dissected the commentary. The man watching from the pillow had a huge face, lined vertically so that the features tended to melt toward the chin. A stranger, unconscious of the true state of affairs, might assess the scene as a middle-aged man getting bad news from a consultation of young doctors.

On screen, they saw a skinny Texan with a big grin mounting the steps of the Capitol, a Congressman with a message. They saw him campaign at home for a seat in the United States Senate and go down to defeat by a fistful of votes; he made speeches in the Senate well; he posed with Franklin D. Roosevelt, his idol; he was with bald Sam Rayburn, the teacher; he worked in an office with his dark and modest young wife beside him; he took a dangerous step for a Texan by espousing civil rights legislation; he became a majority leader, the youngest in history; a myocardial infarction brought him down, and he was shown in Bethesda Naval Hospital; the mean and vicious fight between Kennedy and Johnson for the Democratic Party nomination; the need of each for the other on a winning election ticket.

All of it evoked memories. The man on the pillow was silent. The cameras were now at Hyannis Port, in Massachusetts,
a hedge and some homes on the edge of a surly sea. Had the
President's father been told? No one knew. His mother knew.
She said she would attend early Mass tomorrow. The camera
switched to that cold day in January when the young man announced, at his inauguration, that the torch had been passed to
a new generation. With confidence, he prepared to lead mankind to the stars.

The recollections had run out. The conversations were desultory.
At this hour no one could think of anyone to call on a telephone.
The Attorney General was relieved when word reached the seventeenth floor that the autopsy was over. He asked about the
medical findings but was told that they were tentative, mostly
involving a big head wound and a shot in the back of the neck.
The White House already had the news.

Shriver reminded Robert Kennedy that the family had to
go to Gawler's and select a casket and bring an embalmer to
Bethesda. This was an integral step which had been overlooked.
Mrs. Kennedy saw her brother-in-law approach, and she must
have known that another decision would have to be made. He
began by reminding her that the Secret Service had damaged a
handle on the Oneal casket. She said that she had no intention
of using that casket anyway. Mrs. Kennedy did not want to be
reminded of Dallas. She would not use that casket; she could
still see herself running after it, holding her fingertips on the
top, as official Dallas shouted that the President's body would
have to remain for an autopsy. The terror had been lodged
within her from the sound of the first shot, and nothing since
had lessened the pain.

Kennedy told her that Shriver had asked about funeral directors, and three names had been submitted as establishments
of good taste. Sarge had selected Gawler's, and, if she agreed,

someone would have to go there and select a casket. Did she have any ideas? She did not. Did she want any special person to select it? Before she could answer, the Attorney General said: "Kenny and Larry and Dave were very close to him. Why not send them?" Mrs. Kennedy agreed.

The three men were ready to leave. All they asked was some guidance on what kind of casket the family would like. And how about that embalmer? They were told not to worry; the embalmer should be waiting. He had been on notice for a couple of hours. They asked Clint Hill to have a car ready at the front entrance and have the driver find out how to get to Gawler's.

11 p.m.

Time always saps the excitement of the game in the mind of the winner. It was one hour past Lee Harvey Oswald's bedtime. In his opinion, he had won whatever contest the law had projected in this glassy office. Wit had been honed with sparks against wit for nine hours. They had sent their best—Fritz and Hosty and Kelley and Clements, postal inspectors and FBI and Secret Service and Dallas detectives. Their best had failed to crack his contention that he was guilty of nothing more than carrying a revolver into a motion picture theater. They had battered at the wall of his will and hurt their hands.

Most of the time the office had been filled with the authority of the law, standing, sitting, smiling, frowning at him. Lee Harvey Oswald had taken them on one at a time and all together. He had been smart enough to concede that he deserved the punch in the eye for resisting policemen; he had been bright enough to protest that, in spite of all their protestations, he had no lawyer, no trained opinion to counsel him, nothing but his own magnificent intellect to keep the hounds baying at the foot of the tree.

Long since the questions had become repetitious and Mr. Oswald tired of foiling them. Detective Adamcik walked into the little office and studied the skinny body, the thinning hair of indeterminate color, the lean face melting toward the long neck, and he said: "Where were you at the time this assassination occurred?" The prisoner did not deign to look up at the new opponent. He stared straight ahead, moving his wrists in concert with the handcuffs. Adamcik waited for a response. All he heard was the clamor of the press in the hall. The detective sat to guard the prisoner.

Will Fritz, having arranged a press conference without the consent of the prisoner, returned to ask a question. The captain had rank; he deserved a response. "You took notes," Oswald said insolently. "Just read them for yourself if you want to refresh your memory. . . . Now you know as much about it as I do." The captain could accept an affront. He blinked with big hyperthyroid eyes and proceeded with the details of the show to be staged in the assembly room. Neither the question nor the answer were, at this hour, of importance. He would like to have cracked this young man, but it was not a requirement to the prosecution of the case.

If the situation could be related to a card game, Fritz had the winning hand. He realized that nothing his adversary had could trump the cards held by the police. If Fritz felt disappointment, it was in the cautious play of Oswald. The game would close with the prisoner holding onto a few chips, and Fritz would like to have seen them all on the table as stakes. Oswald would have enough in reserve to play again tomorrow, and maybe another day. He might be turned over to Sheriff Decker with sufficient resources to continue the game in court.

The questions became fewer. The hour was late; Homicide and Robbery had been on duty since 8 A.M. It was becoming difficult for the men to concentrate. Some of the conversation between detectives was punctuated by "I don't know . . ." Men asked each other; "Did you take that statement?" and heard "Could be" or "I forget." Most of them had lost a concept of time. In penning reports, they had trouble trying to recall if something happened at 2 P.M. or 4 P.M. There was the additional difficulty of trying to recall the names of other detectives who may have been present. They lost patience with each other, the final fatigue.

Lee Harvey Oswald shook his head slowly, as though he was dealing with children. "How could I afford to order a rifle on my salary of a dollar and a quarter an hour," he shouted, "when I can't hardly feed myself on what I make?" He bent forward and his elbows perched on his knees. A detective asked the prisoner to explain the difference between a communist who is a Lenin-Marxist and one who is a pure Marxist. "It's a long story," he said sullenly, "and if you don't know, it will take too long to tell."

It was late for a phone call, but the ring at 1316 Timberlake in Richardson was monotonously insistent. Gregory Lee Olds decided that it was easier to succumb to the demand than to count the rings. The man on the other end of the wire was a board member of the American Civil Liberties Union. He asked if Mr. Olds knew whether the civil rights of Oswald had been protected. Olds said he knew nothing about the case. He was, as a businessman, editor of a Richardson weekly newspaper; as a militant citizen he was president of the Dallas Civil Liberties Union.

Well, the caller said, the president of the Austin affiliate had phoned long distance. He said that the ACLU should be concerned about the prisoner. In Austin there had been some television pictures of Oswald in a hallway somewhere, in a crowd of intense faces, holding his handcuffs up and shouting hoarsely that he wanted a lawyer and had not been given the opportunity to get one. He knew his rights, he said, but he wasn't getting them. Olds, who was ready to retire for the night, said that he would check into the matter at once. He admitted that he knew no more than that the President had been shot and a policeman killed. Some young fellow from Irving had been arrested.

Olds was a patient man. He was also persistent. He could be deflected but not stopped. He phoned the Dallas Police Department and asked to speak to the chief of police. The chief was busy. He would speak to a deputy. No one knew where they were. Would he want to speak to a detective? No, he would not. Mr. Olds said he was president of the Dallas Civil Liberties

Union and he would speak to the man in charge of the assassination case and no one else. He was given one or two men who informed him politely: "Captain Fritz isn't available, but you can tell me . . ." Gregory Olds slammed his teeth tight and said that he would wait.

Will Fritz got on the phone. He was asked if he was in charge of the Kennedy case, and he said yes. Olds explained his position and said that the Civil Liberties Union was anxious to see that Oswald had whatever legal representation he desired. The captain was a figure of bland unction. The question had come up, he said, and the police department had been at pains to detail Oswald's rights, but he had declined assistance. Fritz himself had given Oswald the right to phone counsel of his choosing, had ordered the jailers not to hinder the prisoner, and, in fact, Oswald had made a couple of phone calls. The best that the captain could tell Mr. Olds was that Lee Harvey Oswald had not made any requests to him.

Olds hung up. He knew that there are prisoners in every city who refuse the services of an attorney on the assumption that one is not needed. The ACLU had considerable experience in this field; some defendants decided to represent themselves in court; a few, tragically, equated innocence with acquittal. These often paid in prison. The easy way for Olds would have been to retire, safe in the knowledge that he had the word of the Dallas Police Department that all the legal safeguards had been offered to Oswald and he had spurned them.

The editor decided to take a more difficult course. He phoned the board member and suggested that he call a few other ranking members and that they meet in the Plaza Hotel lobby at once. There would be a conference about Lee Harvey Oswald. The men sat on a couch and discussed the case. The guilt or innocence of Oswald did not come to the surface; it seemed incredible that, considering the trouble he was in, he didn't want a lawyer. Someone suggested: "Call the mayor."

If there is one thing to which Gregory Lee Olds was accustomed, it was disappointment. Chiefs of police, mayors, and prosecutors regarded an assortment of questions from the ACLU as a personal challenge. Olds got on the phone again and asked for Mayor Earle Cabell. Who was calling? He gave his name and rank and was told that the mayor was busy. The editor wondered what could keep a mayor busy after 11 P.M.

The best thing, he told his confreres, would be for the body of men to walk across the street to police headquarters and ask the questions directly. They might even meet Lee Harvey Oswald himself. Surely, if the police department was justified in its position, it would have no objection to a group of men saying: "We're from the American Civil Liberties Union. Do you need a lawyer?" If he said no, the case was closed and they would go home. Curry and Fritz would probably be happy to be rid of them.

They were directed to the third floor and the elevator lifted them up and spewed them into the madness of the corridor. Men and cameras curled around them in a stream flowing in the opposite direction. Olds saw a man he knew: Chuck Webster, a professor of law. They explained the problem. Mr. Webster said that he had been around headquarters practically all evening and he thought he knew the man who might reassure the ACLU. Webster escorted them to the other end of the hall and introduced them to Captain Glen King. The captain was a gracious man. He said that Oswald had been charged with the murder of Officer Tippit and was, at this moment, being formally charged in the assassination of the President.

Olds said that this was not their concern. All they wanted to know, in words of one syllable, was whether the prisoner had been advised of his right to counsel. That's all. Glen King said that, so far as he knew, Oswald had not made a request for counsel. That's an edge of an answer, but it lacks body. Had the police department advised Lee Harvey Oswald—never mind what he asked for or hadn't asked for—had they told him that he was entitled to a lawyer, that he could have one right now or at any time throughout this interminable day?

The captain thought that the man best equipped to answer the question, to put all minds at rest, would be the J.P. Mr. Johnston was down in the basement at this moment. Why not run down there and ask him? Mr. Olds remained upstairs. The ACLU men went back down the elevator. It seemed awkward that such a simple question required the affirmation of so many officials. Each in turn was certain that Oswald was protected, but no one was certain just how. The best source, they thought, would be the justice of the peace.

The fact was that, consciously or unconsciously, Oswald's legal rights were in jeopardy. Shortly after 2 P.M. the police department of Dallas had told him that he didn't have to answer questions, that anything he said could be used against him in court, that he did not have to pose for press photos or answer questions from the newspapermen or television people. Up to that point the law had blinded him with the brilliance of justice. Beyond that point legal darkness had descended on the scene. He had asked again and again for a lawyer. He had requested the services of John Abt of New York and, when Oswald had reminded the police inoffensively that they had taken his thirteen dollars away from him, he was told to make the phone call collect. This gave him an unnecessary hurdle to clear, and he had failed it.

On at least one other occasion, Oswald had told the officers that, if he could not locate Abt, he would consult the American Civil Liberties Union. He had also declared that he was a member of the ACLU. Will Fritz, surprised, asked how much Oswald had paid in dues, and the prisoner told him five dollars. If the department had a desire to protect the rights of the prisoner, it would have been a small gesture to have phoned the Dallas branch of the American Civil Liberties Union and said:

"We have a fellow over here who says you're his second choice to help him. Would you like to send someone over to have a talk with Oswald?"

Instead, almost the entire day and evening had been spent inundating the prisoner in a bowl of hostile faces. He faced their combined cunning, analyzing the innocent questions before responding to them, glorying in the attention he drew from the world, perhaps even exultant at the opportunity to hold them off in the swordplay of the tongue.

Two men of the Olds group met Justice of the Peace David Johnston. The judge appreciated the nobility of their cause at once. The ACLU men returned to Olds and said that the judge had assured them that Oswald's legal rights had already been explained to him and he had "declined counsel." The law had done its duty. The American Civil Liberties Union had done its duty. It was time for good men to go to bed.

Around Times Square, the taxicabs waited with dimmed lights on the side streets. The theater crowds had gone home. The pancake restaurants were bright with young faces, happy to be out of the chill wind at seventy-five cents per face. A ribbon of news walked in yellow lights around the old Times Tower, telling the story over and over. On the corners, men in stockinged hats sold the morning papers.

The hands of the clock met at twelve, and the big one moved a minute past to inauguarate another day in New York. *The New York Times* spread the assassination headline across the top of page one as though somewhere it might electrify someone who had not heard. Ten miles outside of Norwalk, Ohio, eighty-four inmates of the Golden Age Nursing Home fretted over the loss of one so young. In twelve more hours sixty-three of them would burn to death. In Los Angeles, Aldous Huxley, a littérateur, was dead of cancer. The paid obituary notices in *The New York Times* were drenched with the tears of the maudlin. There were twenty-two notifications that John F. Kennedy, President of the United States, had passed away. The American Booksellers Association, Igor Kropotkin, president, mourned his death and "while honoring all his great accomplishments, single out with gratitude his long devotion to the world of books." Panic and remorse were still dominant and many people sat the long vigil before the television set to subject themselves to the redundant hammer blows. The flagellants and the oratorical sadists were betrothed. The parade was begun over and over and over. Mr. Kennedy smiled and waved his hand, and Mrs. Kennedy flicked the dark lock of hair from her eye and smiled vaguely in an ocean of strange and curious faces.

Everybody knew that a scene would come when people would fall on grass, scrambling to safety; when the President would begin to topple; when the young lady in pink would begin to climb out on the trunk; when a motorcycle policeman would ram his vehicle into the curb, draw his gun, and glance helplessly at windows and railroad tracks. Everybody knew this. The lacerated mentalities had to witness it once more, as though this time the story would end happily.

In time, David Brinkley of the National Broadcasting Company tired of the blows. "We are about to wind up," he said slowly, "as about all that could happen has happened. It is one of the ugliest days in American history. There is seldom any time to think anymore, and today there was none. In about four hours we had gone from President Kennedy in Dallas, alive, to back in Washington, dead, and a new President in his place. There is really no more to say except that what happened has been just too much, too ugly, and too fast."

Too fast to some, too slow to others. In Washington, Muggsy O'Leary drove down Wisconsin Avenue and turned into the big parking lot at Harrison. He and the tough, sentimental Gaels—O'Donnell, O'Brien, and Powers—felt that this was the longest, slowest, saddest day of any year. It was a day of so many scores of individually remembered sorrows that no one of them could recall them all. O'Leary, a Kennedy idolator, was a member of the Secret Service because John F. Kennedy endorsed the appointment. The blackbeard, O'Donnell, sat watching the lights of his world flash by the car to explode into the blackness behind. O'Brien, the conciliatory redhead, the onetime bartender, the political mathematician, was doing something that had to be done. He did not relish the task, but he may have been affronted if someone else drew the assignment. Powers, the bald ward leader of Boston, the man who first looked politically upon the tall stuttering son of Joseph P. Kennedy, Sr., the man who first said: "Okay. I'll try to make a Congressman of you"—this is the one who knew the aspirations to be said for the repose of the soul, the spiritual phrases which begged clemency for the sinner.

These went to buy a casket. A saint does not comprehend the finality of death; a sinner does. The four men knew the mystery of death as they knew the zest and joy of political battle. Death was an unlocked front door; the musty odor of flowers; a red vigil light; camp chairs; the ferns; stiff blue knuckles clutching the black rosary; the sonorous voice of the priest, kneeling before the casket, intoning: "Blessed be God! Blessed be His Holy Name . . ."; it was the sobs of the women; the cautious handshake of old enemies; ham and whiskey in the kitchen; old men puffing pipes and remembering him that was in the box and his father before him and that one's father before him.

It was a brand-new building, a Georgian structure with lights in the hedges splashing a glow on the pale brick. This was Joseph Gawler's Sons. In one hundred thirteen years, it had buried three generations of its own, and the fourth waited inside the white doorway. There is no trade, no profession, which stands in such permanent delicate balance as a funeral home. It must be solemn but not doleful; helpful but not cheerful; competent but not morbid; spiritual but not hyper-religious; cordial but not intimate; ready to assist but not overbearing. Joseph H. Gawler understood his function. He stepped forward as the four men came in, and introduced himself and his operations manager, Joseph E. Hagan. The four looked around. The lobby floor was white marble relieved by small black diamonds. There was a circular staircase to the right with ember-red carpet and balustrade; a crystal chandelier hung down from another floor.

O'Donnell began to explain their presence. Mr. Gawler interrupted. He understood. The phone call from the White House had explained everything. The visitors felt relieved. Mr. Hagan, a short man with the air of one who is accustomed to becoming confidential within a short span of time, said that he understood that embalming of the President would also be required. Powers nodded. Hagan just wanted the strangers to know that Gawler's was prepared. He had an enbalming team waiting in an office to the left; John Van Haesen, Edwin B. Stroble, and Thomas Robinson. Gawler, a brown-eyed man with a ruddy complexion, said that they wanted the Kennedys to know that, in spite of the hour, everything could be accomplished to the satisfaction of the family.

The gold-lined elevator moved in silence. The four were in a world of sedate whispers. They passed the second floor, with its array of large rooms furnished in French provincial. At the third floor, the party turned left and Mr. Hagan opened the double doors leading to the selection room. The men stepped into a cool place on heavy beige broadloom. Recessed squares in the ceiling bathed the place in warm light. In an alcove and a main room there were two dozen caskets. The men hesitated, eyes darting. There were gray metallic boxes, grayish suede; there were gleaming metal caskets, some in mahogany, some burled in a blackish wood. A few were open, disclosing the white shirred satin. All stood on carriages hidden by pleated skirts.

The men seemed embarrassed to be in the room. They wanted "something in good taste." Hagan didn't care to remind them that everything in the room was in taste. It required a little coaxing to get them inside the big room, where they could examine the array of merchandise. Mr. Gawler said that his original impression had been that President Kennedy would rest in the funeral home, but that . . . No, they said, he would repose in the White House, where he belonged. They walked slowly around the boxes, none with any knowledge of what Mrs. Kennedy would appreciate.

Someone said that price was not a factor. The Gawler group understood that, but they wanted it understood that there was a standard price on each of these items, that a casket would cost no more for the President's family than for anyone else. This also applied to their services at the hospital. Above all, Gawler's exuded an aura of quiet confidence, and this pleased the four men. Two of them stood beside a polished mahogany casket with ornate silver handles. They thought perhaps that something along these lines . . .

The others joined them. They walked around the box. The half-lid was lifted. It looked rich and solemn. There wasn't a hint of garishness. Walking around it, the men noticed that it picked up arrows of light from the ceiling. "This one," they said. Mr. Gawler said that it would be delivered at Bethesda Hospital within the hour. Yes, O'Donnell said. Please do whatever must be done as quickly as possible.

Hagan had heard on television that the President had sustained a massive head wound. The embalming team might find it necessary to process part of the skull, matching it identically with the real color and texture of the hair. It could have saved time if Hagan or Gawler had asked about these things before leaving for the hospital. Some things are left unsaid. It would require a little more time, but it would be worth it to assess the cosmetic damage themselves and plan the repair work.

The four men were outside in the crisp night air within twenty minutes. They were glad to be out again. It is a triumph to be alive in the presence of death. It is deadlier to be able to walk away from it. The poetry of the sentimentalist is dolorous. As he treads the edge of eternity to do a service for a friend, he too dies.

Lee Harvey Oswald stood. It stirred a turning of heads. He was tired of sitting, he said. The handcuffs hung on his thighs. He arched his back a few times and sat. The prisoner was not told that he was about to star in a press conference. He would be taken downstairs, as though for another lineup, and he would not know, until he saw the cameras and heard the questions, that for a brief time he was being tossed to the press as a sop.

In the outer office, Captain Fritz bent over a desk and told detectives Sims and Barratt to make out an arrest sheet on Lee Harvey Oswald in the murder of one John F. Kennedy. It was to be done at once, before the prisoner left the office. Fritz wanted to sign it. Everything he had pointed to Oswald. There was no other suspect. The captain didn't have a piece of evidence which would lead him to believe that another person might be involved. For the sake of Dallas it would be a good thing to present the assassination as solved to the press of the world. The day could be closed on a note of triumph.

The man with the rumpled suit and face introduced himself to a young policeman in the hall. "I'm looking for Joe DeLong of KLIF," he said, holding a pencil and note pad in view. "Can you page him over the loudspeaker?" "Who?" the cop said. "Joe DeLong," Jack Ruby said. The policeman walked to a corner where a microphone stood. A booming voice echoed through the long hall. "Joe DeLong. Joe DeLong of KLIF. Please report to press information."

The hall was nearly empty. A dozen men lounged in groups. Two men unplugged a thick black cable and followed it to an office window and dropped it to the sidewalk. The cop said: "He isn't here." Ruby said: "I'll wait a minute." A reporter, passing, said he was glad that Curry was putting Oswald on display downstairs. It would serve two purposes, he thought. One would be to give the world a solid look at the man. The other would be to permit the press to ask a few questions. Even if Oswald denied everything, they would have a statement from him.

Ruby said: "Thanks" and started down to the assembly room. He felt a tingle of excitement. Tremendous things were happening and Ruby was there to witness them. He did not want to draw attention to himself, and he would seek a position in the back of the room. Sometimes a young punk cop who did not know Jack Ruby might challenge him—might order him off the premises. The older men, friends of Ruby, might not want to bail him out of a situation like that. So he kept the pencil and pad in view and the gun hidden.

It was a nickel-plated .38. It was inside his trouser belt, between the pants and the shirt. Jack Ruby never used it. Sometimes friendly cops asked: "Why the hell do you carry that thing?" Mr. Ruby said that he carried two, sometimes three thousand dollars hidden in the trunk of his car. The gun was a form of insurance. It was carried, as so many were in Texas, like a 14-carat toothpick, the badge of the male. Sometimes, when Ruby was out socializing, he tossed the gun into the trunk of the car. On other occasions, it was to the left of the zipper on his trousers, with the handle up.

Lewis removed his coat and nodded politely to Wesley Frazier. Sometimes a polygraph operator wonders if his brother officers understand the procedure. They thought that all he had to do was to take a subject like Frazier, ask him questions, and watch a needle jump. Officer R. D. Lewis was a qualified operator. He called Adamcik into the other room to ask a few questions about the frightened boy. The more he knew about Wesley, the better the setup for the test. The office lights were turned up, an armchair was turned so that it faced a blank wall. Lewis arranged the blood pressure cuff for a human arm and looked at the needle tracings on a paper on his desk. Who was this kid? What was his name and what kind of material was Fritz interested in?

The kid's name was Wesley Frazier. He lived less than a block away from Oswald's wife. Frazier worked at the School Book Depository with Oswald and drove him home on weekends. Homicide was pretty sure that Oswald was the man they were looking for, but this Frazier kid was something else. A rifle was found in his house. He could possibly be a party to the assassination. Hours ago he had been questioned in Robbery, but he seemed scared. The kid was halfway home when Fritz got this idea for a polygraph test.

Lewis asked more questions. He needed them as controls. Captain Dowdy beckoned for Frazier to be brought in. The test was explained to him. The first order of business was to sit in that chair and try to relax. He would find that it wasn't as easy as it looked. The best way to relax, he was told, would be to keep reminding oneself that you are going to tell the truth, no matter whom it hurts.

The boy sat. The cuff was wrapped around his arm, and his sleeve was shoved high. He was told to stare at the wall and try to think of nothing. Lewis, at his desk, studied the pulsing of the needle. He was getting steady vertical tracings. The beats were fast; that was nervousness. He waved the others back out of sight and began the test.

It took time to get the control questions and the placidity of the victim juxtaposed so that, on simple interrogations such as: "Do you live with your sister?" the needle would not jump. "Ever fire a gun?" induced a spasm peak. There was nothing incriminating in either question or answer ("Yes"), but Frazier, judging by the needle, bordered on controlled hysteria.

Officer Lewis reassured him several times, told him he was doing fine and not to worry about explanations when responding. If a question could be answered with "Yes" or "No," use the single word. Also, when a question was asked which involved Lee Harvey Oswald, the answer would not necessarily involve Wesley. He might be asked if Oswald worked at the Texas School Book Depository building and the answer should be "Yes" without excitement.

Lewis realized that it would be a lengthy test, but he was a patient man. He expected a jump on the needle when he asked a control question such as: "Ever do anything you're ashamed of?" or "When you were little, did you ever lie to your mother?" There were five police officers in the room and the doorway, and there wasn't one who expected to learn anything from Wesley Buell Frazier. All they had managed to do was to scare the wits out of him.

The charge of murder was complete. It was studied by Assistant District Attorney William Alexander. As he stood beside Wade, they made an imposingly tall team. In the courts, Alexander was known as a vengeful prosecutor, a remorseless examiner, an old-time hanging district attorney. He told Sims to call Fritz out. The captain looked at the document, swore that he was the complainant in this case, and signed it. Henry Wade took it, waved the signature dry, and said: "Well, he's filed on."

Oswald wasn't. The charge had no validity until a bail hearing was held by a "magistrate"—a justice of the peace—and signed by him. The young J.P. was down in the assembly room, sharing excitement with the press. David Johnston was thirty-six years of age, and he had never participated in anything as important as this case. The defendant in the action had yet to be notified.

The document was dubbed F-154 and it announced the contending parties as the state of Texas versus Lee Harvey Oswald. The charge was murder; the defendant's address was listed as "City Jail." Inside the flap, the stereotyped words stated that J. W. Fritz had "personally appeared before me" and being "duly sworn," says he has good reason to believe "and does believe" that one Lee Harvey Oswald, on or about the 22nd day of November A.D. 1963 in the county of Dallas and state of Texas, "did then and there unlawfully, voluntarily, and with malice aforethought kill John F. Kennedy by shooting him with a gun against the peace and dignity of the state." To the left of Fritz's signature, Henry Wade signed.

Except for brief cutaway shots, the television story of the assassination had slowed to an embarrassed repetition. On the streets in some cities, men with microphones asked pedestrians what they thought of the death of the President, and the faces moved up to fill the screen with features which, a moment ago, had betrayed pleasant smiles and which rearranged themselves into grim expressions or open-mouthed horror: "Well, I was just plain shocked . . ." "I mean, he was a family man and I don't know much about politics, but . . ." "I was having a sandwich on the job and this guy walks over and tells me the President is shot and I said: 'This a joke?'" "I called my husband and I said: 'What kind of a place is this Dallas?'" "Did they find out who did it yet?"

The President watched the set on the other side of the bedroom with heavy-lidded eyes, half listening, half nodding to suggestions made by his three young assistants. He held up both hands for quiet. " . . . and now," a commentator said, "we return you to Washington." The vision of faces faded, and Andrews Air Force Base came on screen like a well-lighted insect trap surrounded by darkness. A big glistening plane was rocking its way into the patch of light and an announcer said: " . . . just arriving from Honolulu. This is the plane which carried Secretary of State Dean Rusk and other members of the Cabinet, who were on their way to Tokyo. The tragic news reached them out over the Pacific, and Mr. Rusk ordered the plane to return to Washington."

Mr. Johnson heard the dying whine of the jet engines. The Boeing 707 stopped and a ramp was wheeled to the front of the plane. A large percentage of his government was on this aircraft. The men around the President's bed turned to watch. Valenti turned the dials of the set to get a better picture and better sound. The lugged door swung back and men began to file out, squinting in the light.

Herb Kaplow of the National Broadcasting Company said: "Secretary of State Dean Rusk is first off the plane, followed by Secretary of Commerce Luther Hodges, Secretary of the Interior Stewart Udall, Secretary of Agriculture Orville Freeman [the gentlemen of the Cabinet began to collect at the foot of the ramp, around Rusk], Secretary of the Treasury Douglas Dillon, Secretary of Labor Willard Wirtz, and presidential secretary Pierre Salinger.

"They are being met at the plane by Protocol Chief Angier Biddle Duke and Assistant Secretary of the Navy Franklin D. Roosevelt, Jr. The party is moving to the microphones. Dean Rusk will speak for the Cabinet." Mr. Rusk, a man with a voice like dry breakfast cereal, had no intention of making a mistake. The flight had been long and fatiguing, and there was a new "boss." As the landing gear of the plane thumped downward in air, he had walked down the aisle advising the passengers that he, Dean Rusk, would lead the government officials off. He expected the Cabinet members to follow. If there was any speaking to be done, he would do it.

The Secretary of State saw his deputy, George Ball, in the circle of light and nodded. At the collection of black-fingered microphones, Rusk glanced at a curled sheet of paper in his hand and said: "We have fully shared the deep sense of shock at the grievous loss the nation has suffered. Those of us who have had the honor of serving President Kennedy value the gallantry and wisdom he brought to the grave, awesome, and lonely office of the presidency." Mr. Rusk looked up into the lights. "President Johnson needs and deserves our fullest support," he said.

It was the right thing. The words made an adieu to a departed chieftain and offered an unfettered hand to the new one. The President fluffed the pillows behind his head and returned to the "grave and awesome" tasks at hand. He interrupted the conversation to ask what had become of his daughter Luci. She had retired. Mr. Johnson was a kissing husband and father. He could not believe that he had permitted her to go to bed without a goodnight kiss. It was Luci who had reminded herself that her father was now President of the United States and should not be disturbed on this, the most terrifying night of their lives. Luci planned to spend time reciting some prayers for the repose of the soul of John F. Kennedy.

"You'll get it all back tomorrow," Henry Wade said. The district attorney was ready to defend the Dallas Police Department's paramount right to the evidence, but he could see far beyond the confines of the city and county and he knew that the assassination had federal aspects which ought not to be ignored. Chief Curry listened sulkily. His department, he felt, was doing well. It did not require the expensive equipment of the Federal Bureau of Investigation nor interrogations of the Secret Service. Besides, what could they find out about the evidence which Will Fritz and Lieutenant Day had not already ascertained?

With a big trial coming up, it was not proper to permit the chain of evidence to be broken by flights to Washington. In a sense, everybody in the department was walking on tiptoe to make certain that they had a good case against Oswald, an airtight case. Phone calls had been coming to Curry's office all evening asking him to release the gun, the blanket, the empty shells, the works. The calls came from Wade's office, from the Texas attorney general, Waggoner Carr, from ranking officials. Curry didn't care, really, what Lyndon Johnson had to say about putting the FBI in charge of the investigation. This was still a Dallas County case. If the amassing of evidence proceeded properly, it was the Dallas Police Department which would do it; if it failed, the blame would fall on him.

The big D.A. nodded. As they walked down the hall to Curry's office, he reminded the chief that the case could not and would not be taken away from Dallas County. *He* would have to prosecute Lee Harvey Oswald, not the FBI. There was no percentage in spurning federal assistance. The FBI had offices in places like New Orleans, where Oswald had lived. They had a big network which could trace guns, find people who employed Oswald, or knew him in the Free Cuba movement; they could add to all that Curry now had in his hands.

The chief surrendered. He said that he would order Fritz and Day to hand the stuff over to Washington right away. It must be clearly understood that the FBI would sign a receipt for each item, they would photograph it and send copies to the Dallas Police Department, they would fly it up to Washington tonight, run it through their mill, and have it back in police headquarters tomorrow night. In the name of the Federal Bureau of Investigation, Vincent Drain, the big smiler, agreed. He phoned Gordon Shanklin, still in his office down the street, that Jesse Curry had agreed to the lending of the evidence.

Lieutenant Day was working on the Mannlicher-Carcano rifle when he received the order. He gave it to Drain reluctantly, along with the other material. After telling Drain that a palm print of Lee Harvey Oswald had been located on the stock of the gun, Day forgot to explain that it was no longer there. He had lifted it off *in toto* with adhesive paper.

Shanklin phoned Washington and gave the news to Belmont. "We'll be waiting," he said. The head of the Dallas office then phoned the office of the commanding general at Carswell Air Force Base. "One of our agents," said Shanklin, "is taking evidence in the assassination to Washington. We have a directive that you will help us fly it up and wait for it to come back." Ten minutes later, a K-135 jet tanker on the hardstand cut in pod number three, and a blast of heat thrust the cool night of Fort Worth back and away.

The White House Press Office was quiet. Malcolm Kilduff sat at a desk. The dark hair framed the strong face. This was the hive of government on most days. The press should have been clamoring outside the outer office where Helen Ganss sat. This, of all nights, should have been one of statements and releases, of jangling phones and news tickers noisily walking their black feet across white paper. There was no sound except the sobbing of Christine Camp and Sue Vogelsinger. Andrew Hatcher was in another office.

Down the hall thirty feet was the President's office, It, too, was empty. No highly placed statesmen gathered to debate a decision. No bill was ready for signature, the array of pens standing in a holder like asparagus. The sixteen buttons on the scrambler phone were dark. And, between them, the Fish Room was empty; so was the office of the appointments secretary.

Kilduff, assistant press secretary, made a few phone calls. No one asked him to run down to Ralph Dungan's office and think of names important enough to be invited to a funeral. No one asked his knowledge of Lincoln's funeral. He could, if he chose, straighten a few paper clips, or he could go home. There was no self-pity in his makeup. Mr. Kilduff was at his shiniest when he was fighting for something or against something.

As with so many others in government service, he had been a Kennedy man. Malcolm Kilduff had a weakness: he could not work for someone he did not admire. It was fortunate that in John F. Kennedy he saw the bright idealist, the tough politician, the decision maker, the charisma of eternal youth. The man, the job, the future had exploded today and Kilduff sat in the press office, glancing now and then at the big color photos of Kennedy—the handsome, grinning, we-own-the-world face—

telling himself that it was all shattered and the pictures would start to come down tomorrow.

He had made the solemn announcement of the death of a President. It was his function. The press secretary, Pierre Salinger, had been winging to Japan when it happened. Andrew Hatcher, the other assistant press secretary, had not been assigned. The trip to Texas had been Kilduff's first as acting press secretary. Before he left with the President, he knew that, for him, it would be the last. He had been fired.

The word had come down from Kenneth O'Donnell to Salinger to Kilduff. He was offered a position in another government agency. It was a good, dull post, far enough removed from the excitement of the White House to placate O'Donnell. The matter was placed on a lofty plane: we have to get rid of one assistant press secretary and it cannot be Andy Hatcher because he's a Negro. You are the one.

Kilduff had been offered a "better" post with a skillfully disguised ultimatum: the job would be open three days. The man who sat in the press office alone on this long sad night felt that perhaps the President did not know that Kilduff had been fired. In spite of the proximity of their offices, there was no way that Mac Kilduff could get to see Kennedy without the approval of Mr. O'Donnell. Kenny was the appointments secretary and more. He was the captain of the palace guard; the watchman; hatchetman; keeper of the privy seals.

No one got into the President's office, even casually, without Mr. O'Donnell's knowledge, if not his approval. Kilduff, the fighter and brooder, had to find a way. One morning he walked into the office of Mrs. Lincoln, personal secretary to the President, a dark slender woman who kept fancy dishes of candy on her desk. There was no harm in chatting with Mrs. Lincoln. She kept the door between her office and that of the President ajar about thirty degrees.

Kilduff gave her a hearty greeting and partook of the candy.

He chatted loudly near the door. He and Mrs. Lincoln heard the voice of the President, alone and at work, say: "Is that you, Mac? Come on in." The assistant press secretary, who had accompanied Kennedy on his trip to Ireland, walked inside almost apologetically and told the President that he was being offered a "better" job. The smile on Kennedy's face faded.

"What's the matter?" he said. "Aren't you happy here?" Kilduff said that indeed he was; he would like nothing better than to continue serving the President, but he had been told that one man had to be cut from the roster. "Forget it," the President said. "I'll take care of this when we get back from Texas."

Kilduff sat at his desk wondering who had won this tiny political battle in a world of politics. Kennedy was gone and so, except for the sufferance of Lyndon Baines Johnson, was everyone else. On the plane this afternoon, Malcolm Kilduff had tried to bridge the gap between the rancorous Kennedys and the bustling Johnsons. He had run the errands faithfully, delivering messages and responses to messages.

At one point, he had paused beside the seat of Kenneth O'Donnell. A few feet away was the casket containing Kennedy. The appointments secretary had taken Kilduff by the arm and pointed to the broad back of Lyndon Johnson. "He's got what he wants now," O'Donnell murmured. "We take it back in '68."

The office lights remained on. No reporter buzzed Kilduff or asked for an anecdote to sweeten a bitter story. There was nothing to say and no one to whom to say it. Kilduff got his jacket from a rack and put it on. He glanced at one of those smiling photos and said a silent "Good-bye."

★ ★ ★

The Midnight Hours

The room was twice as long as it was wide and, when the lights went on and the two doors were flung open, it was as though sluice gates had been lifted. A swirling foam of faces poured in. The tide knocked over chairs. A small table was slammed into an abutment. A one-way screen, behind which witnesses might study a suspect, was rolled to one side and came to rest against a wall. Uniformed policemen walked backward before the wave yelling: "Take it easy! Take it easy!" Men with still cameras took positions along the side walls or down front. A television camera in the rear lifted one large dark eye above the crowd.

The voices became a bedlam. The demands and complaints blended into a solid roar of incomprehensible sound. The police assembly room was filled in a minute and the wave seemed to curl backward, shoving those at the doors out into the hall. There was a rostrum up front and Assistant Chief Charles Batchelor stood on the low platform and shouted for order. He could not be heard. In the hall Lieutenant T. B. Leonard, leaving for home with Lieutenant George Butler, stood on tiptoe. Batchelor saw them and motioned wildly for them to come in and assist him. Jack Ruby, pressed against the little table at the abutment, climbed on it and crouched with his back to the wall. He could see over the heads, and he was only eleven feet from the rostrum. The sound was deafening, and the press shouted to itself to keep quiet.

On the third floor Fritz said he wanted all of *his* men in that room, and they dropped their assignments to hurry down. Chief Curry was worried, tossed between duty and a public image. He sensed that the two were not compatible. The dour expression was on the little man. He saw Detectives Sims and Boyd flanking Oswald, and the chief said: "Don't let anybody get near him or touch him. If anyone tries to, I want you fellows to get him out of there immediately." Curry laid good groundwork for his defense. In the third-floor hall twenty minutes ago, when the press conference had been suggested, he had asked Wade: "Do you think this is all right?" The district attorney had shrugged and said: "I don't see anything wrong with it." Curry had held up both hands for the press to be quiet, and he had said: "——anything goes wrong with his being down there—if there's a rush, he's immediately going out and that's it. Now, do we understand each other?" The reporters shouted: "Right! Yes! Right!" One more thing: the chief did not want the prisoner to stand up on the rostrum. "Put him down in front of it, on the floor, and put a guard of men around him."

There was a mutual distrust among the savages. Oswald, standing quietly between the detectives, understood the situation. This was not one more lineup; it was a press conference. It may have been pleasing to him. He uttered no protest. The press conference was the only way in which he could restore contact with the world. He could use it to serve his ends, as the police were using it to serve theirs, and the press hoped it would satisfy theirs.

In the front of the assembly room, Wade sat on a desk, dangling his legs. He was a man almost impervious to danger, but he had an uneasy feeling that there was no way out of this room. It was a mob, pressed face-to-face. Without the pencils, the pads, the portable tapes, and the cameras, the faces might have been cast in a motion picture about a lynching. They were angry faces, and they pressed forward and receded in waves. The district attorney shouted to a policeman: "You'd better get some officers in here to protect him!"

The tall solemn figure of Captain Will Fritz could be seen in the hall. He tried to elbow his way into the room, but he failed. A dozen detectives, preceded by the chief, forced their way into the front of the big room with Lee Harvey Oswald. He was a bobbing figure in a vortex of police helmets and ten-gallon hats. Wade saw them coming and waved the reporters back from the rostrum. He saw one lined face above the crowd, over in a corner, and absentmindedly wondered where he had seen it before. It was probably some local reporter or radio commentator. In the hall, Will Fritz still hoped that Curry would put Oswald up on the stage. He would be out of arm's reach of the press, and besides he could be yanked offstage into the prison admitting office in a trice. Curry told the police to put Oswald down front. The wall of protection around the prisoner kept the reporters three feet away.

The inner circle of policemen began to jostle policemen already in the room. There wasn't sufficient space to step out of the way. A roar of sound enveloped the room as the crowd saw Lee Harvey Oswald. Still cameras were held overhead and aimed in the direction of the protective circle around the prisoner. The place smelled of stale sweat and fetid breath. When the prisoner was in front of the lectern, the press yelled to the police: "Down in front! Down in front! Let's get a look at him. Is this the guy, chief? Did he do it? We can't hear anything. Hey, Oswald, why did you shoot Kennedy? How about a statement?"

Ruby, crouched on the little table, saw the police guard flex their knees and he studied Oswald closely. There was a purplish bruise under one eye. The prisoner had not uttered a word, but the nightclub owner interpreted Oswald's expression as being "proud of what he had done." He thought that the suspect was smirking. Oswald acknowledged the greeting of the mob by raising both manacled hands over his head. Jack Ruby saw it as a clenched-fist communist salute.

In a doorway, Gregory Olds watched. He could hear unintelligible shouts. Near the director of the Dallas branch of the American Civil Liberties Union stood a law professor from Southern Methodist University named Webster. Another authority on law, Greer Ragio, stood nearby. Henry Wade noticed them and cupped his hands to ask Chief Curry about Oswald's civil rights. The chief said that "those people" had been given an opportunity to talk to the prisoner.

"Well, I was questioned . . ." Oswald began. The crowd yelled: "Louder! Louder!" The cordon of policemen around the prisoner began to tense. They looked for a nod from someone to take the prisoner out of the room. "Well, I was questioned," the prisoner said louder, and his voice began to crack with the volume. "I was questioned by a judge." The crowd began to grow quiet. Those who continued to yell "Louder!" were told to "Shut up!"

Every eye was on Oswald. He could read the expressions. They were not friendly to his cause. The objective press was subjective. It was a hanging jury. "However, I protested at the time that I was not allowed legal representation during that very short and sweet hearing." They had no time for his protests or his sarcasm. "Did you do it?" they yelled. "Did you shoot the President?"

"I really don't know what the situation is about," he said calmly. "Nobody has told me anything except that I am accused of—" The voice faltered. "—of murdering a policeman. I know nothing more than that. I do request someone to come forward to give me legal assistance." No one stepped forward. His problems about lawyers were not their concern. What they came for was *the story*. It wasn't Tippit; it was Kennedy. The press was determined to try anew, before this man was yanked offstage.

"Did you kill the President?" It was a simple question and it ran through the room from a dozen lips. Oswald shook his head slowly. "No," he said. "I have not been charged with that. In fact, nobody has said that to me yet." He was in a position to appear aggrieved. "The first thing I heard about it," he said, almost plaintively, "was when the newspaper reporters in the hall asked me that question."

"You have been charged—" "Nobody said what?" "Sir?" said Oswald. "What happened to your eye?" "When were you in Russia?" "Mr. Oswald, how did you hurt your eye?" The press in the rear began to shout to the press in front to repeat the questions and answers. Oswald said: "A policeman hit me." Some of the reporters, crouched low in the front line, began to cramp. A few straightened their knees furtively. The chief nodded to a detective.

Oswald was grasped by the arm. The press conference was over. The cordon was tight around him, and the police began to propel the suspect toward the door. A radio commentator held a microphone to his lips and said: "That was Oswald, Lee Oswald, who was charged with the murder of the President of the United States, although he said he did not know it. He's being taken back upstairs, he's being taken back upstairs for further investigation, as Henry Wade pointed out earlier."

The interview was a failure. The big question had been answered with "No." His plea for counsel was a legal complaint and added no substance to the material of the story. The press might have protested that the interview was a mockery, but they had made a fiasco of it and they were silent. Some left the room running to file late stories. Others remained because they saw Henry Wade remain. It was possible that something could be salvaged by staging a conference with the district attorney.

Wade was accustomed to the give-and-take of reporters. He could handle "the boys." As the chief law enforcement officer of Dallas County, his concern would have to be with possibly prejudicing the rights of the defendant to a fair trial. He could snap "No comment" to any question which held a hint of danger. He lowered his head as he sat on the edge of the desk and swung a big foot off the floor.

"He's been formally charged in Precinct Two of Dallas County Judge David Johnston," he said in the toneless tone of one who has been through this situation many times. "He's been taken before the judge and advised of his rights. He's been charged with both killing Officer Tippit and John F. Kennedy. . . ." The reporters formed a tight scimitar. "Can you tell us any of the evidence against him so far, sir?" The D.A. shook his head. "No. We are still working on the evidence. This has been a joint effort by the Secret Service, the Federal Bureau of Investigation, the Dallas Police Department, the Dallas Sheriff's Office, my office, and Captain Will Fritz has been in charge of it."

All the credits had been pronounced. Some of the reporters wrote them. Some stared dully at the man. "What does he tell you about the killing of the President? Does he volunteer anything or what has he got to—?" "He denies it." "Was he charged with the President's killing?" All of the networks remembered Oswald shouting that he had not heard that charge. "11:26 P.M." Wade said. "11:26 he was charged on the latter charge. . . ."

"Do you have a good case?" "I figure," said Wade, squinting, "we have sufficient evidence to convict him." He could have chosen to ignore the question. "Are there other people involved?" "There is no one else but him." The reporters were still copying the words when the district attorney said: "—he has been charged in the Supreme Court with murder with malice. The charge carries the death penalty, which my office will ask in both cases." For a time it appeared that Mr. Wade could anticipate the kind of material the reporters hoped for and enunciate it without waiting for a question. "Is there a similar federal charge?" "I don't know of any."

The pencils and pens were whirling. "Well," said Wade, "there is a lot of the physical evidence that was gathered, including the gun, that is on its way by Air Force jet to the FBI crime lab in Washington. It will be back here tomorrow. There are some other things that is going to delay this for probably the middle of next week before it's presented to the grand jury." Someone asked about witnesses. "We have approximately fifteen witnesses." "Who," said a reporter, trying to complete Wade's sentence for him, "identified him as the killer of the President?" "I didn't say that," the district attorney snapped. "What did they do?" "They have evidence which indicates his guilt." "Do you have anything to indicate why the man killed the President, if he so did?" "Well," said Wade, "he was a member of the movement—the Free Cuba movement." "Fair Play for Cuba," said Ruby. He had heard it on the radio. "What's the make of the rifle, sir?" "It's a Mauser, I believe."

"Does he have a lawyer?" "I don't know whether he has or not. His mother has been here and his brother has been here all afternoon." Sometimes the leonine face came up, and the D.A. studied the faces around him as a gambler might a marked deck. "Does he appear sane to you?" "Yes, he does."

"Why do you think he would want to kill the President?" Motive is important, but the D.A. decided to forget it. "The only thing I do," he said with exaggerated patience, "is take the evidence, present it to a jury, and I don't pass on why he did it or anything else. We, we're just interested in proving that he did it, which I think we have." The questions dragged on. Some involved vital statistics. Others asked about the gun which was alleged to have killed Officer Tippit.

One, which might have induced laughter but didn't, was: "If he has been formally charged with killing the President, how is it he says there is no connection to it?" Henry Wade stared at the man. "I just don't know what he says. He says he didn't do it." An eager voice shouted: "Was he in Russia? Henry, was he in Russia?" Another voice said: " . . . and he no longer has citizenship to the United States. Is this correct, sir?" Wade said: "I can't verify it or deny it." "Are you looking for any other suspects at all now that you've got—" "We're always looking for other suspects, but we have none at present."

"Henry, do you think this is part of the communist conspiracy?" "I can't say that." "Well, do you have any reason to believe that it might be?" "No, I don't have any reason to believe either way." "What time will you begin in the morning with him?" "Seven or eight o'clock, I would say, roughly." The material was growing weaker. Mr. Wade sat quietly. "Do you have some prints on him?" The question was a wild shot, but Wade electrified the group when he said: "They are on their way to Washington at present." "Who?" "Which?" "What's on the way to Washington?" "The gun. The rifle." "Both guns?" "Both guns."

Justice of the Peace Johnston watched with fascination. The questioning turned a corner when the reporters reminded Wade that Oswald said he didn't know he had been charged with the assassination of the President. Wade said he had been filed on. Which was right? "I do not know," the D.A. replied. "He has just been charged. I know he has been advised of the other and taken before the magistrate." One of the newspapermen put the question to Johnston: "Did he answer that question whether the man had been advised that he's been charged? The man said here that he didn't know he had been, Dave. How about that?"

David Johnston thought it over. "He has not been advised that the charge of the murder of the President, because he is on capital offense on the other." The reporters could not decipher the sentence. "He has not been advised?" one asked. The judge said: "He has not been advised." "When will the arraignment be for the President?" Wade reclaimed his press conference. "I imagine in—tonight sometime." The interviewers could not seem to let go of the question. "He has not been arraigned on the assassination?" "No."

"Have there been ballistic tests made locally on the gun?" "No, sir." On and on, the questions probed. The district attorney said: "Is that all?" but no one responded to that. The men kept asking. One struck ore by inquiring: "Sir, can you confirm the report that his wife said he had in his possession as recently as last night, or some recent time, the gun such as the one that was found in the building?" "Yes, she did." Several voices said: "She did?" "She did, but—" "She did what? She did what?" "She said that he had a gun of this kind in his possession." "Rifle? A rifle?" "Last night?" The district attorney sighed. He was mired in this press conference and he couldn't extricate himself. "Last night," he said. "It's that—the reason I answer that question— the wife in Texas can't testify against her husband, as you may or may not know." It was a peculiar rationale. If a wife couldn't testify about the rifle, Ruth Paine could swear that she heard Marina say that the rifle had been stored in a blanket in the garage. Further, the fact that a wife cannot testify would appear to impose a degree of restraint on what a district attorney can repeat of her admissions.

"Mr. Wade, was he under any kind of federal surveillance because of his background, prior to today, today's events?" "None that I know of. We don't have any knowledge—" "Do you think you've got a good case against him?" "I think we have sufficient evidence." "Sufficient evidence to convince—to con-vict him of the assassination of the President?" "Definitely. Definitely."

The district attorney started to move away from the desk. The last few questions came. "What did she say about the gun?" "She said the gun, he had a gun, a gun of this kind in his posses-sion last night." "Does he give any indication of breaking down?" "No, not particularly." "Are you willing to say whether you think this man was inspired as a communist or whether he is simply a nut or a middleman?" "I'll put it this way: I don't think he's a nut." "Does he understand the charges against him?" "Yes."

The district attorney left with David Johnston. Wade said, "You ought to go up to the jail and have him brought before you and advise him of his rights and his right to counsel and this and that." Someone stuck a hand between them and both men stared at Jack Ruby. "Hi, Henry!" he said. "Don't you know me? I'm Jack Ruby. I run the Vegas Club. Henry, I want you to know that I was the one who corrected you." Ruby kept pumping the hand of the district attorney. Wade introduced David Johnston. The nightclub owner shook hands, and passed a card. It featured a line cut of a nude girl in black stockings holding a champagne glass. The wording read: "Vegas Club. Your Host, Jack Ruby." Wade murmured that Johnston was a justice of the peace. Jack Ruby shook hands again.

He bowed away from the group and asked two strangers: "Are you Joe DeLong?" "No," one said. "Why do you want him?" "I got to get to KLIF. I have some sandwiches." "How about us?" Ruby hurried away. "Some other time," he said. He had trouble getting the night number of the radio station. The doors were locked after 6 P.M. There was no way to bring the sandwiches unless he could get someone to unlock the door, and to do this he required the night phone number.

There was a surging excitement in his chest.* He felt that he had been deputized as a reporter. He was helping the press to get the facts straight. Ruby asked nothing in return. He had a compulsion to be a part of this great story. He had to be "in it." Sandwiches were not enough. Correcting Mr. Wade was not enough. Giving out cards to his nightclubs was not enough. Some recipients smiled, crumpled the card, and dropped it. Others had the effrontery to ask: "What will this get me?" No, it was far better to be part of history than to study it.

Ruby saw a man walking by with a microphone and handed a card to Icarus M. Pappas of WNEW, New York City. Mr. Pappas glanced at it and stuck it in a pocket. Another man carried a portable machine stenciled KBOX. The nightclub owner asked him for the phone number of KLIF and got it. There was a row

---

* In cases where feelings are described, they are culled from later recollections of that person or from testimony of persons who were at the scene. In this case, the feelings of Ruby come from his depositions after shooting Lee Harvey Oswald on November 24, 1963.

of phone booths, and Jack Ruby got into one next to Mr. Pappas, who was phoning New York. KLIF answered the ring and Ruby said: "I'm Jack Ruby. I have some sandwiches and good pickles for DeLong and the night crew. I hear you're working late."

Outside the booth he could see Pappas trying to attract the attention of Henry Wade. "Hold it a minute," said the deputized reporter and brought the district attorney to the radio reporter. He popped back into the booth and said: "How would you like an interview with Henry Wade? I can get him for you." The man at KLIF thought it was fine. Wade was talking to Pappas when Ruby took him by the arm gently and said: "There's a call for you, Henry." Wade went into the booth and was interviewed.

When it was over, Wade held the receiver for Ruby. "Now," said Ruby into the phone, "will you let me in?" The night man said: "All right. I'll leave the door open for five minutes. Just five minutes." The station was a block away. To Ruby it represented a crisis. He could walk the block within five minutes, but the sandwiches and soda were in the car. If he reached the car, he might as well use it to drive to the radio station.

He was up on the main floor, almost trotting, when someone grabbed his arm. "Jack," said a bright young face, "where is everything happening?" It was Russ Knight, radio reporter of KLIF. He carried a portable taping machine. The novice reporter thought about it for a moment. "Come on downstairs," he said. Knight was from the station whose attention he was soliciting. It was important to show as many of these people as possible that Jack Ruby could do things for them that they could not do themselves. In the basement, Ruby said: "Henry, this is Russ Knight of KLIF" and hurried back to the main floor.

In a minute, he was out in the cool midnight air, hurrying to the parked car. On the front seat, Sheba sat waiting. Ruby sometimes referred to her as "my wife." He started the car and moved it quickly to the curb in front of KLIF. There, across the sidewalk, was the door. Ruby parked, grabbed the big bags, slammed the door, and hurried to the building. The door was locked. The reporter had missed his first deadline.

He stood in front of the door, breathing. He waited ten minutes. Russ Knight came down the street, returning to KLIF with a fresh interview with Henry Wade. "I brought some sandwiches and soda for you guys," Jack Ruby said. Knight unlocked the door. They went upstairs to the radio station. The Good Samaritan had bought his way in for $9.60.

The several parts of the funeral had been hammered together. Nothing was "finalized," but Sargent Shriver and his White House "pickup team" could see the outlines and the stages at 1 A.M. The handsome, square-jawed man who directed the Peace Corps could not sit in Dungan's office any longer. Walking was a necessity. He chose to go all the way across the White House to inspect the decoration of the East Room.

As Shriver stood, he beckoned to David Pearson and Lloyd Wright to follow him. He walked sturdily, purposefully through the curving empty corridor of the main floor. Somewhere ahead, he heard John F. Kennedy speaking. It was strange and depressing to hear once more the clear New England accent of a man who would not be heard in this building again. Mr. Shriver continued ahead. In a small office, empty, a television set was on. Pearson and Wright followed their man to the doorway and watched him take an empty chair.

The President, with chin high, was addressing the people of West Berlin. He was stirring and forceful and, beneath the stand, scores of thousands of Germans turned bright expectant faces upward, like cool petals to a warming sun. *"Ich bin ein Berliner!"* shouted Kennedy in limping German and a deep roar of approval came up from the crowd. Sargent Shriver stood. He left the set on, as though he didn't want to be a party to stilling that voice forever.

At the office of the President, Shriver paused again. A United States marine, stiff in dress blues, guarded the empty place. He flicked the office lights on, and the bright interior became the natural frame for the voice coming from a box. Every place, every sound was designed to salt sorrow with recollection. The outside walk, between the swimming pool and the Rose Garden, clicked with the Kennedy heels, the Kennedy chuckles when the children rushed him from ahead, the Kennedy laughter when he was in the pool with Dave Powers and Dave said: "I had to learn the breast stroke because it's the only way I can swim and talk to you." Next to the boxed hedge, the President dozing fitfully behind sunglasses on the lounge, hands behind head, the lean figure as straight as an exclamation point, secretly watching his wife and the children at the swings and the slides.

Shriver was a couple of strides ahead of Wright and Pearson, as the President used to be a couple ahead of O'Donnell and Salinger. The ridiculous recollections could go on and on. Shriver continued along the grand corridors, empty except for a pair of Secret Service eyes at posts along the way. He walked along the main corridor to the East Room and inside. The artist who had studied the Lincoln funeral, William Walton, was lighting a cigarette. Shriver nodded to him and to Richard Goodwin, who had another book in his hand.

"She wants the East Room to be prepared for him," said Walton, knowing that there was only one "she"—"like it was for Lincoln." The visitors had known this for hours. So had Walton and Goodwin and Schlesinger. It is possible that Walton was relating this to himself. "If they are going to get here about two-thirty, I doubt that we can match the Lincoln scene. . . ."

There was an artistic disinclination to do the room à la Lincoln. Goodwin said: "They were pretty rococo in those days." Shriver looked over some of the drawings of ninety-eight years ago and agreed. The catafalque was heavy with black bunting. Mirrors were covered with black gauze, as though the sight of an image alive might bring bad luck. The hanging crystal chandeliers were swathed with circular black; the windows were edged with crepe. It was overdone. To imitate it would be deliberately depressing. Shriver agreed with Walton.

"I think we can capture the right feeling," said Walton, "and yet adapt it a little more to Jack Kennedy." The artist decided to make the room mournful without making it sob. The mirrors would have slender skeins of black around the frames, but not across the glass. Bits of black here and there in the room would establish the proper aura of respect for the dead, without darkening the place and robbing it of life.

Pearson said: "Maybe there ought to be a crucifix." Like everything else thought of or devised that night, it was "sent for." Christ Himself must be in good taste. When a bloody crucifix was tendered, Sargent Shriver declined it and sent for one in his Maryland home. Someone brought up the option of open casket versus closed. The artist was positive that the President would prefer to have it closed. "Jack didn't like to be touched," he said quietly to David Pearson. "I doubt whether he would like to be stared at now."

The elevator door opened on the fourth floor and Deputy Chief Lumpkin and four detectives stepped off. Within the embrasure of heads was Lee Harvey Oswald. He carried his handcuffs forward of his body and he watched the procedure of transfer from Homicide and Robbery Division to the municipal jail with no interest. One of the jailers asked if he had been fingerprinted and mugged, and a detective said: "I think so." The prisoner knew he had not been photographed. The frisking was repeated. A prisoner's card was filled out by the jailers as a receipt for the custody of Oswald.

The party went to the fifth floor. There, Lumpkin turned his man in for the night. The late shift in the prison went through the procedures as though Oswald had not been through it before. This time he was permitted to keep trousers and an undershirt. Two men led him to the old row of three cells and put him in the center one. The cuffs were taken from his wrists. He asked about permission to take a shower and he was told to get some rest. For the second time, he announced that he had "hygienic rights."* The two men who kept the dangling ceiling bulb lit sat for a long night. If Oswald noticed that the Negro had been removed from Cell Number Three, he made no comment.

On the floor below, Officer R. D. Lewis completed his polygraph test on Wesley Frazier. He had nothing to show for fifty minutes of work. Two policemen had been stationed behind a one-way mirror watching the witness and listening to his replies to questions, and they, too, were convinced that, if there was such a thing as an assassination plot, this young man was not a party to it.

Lewis ran through the tape, studying the controls and the intensity of response, and shook his head negatively. One of the policemen phoned Captain Fritz. The captain ordered the three detectives to release the boy and to take him home. The detectives were to report off duty and be available in the morning. It did not assist morale to know that Fritz was more overworked than his men. Stovall knew that he would get home about 2 A.M., after delivering Frazier and Linnie Mae Randall and the Baptist minister to their homes in Irving. The officers would have to be out of bed and ready before 8 A.M. The only man who might sleep straight through the night was Oswald.

---

* The Bar Association of Dallas County did not concern itself with the rights of the prisoner to counsel until the following day. An Eastern dean of a law school phoned a Dallas attorney about the matter of counsel. The Dallas lawyer phoned H. Louis Nichols, president of the Dallas Bar Association. Mr. Nichols phoned a criminal lawyer to "refresh my memory." The criminal lawyer said there was an obligation to appoint counsel after a prisoner has been indicted. Mr. Nichols asked himself if the bar association owed anything to Oswald. Saturday afternoon he visited Oswald in his cell. The prisoner was interested in representation by "John Abt" or someone from the Civil Liberties Union. Nichols did not know Abt or a lawyer from ACLU. The interview was friendly and fruitless.

Robert Frazier was explicit. Evidence was going to continue to come into the FBI's Weapons Identification Section all night long. He wanted each item to be tested exhaustively. New York had just Telexed Alan Belmont that a firm in Chicago called Klein's Sporting Goods was known to have sold the Mannlicher-Carcano rifle by magazine coupon. Chicago had been asked to track down Klein's management and, if necessary, to get them out of bed and try to locate a Dallas or New Orleans order from Lee Harvey Oswald, Lee Oswald, L. H. Oswald, Alex Hiddell, A. Hidell, or A. Hidel.

Handwriting experts would be studying the Oswald and Hiddell signatures; others were working all night in New Orleans to get background and biographical matter; a man was in the State Department working on Oswald's Russian file; Gordon Shanklin in Dallas had a group of agents with the police and the Secret Service; Drain was having evidence photographed before flying it to Washington; Frazier was taking a team to the White House garage to examine the President's car.

The hour was late, but the men, members all of an elite corps of law enforcement agents, wanted to "wrap the case up" before dawn. In Dallas, Chief Curry and Captain Will Fritz thought it was already "wrapped." The FBI considered that Dallas had a good case against Oswald; the bureau wanted to secure the loose ends, such as tracking and tracing the rifle and revolver to Oswald; identifying a bit of metal taken from the President's brain as coming from that rifle; examining that recent trip to Mexico, inch by inch, to find out whether Oswald was part of a conspiracy or acting alone.

Frazier, who looked like a twenty-year bank teller, led Agents Charles Killian, Cortlandt Cunningham, Orrin H. Bartlett, and Walter E. Thomas onto Pennsylvania Avenue on the west side of the building. Two Agents, Sibert and O'Neill, were on their way in. They said they were returning from the autopsy. Supervisor Frazier didn't have time to listen to a summary of the results. All he asked was: "Did you find anything?" They said yes, and displayed a shallow salve jar. The top was unscrewed. On white cotton batting were slivers of lead.

"Here are some metal fragments from the President's head," Sibert said. One weighed 1.65 grains; the other was 0.15 of a grain. Frazier touched them as though they were rare gems. "Is this all you found?" They said that X-ray plates disclosed that there were thirty or forty tiny bits of metal along the edges of the skull and embedded in the brain, but the doctors felt that they were too tiny to locate and withdraw.*

The Frazier group drove to M Street and identified themselves to the Secret Service agents. Deputy Chief Paterni had agreed to an FBI examination of SS-100-X, and they trundled it out from the alcove by hand. The plastic cover was removed; so was the leatherette convertible top. An FBI photographer mounted an aluminum ladder and began a series of shots from above, from the sides, from front and back. A shortwave aerial on the left side was broken off. Robert Frazier guessed it might have happened when Clint Hill made a dash for the back of the car as Mrs. Kennedy tried to climb out. Shots were made of the interior of the trunk.

When this phase was complete, Frazier and his agents moved in for an intense survey of the vehicle. The radiating crack on the windshield was examined, measured, and photographed. The glass was double, fused together by a gelatinous substance. The

---

* Each of the bullets fired at President Kennedy weighed 161 grains: total 483 grains. The final accounting is: (1) The first shot probably missed the car on the right side, ricocheting off the curb and spraying James Tague at the Commerce Street underpass; 161 grains; (2) the bullet on the stretcher weighed 158.6 grains; (3) fragment found in front section of President's car: 44.6 grains; (4) fragment on front floor of car: 21 grains; (5) two fragments from Kennedy's head: 1.65 grains and 0.15 grains; (6) fragment from wrist of Governor Connally; .5 of a grain; (7) fragment from rear rug of car: .9 of a grain; (8) fragment from rear floorboard of car: .7 of a grain; (9) fragment from rear carpet of car: .7 of a grain. Total found: 228.80 grains; first bullet missing: 161 grains. Total: 389.80. Unaccounted for: 93.20 grains.

outside of the crack, at the front of the windshield, was smooth to the touch. On the inside, it felt like a small sharply edged crater. A receptacle was held under it; then Frazier ordered it carefully scraped with a sharp jack knife. Metal fragments, as small as bits of rust, were recovered. The metal content was identical with that of the bits of bullets recovered.

A dent was found in the upper frame of the windshield. This too was measured and observed. Frazier thought that a bit of flying metal might have hit it.* Inch by inch, the FBI men examined the exterior of the automobile, the body, the doors, the wheels, fenders, hood, trunk, even the ribs of the tires. No one hurried. The work was tedious. Frazier opened the back doors, then the front doors. The interior was still laden with petals of red and yellow roses.

Each one was removed and examined to see if any metal content adhered to the flower. The two limousine blankets, sealed in pockets in the doors, were removed and spread on the floor of the White House garage to be felt and dusted. On the back seat and on the rug FBI men picked up dry clots of blood and brain tissue. These scrapings were placed in envelopes, in the same manner that Doctor Burkley's warrant officers had done earlier. Most of these grisly bits were sifted between rolling fingertips for metal.

The metal runner which held the rug down, in the front section as well as the rear, was carefully unscrewed and lifted clear. Some FBI agents held bright spotlights as others began the painstaking task of feeling each inch of the rugs, then lifting them clear to the floor of the garage for additional examination. The floorboards of the car were probed.

Three bits of lead missile material was found on the rug. The rear seat, which operated on a hydraulic lift, was removed.

---

* The dent was found to have been made in a New York garage when someone tried to close the convertible top and hit the aluminum frame of the windshield.

The trunk was opened. In all cases, a visual examination came first; notes were taken. The two sizable bullet ends found earlier in the front of the car were already being tested in the Weapons Identification Section of the FBI. It was considered possible that, as the President fell to his left with his head angling down, the third bullet may have crashed through the back of the skull, tumbled through the brain and out the top of the head, and spun to the inside of the windshield on the driver's side, to inflict a crack in the glass and fall to the floor.

Frazier realized that many suppositions would never be proved. There were possibilities and probabilities and few provabilities.

On a table with dead microphones Jack Ruby made a space and put the bags of sandwiches and soda down. "Well!" he said happily, and then comically slapped a hand over his mouth and pointed to the mikes. The radio announcer on duty, William Glenn Duncan, Jr., pointed to a thick glass section, where a sound engineer was playing a tape. "You can talk," he said. Russ Knight came in. He was going to splice his interview with Wade so that it would be a flawless interview. Ruby waved everyone in to shop for the right sandwiches and drinks.

One of each was brought into the sound engineer, who smiled and nodded his mute thanks. "Isn't it awful," said Ruby. "A weasel like that." He chewed a sandwich and looked around. It was not the first time he had been in a radio studio. On the receiving end, in his automobile, radio stations always sounded loud and imperative and musically busy. In a room like this, a man felt like a dead fish in a vault. Excitement seemed to be one room removed, except when Mr. Duncan cut in with the late news. For a few minutes, Ruby stopped chewing and listened in fascination as the words fled in all directions to be heard in homes and automobiles and restaurants all over the Dallas prairie.

There were times to be quiet and times to speak. Jack Ruby asked Knight if he had seen that "terrible ad" placed in the *News* by someone named Bernard Weissman. It referred to President Kennedy in disgraceful terms, and Jack Ruby was afraid it might reflect on the Jewish community. Knight didn't think so, but Ruby feared that the gentiles would think, with a name like Weissman, that it had been inspired by Jews. Besides, Ruby was sorely afflicted with a feeling that Dallas assessed Jews as being without courage. "Jews have guts," he said frequently. "They're ready to stand up and fight for the things they believe in."

Knight played the interview with Wade. A little later he devised an oral editorial about the assault on Adlai Stevenson and the Weissman advertisement. Ruby said that Russ Knight was one guy who "will go to bat immediately if anything is wrong." Once again, the nightclub owner was manipulating the news in a small way. He was part of a big story, and he would remain as long as he was within it. Russ Knight kindly mentioned Jack Ruby as the man who suggested, at the press conference, that reporters ask the D.A. if Oswald was sane. It was pleasing to know that one's name was flying through the night sky to thousands and thousands of ears.

The men were told to go home, complete whatever they were working on and leave. Captain Fritz knew that some of them had been on duty sixteen hours. He wanted them back in the morning. In the race to close the case, he had an understandable ambition to help Homicide and Robbery to lead it. Fritz, sitting alone in his office, thumbing through the reports, was sorry that all that evidence had gone up to Washington. If Chief Curry had given the captain one full day, he would probably have traced that cheap rifle to some shop in Dallas and from there right to Oswald.

The phone rang. He picked it up and learned from Lieutenant Knight, in charge of the Bureau of Identification, that there were no fingerprint and no mug shots on Lee Harvey Oswald. Fritz thought he recalled prints being made. These were done by Lieutenant Day for swift comparison with latent prints on the rifle and the cardboard box. Knight wanted something more permanent. The captain said that the prisoner had been sent up to the fifth floor for the night and to go get him.

Knight and Sergeant Warren went to the jail and filled in a checkout card at 12:35 A.M. Oswald did not protest. He was lying on a lower bunk, still in his T-shirt and slacks. He sat up and announced that he had been fingerprinted, but it was explained that it had been done with an inkless pad. Now that two felony counts had been filed against him, the Dallas department required permanent prints and photos. Copies of these would be sent to the Federal Bureau of Investigation as a matter of course to find out if the prisoner had a criminal record or if he was wanted by any other police department.

The car stood halfway in the gate at Carswell. Vincent Drain sat in back with his evidence. An Air Force sergeant phoned that he had a visitor without permit who wanted access to the hardstand. Except for the yellow overhead lights at the gate, Carswell Air Force Base was dark. In the distance, a service runway was flanked with deep blue lights. Drain and his FBI driver waited. A car pulled up behind them.

A man got out, walked up, and introduced himself as Winston Lawson, Secret Service agent. "I remember," said Drain. "You were at headquarters this afternoon." Lawson said that Inspector Kelley had heard about the evidence being transferred and had ordered him to accompany Drain. The FBI man had no objection; Gordon Shanklin had agreed to it. The sergeant at the gate was told that there were now two men to get aboard a special plane—one FBI, one Secret Service.

A few minutes later an officer arrived in a staff car. He greeted the agents, apologized for the delay, and took them through the barracks section in the darkness and out onto the air strip. They had to climb a small iron ladder. Lawson said he would help Drain with the packages. The FBI man declined with thanks. He clutched the material because his function was to protect the chain of evidence. Drain knew that when he arrived in Washington, no matter how tired he felt, he would have to remain with this material until Robert Frazier's section had completed its exhaustive tests; then he would have to reboard this jet tanker at Andrews and fly back to Carswell and return the material to Captain Fritz.

The hatch on the K-135 closed, and the two men secured seat belts as the big silvery bird waddled to the head of the runway and received permission to take off for Andrews. When the shrieking of the engines settled down and the craft was airborne, Lawson took out a pencil and asked Drain what kind of evidence he had. It consisted of the following:

(1) A live 6.5-millimeter rifle shell found in a 6.5 Mannlicher-Carcano rifle on the sixth floor of the Texas School Book Depository building; (2) three spent 6.5-millimeter shells, found on sixth floor of the Depository inside the northeast window; (3) one blanket found in garage of Mrs. Ruth Paine, 2515 Fifth Street, Irving, Texas; (4) shirt taken from Lee H. Oswald at police headquarters; (5) brown wrapping paper found on sixth floor, near rifle, believed to have been used to wrap the weapon; (6) sample of brown paper used by School Book Depository and sample of paper tape used for mailing books; (7) fragment of bullet found in the wrist of Governor John Connally; (8) Smith and Wesson .38 revolver, V510210, taken from Lee H. Oswald at Texas Theatre; (9) .38 bullet recovered from the body of J. D. Tippit; (10) One 6.5-millimeter bolt action rifle, inscribed "1940, Made in Italy," serial number C2766. Also inscribed on the rifle was a crown which appeared similar to an English crown. Under it was inscribed "R-E." Also inscribed was "Rocca," which was enclosed in rectangular lines and was on the plunger on the

bolt action on the rear of the gun. On the four-power scope of the gun was inscribed "Ordnance Optics Inc., Hollywood, California, 010 or 010 Japan." Also inscribed was a cloverleaf and inside the cloverleaf was "OSC."

The copilot came back with cardboard containers of coffee. Lawson completed his list and looked out a window. The night was bright with stars. The plane raced them but appeared to stand still. The drone of the engines reduced conversation and increased drowsiness. A radio operator told the agents that they would make Andrews about 4:30 A.M. Eastern Time. Vincent Drain turned his watch ahead an hour. He estimated that he would have this material in Frazier's hands by 5 A.M.

The show was over. The audience had dissipated. Roy Kellerman phoned Clint Hill on the seventeenth floor. "Come on down," he said. "I want you to look at these wounds." The Gawler group arrived. Joseph Hagan introduced himself and his assistants to a Navy enlisted man. For the embalming, he had Mr. John Van Haesen, Mr. Edwin Stroble, and Mr. Thomas Robinson. They would not begin their labors until the autopsy team signified that its work had been completed.

It had. Brigadier General Godfrey McHugh remained in the room with the body, as he had vowed to do. Dr. George Burkley got to his feet, walked over to the sheet-covered remains, and worked on the left hand. The wedding ring came off. The doctor, who felt emotional about the death of his patient, took the gold band and went to the seventeenth floor with it. The Attorney General came to the door and held out his hand. The graying man said he would like to deliver it "personally" to Mrs. Kennedy. There was no reason to doubt that Robert Kennedy would give it to her. Still he held the door open and Burkley went inside and handed the ring to Jacqueline Kennedy. The rear admiral made a touching speech about his feelings. For a moment, grief matched grief in mutual appreciation.

In the autopsy room, the sheet was removed from the President's body, and Kellerman ordered Clint Hill to make his observations. The body was turned face down, then returned to its original position. Hill was stoical as he noted a bullet puncture at the base of the neck in the back and a small hole in the rear of the head, in addition to the big rent in the middle of the head. He was sent back to the Kennedy suite to stand guard and to file a personal report.

The Navy doctors removed their X-rays from the opaque screen and their notebooks from the autopsy table. They were zealous men who had tried to reduce observations and measurements to a precise science and failed. They could and would correct original impressions, but those who enjoy varnishing history with suspicion would not forgive the doctors for misreading a tracheostomy.

The three men were replaced by four. The function of the new men was to restore John Fitzgerald Kennedy to an approximation of serene sleep. In a manner of speaking, this is the most tender and most difficult of services. It is normally performed in secrecy. For Joseph Gawler's Sons, it would have been easier to take the body to their establishment. The instruments and material would be at hand, and the body could have been returned in ninety minutes.

This was not permitted because Mrs. Kennedy did not want the body taken from the hospital. Understandably the word hospital did not have the note of finality encompassed in "funeral home," "autopsy," and "embalming." It lifted the weary spirit a trifle to think of a Navy officer in a Navy hospital. It postponed a final accounting.

1 a.m.

The shrewd amiability of Lee Harvey Oswald wore off. He had spoken more words to more people on this day than on any other in his life. He had hidden his antisocial attitude, had worn his aura of innocence with authority, and had used the keyboard of the world press to beg for the civil rights which should have been his. The game had become a bore a long time ago. The clock had passed Oswald's bedtime by three hours.

He was not pleasant in the Identification Bureau on the fourth floor. Posing for the mug photographs, he obeyed directions like an automaton; the teaspoon edge of the lower lip began to emerge in the sullen pout which had been his badge from the age of two. Oswald would not roll his fingers on the print pad; he dragged the fingers across it until a policeman took the digits one by one and rolled them properly. Again he was asked to sign his name at the bottom of the card and he said: "No."

The police did not remonstrate. They tossed a dirty cloth, damp with benzine, for him to wipe his fingers clean. At ten minutes past one, the negatives and prints were declared satisfactory, and two jailers escorted him through the little foyer with the file cabinets and back up to the fifth floor. He noted that one guard sat on a chair propped outside the alley of three maximum cells; one was inside. There were no greetings either way. He went to his cell—the one in the middle—washed his hands in the chipped sink, urinated in the sloping basin built into the floor, and settled down for the night.

He was on his back on a lower bunk, hands clasped behind his head, eyelids almost closed and glistening at the naked ceiling light. He kept his shirt and trousers on this time, kicked the loafers off, and crossed ankles. The guards could not tell whether he was awake or sleeping. The respiration appeared to be slow.

The guard who sat opposite Cell Two would surmise that Lee Harvey Oswald fell asleep quickly. He could have been wrong, but after a few minutes the toes did not twitch, the eyelids were closed, the features appeared to fall into the relaxed aspect of a weary child. No man who had a grave crime on his mind could relinquish it so easily. No innocent person, charged with so heinous an offense, could sweep the terror from his mind and lapse into sleep.

There was nothing on the conscience of Jack Ruby, and yet he could not entertain the thought of sleep. The two men were only a few hundred yards apart, but Ruby burned brighter as the hours faded toward dawn. He could not let go of his part in this story, small as it was. He asked the men at KLIF how they liked the Exotic Cola. They said fine. Was the interview with Henry Wade satisfactory? Indeed it was, and Glen Duncan and Russ Knight were thankful to Jack Ruby for setting it up.

He wasn't boastful. Nor was he feverishly excited. The nightclub owner coaxed endorsement from the professionals. He spoke to Russ Knight—known to his fans as Weird Beard— about the techniques of conducting an interview and of how the tapes are cut and spliced to eliminate the weak and irrelevant sections. The mood, which was exuberant, dropped to resentment when Ruby thought of the man who had no name, "you know, the creep." Jack Ruby had not the slightest doubt that the prisoner was guilty. Dallas had the right man.

Knight was working and listening. Ruby, still the editorial strategist, asked why Gordon McLendon, owner of KLIF, could not deliver a sermon on radio against the forces which spawn "creeps" who placed treasonable advertisements in newspapers, pelt statesmen like Adlai Stevenson, and shoot Presidents who are guests of Dallas. The city was violent. The radio reporter was not too familiar with political terms, but he sensed that Ruby expected that someone "with guts" would speak out against the extreme right wing in Texas. Those who knew Ruby were aware that he had a chameleon character which could switch colors, from hilarity to resentment, from generosity to fisticuffs, from charitable impulses to tears, without changing emotional gears.

Duncan got the impression that the entrepreneur dropped the role of crusader abruptly and, far from grieving for President Kennedy, appeared to relish the personal contacts he had established in headquarters. Out of great tragedy had come a small measure of stature.

He wandered around the studio as Duncan prepared the 2 A.M. newscast and struck up a conversation with a young man. Ruby gave him a card to the Carousel Club. He found another employee, Danny Patrick McCurdy, and confided that he had announced that he was closing his nightclubs for the weekend. Moodily, Ruby stared at the floor and figured it would cost him between $1200 and $1500, but he would rather lose it at this moment in America's history. Such a loss could easily be translated into a mark of respect for the President.

The shine began to wear off KLIF. Jack Ruby, rumpled and righteous, had told everyone how he felt about everything. He was on the side of the angels. The broadcasts would continue all night long, but there was nothing to be gained in the studio. The paper bags were empty. Everyone had been thankful. Ruby was sure he could gain entrance at night again. He had established a rapport with KLIF. Besides, a new idea had crossed his mind. He had seen the *Dallas News* ads at Phil's Delicatessen, and in capital letters they said: CLOSED under the names of his nightclubs. The *Dallas Times Herald* would be wide awake at this hour, and he could drive over there and make certain that those small one-column ads shouted "CLOSED." Also he could see the advertisements of his competitors. So far, those gentlemen showed an elemental lack of "class" by remaining open and making a profit.

The final abuse of the body was under way. Pumping leads were established under the armpits. One forced a formaldehyde compound through the arteries of the body as a tube on the opposite side accepted the last of the body's blood. Gawler's men were efficient and almost silent. The four maintained their separate tasks. This time it was difficult to keep the hands from trembling. All of the four had lived in and around the capital with this charmer, this buoyant President. When the sheet was curled off the body, the professionals looked at what was left. Each man kept his features immobile, but each felt the depression of death.

A cosmetician studied the bloated face. Roy Kellerman got to his feet, walked over, and whispered: "How long?" The answer, whispered, was "Not long." He asked again: "How long?" An embalmer looked at his wristwatch. The time in Washington was 2:30 A.M. "An hour," he said. "An hour and fifteen minutes." Roy Kellerman strode back to his witness chair and phoned Clint Hill. "Tell the Attorney General we leave about 3:45," he said. "Tell the White House too."

The art of making a body presentable is no favor to the dead. It is designed to please the next of kin, to assure the living that he sleeps. The ultimate hypocrisy is jamming shoes on the dead. All of it is countenanced, expected, and paid for by the living. Gawler's, known for its discretion and good taste, knew from experience that the restoration of those who die by violence—especially with head or face wounds—is particularly difficult. Joseph Hagan walked around the body, noting the lacerated areas, and snipped a small bit of hair from behind the President's head.

"Go back and match this," he said to one of his men. "Bring enough to cover this open section on the head." He looked toward the bench where Kellerman and Greer and Burkley sat like dazed dolls. "And hurry," he said. A slight curved mesh was fashioned for the missing part of the head. It was a malleable fabric. The scalp would be pulled tight over it.

The Dallas casket, which had reposed against a wall, was taken away.* The dark mahogany one was wheeled in on a trolley. The President's clothing was placed on a chair, the creases neatly folded. Deft touches of a compound were placed on his eyelids to keep them closed. The lashes were brought down. White shorts were brought up over the legs. Black sox were peeled upward over the feet and ankles. The unresistant body began to take on the hue, the composed expression of John F. Kennedy.

Justice of the Peace David L. Johnston was ready. There was nothing the young man had to do except to read the charge to the prisoner, advise him of his rights, and tell him that murder in the first degree was a nonbailable offense in Texas. The magistrate had been in the room when Fritz signed the document at 11:26 P.M., and he could have read it to Lee Harvey Oswald at that time. There was no room for an exercise of options by the judge. Anyone could have read it and disposed of the hearing within ten minutes.

At 1:30 A.M. he was ready. Lee Harvey Oswald was sleeping. The judge told Chief Curry to get the prisoner and find a safe place for the arraignment. The chief thought that the jailers could bring him down to the I.D. Bureau on the fourth floor, the same place where Lieutenant Knight had made the fingerprints.

---

* One of the small mysteries is what happened to the Dallas casket. The Department of the Navy professed ignorance. Bethesda Hospital, at that time and for several years thereafter, was under the command of Captain R. O. Canada. In May 1968, Canada sent word to the author that he would not be permitted to see the empty autopsy room, unless he had an "okay from the White House." The public relations officers professed to know nothing of the disposal of the Dallas casket.

Phone calls were made. Superior officers of the police department were ordered to be present. This, they were told, was the big charge—assassination of a President.

Fritz slammed a desk drawer closed and took an elevator up. Chief Batchelor walked up. So did Chief Stevenson. Assistant District Attorney William Alexander represented Henry Wade. Another assistant D.A., Maurice Harrell, was present. On the third floor, most of the reporters had left for the night on the assurance that nothing exciting would occur until morning. Many of them were in hotels, tapping out final recollections of a bad day. A few, notably wire service men, remained in the corridor, and asked why top-ranking officers were hurrying to elevators and stairwells.

It was "nothing." None of them guessed that it would be a hearing on the assassination because the press had been assured that Oswald had been arraigned on that charge before midnight. Some in the hall surmised that the excitement was caused by a possible attack on a jailer by Oswald. Or maybe he had committed suicide. Or tried to. On the other hand, it could be that he had cracked and wanted to confess the crime. He could be bargaining for a second degree plea.

In the nearby hotels, out-of-town reporters were chagrined to learn that Dallas room service stops at 10 P.M. In New York and Washington, they told their hotel operators, there were places where a man could phone for a steak and a bottle of liquor at any hour. They were reminded that they were in Dallas, where decent people are sleeping at this hour. Besides, liquor was out of the question. The chief of the *Los Angeles Times* bureau, Robert J. Donovan, asked if it would be possible to pay a bellboy to go out and get some food.

An old Negro arrived. Donovan looked at him. The man was an anachronism, even for Dallas. The black shiny skin was furrowed; the hair was white along the sides. He wore a jazzy uniform with bell-bottom trousers and a gay pillbox hat on his head. "We are a bunch of newspapermen," said Donovan. "We haven't had anything to eat since breakfast in Fort Worth. We would appreciate it if you could go out somewhere and get us something to eat." The old man looked at the tired faces and nodded. It could be done.

"Now," said Donovan, "we can pay you a little extra if you can get us a jug. We know we can't get liquor in Dallas. We figure that you would know a place." The dark compassionate eyes of the old man swept the room. "No," he said. "I can't do that." The voice was deep enough to come from the throat of a prophet. "Ain't there been enough laws broken in Dallas today?"

Two jailers awakened Lee Harvey Oswald in his cell and he demanded to know what was going on. Sergeant Warren said: "I got my orders from the chief. He says to bring you down to I.D. again." If the prisoner silently debated the notion of refusing, he gave it up. For a bond hearing, they would have carried him. Docilely he held his wrists out and the handcuffs were snapped. He was sandwiched between guards and left the maximum security area, half sleepy, half angry, and was taken down a flight of stairs.

On the fourth floor there was a heavy metal prison door. This was unlocked, and Lee Harvey Oswald found himself in the presence of a considerable company of men. Normally this hearing would have been held in one of the courts on the first or second floor of this building. It was not put on the statute books to become an act of secrecy. This is what it was. The little foyer with the file cabinets was a chosen setting. The prisoner could holler as loud as possible for counsel—or shout his innocence—and no one could hear him except those police officers who believed him guilty. The outcome of the hearing would have remained the same in any case, but there was a desire on the part of the law to recite both charges—Tippit and Kennedy—in a private place. There could be but one reason for this: to keep Oswald's mouth shut.

Judge Johnston stood behind a check-in counter. He had the formal charge before him. Oswald glanced at all the faces. He saw no friends. This could not have hurt him because he never had one. "Is this the trial?" he said sarcastically. Oswald knew better. The justice of the peace held the affidavit up. "No," he said. "I have to arraign you again on another offense." Oswald did not want to be a party to it. The judge appeared to be nervous.

He began to parrot the printed form before him, stating that he was arraigning Lee Harvey Oswald for the murder with malice of John F. Kennedy, cause Number F-154, the state of Texas versus Lee Harvey Oswald.* The malcontent appeared to be offended. "Oh," he shouted at the assemblage, "this is the deal, is it?" The judge kept talking, saying that this charge had been filed at 11:26 P.M. on November 22, 1963, by Captain J. W. Fritz of the Homicide Bureau of the Dallas Police Department, and that Fritz, as the complainant, had signed it.

Across the bottom of it, Judge Johnston penned: "1:35 A.M. 11-23-63. Bond hearing—defendant remanded to Sheriff, Dallas County, Texas. No Bond—Capital offense." Oswald watched, and said: "I don't know what you're talking about." He began to ask about John Abt. He spelled it for the judge. Oswald's plea was that if he had constitutional rights, then one of them included the services of a lawyer. He had asked for John Abt of New York almost all day. In addition, he told the judge, he had said that if Abt was unavailable he would accept the services of a Dallas American Civil Liberties Union lawyer.

"You will be given the opportunity to contact the lawyer of your choice," Johnston said blandly. The prisoner was irritated;

---

* In testimony before the President's Commission on the Assassination of President Kennedy, J. P. Johnston swore he apprised Lee Harvey Oswald of his constitutional rights "again." Chief Jesse Curry, a witness in the same hearings, swore: "I do not recall whether he did or not."

this is what he had heard all day. He had pleaded for legal assistance for the past eight hours. He had begged for it at the press conference. He had phoned for it. No one had stepped forward. Some of the police officers who now stood silently behind him knew that the American Civil Liberties Union had contacted the police to protect Oswald's rights. The law lied when it said he had declined the services of a lawyer.

He was boxed in firmly and in inquisitorial secrecy by men who proclaimed themselves the upholders of the law. He cannot have hoped to escape the charge of assassination: there were too many witnesses; he had hidden his gun between cartons; the dead shells were still on the floor when he departed; he must have known that the curtain rod fable could be checked and that even his naïve friend, Wesley Frazier, would not believe it. Lee Harvey Oswald knew, once he made up his mind not to flee Dallas, that he would be caught and charged with the assassination.

Whatever grand design he had in mind for himself involved the use of an attorney. There can be no doubt that, considering the obvious trail of evidence he left behind him, that he would probably have been convicted in a reasonable trial. The mirror maze of thinking in which he involved himself was not deficient in simple logic. Arrested—Tried—Convicted must have been arithmetical progressions to him. Above all, he required a forum, a debating pedestal. He could have made a bid for fame in a Marxist speech at a trial. Or he might have penned a runaway best seller in prison. He was a bookish man. In his extreme penuriousness, he had once spent hard-earned dollars to have a public stenographer in Fort Worth pen the "history" of his visit to the Soviet Union.

Had he listened to Judge Johnston, Oswald might have noted that this was the second time he had been "remanded" to the custody of the sheriff, Bill Decker. He had the right to demand the transfer "forthwith." It would have embarrassed Fritz and Curry and perhaps the judge.* The hearing was over. Curry nodded to the jailers. They took their man back through the iron gate and up to the fifth-floor cell. He would be permitted six hours of sleep.

Sheba was patient. She dozed on the front seat for hours. The noise of pedestrians seldom disturbed her. The little dog recognized the voice, the cough, and the step of her master. When Jack Ruby came out of KLIF, Sheba was alert, standing on the front seat. He slid in on the driver's side, made the cooing expressions he always did when he returned to one of his dogs, and put the car in gear.

He was driving west. The streets appeared bright with light, but there was no one on them. Ruby could look down a mile of broad avenue and watch the diminishing rows of traffic lights switch, like an array of colorful soldiers, from orange to red to green. The trouble, as far as Ruby was concerned, was that he had time. He was a night person. The Carousel and the Vegas were closed, but his mind was alert and open. It would not shut down for sleep.

Since childhood he had had a fear of being alone. He spent as little time as possible in the apartment. He asked his friend George Senator to share the place with him so that there would be someone to talk to. When there wasn't, Ruby read the papers quickly, tossed them on the bedroom floor, and reached for the phone. The hour didn't matter: "Did I wake you? I'm sorry. Listen, I just thought of something. . . ."

Even the dogs were there for the monologue. He had several. They were small, like Yorkshires, and they stared moodily at Ruby from overstuffed chairs and beds. By day, when he had no work to do, he could stop in the bank to make a deposit or a

---

* Had the plea been granted, it would have saved the life of Lee Harvey Oswald. Thirty-four hours later, he was shot to death in the basement of police headquarters by Jack Ruby. In a prison ward, Ruby died of cancer three years later.

withdrawal, kill an hour in a newspaper advertising department, stop in at police headquarters and talk crime with the police, visit Eva for an hour, work on ideas to peddle a notion he had, called The Twistboard, shop for groceries, visit a stripper, hand out free cards to his nightclubs, kill time in Phil's Delicatessen, or glue himself to the phone.

He had no real yearning to visit the *Times Herald*, except to check his advertisements. He might find someone there to discuss the assassination with him, and he could tell how he nailed Henry Wade for one interview after another, and of how he corrected him—politely, of course—about the Free Cuba group. Jack Ruby knew what a big shot was. He also knew he wasn't one.

2 a.m.

Dallas was so quiet that Jack Ruby could listen to the continuous kiss of his tires on Jackson Street. The night air was chill and sweet and clear. On his left a big sign flashed hope to the emptiness: "Life Building." At a red light near Field Street, if a man sat quietly, he could hear the summer thunder of the freight yards at Lamar Street rippling the length of a hundred cars or more. It was a noise similar to the one President Kennedy had heard from his pillow at the Hotel Texas last night. Another town; another railroad; another man.

Ruby turned right on Field. A car flashed lights at him. A horn honked. A policeman yelled: "Hello, Jack." Ruby followed the lights into a parking lot diagonally across the street. The policeman, Harry Olsen, was a friend. There was a woman with him. She worked for Jack. She was a British girl with a lilting accent. He called her Kathy Kay, but her name was Kay Helen Coleman.

The nightclub owner had trouble understanding people who were in love. It embarrassed him. Olsen was in love with Kathy and was going to marry her in a month. But Ruby could not see it that way; he thought of Olsen as still married to another woman, going out on dates with this English girl. Mr. Ruby thought of this liaison as "secret"; Olsen was a man with "marital problems."

The policeman and the nightclub girl had met at 11 P.M. and they had been drinking beer and conversing about the assassination. The more beer, the more emotional the event became. The three met near the shack of the parking attendant, Johnny.

There were vast level empty spaces, and a few automobiles left overnight. The four asked each other: What do you think? Wasn't it terrible? The standard question.

Sheba was permitted to sniff around the parking lot. When she wandered too far into the silent darkness, her master called her with a chirrup. The dog returned happily, scampering around the legs of Jack Ruby, as though daring him to try to catch her. This once, the nightclub owner found two persons who were more emotional than he. He would like to have told Olsen about Jack Ruby and the district attorney, but the couple had been wound up tight, and they had opinions which must be vented at once. "If he was in England," Kay said in that prim lilting accent, "they would have dragged him through the streets and would have hung him." Olsen thought that Oswald should be cut, inch by inch, into ribbons.

The nightclub owner listened to them and to Johnny, the parking attendant, and he decided to leave. They grabbed his arm and held him. Jack Ruby was told that he was the greatest guy in the world. Just the greatest. They did not want him to go. The conversation had the quality of a snowball running down-hill; it became bigger, more magnificent as the mood plumbed the depths. At times, any two of them would burst into tears.

No one asked Olsen why he had not been on duty as a Dallas patrolman that day. He had fallen and broken a kneecap weeks before. The leg was still in a cast and he had been assigned to "light duty." Ruby's face sagged. When he found room in which to elbow the conversation, he kept saying: "What a wonderful man President Kennedy was. You know, I feel so sorry for Mrs. Kennedy and those little children."

Mrs. Coleman told Ruby that, since Harry broke the knee-cap, they seldom got out. They were so depressed this evening that she had driven him to the Sip and Nip on Commerce Street. They had some beer, and they expressed their solemn sentiments to Lee, the bartender. The place closed at midnight,

JIM BISHOP

so they had driven around, looking for places in which to exercise the catharsis of conversation.

"I took some sandwiches down to the boys," Ruby said, lying to the policeman. This was another generous gesture that made Ruby the greatest guy in the world. The parking attendant listened, but no one gave him room for opinion. He knew Ruby was the owner of the Carousel, a block away at Akard and Jackson. Right now Mr. Ruby was sitting in the driver's seat of his car with the door open. Olsen was leaning on the open windowsill. Kay Coleman noticed that her boss had his "starey, wild-eyed look." She soon found out what had upset him. He was angry because his striptease competitors at the Theatre Lounge and the Colony Club had remained open, doing business, earning profits after he, Jack Ruby, had shown the proper manner in which to show respect for a dead President.

Olsen discussed Oswald, and Ruby quietly played his trump card. "I saw him." He waited for the proper reaction. They said: "You did? What did he look like?" Ruby said that he had seen him taken down at the assembly room and had been close enough to Lee Harvey Oswald almost to touch him. "What do you think, Jack?" said Olsen. The nightclub owner, in a clearly superior position, bent his lip into a sneer. "He looked just like a little rat. He was sneaky-looking, like a weasel."

His friends nodded sagely. That was the impression they had. It was a sad day for Dallas that a thing like this had to happen because some crazy creep had a gun. "It's too bad that a peon like that can get away with something like that," Ruby said. The group nodded agreement. They talked on, Ruby calling Sheba back into the car, stroking her head and calling Oswald a "son of a bitch." Patrolman Olsen congratulated Ruby for closing up for the whole weekend. The conversation became repetitious, but the foursome continued; excited thoughts were flung to the night air, hardly listened to in passing, and others spurted upward, like an erratic fountain of thought.

Ruby decided that, no matter how late this agreeable debate lasted, he was going to stop at the *Times Herald* before taking himself and Sheba back to the smelly little apartment and throwing himself on a bed.

The four men were no longer sharp of mind. They studied the supine President as Gawler's men studied another one, except that this one was alive and new. Jack Valenti, so fatigued that he sat too erect, riffled through the pages of notes and suggestions. Cliff Carter, as big as Lyndon Johnson physically, was crushed by the weight of the hours. Bill Moyers, the facile student of government, kept a vigil on the President's eyes.

The lids were heavy. They dropped a little, then lifted suddenly and stared at the hypnotic screen. A station was running a hastily wrought biography of "Lyndon Baines Johnson, 36th President of the United States." The subject placed the bedclothes up under his arms and fought sleep. The men around him had no further words, no suggestions. Johnson felt a confidence in each one, but none of them were politicians. This was a "pickup" team. It is doubtful that, collectively, they knew more about the seat of power than its title. Their loyalty to him was touching but temporarily unproductive.

Lyndon Johnson understood power and its uses. Like a slow and ambitious student, he had studied twice as hard as his competitors in the Congress and at the White House. He had a more practical feel for government than his predecessor because he had watched the wheels of congressional committee spin eccentrically, and he understood the relationship—the true relationship—of the men on the Hill with the men on the High Bench, in relation to the Man in the White House.

Each of these was the right man for his time. The country had been aroused by the youth and exuberance of John F. Kennedy, who admonished citizens and legislators to execute his plans, not because it was politically right, but because it was good for America. The new frontier, as was true of the new gim- mickry, was in outer space. Kennedy could draw more attention shooting his cuffs than Johnson could declaring war.

The country was not prepared to receive Lyndon Johnson. To retreat a step, it was not prepared to lose its Galahad in Texas. In pain, the people acquired guilt. They had felt it, in a similar situation, three times in the past hundred years. This time the scars would be deeper because of the almost instant commu- nication of television. They *saw* what happened in Texas. They saw it again and again, as a repentant slayer relives his crime.

Those who had opposed John F. Kennedy were now pre- pared to receive him. The people who voted against him wept. The Congressmen who had disarrayed his program, stifled his progressive measures in committee, beat him on the floor of House and Senate with loud "Nays," worked hard this night on speeches of sanctification and superlatives to be laid reverently in the back of *The Congressional Record*.

The mass of people, like lovers who have strayed, desired to touch him. They would pray for him, defend him, buy a color photo of him or an ashtray with his name on it, adore his widow and vote for his brothers, change the names of boulevards, schools, hospitals, capes, and stadia to Kennedy. The people would resurrect him and they could not do this without spiritu- ally rejecting Lyndon Baines Johnson.

The new face was older, tougher, earthbound. The features were not intended for flights to the stars. The accent was Texas. Ironically none of the Texans around the bed could hear it. The new man understood the phrases: to think; to do; to produce. He knew more about them than Kennedy, and he had just com- pleted the first eleven hours of earning his salary. And yet effi- ciency was not good enough. Grief is not an intellectual exer- cise. The national heart was depressed; the country was taking its pulse.

"It's getting late, Mr. President," Cliff Carter said. It was a hint to close the book for the day. If he said: "Stay," they would remain seated. The brown eyes opened wide, moving from man to man. "It is," he said. The President propped himself up on one elbow. "Now you all go to bed and get some sleep." He looked at the little clock on the night table. It was nine minutes past three. "We'll be leaving here at eight in the morning." If there was shock, none showed it. They would be up at seven. Moyers asked if he should shut the television set off. "No," the President said. "I'll take care of that."

They said "Goodnight, Mr. President." As they left, he was still awake, still looking.

To the witnesses, the morticians were, in a manner of speaking, magicians. They had been given a broken shell of a man, and they had walked around him many times, whispering incantations to each other, applying the laying on of hands, and the shell began to look more and more like John F. Kennedy. The brows, the cheeks were smoothed outward and downward. The natural complexion of the President seemed to return. Thatches of thick chestnut hair were applied to the head and were combed out.

It had not hurt as much for the Greers and Kellermans and Burkleys and McHughs to stare at him when he was torn and broken. It hurt now. An undershirt, a pair of trousers, a white shirt were put on the unresisting frame and, when the tie was knotted, he had everything but breath. Greer turned away. Kellerman found it difficult to believe. Perhaps McHugh or Burkley might have fleetingly remembered the old story about the genie who could grant but one wish.

The entire evening had been morbid and gruesome, but the government had insisted on having witnesses. Death, in this case, was not a personal matter but an affair of state. From the moment the first shot in Dealey Plaza split the sky until the last volley drifted across the green hill at Arlington Cemetery, everything that happened to this man, everything that was noted, surmised, or conjectured, every conclusion for good or ill would become history.

The dark jacket was put on. It was buttoned and the hands were entwined across the front. The shoes were put on. Hagan and his men walked around the body again and again, tugging at wrinkled cloth, smoothing the hang of the suit of clothes, studying the serene features from the sides, the front, and even from the back of the head. All of it would have to be done again, when the body left the table, but the men wanted to be reasonably sure that they were satisfied with their work.

The long night was running out. There was a morning chill on the stone of Saint Peter's Basilica in Rome. The amber rays from the high windows crossed swords inside the nave. Paul VI knelt at an ornate *prie-dieu* before the main altar, the dark baldachinos lifting a canopy over the Host as the assembled monsignori carried the Pope's heavy brocade vestments forward to cover the prayer bench.

Thus began the pontifical requiem Mass for the repose of the soul of John F. Kennedy, a supplicant Roman Catholic, a sinner. The Pope clasped his hands, the fingertips touching each other, and he began his entreaty to Almighty God by asking forgiveness of sin. The early communicants at St. Peter's Basilica, pious Romans and inquisitive tourists, saw the Pope and knelt on the marble floor in astonishment. Prayers had been assaulting the gates of heaven for John F. Kennedy in many tongues and many temples, but if the credos of the Catholic Church are to be accepted, the real authority, the valid plea for the soul came from the lips of Father Oscar Huber, the little priest who had never seen a live President.

Thousands of miles to the west, Vincent Drain rested his eyes. If he had a prayer to say, it had been said. The tanker was high in the sky and a little smile crossed his lips. The FBI agent approached the ladder laden with packages of evidence, the military personnel had saluted him.

As he was smiling through the closed eyes, he felt something move at his feet and looked down to find a sergeant removing his shoes. "Hey," the FBI man said, "what are you doing?" "You look like you need rest, mister." The shoes came off. The jacket was hung. Then sleep fled, and Vincent Drain went forward with his evidence, to sit behind the captain and await the pink flush of dawn on the flight deck.

In London, Sir Alec Douglas Home knelt in Westminster Cathedral, peering like an emaciated owl over the tops of his spectacles. The first British services for Kennedy had begun. In Paris, groups of blue-clad laborers stood before the window of a television shop, studying the animation of a face which was no more. Down the Kurfürstendamm, past the Kaiser Wilhelm Kirche and the zoo paraded the remnants of the torch carriers of Berlin.

There were torchbearers in Bern, too, but the Swiss started before dawn. In the black velvet of the hills, they had made a glacial river of fire descending toward the old city. Thousands of Britons braved the needle veils of morning rain to stand in reverence before the United States embassy in London. A long time ago, the dead President had called this home. He and young Joe had come here to study and to be at the side of their father, Ambassador Joseph P. Kennedy, Sr.

The Kennedy boys had complained at wearing homburgs and carrying furled umbrellas, but they had come to know the ambassador as an absolute authority. They had made fun of each other in the embassy when the hats went on. The people who braved the morning drizzle knew nothing of these things, but they paid their homage in rain because they felt they had lost "well, a cousin of sorts."

Everywhere the hearts of the multitude felt regret at the passing of a man. Governments were different. Official grief was stereotyped: a wreath; a warm cablegram to a widow; signing a book in an embassy foyer; an emissary at a funeral. Governments are concerned with the living. The morning reaction would be coming in from everywhere. Britain understood Kennedy, but could the United Kingdom depend upon Johnson? Germany believed that the format and policy of the Kennedy regime would remain intact, but would the spirit be alive? Argentina was worried. Italy would appreciate assurances of continued support.

France, well, France would send De Gaulle to find out for himself. America, too, was fearful. It had gone to bed with its confidence shaken. How much did it know about Lyndon Johnson? Didn't he have a heart attack? Who succeeds Johnson? Well, there was old reliable John McCormack, seventy-one years of age, Speaker of the House. He was an excellent politician, an ideal lieutenant—but a captain? Politically and philosophically, he was hardly sophisticated. Behind him, in line of succession, came Senator Carl Hayden of Arizona. He was eighty-six years of age, a man revered by his confreres as a great American. The mileage on his mind was insuperable. In a day, America had lost its youth.

The casket was wheeled sideward toward the table. They touched. The witnesses offered assistance, but Hagan declined. He and his men worked easily. Both lids of the casket were open. On a signal, they lifted the body, kicked the autopsy table away, and held the body over the casket and lowered it gently. The final tender service to the flesh had begun. The men walked back and forth around the box, straightening feet and seams and shirring. The tie was firmly held under a clasp, and the jacket was draped softly downward.

The hair was carefully combed once more. A rosary was carefully laced through the fingers. The morticians examined their work from every side. They rubbed cloths over the dark mahogany, where hands had touched it. Joseph Hagan looked across the room at the witnesses. "We are ready," he said softly. The lid came down.

The word reached the seventeenth floor. The Attorney General looked at his watch. It was close to 4 A.M. Relatives and friends hurried to get coats, women to powder. A pall of cigarette smoke hung in the sitting room and it undulated as the guests made ready. They had served the widow well. Conversations of many hues had diverted her mind from the permanent shock. She had been forced, for a time, to think of other things. She had lapsed into a staring, dull expression, but there was always someone present to ask a question, tell an anecdote, make a suggestion.

Robert F. Kennedy had been heroic. He sublimated his crushing sorrow to serve hers. It is possible that Robert Kennedy had more courage than either Joe or Jack or Ted. Everything affected him more and showed less. He often felt pity, rancor, indignation, contempt, and sorrow—and denied them. His likes were as fierce as his dislikes and as possessive. He was small and tough and shyly sentimental. It was the Attorney General who originally divided the world into "them" and "us."

Mrs. Kennedy spoke this evening of her husband as though he were living. Bobby spoke of him as dead. He had spent time in the kitchen with McNamara and others discussing the Kennedy regime, the Kennedy policies, as though they had been killed in Dallas, and he wondered aloud if Lyndon Johnson would try to resurrect them. He had loved his brother slavishly, so that if an error was committed, he wanted to take the onus upon himself. When Robert irritated men of lofty station, they complained to the President, and he smiled and said: "I can handle Bobby."

He could. Jack was dead, and old Joe could not speak, and for a time Robert F. Kennedy would wander in the fields of McLean, Virginia, with his Newfoundland dog, wondering about himself, asking himself if the attainment of power was worth the result, brooding over Jacqueline's brooding, worried because the face in that majestic office wasn't Jack's. There was no one to handle Robert Kennedy. For several months, he would not be able to handle himself.

At the stone dock, a Navy ambulance was backed tight. Limousines sat in the dark. Ranking naval officers, summoned by subordinates, appeared. The witnesses who had waited all night, stood for a moment. Kellerman met Hill and Landis in the corridor, and led the party to an anteroom. The Attorney General, head down, took his sister-in-law's elbow. Behind them, in slow procession, were the Kennedy sisters, Ted Kennedy, Powers, O'Donnell, O'Brien, McNamara, the old friends, the new Cabinet members—they filed on in greater numbers than anyone had thought.

Naval personnel stood around the casket. Kellerman smiled briefly. "We'll take care of that," he said. The Secret Service agents wheeled it out onto the dark dock and into the ambulance. The trolley was taken from underneath and pulled back to the autopsy room. Roy Kellerman made the arrangements. He held a whispered conversation with the Attorney General and came back to the dock. "Bill," he said to Greer, "you drive. I'll sit up with you. Mrs. Kennedy and the Attorney General are going to ride in back with the body. Clint, take the second car."

There were motorcycle policemen in a wedge on Palmer Road. Kellerman went out and told them the route back to the White House. He wanted no sirens, no noise. They would aim for about thirty miles per hour and hold it. The cops could get the cortege through the red lights. They would go down Wisconsin and then left to the White House. At the northwest gate on Pennsylvania Avenue, the motorcycles would turn away, permitting the ambulance and the automobiles to go through.

Kellerman hurried back to the dock. He saw the widow stooping to get through the back, where, as before, she sat on one side of the casket and Robert sat on the other. They drew the shades. The doors of the limousines were slamming. Drivers put their lights on. Officers of the United States Navy stood on the dock at salute. The faces, the uniforms, the rank were blurred in the night light. A few enlisted men had put white caps on and stood on the ground, saluting.

Greer took the order from Kellerman. The ambulance lights went on. The rotating red beacon on top swept the night air with carmine crayon. The motorcycle cops forced their weight down on their starters and the racket of exhausts was like gunfire. They moved out onto Palmer, behind the giant monument of Bethesda.

There were houses of naval personnel here. The hedges and lawn flowers were stiff and dead, but the occupants were soft and alive. In the dark, they stood on their porches, watching the cortege come toward them, the sweeping beacon magnetizing the eyes of parents and children. They saw it coming toward them, the phalanx of police a spearhead, and they saw it pass, and they watched it go away, up the climb of Palmer toward the front drive, where the grass was velvet green. Commanders in pajamas shifted sleepy youngsters from one arm to the other and came to stiff salute. Some women watched and wept. Some shook their heads perplexedly. A few blessed themselves and wished him a long calm voyage to whatever haven sailors seek.

On shortwave, the word went to the White House. The President of the United States was on his way home. Hurriedly, a triangular piece of crepe was hung on the front door of the main entrance. Sleepy honor guards were whipped by words to attention. Officers strode up and down in the darkness, sabers against shoulders, the mouths forming the same truculent O which was so common to the face of Lee Harvey Oswald. The final tape of black was tacked to the bottom step of the big catafalque. In the darkness, saffron lights shone from the second floor. The great fountain on the lawn tossed a stream high in color, and it diffused in veils as it fell.

The house waited for him as it had waited for others. The sound of sobs had been heard here before. The house had heard the rattle of death, the smacking kiss of the bridegroom, the cry of an infant. This man had no more days to go. The house must live on, to endure what time would bring.

3 a.m.

Stage center was empty. The actors were resting. A police department cleaning employee pushed a broad broom down the third-floor corridor. It held a sizable assortment of cigarette butts, bits of film, and used photography bulbs. The hall was empty of reporters. In the press room, a typewriter clacked, but no one could identify the young man who hit the keys with one finger. A detective sat, with hat off the edge of his forehead, in Forgery.

A building employee moved from door to door, snapping lights off. Captain Will Fritz was gone. The chief had called it a night. A deputy chief sat in the front office reading the morning paper. It was so quiet one could hear the pulsing refrigerant in the soft drink machine. Paper cups filled trash baskets and the most recent ones rolled on the floor. Lieutenant Day's bureau was locked.

A radio operator listened to a DWI—driving while intoxicated—report coming in from a squad car somewhere in the night. Phones rang, and no one answered. On the basement floor, a prison admission clerk read a soft-cover novel. On the fifth floor, a maximum security guard got off the chair and walked a few paces up and down to keep awake.

Lee Harvey Oswald slept. The mouth was slightly open; the respiration was slow. He could not have been fearful. Sleep had come naturally and swiftly. If he had dreams, they were secret. He murmured no words. The limbs did not twitch. The sleep was restorative, and he would be ready for Fritz after a hearty breakfast.

Dallas stood tall and cool in the night. The Texas School Book Depository building, except for the night-light on the ground floor, was dark. The Hertz sign on the roof flashed: "3:08" then "62 degrees," but it did not disturb the pigeons, who slept with their heads under their wings. The switch engines made their short hauls across the overpass, coupling and uncoupling strings of cars. The brightly lighted building behind the railroad station was the *Dallas News*. The diesel trucks on Stemmons Freeway made pulsing sounds as they rushed through Dallas and out the other side. A novelty shop, closed with all lights on, featured a dinner dish with a portrait of Mr. and Mrs. John F. Kennedy for a dollar ninety-eight.

Jack Ruby looked at his watch and said he had to go. The policeman and his bride-to-be said okay. Sheba, curled on the front seat, had no opinion. The shame of the whole thing, said Jack Ruby, was that poor little woman who would have to come back to Dallas for the trial. Every time he said it, the sentence took on the aspect of a whole new thought. It was difficult to ascertain whether he was depressed by the assassination or exhilarated with his role as public relations adjudicator.

Marina Oswald slept with hair rollers wound tight. Marguerite knew how to foreclose a crisis to get proper rest. Trauma One, at Parkland Hospital, was dark. Even with sedation, Governor Connally slept restlessly. Lacerated flesh and broken bones had returned from shock to protest. Roy Truly, the manager of the School Book Depository, had grimly told himself: "I'm not going to get an ulcer over this," and he wasn't. It had taken some time for Wesley Frazier to fall asleep, but he made it. The few who were still wakeful were those whose night duties required it; those who were chronically ill and could not sleep; those whose dull lives had been improved with a new topic and whiskey.

The automobile appeared to be in sections. The rear seat was on the floor of the garage. The bubbletop was sitting by itself. The floor rugs, the metal stripping were in a separate group. Even the interior of the trunk had been taken apart. Whatever

SS-100-X could tell the FBI had been told. The dream car was a nightmare. Robert Frazier ordered his men to put it together. Two Secret Service men watched. When the job had been completed, the car was tested and found to be in running order. The ignition was shut down, and the men pushed the car back into the alcove at the rear.

The New York office of the Federal Bureau of Investigation had traced a big shipment of cheap Italian military rifles to Crescent Firearms, which sold in lots to mail order distributors. Early in the evening, the Dallas office had notified Washington that the rifle found on the sixth floor of the Texas School Book Depository building was a 6.5-caliber Mannlicher-Carcano with the serial number C2766 stamped on it. Alan Belmont had passed this information on to all field offices. The New York group, contacting one gun house after another, found that Crescent had them.

The company had cooperated in keeping the office open as the FBI agents watched employees run through the files. The records were not overly precise, but they indicated that C2766 had been sent to Klein's Sporting Goods, Incorporated, at 4540 West Madison Street, Chicago. The Chicago office of the FBI was alerted and, late at night, found William J. Waldman, vice president of Klein's, at his home, 335 Central Avenue, in Wilmette, Illinois.

Mr. Waldman agreed to accompany the FBI agents to the office. It would be reopened, and he might need some of his own personnel to help run through the records. Lights were turned on, and file cabinets were ransacked. The first order of business, Mr. Waldman said, was to check up on the purchase records of the company. It was not a simple matter because Klein's purchased a lot of sporting goods, a variety of merchandise, of which guns was but one.

It was after midnight when an invoice was uncovered from Crescent Firearms of New York. It was dated February

7, 1963, and cited a shipment of Mannlicher-Carcano 6.5's, amounting to a hundred rifles per box. It was shipped by North Penn Transfer-Lifschultz and there were ten cases. Waldman showed a receipt, indicating that the guns had been paid for on March 4, 1963.

The guns had been advertised in hunting and sporting magazines. Purchasers filled out a coupon and remitted money orders. The next move, Mr. Waldman said, would be to start hunting through the microfilmed photostats which were kept on file. This would take some time because they were dropped into the files as they were received. There was no specific order of filing. Locating C2766 was going to be tedious rather than difficult.*

The drive before dawn was one of reflection. It was not planned that way. From the ambulance back through the cortege of six cars, conversation was slow or brief, and each of the mourners had time for his personal thoughts. It had been a long, long day. They were in another day but, not having had the blessing of sleep, it was considered to be the same one. In an hour the first streaks of light would be coming up behind the Capitol dome.

Death was not devised for simple meditation. The mind must be ordered to think about it, to absorb the crushing finality of it, to accept it, to plan logically. The sorrowing mind will refuse. This often leads to a carousel of thoughts which follow in sequence leading to no conclusions. William Greer, for example, was driving the ambulance slower than he or Kellerman had expected and deep within Greer he could hear the President reciting the lonely lines of Robert Frost:

---

* At 4 A.M. Dallas time, the order for C2766 was located. It was a coupon clipped from the *American Rifleman* of February 1963. It was ordered by "A. Hidel, P.O. Box 2915, Dallas, Tex." It was paid for by "A. Hidell," a name Lee Harvey Oswald told Captain Fritz he used now and then. The price of the rifle, with four-power scope on it, was $19.95. There was an additional charge of $1.50 for postage and handling. The Mannlicher-Carcano had been mailed to the purchaser on March 20, 1963.

The woods are lovely, dark and deep,
But I have promises to keep,
And miles to go before I sleep,
And miles to go before I sleep.

It is not a refrain which can be shut down. "And miles to go . . ." The route from Bethesda was nine and a half miles to the White House. It was the last time Greer would drive the President. He would take his foot off the accelerator, watch the police escort move ahead, then see them turn and slow down. A cab driver, dozing on a street corner, would watch the procession, then sit upright and put his car in gear and try to keep pace with those who wept with reason. A milk truck joined the line of cars, and a motorcycle patrolman was dispatched to go back and turn it away.

Each one had his separate memories, and if they crossed the line of thought of someone else in those cars, it was accidental. The thoughts raced from comedy to solemnity, and some tried to recall whether John F. Kennedy had discussed his death and if, in doing so, displayed a divination. Some wondered what he would be remembered for; he was here so short a time. He was the second man to free the Negro and, like the first, he departed, leaving the Negro to fight for it.

He had driven the Russians and their ugly missiles from Cuba; he had banned the testing of nuclear bombs in the atmosphere; found time to impart his benevolent blessing to the arts; what else? What else? He had inaugurated a new and less paternalistic attitude toward his brethren in South America; he had stopped Big Steel from contributing to the spiral of inflation; Castro had stopped him at the Bay of Pigs. He had made America feel excited about itself, a thing which is not measurable in tangibles.

Did anyone smile ruefully at the darkness remembering the time the President had asked David Powers to stay with

him while Mrs. Kennedy was in Europe, and Powers said to
the President: "This has got to stop. My family calls me John's
Other Wife." The young widow, flipping through a mental al-
bum of portraits, remembered that, late at night, he often asked
her to play the record from the play *Camelot*. He would sit at his
"night work," barely listening, until the last part of the last song.
Then he dropped his pen as the voice sang:

> *Don't let it be forgot,*
> *That once there was a spot*
> *For one brief shining moment*
> *That was known as Camelot.*

He was moved by those words, and he would stare at his
wife intently, smile, and pick up the pen. Mrs. Kennedy thought
about it, this time with poignancy. "There will never be another Camelot," she whispered. No one had to discipline Kenny
O'Donnell's emotions. The champ was up front under that rotating light. He was dead and would never be seen or heard
again. For a man who exults in battle there is no solace behind
a hearse. If Robert Kennedy was able to lift his thoughts higher
than the grave, he realized that there was a future for the Kennedys. There were always graves for young Kennedys, and there
was always the future borrowing a tinge of Rose over the next
hill. It was, in a way, as though some school bully had beaten his
brother. The boy would have to take Robert on next.

Kellerman was all cop. He could keep his mind on that gro-
tesque autopsy for hours—if so assigned—and he could concentrate on the motorcycle cops and the pace of the cortege.
Duty divorces emotion. Mr. Kellerman glanced up and down
each passing side street as though he was still looking for a
threat. The political philosopher, Larry O'Brien, would remain
in the arena where the action was. He did not know Lyndon
Johnson well, but he knew that Johnson had long ago assessed

him. If the President asked him to stay, O'Brien would stay. If he was asked to leave, he would go home to Massachusetts. There were high-ranking government officers who could not sleep tonight trying to remember whether they had been cordial or cold to Lyndon Johnson. Men and their wives sat recounting every meeting, trying to remember every word. The conundrum resolved itself to: "Never mind my impressions of him. What were his impressions of me?"

The cortege passed Georgetown University and turned left. Greer was in no hurry. He was moving slowly and he looked at the familiar houses and shops on M Street and then he curled the ambulance around and onto Pennsylvania Avenue. There was a slight slick on the pavement as though a drizzle had dampened this part of Washington.

The ambulance kept to the middle of the street. A few people were out in front of Blair House, one of them a woman with her knuckles in her open mouth.

The motorcycle police approached the West Gate. Across the street in Lafayette Park, the people stood stonefaced as they had on the morning of April 15, 1865. They did not weep because many of them could not yet believe. The ambulance came to a stop. The police escort wheeled away from it like a fleur-de-lis. Inside the main door of the White House, the men who had worked so hard all night said: "Here he comes" and, in a body, they retreated.

Sargent Shriver went out to the front porch and stood waiting under the big light. His skin was pale. A Negro usher began to open the great front doors, the latches snapping loudly. Two White House policemen opened the double west gates. On the driveway, a double row of servicemen heard an officer shout an order and it rang off the front of the mansion. Greer bent low behind the windshield, nodded to the policemen, and started up over the sidewalk slowly and onto the big curving driveway. The soldiers, the marines, the Air Force recruits flanked the ambulance and marched beside it, their officer in front of the bumper, lifting his legs high, the saber leaning on his shoulder. The thump of boots hitting the pavement in cadence could be heard inside the ambulance. They could be heard in the silence across the street. The lights of the cars passing slowly onto the grounds washed the dark old bark of the big trees.

Everywhere he went, this man had heard "Hail to the Chief." This time there were the muffled drums of feet. Roy Kellerman glanced at his watch. The time was 4:24. Greer saw an honor guard of United States marines ahead and he pulled slowly abreast of them and stopped. ". . . and miles to go . . ." He pulled up the hand brake and turned the ignition off. ". . . and miles to go . . ."

The cars behind them, lights blazing, came to a halt. Except for a hoarse military shout now and then, and a car door opening or closing, there was no sound, no conversation. The relatives and friends who had kept the vigil emerged, walking hesitantly up the line of cars to the ambulance. Kellerman and Greer opened the back. They helped Mrs. Kennedy and Robert Kennedy down. There were no tears. They looked around and up toward the portico, like people who are surprised to find that they are here.

Military men in dress uniforms slid the dark casket out on its bearings. At a whispered command, each man grasped a silver handle with both hands. At another command, they turned to face front and held on with one hand. This was the darkest hour, and the light from the portico was brighter than ever. It scintillated along the curving lid of the casket as the young men began their slow steps through another long guard of honor onto the porch.

Mrs. Kennedy and her brother-in-law followed. The others fell in awkwardly behind them. The casket tilted as it was carried up the few steps. The guard of honor extended through the doors and on into the main lobby. Across the street, a woman wailed aloud. The group walked slowly through the main doors into the cool brightness of the marble reception hall. An honor guard of marines stood at attention. The sound was new, the squeak of shoes. Sargent Shriver kissed his sister-in-law, whispered a consoling word, and took his place on the opposite side.

The pallbearers turned left and into the great East Room. There, for the first time, Mrs. Kennedy could see what had been done. She saw the great black catafalque and the bits of crepe on the chandelier and the frames of mirrors. She nodded to herself, satisfied. The casket was lifted high, centered, and lowered on the dark dais. Robert Kennedy looked and then looked away.

The men who had carried the body stepped back. They looked furtively at each other. No one had told them what to do next. A marine officer stepped smartly to the center of the parquet floor, clicked his heels, and whispered orders as a priest in cassock and white surplice walked to the head of the casket, sprinkled holy water, and murmured prayers. The marines marched off, the boots thumped in unison, and the little wall bracket electric lights shivered.

There was an altar boy; the priest whispered to him to light the taper and light the four candles surrounding the catafalque. The boy was nervous. Two officers snapped an American flag open and draped it over the top of the casket. David Pearson watched the altar boy fumble. So did another former altar boy: Robert Kennedy. Someone noticed that Mrs. Kennedy was no longer in the doorway. A moment later, the Attorney General disappeared.

The priest knelt. His prayers were his own. Some, in the doorway, knelt. Others stood. Stuart's portrait of Washington stared out in pride and did not see the thirty-fifth man to the office. An officer inspected the guard of honor—a man from each branch of service. His face was hard and stern, and he turned heel and toe away from each man to march to the next.

At the conclusion, he muted a barked command and the honor guard lapsed into parade rest.

General Godfrey McHugh stood straight and silent. He was near the casket, thumbs on the seams of his trousers. Beside him stood Secret Service man Clint Hill. The East Room was so quiet that men could hear each other breathe. An usher came into the room, strode over to Hill, and whispered that Mrs. Kennedy said that she would be downstairs in a minute. She wanted the casket opened.

The flag was folded and removed, to be draped over the forearm of an officer. The general and the Secret Service man stepped up on the side of the catafalque to try the catch. They fumbled. Then it snapped, and the lid came up. They did not want to be fumbling when Mrs. Kennedy arrived. Clint Hill lifted it wide. He looked inside.

He studied the face more boldly than he had before. The President looked composed. The jawline seemed a little broader. The thick chestnut hair looked darker against the white satin pillow. Carefully they lowered the lid without closing it. McHugh looked back across the room. Mrs. Kennedy stood in the doorway, on the arm of Robert Kennedy.

For the first time, she looked exhausted. The feet were a bit too wide apart. The head was slightly down, the mouth hung open. The eyes held the haunted look of the long day. Robert Kennedy held her elbow and whispered to her. They started slowly across to the center of the room. General McHugh barked an order: "Honor guard, leave the room!" There was a hesitation. Each man did an about-face and started to walk away.

"No," Mrs. Kennedy said, holding up a hand. "No. They can stay." They stopped but did not turn back. One man was in midstep, and remained in that attitude. Robert led her to where Clint Hill stood. The Secret Service man lifted the lid high and stepped down. The Attorney General helped the lady up the step. She stood looking in, still wearing his dried blood on her strawberry dress and on her stockings. She stared at the image and asked for scissors. Hill got them. She reached in and snipped a lock of hair. Robert Kennedy glanced at his brother and turned his glance down. Mrs. Kennedy held the snip of hair and the scissors.

Then she turned away. "It isn't Jack," she said.

EPILOGUE

Looking back over the shoulder, one gains perspective. The further back, the more the mind focuses on infinity. Five years after the assassination of President Kennedy, it is clear that America immersed itself in an emotional bath on November 22, 1963. It soaked a long time but emerged no cleaner. In June 1968, President Kennedy's brother, Robert, was assassinated. Between those events, other crusaders had been killed. Others will be killed.

These sorrows are not the symptom of a sick society. To the contrary, the health of the community is displayed by the increasing amount of mass shock which follows each assassination. America is deluded by a veneer of gentility and sophistication. The country feels that it is "above" violence. No culture, no country is. Man husbands hate as he does love. It is mundane for one person to wish another person dead. Some kill symbolically with an anonymous letter or a threatening phone call. A few stick pins in dolls. The coward misses on purpose. The fanatics, the sick, transform their hate and frustration into a final, physical act.

This book was written for two reasons. One is that, for a number of years, the minute-by-minute account of an event has been my forte, and that day in that city lends itself to this kind of writing. The second is, as the list of source material will show, a great number of writers have spent a lot of energy bending these events to preconceived notions. And yet I cannot claim complete and unqualified accuracy for myself because I have never written anything which, in the final analysis, is exactly as I had hoped.

The nonfiction writer, unlike the novelist, is stuck with facts. They can be, collectively, undramatic and antidramatic. Solid facts have ruined good scenes.

My feelings about the people in the book and those, like the members of the Warren Commission, on the perimeter of events are turbulent. There are liars and second-guessers in the cast, and self-hypnotists too. I cannot believe that Mrs. John F. Kennedy said: "I love you, Jack!" as her husband fell dead in her arms. She doesn't even remember crawling out on the trunk of the car. Those riding with her can recall everything they heard her say. "I love you, Jack!" is not one of them.

Will Fritz was a good, plodding cop, but I do not admire him as Inspector Kelley of the Secret Service does. Fritz kept no record of his interrogations of Oswald. A pool of police stenographers was available to him. Also, for a few dollars, he could have rented a tape recorder from a shop near police headquarters. His questioning was cautious and repetitive. I am left with the impression that Fritz was afraid of alienating his prisoner. The book does not reveal him in this light because the known facts are counter to it.

I felt, at times, that Chief Jesse Curry wanted to hide from the assassination. He remained at Love Field until 4:05 P.M. although Oswald was arrested at 1:40 P.M. The chief spent much of his time in his office worrying about the press and the possible indictment of his department. The district attorney had the impression that the chief did not know a great deal about the case. Nor do I believe that Fritz and Curry were solicitous of the rights and welfare of Lee Harvey Oswald. Every time the prisoner entered or left the office of Captain Fritz he ran a gamut of vicious reporters. He also pleaded for "John Abt or a lawyer from the American Civil Liberties Union." The police department told the ACLU that Oswald had declined the services of an attorney.

Fritz has said that if his men were close to the President when the shots were fired, they might have picked off the assassin in the sixth-floor window or, at the worst, sealed the building against escape. When the shots were fired, Dallas policemen fell off motorcycles, drew revolvers, and ran in diverse directions as the echo chamber of Dealey Plaza tossed the explosions back from three walls. In addition, they were unable to seal the Texas School Book Depository building until at least 12:34 P.M., at which time Mr. Oswald was sauntering back up Elm Street to board a bus.

The press was abominable. Reporters were demanding, hysterical, and abusive. It was within the power of Captain Glen King and his superiors to seal the third floor against cameras and journalists at any time. The reporters could have been evicted to the first floor where, once an hour, Captain King could have appeared before them with a typewritten report of progress. The police were afraid of the press. The midnight conference with Oswald was a mockery of justice. The lion was thrown to the Christians.

A good case can be made out for any theory about the three shots. As a mediator on television, I have listened to some which induced laughter. The best procedure is to work backwards. The vast majority of witnesses agree that they heard three shots. Zapruder's film proves that the third shot blew the top off of John Kennedy's head. This leaves two for accounting. Governor John Connally, who remained alert and conscious through the ghastly scene, is a hunter. He heard a rifle shot and swung toward his right, then toward the left to look at the President. Mr. Kennedy was lifting both hands upward. A second shot rang out, and the President grasped the throat area and began to fall toward his left. At the same time, the Governor felt as though someone had slammed him in the back. This would indicate that the second shot hit Mr. Kennedy, furrowed between the strap muscles of the neck, nicked his tie, emerged pristine, having hit no bone, punctured Connally's rib cage, and emerged exactly where the films show Connally's right wrist to be—

coming up toward his chest. It hit the wrist, fractured it, and was spent in a shallow furrow on the left thigh and remained there until it fell off a stretcher.

If there is a mystery—and I don't think there is—it lies in the first shot. A direct line from Oswald's window, down to the position of SS-100-X, and straight to the underpass at Commerce, will show that this is the one which hit the pavement on the right side of the car, sending up a shower of gravel from the pavement. A woman on the curb opposite the car was hit by a "spray." President Kennedy's seat in the car was elevated three feet higher than the Connally jump seats. Undoubtedly Kennedy heard the first shot; undoubtedly he felt the spray of concrete and realized that someone had taken a shot at him. The bullet is believed to have tumbled upward off the pavement, nicked a curb, and sprayed the face of James Thomas Tague standing beside his car at the underpass—on a straight line from that sixth-floor window.

Marina Oswald has my sympathy, not my esteem. She was ready to tell about her husband's rifle and the blanket in which it was wrapped. Why did she not volunteer the information she had about her husband's attempted assassination of Major General Edwin Walker? Why not tell about how she locked him in the bathroom because he vowed to kill the Vice-President on a visit to Dallas? I cannot quarrel with the posture of a mother who wishes to protect her babies and herself, but if she proposes to reveal some truths and withhold others, she does not qualify as reliable.

Lee Harvey Oswald's mother makes the perfect portrait of the permanently aggrieved woman. When I was in Dallas and Fort Worth, she was one of the few persons I could not locate. And yet I feel that I know her. She has one-way eyes and a mind to match. Marguerite Oswald formulates logic so illogically that it becomes predictable. As long as she lives, Lee Harvey has a friend at court.

My conversations with Judge Joe B. Brown and the host of anonymous men he sent to my hotel suite at the Dallas Statler Hilton were the most revealing. Brown was the judge who presided at the trial of Jack Ruby for the killing of Lee Harvey Oswald on Sunday, November 24, 1963. He had sustained several heart attacks and had an ardent desire to write a book defending his conduct of the trial. He sat smoking a pipe, a man with an excellent memory for names and events. The people he sent remained "anonymous" because many of them—for example, police department employees—had been ordered not to discuss the assassination with anyone. The material given me by the judge and his faceless friends filled many of the blank spots of that bad day in Dallas. The judge took me to the jail one night so that I could see the "old" maximum security cell in which Oswald reposed for two days.

The management at the Hotel Texas in Fort Worth was kind in permitting Mrs. Bishop and me to occupy Suite 850 which had been occupied by the Kennedys on November 21 and 22, 1963. We were in it on November 21 and 22, 1967, the fourth anniversary of their stay. In all cases, I photographed everything—interiors, furniture, lobbies, exteriors, the parade route—all in color. The only time I was forbidden the use of a camera was when Mr. Roy Truly of the Texas School Book Depository took me up to the window used by the assassin. He pointed to the Nikon and said: "Leave that thing down here."

I am certain that, without the unqualified assistance of Chief James Rowley and the Secret Service, and Cartha de Loach and the Federal Bureau of Investigation, this book would have been a guessing game. The agents of both services who worked on this case in Dallas or in Washington were made available to me for individual questioning. Their report sheets showed exact times on events, even small ones. At the National Archives, I hefted the cheap rifle Oswald used, examined the revolver used to kill Officer Tippit, and counted the bits of bullets which have been recovered.

I am grateful to President Lyndon Johnson for a private interview on the assassination. It was the first time he had discussed it and, from the manner in which it affected him, it may be the last. Mrs. Lyndon Johnson, always gracious, is the only person I interviewed who wept. Malcolm Kilduff, assistant press secretary, and Jack Valenti, President Johnson's most confidential aide, cudgeled their heads to recall every scrap of pertinent information.

William Greer, who drove SS-100-X, has retired from the Secret Service. I visited him at his home in Maryland. His wife was ill and it was not a time to badger a man with ugly memories, but he sat and said: "Go ahead. It will take my mind off other things." The men of Gawler's Sons were discreet and ethical. Cliff Carter, who sat with President Johnson that night at The Elms, has a long and accurate memory.

Father Oscar Huber would not have seen me except that he was so angry at an earlier book about the assassination. This was also true of Roy Truly and others—some of whom assert that they were listed as having been interviewed but weren't. Father Huber, a spiritually complacent man, becomes feverishly angry when he considers an author who claims that the priest, leaving Parkland Memorial Hospital, said: "He's dead." "I did not!" Father Huber says, "and I wrote that guy a letter and offered to pay his airfare back to Dallas to prove it to him. He never answered my letters."

All of the interviews helped to add chips and bits to the research. But the 10,400,000 words of the so-called "Warren Commission Report" is and must remain the primary source of all material on the assassination. It is often repetitious and disorderly, and it required two years for me to read and annotate, but it was worth it. Two sets of the twenty-six volumes were used for cutting out affidavits and placing them in the right minute of the eighteen loose-leaf notebooks I kept on November 22, 1963.

Others who helped to make this book as complete as time and diligence can make it are: My wife Kelly, who helped with interviews, stenographic notes, copying documents, and retyping the manuscript; my daughters Karen and Kathleen, who helped to paste notes in the proper book, sometimes placing a "2:05 P.M." note in the "2:05 A.M." book; Mrs. Deloris Goldaker, who typed notes off and on for four years; and Miss Millicent Harrison, who separated the originals from the four carbon copies.

To assist the future researcher, following is a list of the sources used in researching and writing this book.

Source Material

Following is a list of the sources used in the researching and writing of this book.

1. *The Warren Commission Report* (Condensed Version). Washington, D.C., U.S. Government Printing Office, 1964.
2. The Editors of *The New York Times* and Viking Press, *The Kennedy Years*. New York, Viking Press, Inc., 1964.
3. Warren Leslie, *Dallas City Limit*. New York, Grossman Publishers, 1964.
4. *Miami Herald;* editions 1963–1968.
5. *The New York Times;* editions 1963–1968.
6. The Associated Press; 1963–1968.
7. United Press International; 1963–1968.
8. Pierre Salinger and Sandor Vanocur (eds.), *A Tribute to John F. Kennedy.* Chicago, Encyclopedia Brittanica, Inc., 1964.
9. Bill Adler (ed.), *The Kennedy Wit.* New York, The Citadel Press, 1964.
10. *Four Days.* Compiled by United Press International and *American Heritage* magazine, 1964.
11. *JFK Memorial Book.* Special edition. *Look* magazine, November 17, 1964.
12. Paul Ballot, *Memorial to Greatness.* Aspen Corp., 1964.
13. John W. Gardner, *To Turn The Tide.* New York, Harper & Row, 1962.
14. *Hearings Before the President's Commission on the Assassination of President Kennedy,* Volumes 1 through 26. Washington, D.C., U.S. Government Printing Office, 1964.

15. John F. Kennedy, *Profiles in Courage*. New York, Harper & Row, 1955.

16. G. Lieberson and J. Meyers, *John Fitzgerald Kennedy . . . As We Remember Him*. New York, Atheneum Press. Columbia Records, 1965.

17. James MacGregor Burns, *John Kennedy: A Political Profile*. New York, Harcourt, Brace & World, Inc., 1959.

18. *The Speeches of Senator John F. Kennedy. Presidential Campaign of 1960*. Washington, D.C., U.S. Government Printing Office, 1961.

19. *Public Papers of the Presidents of the U.S. John F. Kennedy, 1963*. Washington, D.C., U.S. Government Printing Office, 1964.

20. John Hersey, "Survival"; reprinted in *Here To Stay*. New York, Alfred A. Knopf, Inc., 1963.

21. Jacques Lowe, *Portrait: The Emergence of John F. Kennedy*. New York, McGraw-Hill, Inc., 1961.

22. William Manchester, *Portrait of a President*. New York, Little, Brown and Company, 1962.

23. Joseph McCarthy, *The Remarkable Kennedys*. New York, The Dial Press, Inc., 1960.

24. Hugh Sidey, *John F. Kennedy, President*. New York, Atheneum Press, 1963.

25. Jim Bishop, *A Day in the Life of President Kennedy*. New York, Random House, Inc., 1964.

26. Anne H. Lincoln, *The Kennedy White House Parties*. New York, The Viking Press, Inc., 1966.

27. NBC News Staff, *Seventy Hours and Thirty Minutes*. New York, Random House, Inc., 1966.

28. Fred J. Cook, "The Warren Commission Report": Part I, "Some Unanswered Questions"; Part II, "Testimony of the Eye Witnesses." *The Nation*, 1966.

29. *Photoplay* magazine; editions 1963–1968.

30. Charles Roberts, *The Truth About the Assassination*. New York, Grosset & Dunlap, Inc., Publishers, 1967.

31. Arthur M. Schlesinger, Jr., *A Thousand Days*. Boston, Houghton Mifflin Company, 1965.

32. William Manchester, *Death of a President*. New York, Harper & Row, 1967.

33. Theodore C. Sorensen, *Kennedy*. New York, Harper & Row, 1965.

34. Sylvia Meagher, *Accessories After the Fact: The Warren Commission, The Authorities and The Report*. Indianapolis, The Bobbs-Merrill Co., 1967.

35. Jean Stafford, *A Mother in History*. New York, Farrar, Straus, Giroux Inc., 1965.

36. *Esquire* magazine, December 1966.

37. Richard J. Whalen, *The Founding Father*. New York, The New American Library, 1964.

38. Penn Jones, Jr., *Forgive My Grief*. Midlothian, Texas, Midlothian Mirror, 1964.

39. Gore Vidal, "The Holy Family." *Esquire* magazine, April 1966.

40. *Inaugural Spectacle*. Special edition, *Life* magazine, 1961.

41. Harold W. Chase and Allen W. Lerman (eds.), *Kennedy and the Press*. New York, Thomas Y. Crowell Co., 1965.

42. William H. A. Carr, *JFK, An Informal Biography*. New York, Lancer Books Inc., 1962.

43. Deane and David Heller, *Jacqueline Kennedy*. New York, Lancer Books Inc., 1962.

44. Mark Shaw, *The John F. Kennedys*. New York, The Noonday Press, 1959.

45. Arnold Bennett, *Jackie, Bobby and Manchester, The Story Behind the Headlines*. New York, Bee-Line Books Inc., 1967.

46. Edward Hymoff and Phil Hirsch, *The Kennedy Courage*. New York, Pyramid Books, 1965.

47. Lonnelle Aikman, *The Living White House*. Washington, D.C., National Geographic, 1966.

48. Stanley P. Friedman, *The Magnificent Kennedy Women*. New York, Monarch Books Inc., 1964.

49. Evelyn Lincoln, *My Twelve Years with Kennedy*. New York, David McKay Co. Inc., 1965.

50. Theodore H. White, *The Making of a President*. New York, Atheneum House, Inc., 1961.

51. *Assassination of a President. The New York Times,* special edition, 1963.

52. Robert J. Donovan, *PT 109, JFK in World War II*. New York, Fawcett World Library, 1961.

53. Paul B. Fay, Jr., *The Pleasure of His Company*. New York, Harper & Row, 1966.

54. Mark Lane, *Rush to Judgment*. New York, Holt, Rinehart & Winston, Inc., 1966.

55. *Ramparts*, November 1966.

56. NBC News Staff, *There Was a President*. New York, Random House, Inc., 1967.

57. *Time* magazine; editions 1963–1968.

58. Sylvan Fox, *The Unanswered Questions About President Kennedy's Assassination*. New York, Award Books, 1965.

59. *U.S. News and World Report;* editions 1963–1968.

60. Harold Weisberg, *Whitewash*. New York, Dell Publishing Co., 1965.

61. Harry A. Squires, "Will the Spell be Broken?" Southland Supplement, *Long Beach Independent Press-Telegram*, 1963.

62. Lawrence Van Gelder, *Why the Kennedys Lost the Book Battle*. New York, Award Books, 1967.

63. Maude Shaw, "White House Nannie." Southern News Services Ltd., 1965.

64. Thomas G. Buchanan, *Who Killed Kennedy?* New York, G. P. Putnam's Sons, 1964.

65. Pierre Salinger, *With Kennedy*. Garden City, N.Y., Doubleday & Company, Inc., 1966.

66. Newspaper columns by Drew Pearson and Jack Anderson 1963–1968. Copyright by Bell-McClure Syndicate.

67. *Newsweek;* editions 1963–1968.

68. Articles by Peter Lisagor. *Miami Herald-Chicago Daily News,* 1967.

69. Robert Oswald and Barbara Land, *Lee, A Portrait of Lee Harvey Oswald.* New York, Coward-McCann, Inc., 1967.

70. Jim Matthews, *Four Dark Days in History.* Carmel, Calif., Special Publications Inc., 1963.

71. Interviews by Jim Bishop.

72. R. B. Denson (ed.), *Destiny in Dallas.* Dallas, Denco Corp., 1964.

73. Josiah Thompson, *Six Seconds in Dallas.* New York, Bernard Geis Associates, 1967.

74. John Connolly, "Why Kennedy Went to Texas." *Life* magazine, 1967.

75. Article by David Pearson. *Miami Herald,* November 22, 1967.

76. Paul Ballot (ed.), *The Thousand Days.* New York, The Citadel Press, 1964.

77. *Look* magazine; editions 1963–1968.

78. Frances S. Leighton, "First Lady's First Day." *This Week* magazine, United Newspapers Magazine Corp., 1964.

79. *The Daily News;* editions 1963–1968.

80. *Life* magazine; editions 1963–1968.

81. Jimmy Breslin, "Death in Emergency Room One." *The Saturday Evening Post,* 1963.

82. Radio Free Europe; transcripts 1963–1968.

83. Mrs. John Connolly, "Since That Day in Dallas." *McCall's* magazine, 1964.

84. Jessamyn West, "Prelude to Tragedy, The Woman Who Sheltered Lee Oswald's Family Tells Her Story." *Redbook* magazine, 1964.

85. Dr. Renatus Hartogs and Lucy Freeman, *The Two Assassins.* New York, Thomas Y. Crowell Co., 1965.

86. *Encyclopedia Brittanica,* 1965 edition.

87. *The Warren Report.* Published by Associated Press, 1965.

88. *Operating Room Nurses Journal,* November 1967.

89. Jim Bishop, *A Day in the Life of President Johnson*. New York, Random House, Inc., 1967.

90. Relman Morin, *Assassination. The Death of President Kennedy*. A Signet Book. New York, The New American Library, 1968.

91. Evelyn Lincoln, *Kennedy & Johnson*. New York, Holt, Rinehart & Winston, 1968.

92. The National Archives.

INDEX